ENERGETICS OF GEOLOGICAL PROCESSES

ENERGETICS OF GEOLOGICAL PROCESSES

Hans Ramberg on his 60th birthday

EDITORS

Surendra K. Saxena

and Somdev Bhattacharji

ASSOCIATE EDITORS

Hans Annersten and Ove Stephansson

SPRINGER-VERLAG

New York Heidelberg Berlin

EDITORS
SURENDRA K. SAXENA
Brooklyn College of the City University of New York
Bedford Avenue and Avenue H
Brooklyn, N.Y. 11210

SOMDEV BHATTACHARJI
Brooklyn College of the City University of New York
Bedford Avenue and Avenue H
Brooklyn, N.Y. 11210

ASSOCIATE EDITORS
Hans Annersten
Geological Institute
University of Uppsala
Uppsala, Sweden

Ove Stephansson
Department of Rock Mechanics
Luleå University
Luleå, Sweden

Designer: Robert Bull

Library of Congress Cataloging in Publication Data

Main entry under title:

Energetics of geological processes.

Includes index.
1. Geodynamics—Addresses, essays, lectures.
2. Geochemistry—Addresses, essays, lectures.
I. Saxena, Surendra Kumar, 1936– II. Bhattacharji, Somdev, 1932–
QE501.3.E53 551.1 76-30859
ISBN 978-3-642-86576-3 ISBN 978-3-642-86574-9 (eBook)
DOI 10.1007/978-3-642-86574-9

Softcover reprint of the hardcover 1st edition 1977

987654321

a volume dedicated
TO HANS RAMBERG
for his continuing achievements in science

FOREWORD

Hans Ramberg is working in an area of geology where 60 years are a short, often negligible period of time. This is not so in the lives of men. For us it is a time for evaluating past accomplishments and a time for friends to express their appreciation and admiration.

Some universities have become famous for this ability to foster eminent scientists in one or several fields. The success of Cambridge University in physics is a well-known example, but if we ask ourselves whether the success of Oslo University in earth sciences is not equally astonishing, then we see that Hans is yet another example of this process; but it is not the whole story.

There were certainly promising prospects when he started his studies in geology: V. M. Goldschmidt had just come back from Göttingen in Germany and Tom Barth had returned from the Geophysical Laboratory in Washington, D.C. Two leaders in geochemistry and petrology at the same time! Hans became a student of Barth, specializing in metamorphic rocks and their problems; but soon the situation changed. Norway was occupied by the Germans and the possibilities for university studies almost vanished. However, in spite of all difficulties he obtained his Ph.D. in 1946 and began participating in the geological mapping of Greenland. In 1947 he went to the University of Chicago and stayed there until 1961 when he came to his present position in the University of Uppsala, Sweden.

From this synopsis of his career it is clear that he has spent practically his whole working life out of Norway. Scientifically, Hans has also lived far away from the mainstream of Norwegian geological research; but his heart is in Norway, in the Trondheim area with its Caledonian mountains. This is the place to which he returns: where he was born, where he started his research and where he still derives inspiration for new developments.

His scientific work has centered around the application of basic physics (in his case thermodynamics and mechanics) to the problems of petrology and tectonics. It is as easy to define these interests as it is difficult to put them into practice. Natural systems are so complex that our models of them easily become unrealistically simplistic. Hans started along the thermodynamic line and his contributions were of major importance to the development of petrogenesis. Even so the great success came when Hans started his famous experiments in centrifuges; in this field his contributions are fundamental.

My first knowledge of Hans came during the War through Norwegian student refugees in Stockholm; he was described as a most promising student. Then his first papers arrived and I was impressed by his fresh approach to the classic problems of metamorphism and metasomatism. I first met him in Oslo in 1946 during a lecture tour. We had little time for discussions so my most vivid recollection is that he was good at coasting when we ran down from Holmenkollen to Oslo! When we spent half a year together in Chicago in 1947–48 I began to appreciate his great gifts fully. Since then this impression has only been strengthened.

Hans should not be surprised that we have found out his age since after all, we have succeeded in determining the age of such a reluctant lady as Mother Earth. The result is this book, with which a score of his friends congratulate him on his sixtieth birthday!

May, 1977 Frans E. Wickman
University of Stockholm, Sweden

FOREWORD

Students who are provided a surfeit of definitive answers and final solutions are taught that it is not necessary to question; the answers are there and will be provided; all the students need do is acquiesce in the assimilation of them. These students may also be taught to regard a knowledge vacuum—of whatever extent—as intolerable, and to feel more anxious to have their knowledge-space filled with something than to review critically whether that something fills the space appropriately. An authoritarian assertion, said to be appropriate and correct, is to be accepted without attempting any assessment of its appropriateness and correctness by application of one's own critical capacities, and may even be welcomed as absolving one of any responsibility should the assertion prove to be wrong. By acquiescing to authority the hope is that the responsibility for being wrong will rest elsewhere. What one dutifully does is to perpetuate what one has been told is right and to contribute by filling interstices with bits of information—a safe way to live, or so it seems. What this approach ignores is that the responsibility for error is not avoided, but is assumed and remains fully shared; one is never absolved of the responsibility to apply one's own critical capacities to the assessment of the appropriateness and correctness of the fabric and structure and of the guiding principles and concepts of one's science.

Consequently, one of the important qualities of a teacher of geology—or of any science—is to stimulate and facilitate questioning and testing by the minds of students. Answers may be suggested, but only in order to be tested; dogma is rejected as such, but its assertions are taken up for examination and dispute. Imaginative hypotheses are prized, even if they turn out to be wrong, because they require scrutiny of possible relationships between apparently disparate facts and principles. Dispute is good because it forces examination; a "wild idea" may be defended tenaciously until it has been demonstrated to be untenable, and then it is happily abandoned. For those who genuinely warrant the designation *student,* regardless of age or experience, honor is not bound up with the positions which are supported during evaluation; rather, to the extent personal honor is involved it is in the clear devotion to the advancement of insight and knowledge. Honor becomes attached to process, not to specific outcomes, to intellectual approach and aspirations, not to personal theories.

In thinking about Professor Hans Ramberg's role in the geological sciences, I am reminded of an occasion during the summer of 1955 when he and I went fishing in Åsenfjord, an inner arm of Trondheimsfjord. His role on that fishing trip in relationship to me was analogous to his role as a geological scientist in relationship to his students, the geological community, and his science.

As we started that trip, both having good eyesight, we had the opportunity to see the same things. However, I, a Chicago-raised nonfisherman, discerned no meaning in certain observations—Hans did. So he guided us to a place in the fjord, he provided me with the necessary equipment, told me how it could be used, and steered our boat, and I proceeded to haul in saithe as fast as I could get a multiple-hooked device into the water, back into the boat, and get the fish off the hooks. That fish story is, of course, one I shall never forget; it was tremendously exciting and it was enjoyable and rewarding to be so successful, but I would not have had anything like the catch I did—several score fish (really!)—had I been alone about it.

What analogy do I see? If not perfect—as perhaps no analogy is—I would note that in the initial observation he demonstrated the ability to discern what would likely be a promising area (he formulated the problem in such a way that success was greatly increased); he then gave me the appropriate fishing equipment (he provided what would be necessary to address the problem—intellectual tools, analytical technique, theoretical framework), told me what to do with the equipment (gave instruction, interpretation, and guidance so that the problem could be effectively addressed by me); and he steered the boat (kept us to the problem area that would be most fruitful). That is, his role has been to question and help others to question, to develop what was needed to address the questions and provide assistance in using his additions to our intellectual arsenal, and continually keep us focused on what would be productive. What Hans did *not* do on our fishing trip—which would also have been little fun for me—was fish all the fish and dump them in my lap! He did not do the job for me and then hand me the result. My little analogy breaks down badly in at least one respect; on that fishing expedition I caught most of the fish.

Hans Ramberg, explicitly and by example, is a teacher of the probing inquisitive study of science rather than of the authoritarian direction of information gathering. Whether one experiences him in the classroom, where his humor and devastating critiques of authoritarian positions are allowed to unfold, or in the colloquium discussion, where he gently explores weaknesses of hypotheses and stimulates to further inquiry, or in his formal publications which seriously and systematically set forth the results of research and thought as they exist at that stage of the continuing evaluation of knowledge, one is compelled to respect his devotion to the improvement of our science by thrusting ahead into the unexplored and by constantly ridding us of the "baggage" that no longer contributes to the improvements of our understanding that we seek.

Hans Ramberg has been an innovative leader in the exploration of new directions and the assessment of new approaches in the study of earth crustal phenomena. His career has been especially characterized by (1) the application of the principles and theory of physical chemistry and thermodynamics to the theoretical evaluation of hypotheses concerning the evolution and genesis of plutonic rocks; (2) the development of experimental determinations of fundamental thermodynamic properties of geologically important materials so as to constrain the evaluation of hypotheses; (3) the careful observation of rocks in the field so as to determine the geological factors that must constrain genetic theory; (4) the development of experimental methodology for the simulation of small- and large-scale structural phenomena of the crust, especially emphasizing scale model tests of the mechanical properties of segments of the crust and the influences of variations of those properties; and (5) the application of mathematical models to the understanding of the mechanical behavior of the crust.

Looking at a recent list of Professor Ramberg's publications, it is clear that the first half, up to 1961, demonstrates a clear emphasis on the application of chemistry to the solution of geological problems, while the second half demonstrates a shift of emphasis, that began in 1955 and became dominant in 1961, toward the application of physics to the solution of geological problems.

It is noteworthy that all through his career, while occupied with theoretical and modeling studies, Professor Ramberg has continued field work. His philosophy has always been that theoretical considerations must be tied to observed geological field relations, structures, and mineral assemblages. The interplay between field geology and the laboratory/theoretical geology conducted indoors generates new ideas and revises or rejects old ones that prove to be invalid. Such a synthesis is able to be most fruitful when performed by a single individual making both approaches. Over the years he has generated many ideas. His book, *The Origin of Metamorphic and Metasomatic Rocks,* and many of his papers both before and after, treated many problems in mineral chemistry related to the structures and assemblages of metamorphic rocks and what the petrogenetic significance of these are. Throughout his work he attempted to emphasize quantitative approaches and point the way to solutions that could be found by experimental and modeling studies. Professor Ramberg has been an outstanding source of ideas for researchers in thermodynamic mineral chemistry and petrology over three decades.

Since the early 1960s Professor Ramberg's work in the mechanics of geological processes has won wide respect and represents a new approach to geological problems. It is valued as being correct in terms of its mathematics and physics. He has used this work to question "fad-ism" with regard to plate tectonics, but I know from personal conversation with him that his motivation is to cause us to pause and think about problems and questions, not to deny the real power of the plate tectonics concept. This work in geodynamics is his most unique research. In his geochemical work he has achieved a position among the leaders, but his geodynamical research has achieved for him a position of splendid isolation. He has synthesized research based upon experimental models, mathematical models and analysis, and detailed and regional field studies in forging ahead to reach new levels of understanding of small-scale and large-scale geological structures. His work has been prized by those engaged in extraterrestrial studies as well as engineers investigating the responses of materials at the earth's surface, and not least by his intended primary audience, the geologists engaged in the study of the deformation of the earth's crust.

Hans Ramberg is a buoyant person, lightness rather than great seriousness being the essence of his nature. These words are used in the definition of *levity,* his personality being in striking contrast to his constant concern for *gravity* as a phenomenon to which attention must be drawn in the interpretion of geological features. Since his very earliest published works, and in the context of both geochemical and geodynamical problems, the effect of gravity has been continually noted and has, of course, been central to his innovative model experiments using a large centrifuge. It may be that it will be through his contributions to our understanding of gravity-driven processes that Hans Ramberg will be longest remembered.

Professor Ramberg's contributions have not been limited to what is represented by the record of his publication list. He has characteristically scrupulously refrained from adding his name to the publications of research students and post-doctoral fellows who have worked under his guidance while a professor at the University of Chicago (1948–1961), a professor at the University of Uppsala, Sweden (1962–present), and a part-time professor at the University of

Connecticut (1970–present), unless he had personally made a major contribution to the day-to-day work of the research and to the written report. He has not claimed co-authorship where his role has been in conception, inspiration, definition, encouragement, and critical appraisal; but his contributions through his students have also been large and have continuously emphasized the application of rigorous theoretical and experimental chemistry and physics to the solution of geological problems.

Professor Ramberg has consistently risked unpopularity by questioning dominant paradigms rather than "falling into line" to get the most quick mileage possible out of what was most in vogue at the moment. His independence of thought and his devil's advocate roles have helped to keep physical and chemical inquiry focused on soft spots in geological hypotheses and theories as they have been developing. This is a contribution of great value for which the geological community owes Professor Ramberg recognition and gratitude. At times, positions that he has inquired into, and perhaps therefore been identified with, and advocated in order to achieve the test of controversial assessment have finally proven to be wrong. Even so, the test of one position against others has proved to be an important service. But more, he has been a leader in invoking new approaches to inquire into unsolved problem areas. His work has stimulated both those closely associated with him and those who have become aware of his work, to seek deeply and more widely; he has catalyzed geochemical and geophysical research for 30 years, half of the time working in the United States and half of the time in Europe. I am proud to have had some professional exposure as well as personal association with one of the major post-World War II contributors to the movement in geological science to apply chemistry and physics to the solution of geological problems.

May, 1977 Paul H. Reitan
State University of New York at Buffalo

PREFACE

This volume is dedicated to Hans Ramberg on the occasion of his 60th birthday by a number of admirers, friends, colleagues, and former students as a token of their high esteem and affection. The title and division of this volume into two broad categories, geodynamics and geochemistry, illustrate the important fields in which his influence as a pioneering researcher and teacher has had its greatest scientific impact.

It was not an easy task for the editors to select a limited number of authors for this volume from the large world-wide circle of admirers, friends, and former students of Professor Ramberg. The editors are painfully aware that many of Professor Ramberg's former students and admirers who have gained eminence and reputation as geoscientists could not be invited to contribute to this volume because of the limitation on the number of pages.

In this volume we have compiled original researches, reviews, and ideas of the contributors in the fields of geodynamics, structural geology, geochemistry, and petrology, a cross-section of the areas in which Professor Ramberg's own original contributions have been prolific. Because of the diverse topics of the papers for this volume, it was not possible to arrange them in any logical order beyond the broad subdivisions geodynamics and geochemistry. However, readers may notice that the topics of wider scope appear earlier in the book.

Throughout the preparation of this volume, we have received valuable secretarial and clerical support from the Department of Geology of the Brooklyn College of the City University of New York. We are deeply grateful to Mona Saxena and Lee Bhattacharji, for their support, encouragement, and patience. The efficient and skillful editorial help of Ruth Adams and her deep interest in this volume greatly expedited its publication. Finally, we would like to thank the contributors for their promptness in sending the papers and in dealing with editorial requests.

May 1977
City University of New York

S. Bhattacharji
S. Saxena

CONTENTS

LIST OF CONTRIBUTORS

V. V. Beloussov, Institute of Physics of the Earth, Academy of Science of the U.S.S.R., Moscow, U.S.S.R.

J. Sutton, Department of Geology, Imperial College, London, U.K.

J. Watson, Department of Geology, Imperial College, London, U.K.

B. Bayly, Department of Geology, Rennselaer Polytechnic Institute, Troy, N.Y. 12181, U.S.A.

H. Koide, Geological Survey of Japan, Tokyo 162, Japan.

S. Bhattacharji, Department of Geology, Brooklyn College, City University of New York, Brooklyn, N.Y. 11210, U.S.A.

S. M. Schmid, Geologisches Instiut des ETH, Sonneggshr 5, 8006 Zurich, Switzerland

M. S. Paterson, Research School of Earth Sciences, The Australian National University, Canberra, A.C.T. 2600, Australia.

S. K. Ghosh, Geological Institute, Department of Geology, Jadarpor University, Calcutta, India.

P. J. Hudleston, Department of Geology and Geophysics, University of Minnesota, Minneapolis, Minnesota 55455, U.S.A.

J. L. Roberts, Department of Geology, University of Newcastle-upon-Tyne, Newcastle-upon-Tyne, NE 1 7RU, U.K.

P. F. Williams, Geologisch en Mineralogisch Instituut, Der Rijksuniversiteit-Leiden, Nederland.

H. J. Zwart, Geologisch en Mineralogisch Instituut, Der Rijksuniversiteit Leiden, Nederland

A. Miyashiro, Department of Geological Sciences, State University of New York at Albany, Albany, N.Y. 12222, U.S.A.

K. S. Heier, Norges geologiske undersøkelse, Box 3006, Trondheim, Norway.

G. Grønlie, Institutt for geologi, Universitet i Oslo, Oslo, Norway.

H. K. Mao, Geophysical Laboratory, 2801 Upton St. N.W., Washington, D.C., U.S.A.

P. M. Bell, Geophysical Laboratory, 2801 Upton St. N.W., Washington, D.C., U.S.A.

J. Ganguly, Department of Geosciences, University of Arizona, Tucson, Az 85721, U.S.A.

L. L. Perchuk, Institute of Experimental Mineralogy, U.S.S.R. Academy of Sciences, 142 432, Chernogolovka, Moscow, U.S.S.R.

A. Lasaga, Department of Chemistry, Harvard University, Cambridge, Mass. 02138, U.S.A.

S. M. Richardson, Smithsonian Astrophysical Observatory, Cambridge, Mass. 02138, U.S.A.

H. D. Holland, Department of Geological Sciences, Harvard University, Cambridge, Mass. 02138, U.S.A.

P. J. Wyllie, Department of Geophysical Sciences, The University of Chicago, Chicago, Illinois 60637, U.S.A.

E. J. Olsen, Field Museum of Natural History, Chicago, Illinois 60605, U.S.A.

S. K. Saxena, Department of Geology, Brooklyn College, City University of New York, Brooklyn, N.Y. 11210, U.S.A.

I

ENERGETICS OF GEODYNAMIC PROCESSES

1

V. V. BELOUSSOV

Abstract The development of the structure of continents in all its manifestations and at all stages could be associated with the phenomenon of gravitational instability as the principal cause of endogenous processes. Gravitational instability on a general scale of the Earth appears both in the process of primary accumulation of the protoplanetary material and in the consecutive changes in its phase state. These changes occur as the result of heating of the Earth's interior largely under the influence of radioactive decay. Gravitational instability appears on different levels within the globe and on different scales. The asthenosphere is an important level of instability. Near the surface, on a smaller scale, the instability appears in the process of sedimentation and as the secondary product of tectonic displacements. This article describes in a qualitative form the general scheme of processes in the Earth related to gravitational instability. Only parts of the scheme might be considered proven; other parts remain suppositional. When proposing suggestions, the author has sought to formulate the most logically secure concept which can serve as a guide for further studies and, to a certain extent, as an alternative to the "new global tectonics."

GRAVITATIONAL INSTABILITY AND THE DEVELOPMENT OF THE STRUCTURE OF CONTINENTS (attempt at a synthesis)

1 DIVISION OF THE EARTH INTO GEOSPHERES

In the light of modern cosmogenic concepts the division of the Earth into the core, the lower and the upper mantle and the crust took place by the process of differentiation of the Earth's matter. The Earth's matter accumulated from the particles of the protoplanetary cloud (Levin, 1972). Since accumulation occurred largely at random, the distribution of the Earth's matter was in the first stage quasi-homogeneous, i.e., homogeneous in large masses and inhomogeneous in small ones (Safronov, 1964). This very inhomogeneity created gravitational instability, which resulted in the tendency of particles to move in the direction leading to the decrease of potential energy, i.e., to the redistribution of particles by density with the resulting accumulation of relatively heavy particles in deep geospheres and of relatively light ones in the outer geospheres.

Gravitational differentiation was promoted by the heating of the Earth caused by radioactive decay, energy of particle collisions, and tidal friction. Heating reduced the viscosity of material, thus facilitating displacement of particles. As is known, some researchers hold that in the early history of the Earth there was a stage of complete, or almost complete, melting when differentiation occurred at a more rapid rate (Wetherill, 1972). Gravitational differentiation has lasted during the whole of geological time, and is still active. The level of the initial division of the heavy and light matter in the Earth, already divided into geospheres, apparently lies near the boundary between the mantle and the core (Artjushkov, 1968, 1970). The upper mantle composition is, evidently, close to the average composition of the Earth. This is indirectly confirmed by the fact that the most reliable average calculated density of matter in the foundation of the lower mantle (about 5.6 g/cm^3) is close to the mean density of the whole Earth (5.52 g/cm^3). The study of the eigen-oscillations of the Earth reveals substantial inhomogeneities in the layer 200–300 km thick near this same boundary (Jarkov et al., 1974). These inhomogeneities could be associated with the primary division of matter.

The division of the Earth into geospheres and, especially, the separation of the crust from the upper mantle, should not be regarded, as a purely mechanical process. Physico-chemical processes also played an important role in this differentiation; their meaning is still far from being obvious. It is of no concern to us now, however, since only the final result is essential, i.e. differentiation of matter by density. Calculations show that in this process the energy escape reached 1.5×10^{35} erg during the existence of the Earth.

2 ASTHENOSPHERE

An extremely important outcome of differentiation at an early geological stage due to partial melting and under the combined effects of material transfer, pressure and temperature, was the formation of asthenospheric layer roughly at depths between 100 to 400 km.

The upper mantle under the continents, evidently, largely consists of garnet and spinel peridotite with local eclogite inclusions (Luts, 1974; Sobolev, 1974). The partial melting of these rocks, taking place at the indicated depths and at about 1400°C, causes the melting of basalt, which probably forms films around crystals (mostly olivine) that remain solid.

An essential feature is the fact that the basalt liquid of the asthenosphere can accumulate into large masses and rise to the surface only under certain specific conditions, the most important of which is high heating causing convective movement. Usually the basalt remains for a long time in the asthenosphere in the form of small scattered inclusions. The amount of such scattered liquid in the asthenosphere can vary from negligible to, perhaps, 30%. The presence of liquid decreases not only the viscosity, but also the mean density of the asthenosphere. When there is 15% of liquid in a given volume, the asthenospheric matter reduces its density by 0.1 g/cm^3. In this case the density of the material in the roof of the asthenosphere becomes lower than that of the overlying solid upper layers of the mantle (the so-called substratum), and this situation produces the state of gravitational instability. This level of instability plays an extremely important role in endogenous processes.

3 INHOMOGENEITY OF HEAT FLOW AND OF THE STRUCTURE OF THE TECTONOSPHERE[1]

The mean value of heat flow is $1.5 \cdot 10^{-6}$ cal/s·cm^2 and the most frequent value of the heat flow is $1.1 \cdot 10^{-6}$ cal/s·cm^2; however, considerable variations in the flow are observed in different places (Polyak and Smirnov, 1968; Lubimova and Polyak, 1969; Lee, 1970). These oscillations correlate with the regions of various present-day endogenous regimes. On continents the lowest flow is observed on ancient crystalline shields (less than $1 \cdot 10^{-6}$ cal/s·cm^2). In the zones of weak orogenesis, for example, in places of Paleozoic folded regions, it rises to 1.5 in the same

[1]The tectonosphere is understood as the combination of both the lithosphere and the asthenosphere.

unit. But in the Tien Shan, where strong recent tectonic activity is observed, it increases to 1.8. The mean flow values are still higher (2.0) in rift zones. Finally, the highest heat flows on the continents are connected with regions of modern volcanism (3.6). These regions, as thermal regimes, could probably be analogous to the ancient geosynclines. This similarity is confirmed by the paleo-temperature measurements, which showed that in the areas of regional metamorphism during its development the heat flow should be 3–5 times higher than the normal (Billings, 1937; Bederke, 1947; Verhoogen, 1961; Zwart, 1962). Since regional metamorphism is developed in geosynclines, these figures should be regarded as characterizing geosynclinal zones during the development of regional metamorphism. They are, therefore, close to the mean heat flow of present-day volcanic regions.

Though the data on the depth of occurrence are still limited, the present data on thickness and properties of the asthenosphere point to considerable inhomogeneities in its distribution and to the fact that these inhomogeneities on the whole correlate with the differences in the heat flow values (Cook, 1962; Fedotov et al., 1964; Tarakanov, 1965a,b; Lukk, 1966; Fedotov, 1966; Tarakanov and Leviy, 1968; Brune, 1969; Dorman, 1969; Herrin, 1969; Archambeau et al., 1969; Ansorge et al., 1970; Zorin, 1971; Knopoff, 1972; Tuneto Kurita, 1974). Under ancient shields the asthenosphere is very weakly developed and, possibly, is absent in some places. Where its existence is certain the asthenosphere starts at the depth of more than 100 km (120 km on the Canadian shield) and its thickness is only several tens of kilometers. Under the platforms with the sedimentary cover the roof of the asthenosphere is at a depth of about 100 km, while the thickness reaches 150 km. Under the regions of recent orogenesis (for example, under the Alps) the roof of the asthenosphere rises to the depth of 80 km; it becomes thicker and its influence on seismic waves velocities increases. Still more obvious is the low velocity layer under the rifts: here its roof is 60–70 km from the surface and, probably, it joins in the upper part with the lenses of layers of intermediate velocities, lying between the crust and the mantle. The velocity of transverse waves in the asthenosphere is on the average equal to 3.7 km/s, as compared to 4.6 km/s in the substratum. Present-day volcanic regions (for example, on the Kuril Is.) show signs that the low velocity layer rises as far as the foot of the crust and its thickness exceeds 200 km.

Thus, in accordance with the available data, a relationship is evident between endogenous regimes, heat flow intensity and the state of development of the asthenosphere. The most active endogenous regimes, which could be called "excited," are connected with high heat flows, and the thicker and nearer to the surface asthenosphere, which greatly affects the seismic waves' velocities. These properties of the asthenosphere are, apparently, connected with more intensive heating of the upper mantle by a high heat flow, in the presence of which partial melting actively takes place. This includes geosynclinal (especially, the ophiolite stage of the eugeosynclinal regime), orogenic, and rift regimes. In these cases the thermal excitation of the upper mantle might be suggested. Quiescent regimes (platform) are associated with relatively low heat flows.

Consequently, there are grounds to suppose that spatial inhomogeneity of the endogenous development of the tectonosphere is caused by a corresponding inhomogeneity of the thermal regime of the upper mantle.

4 ASTHENOSPHERIC DIAPIRISM

The phenomena of diapirism on the boundary between the asthenosphere and the substratum develop under conditions of excited endogenous regimes and may be considered as the direct cause of the vertical crustal movements. The conditions of asthenospheric diapirism were principally considered by Ramberg (1967, 1972). This problem was also studied by Van Bemmelen (1972) and Maxwell (1968, 1969). The actual reflection of this process on the surface should depend on the properties not only of the asthenosphere, but also of the lithosphere as well. The author (Beloussov, 1975) gave, elsewhere, a more detailed description of the possible results of different relationships between the asthenosphere and the lithosphere. If the lithosphere is divided by deep faults into long blocks, which can move against each other in a vertical direction, then it is quite probable that this property of the lithosphere is taken advantage of, and the asthenospheric diapirism results in the contrasting division of the lithosphere into the rising and sinking blocks. Thus the contrasting vertical movements of the lithosphere appear, which is characteristic of the geosynclinal and orogenic regimes. The difference between the first and the second is in the fact that, in the first the subsidences dominate over the uplifts, while in the second, the uplifts dominate over the subsidences. The author attributes this to the differences of permeability in the lithosphere. The geosynclinal regimes appear in places, where the lithosphere has greater permeability, and, therefore, the lithosphere, saturated by basic and ultrabasic magma, after solidification of magma, becomes heavier and partially sinks into the asthenosphere. Under these conditions the movements of the roof of the asthenosphere and the lithosphere are directed towards each other: over the asthenospheric diapir the lithosphere sinks (Fig. 1A).

Orogenic regimes are associated with areas of very high permeability which coincide with individual deep faults. The blocks between the faults are almost impermeable and, therefore, are pushed upwards over the diapirs (Fig. 1C). A change in the character of lithospheric permeability occurs during the middle stage of development of a geosyncline (in

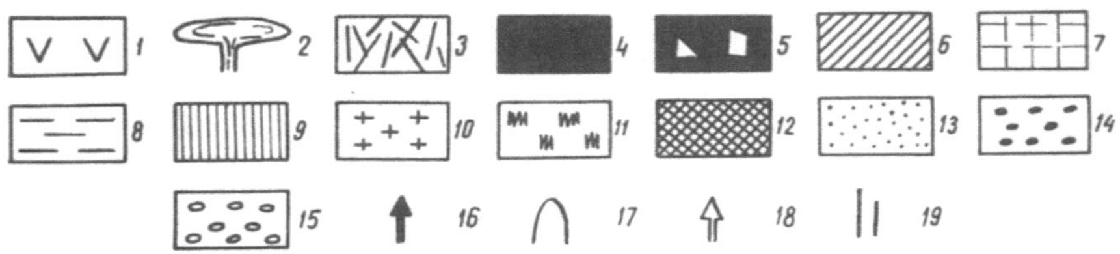

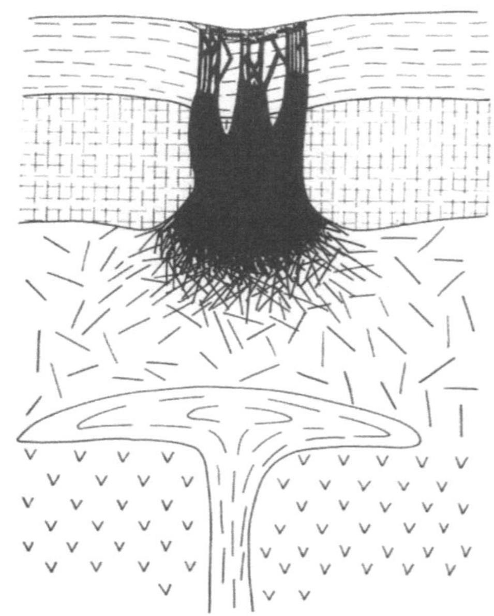

A

Fig. 1A–D. Some endogenous regimes and deep processes (A) Ophiolite stage of eugeosyncline regime; (B) Inversion stage of eugeosyncline regime; (C) Orogenic regime; (D) Rift regime. *1*: Golytzin layer of Upper Mantle; *2*: Heated material rising from Lower Mantle; *3*: Asthenosphere (density of black lines indicates the degree of partial melting); *4*: Melted basic and partly ultrabasic material from the asthenosphere; *5*: Crust-mantle mix; *6*: Eclogites; *7*: Lower part of Lithosphere (substratum); *8*: Continental crust; *9*: Metamorphism of granulite facies; *10*: Granitization; *11*: Metamorphism of amphibolite facies and deep folding (deep crustal diapirism); *12*: Newly recrystallized crust; *13*: Lower terrigenous geosyncline formation; *14*: Flysch formation; *15*: Molasse formation; *16*: Basic volcanoes; *17*: Foci of intermediate and acid magma; *18*: Intermediate and acid volcanoes; *19*: Deep faults

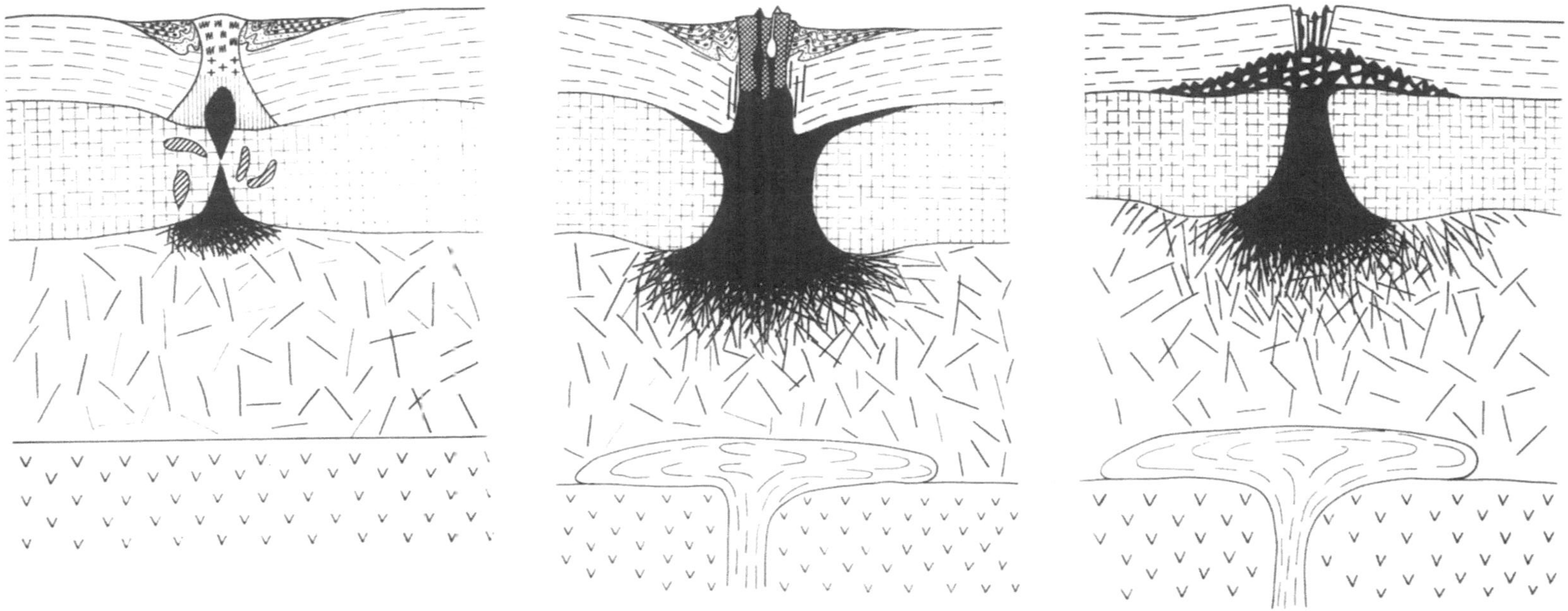

B C D

the so-called inversion stage), when the recrystallization connected with regional metamorphism and granitization "heals" many channels in the lithosphere (Fig. 1B).

If the lithosphere is uniform and monolithic, then asthenospheric diapirism can form a wide domelike uplift in the lithosphere only, which, by stretching and breaking, provides a structural basis for the development of the rift regime (Fig. 1D).

5 DEEP CRUSTAL DIAPIRISM

The next level of manifestation of gravitational instability is the diapirism appearing in the continental crust at the depth where regional metamorphism and granitization are developing.

It has been known for a long time that during granitization the density of rocks is reduced and they acquire the ability to float. What is not taken into account is that temporary reduction of density should also accompany the process of regional metamorphism, even at its earliest stages. Such reduction of density is caused by water escaping both from the crystalline grit and from the adsorbed state during heating of rocks, and it is limited to the time during which the water stays in the pores. Calculations show that in this process the increase of volume and the decrease of density can reach 10–15%. Actually this should lead to diapirism of different metamorphic rocks, including such slightly metamorphozed rocks as the phyllites. The processes of diapirism, however, should stop as soon as the water in pores escapes from the rock and the density of the rock increases. During this same stage of water escape, the viscosity of rocks is reduced and they become amenable to large and complicated deformations. Since the pelitic rocks, dominating the geosyncline in its early stage of development, have low water permeability, this stage of water saturation of rocks can last millions of years.

Deep crustal diapirism provides an explanation for the characteristic peculiarities of deformations of metamorphic rocks (such as complicated disharmony, multistage combination of folds of different sizes, steep hinges, criss-crossing of folds of different orientations) which can be attributed only to the influence of gravity or forces due to volume changes. Ez (1976) expressed the opinion that the structure of the crystalline foundation of the continental crust can in general be explained only by the manifestation of a certain convective mechanism.

However, deep crustal diapirs are the instruments causing folding also in the surrounding normal sedimentary rocks: metamorphic (and also granite and granite-gneiss) diapirs, intruding into such rocks affect them mechanically by pushing them apart and crumpling them into folds (Fig. 2). Such is the explanation of the structure of the axial parts in many folded zones (Beloussov, 1972).

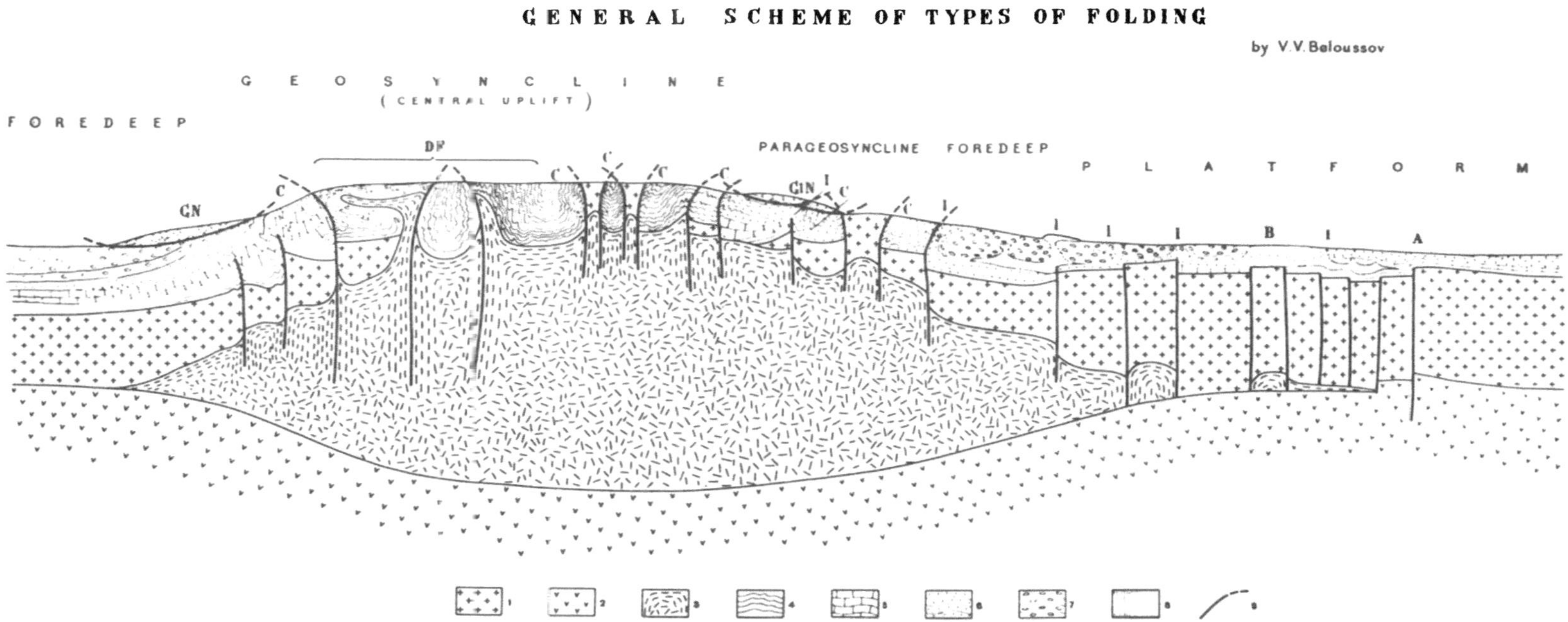

Fig. 2. General scheme of types of folding *1*: Crystalline basement of Upper continental crust; *2*: Lower continental crust; *3*: Rocks under metamorphic and ultra-metamorphic activation; *4*: Lower Terrigenous geosyncline formation; *5*: Limestone formation; *6*: Flysch; *7*: Molasse; *8*: Evaporites; *9*: Faults; *DF*: Deep folding; *C*: Folding of general crumpling; *I*: Injection folding; *B*: Block folding; *A*: Anteclise; *GN*: Gravity nappe

Tectonic nappes of the Pennine type should also be considered as a manifestation of deep crustal diapirism.

In order to understand the process of deep crustal diapirism, it is important to have in mind that in that process floating up of relatively light material is accompanied not only by its squeezing *from under* the heavier material, but also by its compression in the horizontal direction. Gravitational equilibrium is distorted when less dense material occurs not only *under* the denser one, but also *alongside* it, at the same gravipotential level.

If in that case the denser material has sufficient plasticity, then it flows towards the inclusion with lesser density, trying to squeeze it and push it upwards. This mechanism should be important in the development of horizontal compression in the inner parts of folded zones, when they experience deep heating.

6 OTHER PROCESSES CAUSED BY GRAVITATIONAL INSTABILITY

The following processes of structure formation, connected with gravitational instability, are worthy of only a brief notice: (1) Falling apart of the upper parts of lifted crustal blocks, in connection with which vertical faults acquire a tilt in their upper part, the lifted block pushes the surrounding rocks and causes their crumpling; (2) slipping of the Helvetian nappes from the slopes of the tectonic uplifts; (3) squeezing of light and plastic rocks from under tectonic overthrusted heavy massifs; and (4) diapirism in sedimentary layers connected with gravitational instability of sedimentary origin (salt and clay diapirism).

7 THE EARTH AS A THERMAL ENGINE

There is a certain rhythm in the development of the crust of continents, which is expressed in the so-called tectonic, tectonomagmatic, or, more generally, endogenic cycles. The cycles are most conspicuously expressed in the regions of the eugeosynclinal regime, which further changes to orogenic regime. In such regions the initial and the final stages of the cycle are distinguished (i.e. the ophiolite stage of the eugeosynclinal regime and the orogenic regime), and have common features indicating thermal excitation in the asthenosphere, which are the mobilization of the mantle magmas (ophiolites and basic magmas) and the contrasting vertical movements, directly connected, to our mind, with the asthenospheric diapirism. The middle part of the cycle (the inversion stage of development of the geosyncline) is rather unique because of the absence of uplift to the surface of the mantle magmas and because of a certain smoothing out of contrasts in the vertical crustal movements. However, the development of regional metamorphism and granitization, and also the capacity of the crust for large plastic deformations indicate that the crust in this middle stage is heated more than in the beginning and in the end of the cycle.

Since the cycles follow one another, the end of one cycle immediately passes into the beginning of the next and, thus, these two epochs of asthenospheric excitation merge into one. This is emphasized, moreover, by the fact that the ophiolite stage of development of the eugeosyncline in some zones is demonstrated simultaneously with the orogenic regime in the others. Thus, for example, the orogenic regime, closing the Baikal cycle, develops in the period (in the end of the Cambrian), when in other zones the ophiolite stage of the Caledonian cycle begins. The orogenic regime of the Caledonian cycle is superimposed in time (the end of the Silurian–the beginning of the Devonian) onto the ophiolite stage of the Hercynian cycle in other zones, while orogenesis, closing the Hercynian cycle, becomes synchronous (in the Permian and in the Triassic) with the beginning of development of the geosynclines of the Mesozoic cycle. The orogenesis at the end of the Mesozoic cycle coincides in time (in the Late Jurassic and Early Cretaceous) with the beginning of the Alpine cycle (Fig. 3).

Recent tectonic activity, manifested on continents in the Neogene-Quaternary time, is an example of the orogenic regime coinciding with the end of the Alpine cycle. As known, it is manifested not only in places of the Alpine geosynclines, but far beyond their limits, which emphasizes the uniqueness of the orogenic regime. It cannot be regarded as a simple termination of development of geosynclines, though the latter are still the most favorable zones for orogenesis. In a number of places in the neotectonic stage, orogenesis is replaced by the rift regime, which to our mind (see above) is connected with the special properties of the lithosphere.

A similar, but correspondingly less expressed rhythm of endogenous processes can be traced both for the miogeosynclinal and for the platform regimes; for the latter the rhythm is evident in the fact that in the beginning and in the end of a cycle the contrast of the vertical crustal movements is also greater than in the middle of the cycle (Fig. 4).

Therefore, the endogenous rhythm is determined mainly by the periodically occurring thermal excitation of the asthenosphere. Such excitation can be associated only with periodic transfer of heat from the deeper geospheres with the rising material. This uplift of the material from deeper geospheres can be justifiably regarded as a part of the process of general differentiation of the globe. This differentiation does not, however, occur regularly in time, but spasmodically. The light material should, probably, accumulate for some time before it can float upwards.

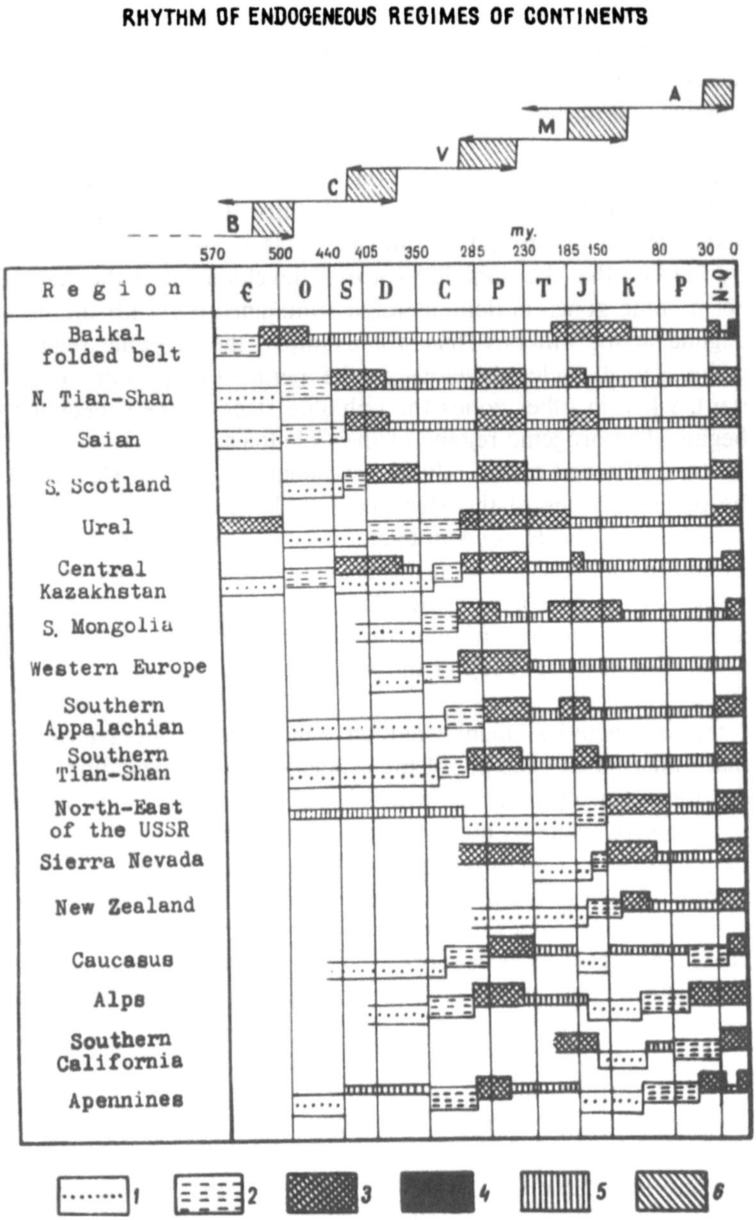

Fig. 3. Rhythm of endogenous regimes of continents *1*: Ophiolite stage and in general the stage of preponderance of subsidences; *2*: Inversion stage of general equality of subsidences and uplifts; *3*: Orogenic regime (stage of preponderance of uplifts); *4*: Rift regime; *5*: Stage of moderate and low vertical movements (mainly platform conditions); *6*: Excited regimes. Endogenous cycles: *B*: Baikal; *C*: Caledonian; *V*: Variscan or Hercynian; *M*: Mesozoic or Kimmerian; *A*: Alpine

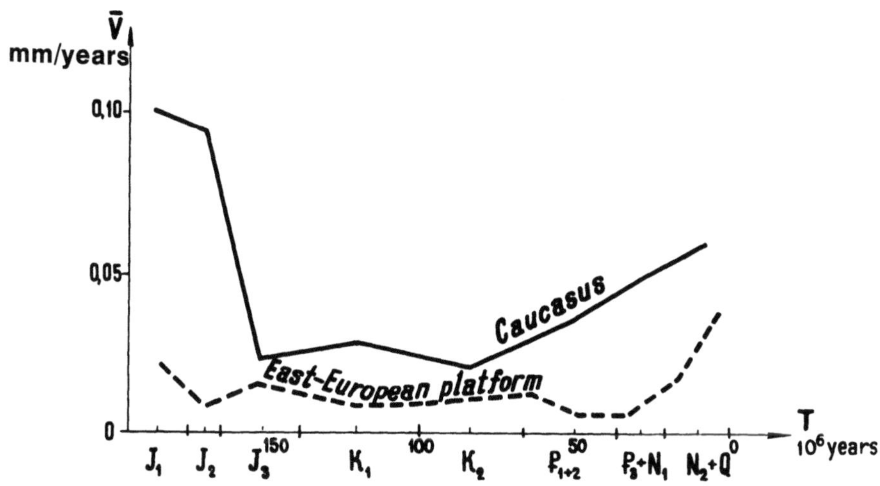

Fig. 4. Variations of average speed of vertical movemens (subsidences and uplifts) during Alpine endogenous cycle for Caucasian geosyncline and East-European platform (after Sholpo, 1969)

The excitation stages of the asthenosphere are divided by the epochs of relative calming down of the processes in the upper mantle, but the same epochs are the periods of time when the crust is heated. We associate this propagation of the heat wave from the upper mantle into the crust with the asthenospheric diapirism, in the process of which the heat reaches the crust with the upper mantle material and gradually warms up the crust.

Finally, the deep crustal diapirism caused by this heating helps to heat the upper layers of the crust, thus increasing their plasticity, which facilitates folding.

Therefore, the Earth might be considered as the thermal engine, in which due to low thermal conductivity the heat accumulates and then escapes by portions through a series of connected mechanisms acting in a sequence at different levels. All these mechanisms are based on density inversion, on gravitational instability and on movement of material tending to reestablishment of equilibrium.

8 HORIZONTAL INHOMOGENEITY IN MANIFESTATIONS OF ENDOGENOUS REGIMES

Different endogenous regimes simultaneously exist on different areas of continents. Beginning with the Proterozoic many of them form linearly elongated zones.

Consequently, the processes of thermal excitation of the asthenosphere described above have irregular development not only in time, but in space as well, mainly coinciding with certain zones and having weaker expression in others. It can be suggested that the floating up of the material bearing heat into the upper mantle from deeper spheres does not take place everywhere uniformly, but is confined mostly within some zones which play the part of super-deep channels. The latter ought to be understood not as gaping fissures, but as zones of high penetrability, filled with low viscosity material. The linearity of tectonic zones reflects the linearity of these super-deep channels.

A general evolution of such channels can be outlined (Beloussov, 1972). In the Archean (3000–2800 million years ago) the channels pierced the entire inner space of the globe below the continents largely at random. This was the reason for manifestations of deformations and magmatism everywhere and for the lack of linearity in the structures of the Archean.

In the Early and Middle Proterozoic (1600–1500 million years ago) for the northern continents and during the whole Proterozoic (to 560 million years ago) for the southern continents, more active endogenous zones (protogeosynclines) and more quiet (protoplatforms) were formed separately. The active endogenous zones are characterized by linearity. This division, however, is unstable and the protogeosynclines and the protoplatforms often interchange places.

The last large stage in the development of the continental crust and the geosynclinal platform starts at the beginning of the late Proterozoic in the northern continents and with the beginning of the Phanerozoic on the southern continents. During this stage the excited regimes are manifested in zones with linear contours. Also linear are the subzones of higher order distinguished within them. In the beginning of the stage the "ancient" platforms on the Pre-Cambrian folded basement are developed. The pattern of further evolution is that, from cycle to cycle the platforms grow, while the geosynclines shrink. This process, however, is not haphazard, but has a definite trend. As a result of such trend at present the excited regimes are concentrated not at random but in two zones: one almost latitudinal—Mediterranean-Central Asiatic; the other, meridional—Circum-Pacific.

The duration of endogenous cycles, which developed on the background of this large evolution, was in the Phanerozoic close to 200 million years. The duration of cycles for the Pre-Phanerozoic time is difficult to establish; perhaps, it varies within 300–600 million years.

How the super-deep channels appeared, what caused their evolution, are problems far beyond the reach of present-day suppositions. One thing is obvious, though, that the continents in their development are connected to great depths, much deeper than the asthenosphere. It is, therefore, impossible to imagine them in the form of thin lithospheric plates slipping over the asthenosphere.

REFERENCES

Ansorge, J., Emier, D., Fuchs, K., Lauer, J. P., Müller, St., Peterschmitt, E.: Structure of the crust and upper mantle in the Rift system around the Rhinegraben. In: Graben Problems. Illies, J. H., Mueller, St. (eds.). Stuttgart; 1970, pp. 190–197

Archambeau, C. B., Flinn, E. A., Lambert, D. G.: 1969. Fine structure of the Upper Mantle. J. Geophys. Res. *74,* 5825–5865 (1969)

Artjushkov, E. V.: Gravitational convection in the Earth's interior. Izv. Akad. Nauk SSSR. Fiz. Zemli, *9,* 3–17 (1968)

Artjushkov, E. V.: The differentiation of the Earth's material according to density and connected phenonena. Izv. Akad. Nauk SSSR, Fiz. Zemli *5,* 18–30 (1970)

Bederke, E.: Über der Wärmehaushalt der Regionalmetamorphose. Geol. Rundsch. *35,* H. 1, 26–32 (1947)

Beloussov, V. V.: Basic trends in the evolution of continents. The Upper Mantle. In: Developments in Geotectonics, Ritsema, A. R. (ed.). Amsterdam: Elsevier, 1972, pp. 95–117

Beloussov, V. V.: Principles of Geotectonics. Moscow: Nedra, 1975, 262 pp

Van Bemmelen, R. W.: Geodynamic Models. An Evaluation and a Synthesis. Developments in Geotectonics. Amsterdam: Elsevier, 1972, 267 pp

Billings, M. P.: Regional metamorphism of the Littleton-Moosilanke area, New Hampshire. Geol. Soc. Am. Bull., *48,* 463–566 (1937)

Brune, J. N.: Surface waves and crustal structure. The Earth's crust and Upper Mantle. Am. Geophys. Union, Mon. *13,* 230–242 (1969)

Cook, K. L.: The problem of the mantle-crust mix: lateral inhomogeneity in the uppermost part of the Earth's mantle. Advan Geophys *9,* 295–360 (1962)

Dorman, J.: Seismic surface-wave data on the Upper Mantle. The Earth's crust and Upper Mantle. Am. Geophys. Union Mon. *13,* 257–265 (1969)

Ez, V. V.: Tectonics of Deep Zones of the Continental Earth's Crust. Moscow: Nauka, 1976, 168 pp

Fedotov, S. A.: 1966. Deep structure, properties of the upper mantle and volcanic activity of the Kuril-Kamchatka island arc by seismic data in 1964. In: Proc. 2nd Volcanological Conf. Moscow: "Nauka," 1966, Vol. III

Fedotov, S. A., Kuzin, I. P.: Velocity section of the Upper Mantle in the region of the Southern Kuril Islands. Izv. AN SSSR, Ser. Geofiz., *5,* 670–686 (1963)

Fedotov, S. A., Matveieva, N. N., Tarakanov, R. Z., Yanovskaya, T. B.: On velocities of longitudinal waves of the upper mantle in the area of Japan, the Kuril Islands. Izv. AN SSSR, *8,* 1185–1191 (1964)

Herrin, E.: Regional variations of P-wave velocity in the Upper Mantle beneath North America. The Earth's crust and Upper Mantle. Am. Geophys. Union, Mon. *13,* 242–246 (1969)

Jarkov, V. N., Dorofeeva, L. N., Dorofeev, V. M., Lubimov, V. M.: Tentative distribution of dissipative function Q(J) in the mantle of the Earth. Fizika Zemli *12,* 3–12 (1974)

Knopoff, L.: 1972. Observaion and inversion of surface-wave dispersion. The

Upper Mantle. Developments in Geotectonics, Ritsema, A. R. (ed.). Amsterdam: Elsevier, 1972, pp. 497–519
Lee, W. H. K.: On the global variations of terrestrial heatflow. Physics of the Earth and planets. Interiors *2*, 332–341 (1970)
Levin, B. J.: Origin of the Earth. The Upper Mantle. In: Developments in Geotectonics. Ritsema, A. R. (ed.) Amsterdam: Elsevier, 1972, pp. 7–29
Lubimova, E. A., Polyak, B. G.: Heat flow of Eurasia. The Earth's crust and Upper Mantle. Geophys. Monogr., Am. Geophys. Union, *13*, 82–88 (1969)
Lukk, A. A.: Structure of the upper mantle of the Earth along the profile the Pamirs—the River Lena. Sov. Geol. *2*, 106–117 (1966)
Luts, B. G.: Petrology of the deep zones of the continental crust and the Upper Mantle. Moscow: Nauka, 1974, 304 pp
Maxwell, J.: Continental drift and a dynamic Earth. Am. Scientist, *56*, 35–51 (1968)
Maxwell, J.: The Mediterranean, ophiolites and continental drift. In: The Megatectonics of Continents and Oceans. Johnson, H, Smith B, (eds.). New Brunswick: Rutgers Univ. Press, 1969, pp. 167–193
Polyak, B. G., Smirnov, Ya. B.: Relationship between deep heat flow and tectonic structure of continents. Geotektonika, *4*, 3–19 (1968)
Ramberg, H.: Model experimentation of the effect of gravity on tectonic processes. Geophys. J. Roy. Astr. Soc. *14*, 307–329 (1967)
Ramberg, H.: Theoretical models of density stratification and diapirism in the Earth. J. Geophys. Res. *77*, 877–889 (1972)
Safronov, V. S.: The primary inhomogeneities of the Earth's mantle. Tectonophysics, *1*(3), 217–221 (1964)
Sholpo, V. N.: Quantitative criteria in evaluation of regime of vertical movements. Geotektonika, *2*, 38–51 (1969)
Smirnov, Ya. B.: Connection of heat field with structure and development of the Earth's crust and Upper Mantle. Geotektonika, *6*, 3–25 (1968)
Sobolev, N. V.: The deep seated inclusions in kimberlites and the problem of the Upper Mantle composition. Novosibirsk: Nauka, 1974, 246 pp
Tarakanov, R. Z.: Travel time curves of P and S waves and velocity section of the upper mantle of the Earth by observation data of the Kuril-Japanese earthquakes. Fizika Zemli *7*, 90–101 (1965a)
Tarakanov, R. Z.: The velocity section on the Upper Mantle in the transition zone from Asia to the Pacific Ocean. Tectonophysics *2*(2/3), 227–237 (1965b)
Tarakanov, R. Z., Leviy, N. V.: A model for the Upper Mantle with several channels of low velocity and strength. The Crust and Upper Mantle of the Pacific Area. Am. Geophys. Union *12*, 43–50 (1968)
Tuneto Kurita: Upper Mantle structure in the Central United States from P and S-wave spectra. Phys Earth Planetary Interiors *8*, 177–201 (1974)
Verhoogen, J.: Distribution of temperatures and heat flow during regional metamorphism. In: Physico-Chemical Problems of Formation of Rocks and Mineral Deposits. Acad. Sci. USSR, 1961, Vol. I, pp. 17–23
Wetherill, G. W.: The beginning of continental evolution. The Upper Mantle. Devel. Geotect. *4*, 31–45 (1972)
Zorin, Yu. A.: Recent structure and isostasy of the Baikal rift zone and the adjacent territories. Moscow: Nauka, 1971, pp. 168 (in Russ.)
Zwart, H. J.: On determination of polymetamorphic mineral associations and its application to the Bosost Area. (Central Pyrenees). Geol. Rundsch. *52*, 38–65 (1962)

2

J. SUTTON AND J. WATSON

Abstract Although the structure of the continental crust as a whole has resulted from the deformation of preexisting rocks, growth structures formed by the addition of magmatic material derived from the mantle can also be recognized. In some instances direct relationships with bodily motions of continental masses have been established; where the new material is largely granitic, the resulting structures are dominated by the effects of relative vertical movements resulting from changes in density distribution. Many Archaean provinces which received large accessions of material over short time intervals show distinctive tectonic patterns resulting from deformation during periods of volume-increase.

INTRUSION INTO MOVING CRUST

1 GROWTH STRUCTURES IN THE CRUST

The characteristic structural patterns of the continental crust are those produced by the deformation of pre-existing rock material. What may be called growth structures developed by the progressive addition of igneous material *pari passu* with the regional displacement of pre-existing rock units are of restricted development in the continental crust, though they are, of course, the dominant structures of the oceanic crust. This contrast reflects the fact that plates of oceanic crust have a life-span of only a few hundred million years, whereas continental crustal units may remain near the Earth's surface for two or three thousand million years. Under present-day tectonic regimes, additions to the continental crust are made principally near plate boundaries which are also the sites at which new distortional structural patterns are generated. In Archaean and Proterozoic times, however, the continental crust appears to have received larger and more widely distributed accessions of material from the mantle; the distinctive structural patterns of certain Precambrian provinces may, therefore, represent distorted growth structures. Our object in this paper is to examine some aspects of igneous intrusion with the aim of following up this possibility.

2 DIKE-SWARMS AND CENTRAL INTRUSIONS

The emplacement of a basic dike swarm fed by magma derived from the mantle produces a growth structure in which the relationships between the distribution of dike fractures and the large-scale crustal movements controlling their opening are fairly clear. The percentage volume of new material added to the continental crust is small. Some regional Proterozoic dike swarms amount to about 4% of the crust (e.g., Watson and Lisle, 1973), while swarms adjacent to Cenozoic central intrusions in north-west Scotland seldom rise to more than 20% of the crust (Skelhorn et al. in prep). Plateau lavas fed by such swarms seldom exceed 5 km or roughly 15% of the average thickness of the continental crust.

The majority of large Phanerozoic basic dike swarms are located near major fractures associated with rift valleys or continental margins. The alignment of these swarms and the positioning of related central intrusions can in many instances be interpreted in terms of the displacement of adjacent continental slabs in response to large-scale plate movements. Such relationships have, for example, been inferred with respect to the Mesozoic swarms of S.E. North America and N.W. Africa (May, 1971). They imply a temporal connection between crustal motion and magmatism almost as close as that exhibited at the mid-oceanic ridge system.

Regional dike swarms and comparable intrusions of Proterozoic age traverse the interiors as well as the margins of continental masses. The

Great Dike, an early intrusion possibly related to extension, bisects the Rhodesian craton. The ~1200 Ma Mackenzie swarm of Canada penetrates to the central part of the mid-Proterozoic North American continent, invading an area not less than 500 km in breadth and 1000 km in length. Its development implies an appreciable extension of the crust on roughly NE-SW lines, and although it is not directly related to a definite fracture system, the motions connected with emplacement of the magma may have been broadly synchronous with those responsible for the development of the Keweenawan rift system and midcontinent gravity high.

The effects of intrusion in a moving crustal framework are illustrated from another part of this mid-Proterozoic rift system by Stephenson's analysis (1976) of ring intrusions in the Gardar igneous province of south Greenland. The ring complexes, dike swarms and lava fields of this province have a well-established spatial connection with large sinistral wrench faults and graben of east-west trend. At the relatively high crustal level revealed by erosion, adjustment of the country rocks to displacement was achieved mainly by brittle deformation. Some of the ring-intrusions on the other hand appear to have responded to the same stresses by ductile deformation. Their prevailing elliptical forms are explained by Stephenson in terms of simple-shear deformation of hot and partially fluid bodies which were initially of circular cross-section; the sum of ductile and brittle shear strain recorded by the intrusions is roughly equal to the sum of sinistral fault displacements in the country rocks.

The late Archaean to early Proterozoic basic dike swarms which traverse the formerly-united continental masses of Canada, Greenland, and northwest Britain are comparable in scale with the Mackenzie swarm. In the region of Søndre Stromfjord, West Greenland, two swarms intersect in the vicinity of an old-established deep crustal shear zone, the Nagssugtoqidian front. Geometrical analyses of individual dikes, together with a consideration of their relationships with the Nagssugtoqidian structures, have led Escher et al. (1976) to conclude, firstly that the swarms were emplaced in conjugate zones of transcurrent motion and, secondly, that relative movements on at least one of these zones took place before, during and after the period of dike intrusion. The inferred senses of movement connected with dike emplacement are considered by Escher et al. (1976), to be compatible with distortion of the entire North Atlantic continental mass with principal stress direction (σ_1) on NNW-SSE and σ_3 on ENE-WSW lines. It is significant that distortion took place without rupture and without preferential development of fractures perpendicular to the direction of least stress. The complex primary form assumed by many intrusions within the main Nagssugtoqidian shear zone (e.g., Watterson, 1974), and the indications that metamorphic reconstitution took place in the Scottish part of the swarm soon after dike emplacement (e.g., O'Hara, 1961; Tarney, 1963), suggest that, at the deep levels revealed by erosion, the crustal country rocks were hot and capable of ductile deformation during the period over which dike intrusion took place. Taken together, these observations suggest that the growth structures defined by the late Archaean/early Proterozoic dike swarms of Greenland and north-west Britain were developed within a large continental plate at high temperatures and under conditions of moderately high ductility. The distortion with which their emplacement was associated affected a moving crustal mass of continental dimensions (Sutton and Watson, 1974).

3 GRANITE EMPLACEMENT

Plutonic rocks which are, in a broad sense, of granite composition (granites, granodiorites, tonalites and their gneissose derivatives), have been characteristic components of the continental crust since the start of the geological record. Some of these rocks are shown by isotopic dating techniques to have remained within the crust for more than three thousand million years. An unknown proportion of granites which yield younger isotopic ages have been derived from preexisting granitic rocks by deformation, heating and partial melting. The emplacement of such recycled granitic material would involve redistribution of crustal material and would not lead to the development of growth structures in the sense defined in the first section of this paper. On the other hand, the emplacement of granites which were directly derived from sources beneath the continental crust provided genuine additions to its bulk. Over the last decade, the results of isotopic studies in strontium and lead which provide evidence as to the derivation of magmas have lent support to the view that substantial volumes of granitic material have been added to the crust in this way, especially during Archaean times.

Comparison of the volume and distribution of granites in tectonic provinces of different ages suggests that there have been significant changes in the distribution of added granitic material. Later Phanerozoic granites are concentrated near converging plate boundaries and are almost confined to orogenic mobile belts. Even in this setting, they seldom occupy more than about 20% of the surface area at the present level of erosion and some very large batholiths are thought to be effectively confined to the upper crust—the Peruvian and Sierra Nevada batholiths, for example, appear to form long narrow strips some 10 to 15 km in thickness which are roofed by piles of related volcanics, but floored by nongranitic rocks (Hamilton and Myers, 1967; Cobbing and Pitcher, 1972; Myers, 1975). At the other end of the time-scale, granites, together with gneisses and granulites of granitic parentage, usually form well over 60% of the surviving Archaean provinces (Talbot, 1973; Bridgwater et. al., 1973).

The abundance of granitic material, furthermore, is much the same in granite/greenstone-belt provinces which represent relatively high-level crustal sections and in provinces of granulite and gneiss which appear to represent sections at depths of 20 to 30 km. There is little evidence that Archaean granites are confined to converging plate boundaries.

The classic distinction between syntectonic and post-tectonic granites in orogenic mobile belts has generally been applied with reference to locally determined sequences or orogenic disturbances. In a wider context it is becoming apparent that many granites which are post-tectonic with respect to the structural development of their crustal envelope were nevertheless emplaced while relative motions of adjacent plates continued. This phenomenon is well illustrated in sections of the circum-Pacific mobile belts, where consumption of oceanic crust continued over a long period. By dating large numbers of granites classified as syntectonic and posttectonic in the Tasman belt of eastern Australia, Evernden and Richards (1962), showed that there was little correlation between their age and their inferred relationships with tectonic disturbance. Many of the granites of post-tectonic aspect were emplaced long before eastern Australia ceased to override the floor of the western Pacific.

Along the eastern Pacific margin the structure of the oceanic plates and adjacent continental margins suggests that the American continents have been overriding the sea floor for many tens or hundreds of million years, a process which continues today along some sectors. The batholiths of late Mesozoic age were emplaced well within the period of advance of the Americas, and in a broad sense are essentially syntectonic. Their form and relationships with the rocks of their envelope provide very little indication of tangential shortening related to this advance. Hamilton and Myers (1967) emphasize the role of vertical rather than tangential movements in the emplacement of the Sierra Nevada and Boulder batholiths of North America. A similar picture emerges from recent accounts of the Peruvian batholith. Vertical block movements are recorded in the envelope and the granites of the batholith were emplaced by vertical movements relative to the rocks of the envelope which, at least at high levels, descended by cauldron subsidence or large-scale stoping (Cobbing and Pitcher, 1972; Myers, 1975). From Pitcher's estimate of about 70 million years as the time taken to emplace the batholith (1975), it seems that the leading edge of the South American continent could have advanced by more than 1000 km over the Pacific floor during the period over which the granite (and the related volcanics which roof it over), entered the crust. The geometrical relationships of the batholiths, unlike those of the dike swarm considered earlier, have no direct connection with the large-scale crustal motions to which the generation of the parent magma was due, but are strongly influenced by the buoyant nature of the granite material. The resulting structures provide a field of study which Ramberg has made his own (1967, 1970).

The mid-Proterozoic suite of intrusive igneous rocks which includes the rapakivi granites and their monzonitic, noritic and anorthositic associates in Fennoscandia and Southern Greenland is of special interest in the light of these considerations. Rapakivi granites dated at ~1750 million years in Southern Greenland form long, narrow strips, roofed partly by related volcanics and narrowing downward to vertical "stalks" against which structures in the envelope turn sharply downward. Bridgwater et al. (1974) consider that the emplacement of these bodies was achieved by vertical displacements of the surrounding rocks perhaps analogous to those associated with the rise of the Western American batholiths.

A mid-Proterozoic rapakivi suite usually accepted as consisting largely of mantle-derived igneous material is widely distributed from south-east Europe through Fennoscandia and Greenland to Canada and the western United States, defining a broad tract through what may have been a single supercontinent. The tectonic provinces penetrated by intrusions of this suite invariably show evidence of early Proterozoic (>1800 million years) tectonic activity, granite emplacement and metamorphism which, in many localities, was of the type associated with high geothermal gradients. Although the ages of the intrusions vary somewhat, they agree in being post-tectonic with respect to the locally recorded sequence of structures. Bridgwater and his colleagues (Bridgwater and Windley, 1973; Bridgwater et al., 1974), point to evidence which suggests that the tract within which the rapakivi suite is distributed remained a zone of abnormally high heat flow for a period of the order of 400 million years. Prior to the emplacement of the suite, tectonothermal activity led to the development of regionally metamorphosed rocks characterized by effects of polyphase penetrative deformation and often by abundant granite. In south Greenland, where the structural style is relatively well known, the tectonic patterns developed during this period do not suggest significant tangential shortening (Bridgwater et al. 1973a). During and after the intrusion of the rapakivi suite, the country rocks suffered neither regional metamorphism nor penetrative deformation and the characteristic structures were fracture systems defining sediment-filled graben such as those which contain the Jotnian of Fennoscandia, the Gardar Sandstones of Southern Greenland, and the Keweenawan of the Great Lakes. The later igneous assemblages associated with these fracture systems are predominantly basic and alkaline.

Over a long time span, therefore, the hot tract defined by intrusions of the rapakivi suite passed from a stage in which the rocks and structures produced were of types normally ascribed to the effects of orogenic crustal activity to a stage in which the rocks and structures produced were those typical of cratonic regimes. Bridgwater et al. (1974) suggest that the prolonged phase of abnormal thermal conditions in Southern

Greenland was related to the development near the base of the crust of a magma "cushion" of the type which has been thought to underlie Phanerozoic rift structures (e.g., Illies, 1970). They infer that the emplacement of the distinctively shaped rapakivi bodies was linked in time with the formation of a crustal arch above the magma "cushion" and with the onset of rifting at the extended crest of this arch. The processes envisaged are those which in Phanerozoic times have been characteristic of the interior parts of continental plates and of regimes involving limited extension of the crust. On the other hand, Bridgwater and his colleagues also see analogies with the long-term development of the Pacific margin of the American continents in Phanerozoic times. The entire tract characterized by intrusion of the rapakivi suite follows the length of the Proterozoic supercontinent as reconstructed from palaeomagnetic evidence by Piper (1976), but is so broad that some intrusions lie far from the long edges of the supercontinent. There is, therefore, no clear definition of a leading edge at which granite intrusion was concentrated.

The assemblages referred to above are concentrated in provinces which were tectonically active in early Proterozoic times. The location of other provinces of the same age, and that of certain younger Proterozoic provinces, with respect to the boundaries of continental plates have been debated in terms of two alternatives—that all such provinces mark sutures between formerly separate plates, and that some were essentially intracontinental or ensialic (compare, for example, Burke and Dewey, 1973; Hurley, 1973; Sutton and Watson, 1974; and Shackleton, 1976). Some of the disputed provinces, like the Nagssugtoqidian belt of Greenland are poor in granite, although they are characterized by effects of regional metamorphism and penetrative deformation. The Laxfordian complex of north-west Britain has a tectonic pattern which may have been initiated by ductile extension (Watson, 1976), and by relatively small sheets of potassic granite which tend to be concentrated near old lineaments, and which could represent the roots of formerly extensive high level intrusions. The western Churchill province is rich in granite, but as yet (Davidson, 1972), neither the proportion of this granite which is Proterozoic as distinct from Archaean, nor the source, distribution or intrusion form of the Proterozoic fraction have been clearly established. Hurley (1972, 1973), has explored the idea that in parts of the Pan-African system of belts granites with low initial strontium isotope ratios were derived from sources in a zone of hot mantle and that such belts may be sited on lines of mantle upwelling. Shackleton has arrived at a similar view from different reasoning (1969). In the present context the view implies that Proterozoic granite may have been added to the crust at intraplate sites, and not simply, as under the later Phanerozoic regimes, near destructive plate margins. The resulting patterns are everywhere dependent on the buoyancy of the newly emplaced granite.

4 THE ROLE OF ARCHAEAN MAGMATISM

In provinces where the structures date back to Archaean times, (>2400 million years), the principal rock assemblages vary between a low-grade type exemplified by the greenstone-belt/granite association of the Kapvaal or Rhodesian craton and high-grade type typified by the granulites and gneisses of the North Atlantic craton. The proportion of granite in these provinces is much greater than that in younger mobile belts, and does not appear to vary greatly with depth. Isotopic studies in several provinces have suggested that much of this granite was derived from the mantle in Archaean times and does not represent the derivative of a primeval continental crust (for references see Moorbath, 1975).

Two points concerning the source and timing of Archaean magmatism which have emerged from recent isotopic studies are of interest in the present context. Firstly, individual greenstone belts of several provinces and the intrusive granites which flank them have yielded scarcely distinguishable ages, a coincidence which suggests that the predominantly basic volcanism which gave rise to the greenstone-belts and the acid plutonic activity involved in granite formation took place during a closely integrated train of events (cf. Krogh and Davis, 1974; Hawkesworth et al., 1975). Where initial strontium isotope ratios have been determined (e.g., in Rhodesia, Hawkesworth et al., 1975), the low values obtained from many granitic bodies suggest that the entire volcanic/plutonic partnership represents material added to the crust from sources in the mantle. Secondly, the granitic and volcanic partners mentioned above, appear from their isotopic constitution to be sharply distinguished from older granite units, such as the basement tonalitic gneisses of Rhodesia (Hickman, 1974; Hawkesworth et al., 1975), or the oldest gneisses of Minnesota (Goldich and Hedge, 1974). Such older units include deformed granites and gneisses with or without enclaves of supracrustal rocks, and represent remnants of crustal material of continental character several hundred million years older than the adjacent granite/greenstone belt partnership. In a few localities (for example in southern Rhodesia, see Bickle et al., 1975), volcanic sequences of the younger greenstone belts are seen to rest unconformably on a granite basement.

The characteristic domed forms of granites associated with greenstone belts and the evidence (despite the occasional records of large-scale inversion or isoclinal folding) that the majority of greenstone belts occupy large synforms, demonstrate the importance of gravity-assisted vertical movements in the development of the growth structures characteristic of Archaean greenstone-belt provinces. There seems to be little evidence that the provinces have consistently grown laterally by the accretion of successively younger strips of volcanic and granitic material. Indeed, the facts that volcanic sequences in some places rest on granite basement and that (for example in the Superior province, e.g., Goodwin, 1972), basement strips appear to alternate with greenstone belts which are all of roughly the same age [Krogh and Davis' (1974) suggestion that the belts showed progressive southward decrease in age does not seem to be universally supported] suggest that the new additions were made within pre-existing continental rafts. The roughly concordant relationships between the roofs of the domed granite intrusions and the basal groups of the associated greenstone successions suggest that the intrusive granite partner at first interposed itself between the basement and the newly erupted volcanics. The elongate or triangular synforms of the down-sinking greenstone belts record vertical adjustments in response to the rapid addition of large volumes of materials showing two widely different density ranges.

In high-grade Archaean provinces, a partnership between predominantly basic gneisses or granulites derived largely from volcanics and associated intrusives and predominantly acid gneisses derived from tonalitic to granitic plutonic rocks is recognizable. This pair of assemblages has in some areas—notably the Godthåb region of West Greenland and the coastal tract of Labrador—proved to be separable from units of predominantly granitic gneiss which are as much as 800 million years older and which represent remnants of an ancient continental crust (e.g., Black et al., 1971; McGregor, 1973; Bridgwater et al., 1975).

In these respects, the constitution of Archaean gneiss or granulite terrains recalls that of the contemporary granite/greenstone-belt provinces. On the other hand, the geometrical relationships of the component rock groups in the high-grade provinces are distinctive. The granitic material, instead of being gathered in large domed intrusions, is intimately interleaved with the metavolcanics and associated basic-anorthositic intrusives, forming a multitude of roughly concordant sheets ranging from a few centimeters to a few hundred meters in thickness. In West Greenland, where the relationships can be established in three dimensions, it seems indisputable that much of the more acid gneiss originated as intrusive sheets penetrating a preexisting layering and disrupting preexisting structures (e.g., McGregor, 1973). The well-established layered sequence of gabbro-anorthosite complexes in the Fiskenaesset region, for example, is seen to be split open and inflated to a much greater thickness by the emplacement of such sheets (Myers, 1976). The intricate interleaving of the basic and acid partners is invariably associated with the appearance of very large, almost isoclinal folds and with the development of high strains. Primary volcanic structures, which are often very well-preserved in low-grade greenstone belts, survive only locally in the high-grade terrains and then only in strongly distorted forms. The connection between the distinctive intrusion forms of the granitic partner and the development of high strains and complex large-scale folds suggests a causal link between emplacement of granite and development of the structural patterns. Myers in a discussion of the genesis of Archaean gneiss provinces (1976), suggests that deformation was associated with gravity-assisted flow resulting from the thickening of the crust brought about by addition of large amounts of granite. From the abundance of granite in Archaean provinces, at least a doubling of the volume of crust in these provinces seems possible, and the isotopic data suggest that crustal growth in any given region was a rapid process completed within 100 to 200 million years.

5 STRUCTURES RELATED TO CONTINENTAL MOTION

In retrospect, the growth structures which can be recognized in large continental masses are seen to relate both to regionally controlled motions of the crust and to local displacements controlled by density differences. The bulk of growth structures resulting from basic magmatism are geometrically simple and bear intelligible relationships to regional stress systems. Structures resulting from granite emplacement, on the other hand, commonly reflect vertical movements of relatively local significance which were controlled by the buoyancy of the incoming magmas. Such structures bear only an indirect relation to the regional stress systems. Uplift and the consequent redistribution of material by processes of erosion and sedimentation further complicate the geometrical analysis of the manner of growth of the crust.

In Archaean times, the almost simultaneous addition of basic and granitic materials derived from sources in the mantle enlarged certain parts of the continental crust to at least twice their former volume over relatively short time periods. Growth involving granite emplacement on this scale does not appear to have taken place during later periods; the fact that the tectonic patterns of Archaean provinces—both those characteristic of the granite/greenstone-belt partnership and those of granulite or gneiss provinces—are also almost unique suggests that these patterns originated as growth-structures.

In Proterozoic times, additions of both basic and granitic material took place within continental masses as well as in marginal tracts. The distribution of the newly emplaced material can in at least some instances be explained in terms of regional distortion of the crust, in response either to bodily movement of the continent as a whole or to extension above zones of mantle upwelling. The structural relationships of many large Phanerozoic batholiths suggest that although they are located above zones at which crust was being consumed, they entered crustal regions which, like the back-arc basins of the western Pacific margin, were capable of extension. These relationships imply that although the sites at which growth of continental crustal plates has taken place have changed through time they have, as one would expect, consistently been sites at which additional material could be accommodated by some degree of extension.

REFERENCES

Bickle, M. J., Martin, A, Nisbet, E. G.: Basaltic and peridotitic komatiites and stromatolites above a basal unconformity in the Belingwe greenstone belt, Rhodesia. Earth Planet. Sci. Letts. *27,* 155–162 (1975)

Black, L. P., Gale, N. H., Moorbath, S, Pankhurst, R. J., McGregor, V. R.: Isotopic dating of very early Precambrian amphibolite facies gneisses from the Godthåb district, West Greenland. Earth Planet. Sci Letts. *12,* 245–259 (1971)

Bridgwater, D., Collerson, K. D., Hurst, R. W., Jessau, C. W.: Field characters of the early Precambrian rocks from Saglek coast of Labrador. Geol. Surv. Can., Paper *75,* 287–296 (1975)

Bridgwater, D., Escher, A., Watterson, J.: Tectonic displacements and thermal activity in two contrasting Proterozoic mobile belts from Greenland. Phil. Trans. Roy. Soc. Lond. Ser. A. *237,* 513–534 (1973a)

Bridgwater, D., Sutton, J., Watterson, J.: Crustal downfolding associated with igneous activity. Tectonophysics, *21,* 57–77 (1974)

Bridgwater, D., Watson, J., Windley, B. F.: The Archaean craton of the North Atlantic region. Phil. Trans. Roy. Soc. Lond. Ser. A. *273,* 493–514 (1973b)

Bridgwater, D., Windley, B. F.: Anorthosites, post-orogenic granites, acid volcanic rocks and crustal development in the North Atlantic shield during the mid-Proterozoic. Geol. Soc. S. Africa, Sp. Pub. *3,* 307–318 (1973)

Burke, K., Dewey, J. F.: 1973. An outline of Precambrian plate development. Implications of continental drift to the earth sciences, 2, (Ed. Tarling and Runcorn), 1035–46. Academic Press, N.Y.

Cobbing, E. J., Pitcher, W. S.: The coastal batholith of central Peru. J. Geol. Soc. *128,* 421–460 (1972)

Davidson, A.: The Churchill province. In: Variations in Tectonic Styles in Canada. Price, R. A., Douglas, R. J. W. (eds.). 1972, pp. 381–434. Geol. Assoc. Can. Spec. Paper 11

Escher, A., Jack, S., Watterson, J.: Tectonics of the North Atlantic Proterozoic dyke-swarm. Phil. Trans. Roy. Soc. Lond. Ser. A *280,* 529–540 (1976)

Evernden, J. F., Richards, J. R.: Potassium-argon ages in Eastern Australia. J. Geol. Soc. Australia, *9,* 1–50 (1962)

Goldich, S. S., Hedge, C. E.: 3,800-Myr granitic gneiss in Southwestern Minnesota. Nature, (London) *252,* 467–468 (1974)

Goodwin, A. M.: The Superior province. In: Variations in Tectonic Style in Canada, Price, R. A., Douglas, R. J. W. (eds.). 1972, Geol. Assoc. Can. Spec. Paper 11, pp. 527–623

Hamilton, W., Myers, W. B.: The nature of batholiths. U.S. Geol. Surv. Bull. 554-C, C1–C30, (1967)

Hawkesworth, C. J., Moorbath, S., O'Nions, R. K., Wilson, J. F.: Age relations between greenstone belts and "granites" in the Rhodesian Archaean craton. Earth Planet. Sci. Lett., *25,* 251–262 (1975)

Hickman, M. H.: 3,500 m.y.-old granite in southern Africa. Age relationships between greenstone belts and granites in the Rhodesian Archaean craton. Nature, (London) *251,* 295–296 (1974)

Hurley, P. M.: Can the subduction process of mountain building be extended to Pan-African and similar orogenic belts? Earth Planet. Sci. Letts., *15* 305–314 (1972)

Hurley, P. M.: On the origin of 450 ± 200 m.y. orogenic belts. Implications of continental drift to the earth sciences, 2 (Ed. Tarling and Runcorn), 1083–90, (1973)

Illies, J. H.: Graben tectonics as related to crust-mantle interaction. In: Graben Problems. J. H. Illies, St. Mueller, (eds.). Stuttgart: Schweitserbart, 1970, pp. 4–27

Krogh, T. E., Davis, G. L.: Zircon U-Pb ages of Archaean metavolcanic rocks in the Canadian shield. Carnegie Inst. Ann. Rep. 1973–4, 241–242 (1974)

May, P. R.: Pattern of Triassic-Jurassic diabase dikes around the North Atlantic. Bull. Geol. Soc. Am., *82,* 1285–1292 (1971)

McGregor, V. R.: The early Precambrian gneisses of the Godthåb district, West Greenland. Phil Trans Roy. Soc. Lond. Ser. A *273,* 343–358 (1973)

Moorbath, S.: Evolution of Precambrian crust from strontium isotopic evidence. Nature, (London) *254,* 395–398 (1975)

Myers, J. S.: Cauldron subsidence and fluidisation: mechanisms of intrusion of the Coastal Batholith of Peru into its own volcanic ejecta. Bull. Geol. Soc. Am. *86,* 1209–1220 (1975)

Myers, J. S.: Granitoid sheets, thrusting and Archaean crustal-thickening in West Greenland. Geol., 4, 265–268 (1976)

O'Hara, M. J.: Petrology of the Scourie dyke, Sutherland. Mineralog. Mag., *32,* 848–865 (1961)

Piper, J. D. A.: Palaeomagnetic evidence for a Proterozoic super-continent. Phil Tran. Roy. Soc. Lond., Ser. A *280,* 469–490 (1976)

Pitcher, W. S. On the rate of emplacement of batholiths. J. Geol. Soc. Lond., *131,* 587–592 (1975)

Ramberg, H.: Gravity Deformation and the Earth's Crust as Studied by Centrifugal Models. New York: Academic Press, (1967)

Ramberg, H.: Model studies in relation to intrusion of plutonic bodies. Mechanism of igneous intrusion. Geol. J. Sp. Issue *2,* 261–272 (1970)

Shackleton, R. M.: Displacement within continents. In: Time and Place in Orogeny. Kent, P. E., Satterthwaite, G. E., and Spencer, A. M. (eds.) Geol. Soc. Lond. Spec. Publ. No. 3 (1969)

Shackleton, R. M.: Pan-African structures. Phil. Trans. Roy. Soc. Lond. Ser. A *280,* 491–498 (1976)

Stephenson, A simple shear model for the ductile deformation of high level intrusions in South Greenland J. Geol. Soc. Lond. *132,* 307–318 (1976)

Sutton, J., Watson, J.: Tectonic evolution of continents in early Proterozoic times. Nature (London) *247,* 433–435 (1974)

Talbot, C. J.: A plate tectonic model for the Archaean crust. Phil. Trans. Roy. Soc. Lond. Ser. A *273,* 413–428 (1973)

Tarney, J.: Assynt dykes and their metamorphism. Nature (London) *199,* 672–674 (1963)

Watson, J.: Vertical movements in Proterozoic structural provinces. Phil. Trans. Roy. Lond. Ser. A *280,* 629–640 (1976)

Watson, J., Lisle, R. J.: The pre-Laxfordian complex of the Outer Hebrides. In: The Early Precambrian of Scotland and Related Rocks of Greenland. Park, R. G., Tarney, J. (eds.). Univ. Keele, pp. 45–50 (1973)

Watterson, J.: Investigations on the Nagssugtoqidian boundary in the Holsteinborg district, central west Greenland. Rapp, Grønlande geol. Unders. *65,* 33–37 (1974)

3

B. BAYLY

Abstract During deformation, a rock changes in internal structure; it becomes a new material with new properties. It cannot effectively be described as *a material* whose properties are strain dependent or as *a material* whose properties are stress dependent; rather, a rock at one moment is a member of a *suite of materials* that can be changed one into another by deformation. Each member of the suite has its own structure and its own mechanical properties—the properties of the members of the suite are *structure dependent*.

Behavior that is structure dependent cannot well be represented by conventional spring and dashpot models, but can be reproduced in an emulsion model with surface energy. Inspection of such a model suggests that the form $\gamma = A\tau^B$ can be replaced by the form $\gamma = (\mathrm{SF})\tau^B$ where SF is a structure factor. If this change is made, tertiary creep leading to rupture can be embraced under the same formula as secondary or steady-state creep. In addition, primary or transient creep will be seen as partly a nonrecoverable change of structure and only partly as a damped recoverable change. It seems that the inclusion of structure factors could make the constitutive equations for rocks correspond more closely to the phenomena they attempt to describe.

A POSSIBLE USE OF STRUCTURE FACTORS IN CONSTITUTIVE EQUATIONS FOR ROCKS

INTRODUCTION

The paper has two parts: Section 1 shows that as a rock deforms, its structure changes; at a late stage in deformation it is a *different material* from what it was at an early stage. If we try to write descriptions in such forms as

shear strength = some function of normal stress or of confining pressure

or

strain rate = some function of stress,

we fail to allow for the variable structural state in an effective way. It is more realistic to write

shear strength = some function of confining pressure and current structural state

or

strain rate = some function of stress and current structural state.

If the current structural state were simply stress dependent then the sample properties would also be simply stress dependent; if the current structural state were simply strain-dependent, the sample properties would be fixed by current stress and current state of strain. The true situation is worse than these because the current structural state is history dependent; but the situation is not as bad as the last statement suggests because despite an infinite variety of possible histories, there are limits on the list of possible structural states into which a sample can be brought. Section 1 discusses a simple mechanical model where the foregoing ragged cloudy generalizations can be replaced by exact quantitative statements. The model may not be close to any geological reality; its value is the concreteness with which certain ideas can be discussed.

Section 2 starts with the view that deformation depends on localized weaknesses: whatever the scale, from the scale of a single crystal upward, the behavior of a bulk sample depends on the population of local weaknesses inside it. In this view, the structural state noted in Section 1 as an essential factor in mechanical behavior reduces to the fracture state; the sample is seen as a homogeneous matrix whose properties are not history dependent, shot through by a population of fractures or, more properly, weaknesses whose sizes and orientations are history dependent. The possibility that this view could provide *current structure factors* of the type called for in Section 1 is illustrated by an example. The example is quite naive but succeeds in embracing under one formula steady-state creep and creep that accelerates to rupture. The purpose is not to advocate this particular formulation but to illustrate the general point that structure factors might help in summarizing the observed behavior of rocks.

The subject matter lies in the area between the works of Hashin (1964), Wu (1966), Gay (1968) and Kinder and Sternstein (1976) and is rather elementary in comparison with any of these.

1 AN EMULSION MODEL

We begin by recapitulating some fundamentals. When a sample is deformed by external forces, energy enters the sample. Of the energy entering, some is dissipated and some is stored. In the process of dissipation, the amount of structural change can be significant or it can be negligible; for example, in the flattening of conglomerates, anisotropy may develop to a conspicuous degree, but in common experiments with water the difference in structural state of the water early in the experiment and late in the experiment is usually negligible. In the *storage* of energy, by constrast, structural change can never be negligible; it may be small, but structural change is the essence of the storage process.

The next point is that recovery of the stored energy is never instantaneous. A common imaginary concept of ideal elastic behavior involves instantaneous recovery, but in any real material a finite time (or perhaps even an infinite time) is required. If real materials are represented by spring and dashpot models, every spring has a dashpot in parallel with it; and stored energy is never recovered wholly in mechanical form—in the recovery process itself, some energy is dissipated. The foregoing ideas admit the possibility of using, as the element that carries the material's ability to recover, an emulsion globule instead of a spring. Of spring and dashpot models, the emulsion model most closely resembles a Burgers substance as in Figure 1A, but an important difference exists that will be indicated after the emulsion model has been described.

The emulsion model comprises two viscous immiscible fluids; one tends to form globules in the other; a considerable surface energy is associated with the interface so that when a sample is fully relaxed, the globules approach spherical. It is not necessary for the two fluids to have any property other than viscosity in themselves; the combination of viscosity and surface energy gives the emulsion all the properties we need. Thus, under a deformation that is, for example, a progressive coaxial plane strain, the globules will become ellipsoids; the surface area will increase and the tendency toward ever-increasing eccentricity will be opposed by the surface-energy effect; for any steady strain rate, a steady-state eccentricity will be approached, but not reached. Consider a late stage when the emulsion is close to the steady-state configuration. A certain

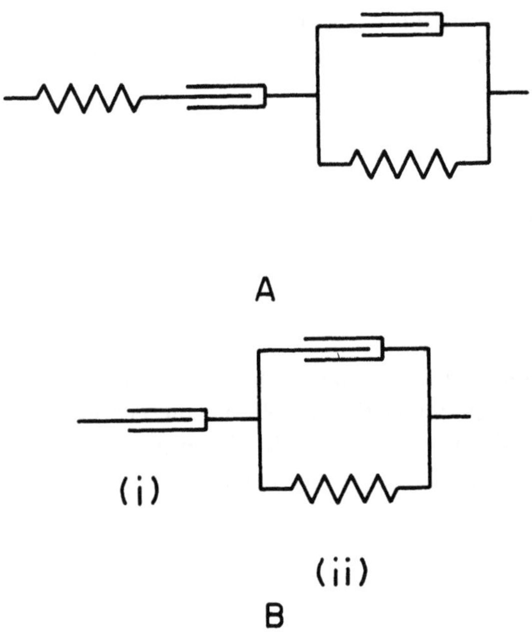

Fig. 1A, B. (A) Conventional representation of a Burgers substance. (B) Representation of a simpler Burgers substance where provision for an instantaneous elastic response has been dropped. As long as it is understood that element (ii) stands for a *series* of elastic responses with different relaxation times, the additional concept of an instantaneous response is an unnecessary complication.

amount of energy will be dissipated in bringing the emulsion to this state; also a fixed amount of energy will be stored, proportional to the amount by which the surface area of the late-stage ellipsoids exceeds the surface area of the corresponding spheres; also a wholly nonfixed amount of energy can be dissipated by continued flow of the matrix around the ellipsoids after these have reached their late-stage configuration. (This last idea requires a sample of infinite dimensions and some postulates about the distribution of ellipsoid centers, that it be random on some scale and homogeneous on another scale so that despite continuous flattening of the whole sample, the statistics of distribution of ellipsoid centers within a stationary grid square does not change. Let us suppose that the distribution of ellipsoid centers has this property.)

The emulsion clearly has the properties sought: upon relaxation of stress, surface energy will drive the ellipsoids back toward spheres, with an accompanying change in the overall dimensions of any representative sample of emulsion, back toward its original dimensions; this recovery will proceed at a rate that can be large or small, and it can amount to a large fraction or a small fraction of the maximum deformation imposed. Thus the emulsion has the essential elements of behavior of any viscous and elastic material. The values of the model are:

(1) Its structural state is very easy to picture and keep track of: a single number, the mean eccentricity of the ellipsoids, specifies the structural state at any moment.

(2) The stored energy associated with each structural state is easy to assess; it is simply the surface energy of the excess area that the ellipsoids have at any moment over the corresponding spheres.

Thus in the equation

total energy put into sample by deformation between time $t1$ and time $t2$	=	energy stored through structural change from time $t1$ to time $t2$	+	energy dissipated between time $t1$ and time $t2$

there is no ambiguity about the magnitudes of the quantities on the right-hand side.

(3) The meaning of the term "instantaneous viscosity" can be inspected rather closely. The bulk viscosity of an emulsion containing spheres is clearly different from the bulk viscosity of the same emulsion reorganized into ellipsoids; thus the bulk viscosity changes during deformation and the question, *What is the sample's bulk viscosity at some instant?* is a real one. In line with point (2) above and using E for energy, we can write

$$\frac{d(E\ \text{total})}{dt} = \frac{d(E\ \text{structure})}{dt} + \frac{d(E\ \text{dissipated})}{dt}$$

If the driving stress field, averaged over the sample, is σ and the strain rate is ϵ so that the left-hand side equals $\sigma.\dot{\epsilon}$, the instantaneous viscosity can be taken as $\sigma/\dot{\epsilon}$, or d(E total)/$(\dot{\epsilon}^2)dt$. Alternatively, some authorities prefer to relate viscosity strictly to that part of the strain that is nonrecoverable; in this view, the instantaneous viscosity equals $d(E$ dissipated)/$(\dot{\epsilon}^2)dt$. Regardless of the usefulness of either form, which depends on one's purpose, the emulsion model makes it easy to see that, in the presence of a changing structure, two definitions of viscosity are available and it shows precisely and quantitatively how they differ.

Up to this point, parallels with a Burgers substance are very close. First, as noted above, we can simplify the Burgers model to the form in Figure 1B—the idea of an undamped spring may in some circumstances be convenient but it is never necessary. The emulsion model then compares with Figure 1B as follows:

(1) The Burgers structural state that corresponds with the eccentricity of the ellipsoids is the length of the spring;
(2) the energies stored and dissipated go respectively into the spring and the two dashpots;
(3) two possible values of viscosity are given by $\sigma/\dot{\epsilon}$ where $\dot{\epsilon}$ is taken either as the strain rate of the whole assembly or else as the strain rate of the viscous element i by itself.

However the Burgers model fails to imitate an essential fact of the real world: *viscous behavior is likely to change with change of structural state*. The Burgers model has no link between the length of the spring and the viscosity of the dashpot element *i*. The emulsion model, by contrast, brings to our attention the fact that increasing eccentricity of the ellipsoids could have a profound effect on the sample's bulk viscous behavior: for example, an emulsion containing spheres will, in the geometry specified, be isotropic whereas if all the spheres are flattened to parallel ellipsoids, the bulk sample will be anisotropic, and the maximum viscosity will be notably increased. This dependence of viscosity on ellipsoid shape brings us to the emulsion model's key property:

(4) The model shows how the mechanical properties of an emulsion at some moment are structure controlled rather than stress controlled or strain controlled. Consider the simplest case where the two fluids involved are Newtonian: then current instantaneous values for the concentration of ellipsoids and their eccentricity lead to current instantaneous values of the anisotropy of the emulsion, its maximum bulk viscosity and its minimum bulk viscosity, without regard to strain history nor current stress. In the absence of a recovery tendency driven by the surface-energy effect, the current eccentricity might be uniquely related to the bulk strain that has occurred since the inclusions were spheres; but *in presence of a recovery tendency, any unique relation between properties and total strain is lost;* a given eccentricity can result from a small strain that grew fast or a larger strain that grew more slowly. Equally, no unique relation of properties to stress can be expected, because the ellipsoids' eccentricity at the moment of interest depends on the history of strain as well as the current stress.

(5) Enlarging on point (4) we can note the following possibilities: (a) the possibility of describing a relaxed emulsion (where the inclusions are spheres) as *a material with strain-dependent properties;* such a description would be partly true but unsatisfactory for the reason just given; (b) the possibility of describing the emulsion as *a material with stress-dependent properties;* this would also be not wholly accurate; (c) the possibility of describing the emulsion as a *series of materials*, differing in their internal structural details, any one of which could be converted into another by strain. Each material in the series would have its own degree of anisotropy, its own maximum bulk viscosity, etc.; and a single material's viscosity could be investigated at different stress magnitudes to see if it were stress dependent, etc. Such a description would seem more satisfactory, because of making possible the distinction of truly stress-dependent effects in isostructural samples from the effects of change of structure.

(6) The model shows that it is not necessary to conclude that the mechanical properties are history controlled, at least not to the extent that one needs to know the history to predict the properties. The number of possible histories for a sample is endless but the number of possible *present structural states* is smaller (possibly one should say a lower order of infinity); and it is the present structural state that, together with the present state of stress, determines the sample behavior.

2 DEFORMATION CONCENTRATED IN LOCALIZED WEAKNESSES

The essence of Section 1 is the idea that a sample becomes a succession of different materials as it strains, each material having its own internal structure and consequently its own set of mechanical properties or responses to stress. The intention was to establish this idea by discussion of hypothetical emulsions whose behavior could be discussed rather precisely. The intention now is to switch to real rocks, and to assert that even though the internal structure is excessively complicated and beyond any kind of precise description, the parallel is sufficiently close; the conclusions drawn in Section 1 carry over; a rock is not well described as a material with stress-dependent behavior but is better described as a suite of materials in which each material has its own mechanical properties and also the property of converting, under strain, into another material in the suite.

The discussion is based on the idea that on almost any scale, the mechanical behavior of a rock sample depends on localized weaknesses. Under this designation, different groups of workers have emphasized

(1) dislocations in crystal lattices
(2) cracks as seen with the scanning electron microscope
(3) sites predisposed to twinning e.g., in calcite
(4) sites predisposed to other lamellar deformation, e.g., in quartz
(5) pressure-solution surfaces
(6) grain boundaries in general
(7) slip surfaces as in clay and soil.

When a material containing such weaknesses converts by strain into another material in the suite, the change in mechanical properties is due mainly to change in the number, size, orientation and cohesiveness of the localized weaknesses. To imitate such weaknesses with an emulsion, it is necessary that the fluid of the dispersed phase, the ellipsoids, be less viscous than the host. Then degrees of freedom that one can use to make the model approximate reality are the eccentricity of the inclusions and their viscosity: in its simplest form, the emulsion model does not admit change in the number of weaknesses, and there are severe restrictions on change in orientation. With real-life weaknesses a possible process is that strain may cause the closing of a crack with one orientation and the opening of a crack of similar properties in a different orientation; such a process could possibly be imitated approximately by an ellipsoid whose orientation could change discontinuously but cannot be imitated by the simple ellipsoid model introduced.

In the discussion that follows, we conceive of a weakness as something more idealized than a grain boundary or electron microscope crack, but less idealized than the ellipsoidal inclusions of the emulsion model. Specifically we imagine a population of weaknesses that are all similar to each other, distributed in such a way that their bulk mechanical effect matches that of the real, highly heterogeneous population of cracks, boundaries, etc. An imaginary homogeneous population of weaknesses of this kind is, on one hand, not too far removed from the real world (or at least not too far removed from the other homogeneous idealizations that are commonly used as mental aids substituting for real granular materials). On the other hand, it is not too far removed from the population of low-viscosity ellipsoids in Section 1. Properties that will be attributed to the weaknesses are:

(1) Some mechanical effect, so that a sample's macroscopic behavior is controlled by the number of weaknesses per unit volume; examples are macroscopic parting strength or macroscopic viscosity;
(2) the property of increasing in number at a rate that is affected by the current shear stress;
(3) the property of decreasing in number, or healing, at a rate that is not affected by shear stress but is increased by compressive stress.

The discussion of weaknesses will dwell more on change in their *number* where the discussion of ellipsoids dwelled on eccentricity. This is a move toward the realm of guesswork and away from the preciseness and completeness of ellipsoidal fluid mechanics, but resemblances persist: for example, an increase in bulk anisotropy could arise in either of two ways—given a fixed number of deformable inclusions, it could arise from change in their eccentricity or, given a variable number of inclusions of fixed shape it could arise from a change in their number. Either way, the main concept remains unchanged, namely, that behavior is fixed completely by the current structural state and not fixed completely by either current stress or current accumulated strain. In addition to this basic similarity, the population of weaknesses resembles the population of emulsion ellipsoids in two more notable ways.

(1) There is some kind of basic state or relaxed state toward which the sample always has a tendency to revert: in the emulsion, it is the state where the inclusions are spheres; in the rock with weaknesses, it is some concentration level of weaknesses that one expects in the natural unstrained state of the rock; it makes no difference to the train of thought whether a zero or a nonzero value is favored for this ground state.

(2) The factor that causes increase in deviation from the ground state is not the reverse of the factor causing decrease. In the emulsion, eccentricity is increased by viscous drag, which is strongly stress dependent, but eccentricity is decreased by the surface-energy effect which is independent of stress. The value of the emulsion model is that it shows the possibility of such a difference; indeed in the emulsion, the difference *must* exist and could be described, if necessary, in a complete explicit quantitative way. In the rock with weaknesses, it is only a postulate that such a difference would exist but the postulate is compatible with what is known about material behavior of such various kinds as annealing and thixotropy.

It is possible to pursue the foregoing ideas in semi-quantitative form. If W_0 is the number of weaknesses per unit volume initially present in a fully relaxed sample, then W, the number of weaknesses per unit volume at a later time in a deformation event, can be written

$$W = W_0 + \int_0^T (\text{weakness accumulation rate})\, dt \tag{1}$$

Also

$$\begin{matrix}\text{weakness} & & \text{weakness} & & \text{weakness}\\ \text{accumulation} & = & \text{formation} & - & \text{healing}\\ \text{rate} & & \text{rate} & & \text{rate}\end{matrix}$$

Considering a compression experiment, the rate of formation of weaknesses would be affected by the temperature and the overall stress field, both hydrostatic and deviatoric components. For given values of these variables, the rate would also be affected by the number of weaknesses already present: with many weaknesses, the overall stress would be concentrated in the small remaining nonweak regions and these would be rapidly torn apart; with fewer weaknesses, the overall stress would be more uniformly dispersed, with fewer severe local concentrations, and the rate of generation of new weaknesses would be smaller. An example of an expression embodying this behavior is

$$\text{formation rate} = \frac{1}{K - \text{concentration}} f_1(\tau, \theta, p)$$

where the first term makes the development of weaknesses self-accelerating to an infinite rate as a finite value K is approached. [$\tau = \frac{1}{2}(\sigma_{max} - \sigma_{min})$, θ = temperature, p = hydrostatic component of stress.]

Turning to healing, the rate at which weaknesses heal is again affected by θ and p, and can be taken as proportional to the number of weaknesses present by analogy for example with radioactive decay; thus

$$\text{healing rate} = \text{concentration} \times f_2(\theta, p)$$

Hence, writing W for current concentration of weaknesses,

$$\text{weakness accumulation rate} = \frac{1}{K - W}[f_1(\tau, \theta, p)] - Wf_2(\theta, p)$$

At constant values of τ, θ, p, this has the form

$$\frac{A}{K - W} - BW \quad \text{or} \quad \frac{A}{K}\left(\frac{1}{1 - \frac{W}{K}} - \frac{BK^2}{A}\frac{W}{K}\right) \tag{2}$$

which, writing w for W/K, behaves like $1/(1 - w) - Cw$; see Figure 2. The critical value for C here is 4. With $C < 4$, the first term dominates and the expression runs to infinity as w runs from 0 to 1; with $C > 4$, the second term dominates and the expression falls to zero before $w = 1/2$. Even without performing the integration in Equation (1), we can see that if the rate of accumulation of weaknesses tends toward infinite the specimen will rupture, whereas if the rate of accumulation of weaknesses approaches zero, steady-state creep supervenes: the rate of healing approaches the rate of formation and W, the number of weaknesses present per unit volume, approaches asymptotically some steady value. In other words, values of C less than 4 lead into tertiary creep while values of C greater than 4 give only unlimited secondary creep.

The behavior just outlined can be compared with phenomena from the real world. First we look in the model for a property comparable with shear strength. C stands for BK^2/A where K is a simple number, the limiting number of weaknesses per unit volume; B is the healing-rate factor f_2 whose value depends on temperature and pressure; A is the formation-rate factor f_1, whose value depends on temperature, pressure and shear stress. At a given temperature and pressure, it is the value of the shear stress τ that makes C less than 4 or greater than 4; the critical value of τ in the model thus corresponds to a material's shear strength in the real world.

A second well-known phenomenon in the real world is the brittle-ductile transition, where apparently one passes from pressure-temperature conditions that permit rupture into pressure-temperature conditions where no rupture occurs however large the stress; or, to put it another way, the

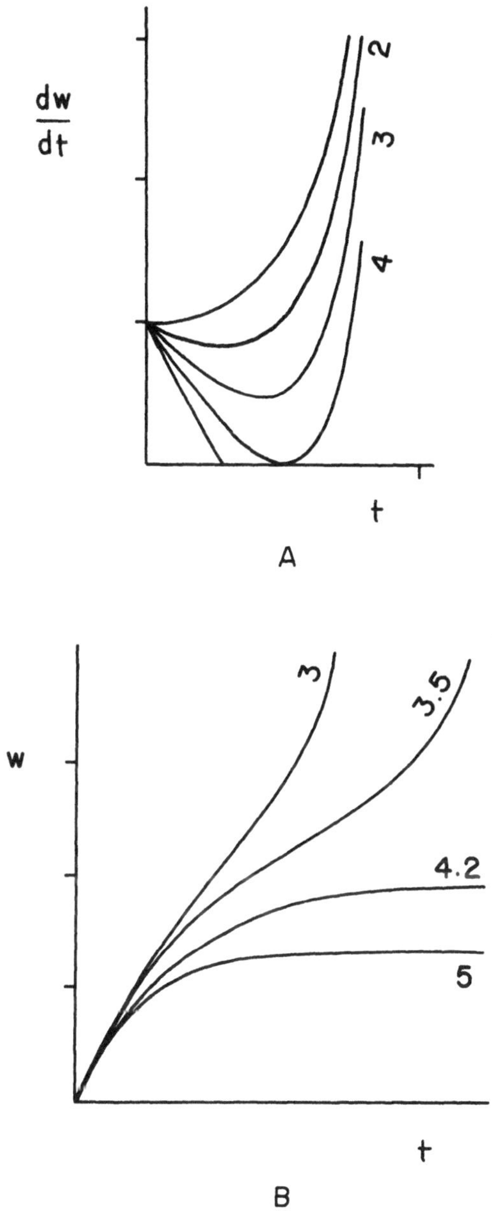

Fig. 2A, B. (A) The behavior of the function $[1/(1-w]-Cw$ for different values of C. (B) Behavior with time of the number of weaknesses in a sample if the rate of accumulation of weaknesses is as shown in diagram A

properties of a ruptured sample become indistinguishable from those of an unruptured sample. This behavior could be imitated in the model if we ask, How weak is a weakness? If the contrast between a weakness and the host continuum is made a function of pressure and temperature such that it becomes negligible in some conditions, the brittle-ductile transition would be imitated. [It would be logical at the same time to insert an exponent thus $1/(K - \text{concentration})^n$ with n approaching zero as the contrast between weakness and host approaches zero.]

The possibility of interpreting secondary creep and tertiary creep of rocks in terms of a changing population of weaknesses has been available for many years, through parallelism with the behavior of metals. The novel point emphasized here is that each new value of W, the number of weaknesses, can be seen as giving a new material. For a given value of W, say W_A, and assuming that each weakness has its own viscosity, parting strength, etc., as in any two-phase composite material, the bulk sample will have a bulk viscosity, a bulk parting strength, etc. As mentioned earlier, the given value W_A can be achieved at different magnitudes of τ and so if necessary the variation in viscosity with τ for the particular material $W = W_A$ could be investigated. If the viscosity turned out not constant, then the creep rate would be nonlinear with τ at *constant* W; but this is quite a different kind of nonlinearity from that normally represented in, for example, expressions of the form used by Weertman, of which discussion follows.

The Weertman form, $\dot{\gamma} = A\tau^B$, is used primarily in connection with steady state or secondary creep (and does not attempt to describe primary or tertiary creep). In concept, a series of experiments at different steady loads $\tau_1, \tau_2, \ldots, \tau_n$ is performed; ideally, a steady strain rate emerges from each experiment, providing values of $\dot{\gamma}_1, \dot{\gamma}_2, \ldots, \dot{\gamma}_n$; then plain numbers A and B are sought to fit the data. It is argued here that such experiments would handle n different numbers $W_1, W_2, \ldots, W_n$ of weaknesses per unit volume in achieving steady-state creep. A more appropriate form to attempt would then be

$$\dot{\gamma} = (\text{some function of } W) \cdot \tau^b$$

The function of W would depend upon τ *among other factors* and might possibly behave, in restricted experiments, approximately like *constant* $\times\ \tau^\beta$. Thus it is suggested that the exponent B in the conventional form combines two contributions (a) the genuine nonlinearity exponent b, that would be discovered in experiments at a given structure state W_A, and (b) the consequences of W changing with τ.

Let us look at the consequences of attempting the different formulation $\gamma = (F_w)\tau^b$. In the idealization above, W is a countable number of weaknesses, to each of which are attributed the same mechanical properties; then the properties of the bulk material are indeed representable as F_w, where the function F involves the properties assigned to the regions of weakness and also those of the host continuum. In the real world, the weaknesses are neither identical nor countable. Nevertheless, a material in which one set of weaknesses has been developed by strain has material properties that we can call F_1; a material in which a different set of weaknesses has been developed has properties that we can call F_2; the difference between F_1 and F_2 is perfectly real in spite of the impossibility of specifying the weaknesses from which the difference results; and F_1 and F_2 derive uniquely from the present properties that either weakness system has, regardless of the history by which the system arose. These features are as true of real-world samples as they are of the idealized models discussed.

Consequences are:

(1) A modified view is taken of the constants in the Weertman formula. For some purposes, it remains convenient to summarize experimental data in terms of A and B but these are not to be taken as material constants belonging to a single material.
(2) The possibility of embracing tertiary creep under the same formula as secondary creep can be pursued, though the attempt necessarily involves some expression comparable withEquation (2) above, containing factors simple enough to be handled algebraically that nonetheless have to try to do duty for the infinite complexity of real-world systems of weaknesses.
(3) The nonlinearity of primary or transient creep need not be attributed wholly to a damped recoverable effect. As soon as due recognition is given to the fact that, even at constant stress, a material's viscosity is likely to change because of changing internal structure, questions arise about the curvature of the transient-creep curve. Part of the nonlinearity is very probably due to damped recoverable effects but almost certainly part is not, and the latter part may be the greater. An indication in this direction appears in the work of Hobbs (1970).

SUMMARY AND CONCLUSION

(1) There are two separate aspects to recovery of a sample after an episode of deformation: the sample may go part-way toward recovering its original dimensions; separately, it may go part way or all the way toward recovering its original internal structure. It is a commonplace that in tectonics, recoverable changes of dimension are usually negligible in comparison with permanent changes, but recoverable change of structure can, nonetheless, be of great importance. By *recoverable change of structure,* we here mean such a phenomenon as the increase and subsequent decrease in the number or size of the local weaknesses in a sample as deformation reaches a climax and declines.
(2) During deformation, the tendency of weaknesses to develop on account of shear stress and to heal on account of compressive stress produces, at any moment, a set of current properties for the weakness population. These current properties determine precisely the behavior of the sample. By contrast, the current stress field gives too little information to determine the behavior; the current strain state equally gives too little information; even knowledge of the two together is too little information. On the other hand, the total history of the sample is too

much information. It is not true to say that we must know the history to predict the behavior: the thing we need to know is the current structural state, together with the current stress field.

(3) The two preceding points can be given precision by considering an emulsion of two viscous fluids with significant surface energy.

(4) Neither the Weertman form of creep law nor the Burgers model as conventionally used makes provision for viscosity to be affected explicitly by structural state; but either could be modified to include such a provision.

(5) Because of the complexity of a rock's internal structure, any attempt to allow algebraically for its variations can only be flimsy. The gap between any model that is algebraically tractable and the real world is distressingly apt to be large. Correlatively, it is hard to use *observation* of the relevant structures as a guide when trying to invent an appropriate piece of algebra, despite recent advances in observational technique.

(6) The fact that a sample is likely to have a different internal structure late in deformation from what it had initially can be emphasized by calling it a *different material.* In this view, a sample during strain becomes a succession of different materials, each of which has its own material properties just as it has its own internal geometry. For the sake of emphasis, this view has been taken in the foregoing discussion. The alternative view is widespread and probably more justifiable, where instead of saying that a sample becomes a series of different materials, we say only that it passes through a series of structural states. No difficulty arises from the latter view if we are willing to write a separate constitutive equation for each structural state; nor does any difficulty arise *in theory* if instead we write one constitutive equation for the material and put appropriate structural-state factors into the equation. It is in practice that the difficulty lies.

(7) Models based on concepts arising in metallurgy and hydrodynamics cannot be expected to be wholly satisfactory for rocks, but the expanding body of work on two-phase composite materials is applicable, as is that on continuum descriptions for materials that contain a static population of cracks. The fact that an early stress increment modifies a material's response to a later stress increment is given serious attention in the literature on polymers, and it is possible that some techniques will emerge in this field that will be of use to geologists.

ACKNOWLEDGMENTS

I have drawn heavily on work from three fields—on different kinds of weaknesses and their role in deformation; on two-phase composites of various materials and configurations; and on testing material samples and summarizing the results in constitutive equations. The ideas used are too general to allow specific attributions, but I wish to acknowledge my debt to the many workers who have helped to build up these fields.

I wish to thank W. D. Means and S. Katz, both for comments on the manuscript and for their other helpful suggestions.

Lastly and very gladly I acknowledge my debt to the man this volume honors. Hans taught us to see the observable facts of geology not as an opaque screen but as a web behind which the true action can be discerned—action that is dominated by transformations of energy. I wish I could see more than a glimpse of the vista Hans surveys with such depth and resolution.

REFERENCES

Gay, N. C.: Pure shear and simple shear deformation of inhomogeneous viscous fluids 1. theory. Tectonophysics *5,* 211–234 (1968)

Hashin, Z.: Bounds for viscosity coefficients of fluid mixtures by variational methods. In: Proc Intern Symp Second-Order Effects in Elasticity, Plasticity and Fluid Dynamics. Reiner, M., Abir, D. (eds.). Oxford-Paris-New York: Pergamon Press, 1964, pp. 434–449

Hobbs, D. W.: Stress-strain-time behavior of a number of Coal Measure rocks. Intern. J. Rock Mech. Min. Sci. *7,* 149–170 (1970)

Kinder, D. F., Sternstein, S. S.: A path-dependent variable approach to non-linear viscoelastic behavior. Trans. Soc. Rheology, *20,* 119–140 (1976)

Wu, T. T.: The effect of inclusion shape on the elastic moduli of a two-phase material. Intern. J. Solids Structures *2,* 1–8 (1966)

4

H. KOIDE AND S. BHATTACHARJI

Abstract Analyses of the San Andreas fault, Neo Valley fault system and many active faults in Japan show that infrequent, high magnitude earthquakes are generated in a bend or gap of a fault system of a reverse step-like pattern characterized by compressional bends and areas of upliftment. On the other hand, more frequent, lower magnitude earthquakes are commonly generated in the extensional zone of strike-slip faults of a normal step-like pattern characterized by en echelon tension fractures, areas of subsidence, extensional bends or basins, which occur in regions of active rifting, high volcanicity, and geothermal activity. Exact distribution of stress due to right or left lateral strike-slip movement of a linear or penny-shaped fault is obtained by an analytical solution of three dimensional theory of elasticity. The results indicate the development of compression and extension on either side of the fault depending on right or left lateral movement producing upliftment in the compressional zone and subsidence in the extensional areas. En echelon arrangements of shear and tension fractures or fault segments in the major fault zone inducing reverse step-like and normal step-like fault system patterns on the crustal rocks are explained by Riedel shear experiments in models. Rock deformational experiments indicate that a reverse step-like pattern is most likely to develop in a fault system under moderate effective pressure in rocks in transition between brittle and ductile failure. The normal step-like pattern is, however, favored under low effective (or high interstitial fluid) pressure in rocks showing brittle failure.

GEOMETRIC PATTERNS OF ACTIVE STRIKE-SLIP FAULTS AND THEIR SIGNIFICANCE AS INDICATORS FOR AREAS OF ENERGY RELEASE

1 INTRODUCTION

A sudden release of strain energy in the earth's lithosphere can sometimes trigger disastrous high magnitude earthquakes. As large earthquakes primarily occur along fault zones, the geometry and physical charcteristics of the fault zone can affect the nature or process of the sudden release of strain energy. It is now well established that the high magnitude earthquakes which occur along the San Andreas fault system in California originate in or near big bends in that system (Allen, 1968; Wallace, 1970; Nason and Tocher, 1970, Scholz, 1973). The main bend in the San Andreas fault system has puzzled many geologists and geophysicists because such a bend in a fault with right-lateral displacement should make it a zone of compressive obstacle. Anderson (1971) has suggested that this bend may have resulted from the rifting off of a piece of North American continent. From his study of the Owens Valley region of California, Pakiser (1960) concluded that if two parallel transcurrent faults with the same direction of displacement are arranged en echelon, and if they extend outward in opposite directions, either a region of relative tension or a region of relative compression will be developed between them. In this paper we will attempt to establish that (1) a bend in a fault or offset of en echelon faults which makes it a region of compression normally generates infrequent but large, high magnitude earthquakes; (2) a characteristic fault geometry can indicate the compressional or extensional region of a fault; and (3) that the fracture or fault geometry in regions of high extensional stress results from failure under brittle condition, while that in regions of high compression is the product of failure in transition between brittle and ductile.

2 THE GEOMETRICAL STEP-LIKE PATTERN OF SAN ANDREAS FAULTS AND GULF OF CALIFORNIA RIFT SYSTEM

The San Andreas fault and Gulf of California rift system which produce right-lateral strike-slip movement, is the boundary between Pacific and North American Plates (Anderson, 1971) (Fig. 1A). The Gulf of California rift system is characterized by en echelon right-lateral step-like strike-slip faults with rhombic shaped extensional basins between them (Shepard, 1950; Rusnak et al., 1964: Lomnitz et al., 1970; Elders et al., 1972). High gravity and heat flow anomalies (Harrison and Mathur, 1964; Elders et al., 1972; Lawver et al., 1975) suggest some sort of igneous intrusion(s) under these extensional basins. Igneous intrusions in extensional fractures at depth under rhombic shaped basins and the connection of these basins by en echelon fractures could produce the step-like pattern of the Gulf of California rift system (Fig. 1A). The San Andreas fault system also shows step-like bends near Los Angeles and San Francisco (Fig. 1). The trend of individual segments of San Andreas fault and the Gulf of California are almost parallel. However, the general trend of San Andreas fault and the Gulf of California rift system shows a major swing due to change in orientation of the step-like pattern of the San Andreas fault from NW–SE to NE–SW in the Gulf of California rift system (Fig. 1). The pivot for the swing is at Salton Sea. The original right-lateral displacement causes at the same time the collision of the wall rocks in the San Andreas fault producing the reverse step-like pattern and the separation giving rise to normal step-like pattern of the Gulf of California rift system.

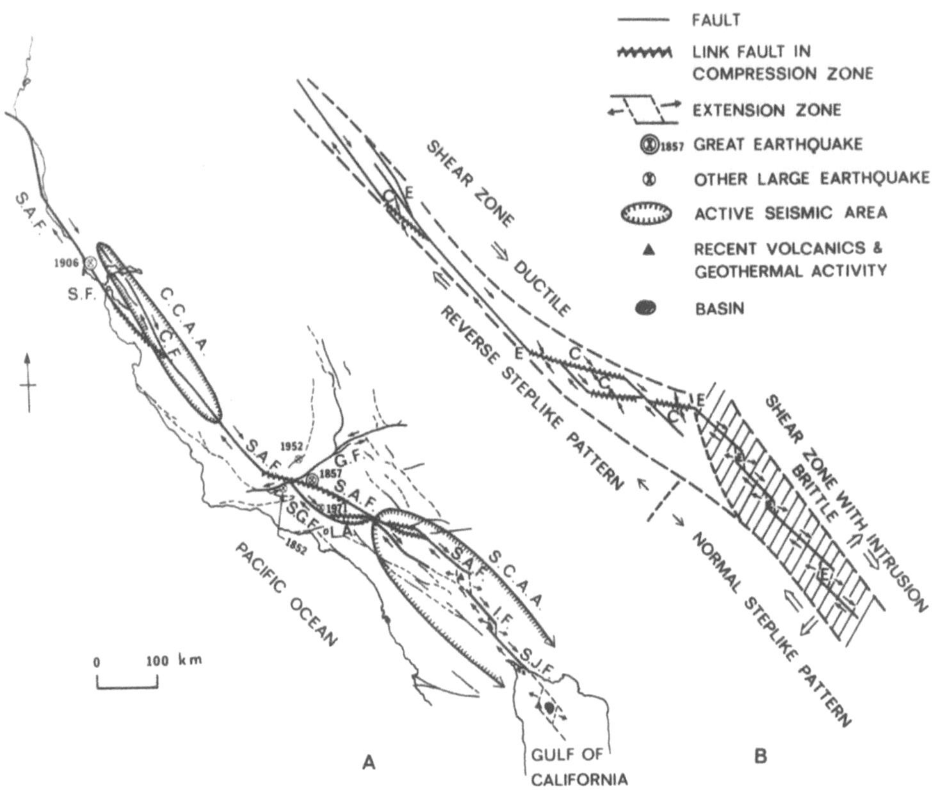

Fig. 1A, B. (A) San Andreas fault system, northern part of Gulf of California rift system and Salton Sea trough. C.C.A.A. and S.C.A.A.: active areas in central and southern Caifornia respectively (after Allen, 1968). These areas have been the sites of many small and moderate earthquakes currently. Note that the large earthquakes occur outside the active areas and near the big bends of the fault system. Recent volcanics, geothermal activity and extensional basins in the Salton Sea trough and Gulf of California rift system are shown (after Elders et al., 1972). S.A.F.: San Andreas fault; C.F.: Calaveras fault; G.F.: Garlock fault; S.G.F.: San Gabriel fault; I.F.: Imperial fault; S.J.F.: San Jacinto fault zone (B) Schematic pattern of the San Andreas fault system, Salton Sea trough and Gulf of California rift system. *C* and *E*: compression and extension zones respectively, *Hatched area:* zone of probable active volcanic intrusion. Normal step-like pattern is formed in the extensional zone of active volcanic intrusion, while reverse step-like pattern is formed in the compressional (shear) zone

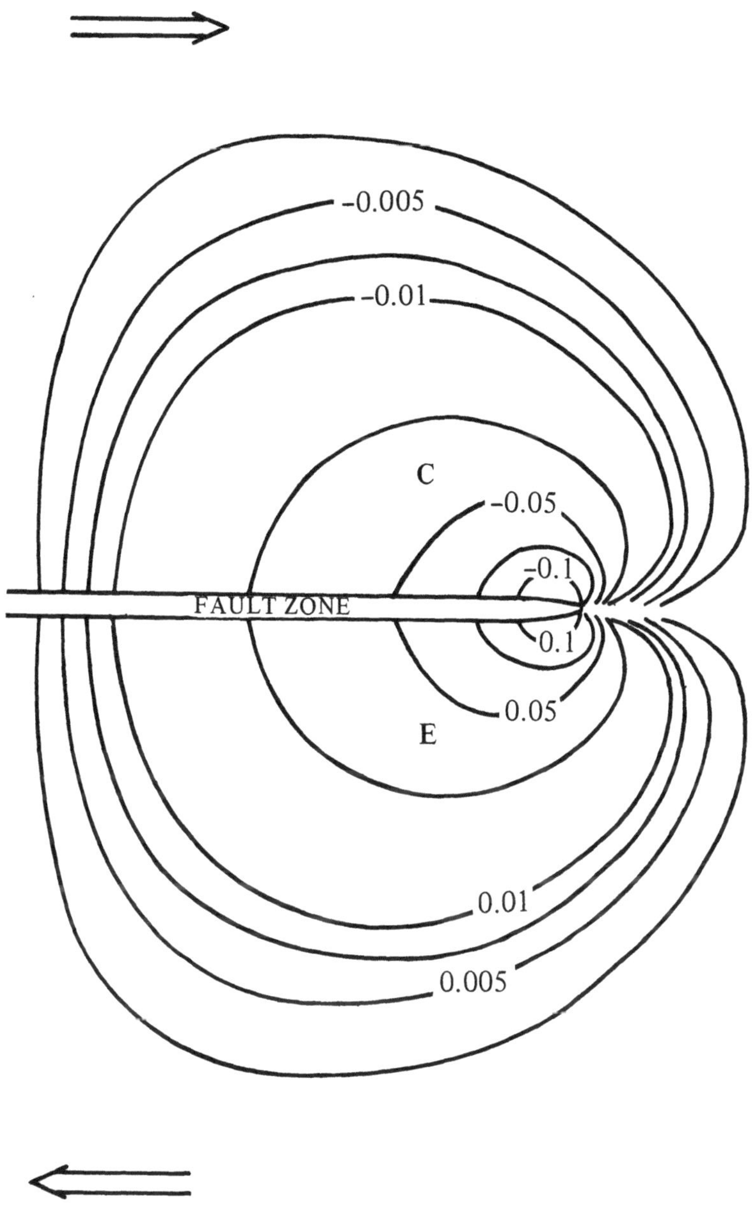

Fig. 2. Contour map of vertical stress around a strike-slip fault zone. Compressional *C* and extensional *E* zones develop on either side of the fault depending on right or left lateral strike-slip movement of the fault. The numbers in the contour show the magnitude of stress when the surrounding rocks of the fault zone are subjected to a shearing stress of 1 kbar parallel to the fault zone. The fault zone is assumed to be a linear or penny-shaped body of aspect ratio 1:100. The host rock surrounding the deep-seated strike-slip fault is considered to be dunite or peridotite with 0.7 Mbar rigidity. The rigidity of the fault zone is considered to be 0.08 Mbar which is about one-ninth that of the host rock. The Poisson's ratio of the host rock is assumed to be 0.28

This relationship between the direction of fault displacement and step-like pattern becomes important for the consideration of the mechanical effects of displacement on such fault geometry. In a normal step-like pattern, the right-lateral fault displacement at a bend would tend to separate the confronting wall rock(s) producing an area of extension, while in the reverse step-like pattern the wall rock(s) in the bend would be pushed against each other thus producing an area of compression or high pressure (Fig. 1B). An analytical solution (Koide and Bhattacharji, 1975a, 1975b) using the three dimensional theory of elasticity has been applied to obtain stress distribution due to right lateral strike-slip movement of a long, elongated or penny-shaped fault. The result (Fig. 2) indicates that both compressional *C* and extensional *E* regions develop on each side of the strike-slip fault. Although the exact magnitude of stress distribution around the fault may vary with the magnitude of shearing stress due to varying strike-slip movement, different aspect ratio of the fault, or for various Poisson's ratio and rigidity of the host rock enclosing the fault, the general pattern of stress distribution remains the same. In the shallower part of the crust, the ground surface around the fault can move readily in the vertical direction. This produces uplift over the compressional region *C* and subsidence over the extensional *E* zone. Progressive development of subsidence and upliftment from the fault boundary can result in a "normal step-like" pattern in the extensional region and "reverse step-like" bend in the compressional zone (Fig. 1). Between the en echelon fault segments, the magnitude of compression or extension becomes high because of stress interaction between fault segments (Fig. 4B). Pakiser (1960) has suggested that volcanic activity is most likely to occur in the extensional zone while failure by shear resulting in thrust or splay faults or folding will be favored in the compressive region.

3 DEVELOPMENT OF NORMAL STEP-LIKE PATTERN OF FRACTURES AND FAULTS

A normal step-like pattern is commonly observed in small faults, especially in those faults which contain hydrothermal veins. It is common knowledge among mining geologists that the thickness of right-lateral vein filled faults increases when their trend turns toward right and becomes narrower when their trend turns toward left (Newhouse, 1942; McKinstry, 1948). It has also been noted in an area of much hydrothermal activity where the faults are normally under high interstitial fluid pressure due to fluid circulation, strike-slip faults of normal step-like pattern are common and generally stable. The hydrothermal solutions carrying ore materials can readily fill the extensional fractures producing gash vein type deposits.

Model experiments (Imai and Koide, 1963; Koide, 1971; see Fig. 3) have shown that a normal step-like pattern of faults develops from a set of en echelon tension fractures (Figs. 3C–D). Tension fractures developed under pure tensile stress can extend for considerable distance in a straight line. However, under compression, extension fractures which are generated from preexisting "Griffith cracks" are characterized by their limited length (Brace and Bombolakis, 1963). Rock deformation experiments have also shown that tensile fractures developed under high confining pressure are relatively small, and their length is controlled primarily by the mechanical properties of the rock and the effective pressure (Koide and Hoshino, 1967; Hoshino and Koide, 1970). However, under increasing axial stress, these fractures interact and concentrate into small shear zones at approximately 30° to the axis of maximum compression, resulting in en echelon arrays of tensile fractures. A shear fracture can develop through the connections of en echelon tension fractures in a single shear zone (Fig. 3). Generally, two types of fault patterns result from the connections of these en echelon tensile fractures (1) step-like and (2) feather-like (Koide, 1968). A normal step-like pattern is most common when the original en echelon tension fractures become connected by shear fractures in shear fault zone (Fig. 3) and are transformed into hydrothermal gash veins or extensional basins (such as in the Gulf of California rift system, Reykjanes rift, etc.). Primary conditions for the development of such a fracture or fault pattern have been found to be the brittle failure of the rock and low effective (or high fluid) pressure (Table 1).

4 DEVELOPMENT OF REVERSE STEP-LIKE PATTERN OF FRACTURES AND FAULTS

Fractures, produced on the ground surface over the San Andreas fault zone in the area of the 1966 Parkfield-Cholame earthquake epicenter (Brown et al., 1967), are of step-like patterns with wide tensional gash segments (Fig. 4). These shear fractures arranged in echelon pattern formed a large shear zone. Ground upheaval indicates compression in the gaps between en echelon shear fractures (Fig. 4). Subsidence, indicating extensional zones, occurred at the periphery of the gaps between shear fractures. As the rupture strength of a rock is greater under compressive stress (Mogi, 1974), fractures are expected to develop more readily in extensional zones than in the compressional gaps. As a result, the fractures normally tend to extend outward from the axis of the shear zone toward the zone of extension (Fig. 3). This outward development of fractures from the fault can produce a concentration of more compressive stress in the gaps between en echelon faults finally resulting in thrust or splay faults in the gaps (Fig. 3E). Connection of large shear faulted

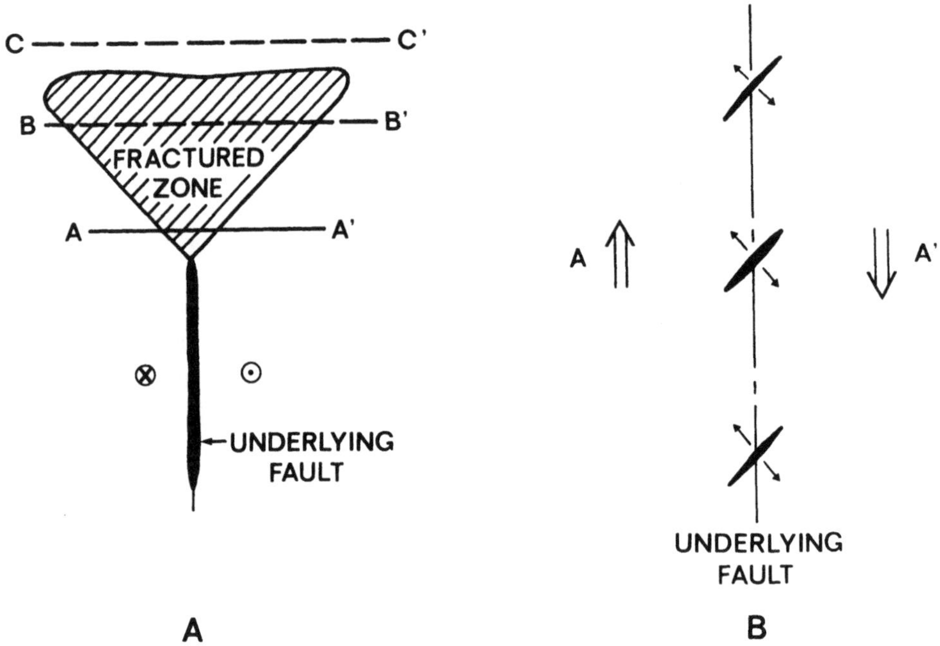

Fig. 3A–F. (A) A diagramatic sketch showing the relationship between an underlying fault and the probable fracture zone development in "Riedel shear" experiment. The effects of right-lateral strike-slip movement with continued displacement on the fracture zone along sections A–A', B–B' and C–C' are shown in (B) to (F); (B) Plan view along A–A'. As the fracture zone is very narrow along A–A', tension fractures can occur across the entire width of the fracture zone. An array of en echelon tension fractures almost parallel to the underlying fault develops; (C) Plan view along B–B'. The major fracture zone here is much too wide for en echelon tension fractures to span the entire width of the fracture zone. However, en echelon arrangement of small tension fractures produces en echelon shear fracture zone(s); (D) Plan view along B–B' with further right-lateral strike-slip movement along the underlying fault. With continued movement along the fault, en echelon tension fractures become connected to form shear fractures. Such shear fractures are also aligned en echelon above the underlying fault. Conjugate faults may also develop in the peripheral region(s) of the shear fractures. Note that the compressional zones C are developed in crustal rock(s) in the gaps between en echelon shear fractures or fault segments directly above the underlying fault. The extensional zones E, however, appear on each side of the compressional zone C, away from the major underlying fault; (E) Plan view along B–B′ at a further stage of strike-slip movement from that in (*D*). Connection of en echelon fractures and fault segments, secondary shear fractures and/or thrust faults develop in the compressional zone C forming a reverse step-like pattern of fault system; (F) Plan view along C–C'. Due to right-lateral strike-slip movement along the underlying fault, en echelon fold system develops close to the surface outside of the possible fracture zone

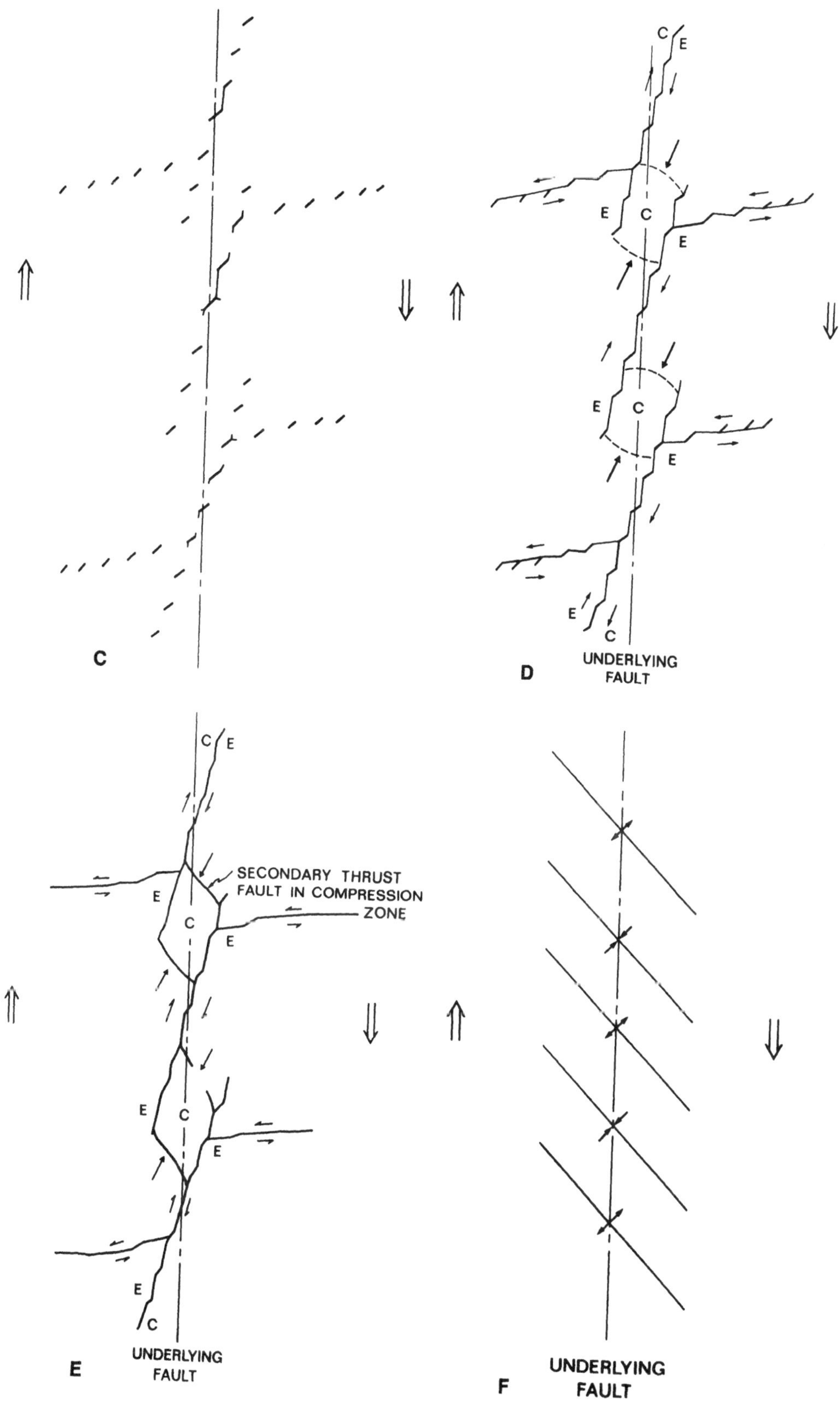

C
C
E
E
C
E
E
C
E
E
C
D
UNDERLYING FAULT
C
E
SECONDARY THRUST FAULT IN COMPRESSION ZONE
E
C
E
E
C
E
E
C
E
UNDERLYING FAULT
F
UNDERLYING FAULT

Table 1.

Fault geometry	Normal step-like pattern	Reverse step-like pattern
Primary pattern	En echelons tension fractures	En echelons shear faults
Size	Small	Large
Fracturing condition	Brittle condition	More ductile condition; transitional between brittle and ductile failure
Stress condition	Low effective pressure, high interstitial fluid pressure	Moderate effective pressure, or low to moderate interstitial fluid pressure
Characteristics of bend	Extensional: tension fractures, tension gashes	Compressional: thrust fault or splay fault
Earthquake behavior	Frequent small or moderate earthquakes	Few large earthquakes
Igneous and hydrothermal activity	Very active	Without or limited activity in peripheral extensional regions
Location	Active rift zone, active volcanic or geothermal area	Transform fault, subduction zone, plate collision zone, intraplate faults
Example	Gulf of California rift system, Reykjanes rift, Iceland etc.	San Andreas fault system, U.S.A.. Neo Valley fault system, Hukozu-Yokosuka fault system, Sagami fault system, Japan

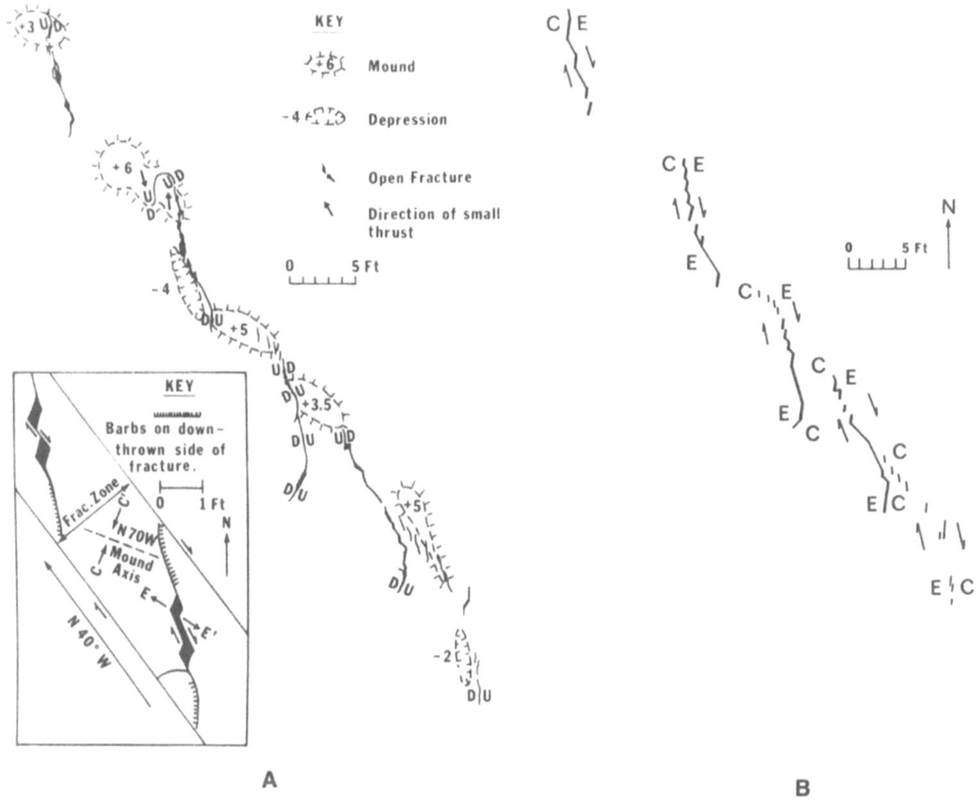

Fig. 4A, B. (A) Diagramatic sketch of surface fractures formed by the 1966 Parkfield-Cholame earthquake along the San Andreas fault (after Brown et al., 1967). Contours (in inch) indicate areas of upliftment and subsidence. *U:* upthrown; *D:* downthrown; (B) Schematic diagram of fracture systems developed as a result of the 1966 Parkfield-Cholame earthquake showing formation of compressional *C* and extensional *E* zones due to upliftment and subsidence

segments by thrusting or faulting in the gaps would produce a reverse step-like pattern as observed in the San Andreas fault system (Figs. 1 and 3).

A reverse step-like pattern of shear fault can be simulated in model experiments (Imai and Koide, 1963; Koide, 1971) by classical "Riedel shear" experiments (Riedel, 1929) in which a cake of plastic clay layer placed on two touching plates are sheared by pushing the plates against each other, thereby imitating a strike-slip fault movement. If the clay layer is thin (Fig. 3A) or saturated with water (Cloos, 1955), only en echelon arrays of tension fractures or gash veins develop close to the line of movement (Fig. 3B) and the shear zone becomes parallel to the line of differential movement. However, if the clay layer is sufficiently thick (Fig. 3A) with differential movement, shear zone on the clay surface becomes very wide (Figs. 3C–E) and the trend of en echelon tension fractures inclines at approximately 30° to the axis of maximum compressive stress (Fig. 3C). This is similar to what has been observed in actual rock deformation experiments under high confining pressure (Koide and Hoshino, 1967; Hoshino and Koide, 1970). The entire system of fractures being confined within a limited width above the underlying fault, becomes practically parallel with it (Figs. 3C–E). The axes of maximum compression and extension become oriented at 45° to the underlying fault (Fig. 5). Thus, the resulting shear fractures (or a set of en echelon tension fractures) orient at about 15° to the underlying fault (Fig. 5). These shear fractures are also arranged en echelon over an underlying fault (Fig. 3D). Connections of en echelon shear fractures would produce a reverse step-like pattern. Such a pattern of fault geometry is most likely to develop under more ductile (transitional between brittle and ductile) condition and at moderate effective pressure. In a reverse step-like pattern of strike-slip faults, displacement induces compression only

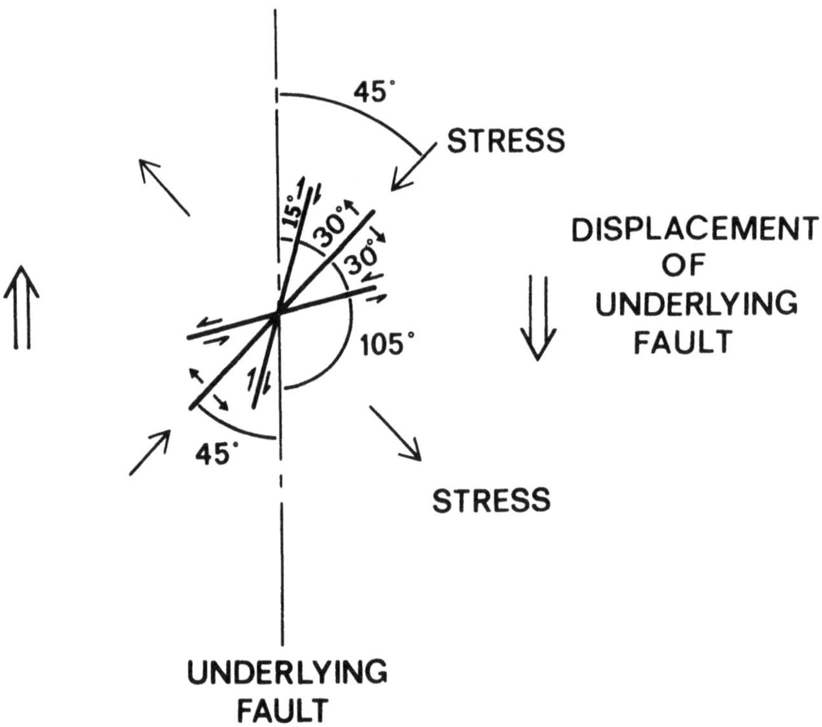

Fig. 5. The relationship of probable orientations of fractures on the crustal rock(s) with the underlying strike-slip fault due to right-lateral movement is shown. The maximum and minimum stress and the tension fractures form a 45° angle with the strike of the underlying fault. An array of en echelon tension fractures or the shear fractures are at an angle of 15° with the underlying fault, while the conjugate shear fractures or faults orient at 105° to the underlying fault

in its bends (Fig. 3E) and no dilatancy. Thus, under high effective pressure or in areas of high compression a reverse step-like pattern of strike-slip fault is expected to be a more stable form than the normal step-like pattern which induces extension and dilatancy. A comparison of normal step-like and reverse step-like patterns of strike-slip faults is summarized in Table 1.

5 REVERSE STEP-LIKE PATTERN AS A CAUSE OF LARGE EARTHQUAKES

Differences in seismic behavior of segments of the San Andreas fault system have been discussed by Allen (1968), Wallace (1970), Nason and Tocher (1970), Irwin and Barnes (1975), and others. Two segments of the San Andreas fault system (Fig. 1) have been characterized by earthquakes only of large magnitude (magnitude of 8 in 1857 and magnitude of 8.3 in 1906). On the other hand, the Gulf of California rift–Salton trough zone and the segments which include the main San Andreas fault from Cholame to near San Juan Bautista in the central California, and the Calaveras, Hayward, and Concord faults (Nason, 1973) are currently the sites of numerous small and moderate earthquakes with considerable creep. Allen (1968) has noted that the two locked zones of the San Andreas fault system are characterized by marked bends and relatively simple fault traces (Fig. 1). The active areas, on the other hand, show considerable evidence of fault trace complexities such as branching, splaying, and many en echelon breaks (Fig. 1). The two locked fault zones are typical examples of compressional bends in a reverse step-like pattern (Fig. 1). The rupture strength of rocks in these compressional bends should, therefore, be high. If previous fractures exist in such rocks, high friction under compressive stress would prevent sliding along the fracture surface. Sliding along other (extensional) segments of fault would build up high stress concentration in the adjacent compressional bends (Fig. 3D), and the final yield in the compressional bends would lead to the generation of large earthquake(s) producing intersegmental sliding. The presence of several thrust faults is indicative of compression around the bend near Los Angeles (Fig. 6). It is also important to note that the 1952 Kern County earthquake and the 1971 San Fernando earthquake were generated by thrust faulting in the compressional bend (Anderson, 1971). The upliftment prior to and during the San Fernando earthquake (Castle et al., 1975) indicates that the accumulation of stress was the actual cause for this earthquake. Recent upliftment along the main San Andreas fault (Castle et al., 1976) indicates accumulation of compressive stress in the bend near Los Angeles and possibility of a large earthquake in this area (Fig. 6). Our theoretical analysis (Fig. 2) of stress development around a strike-slip fault also

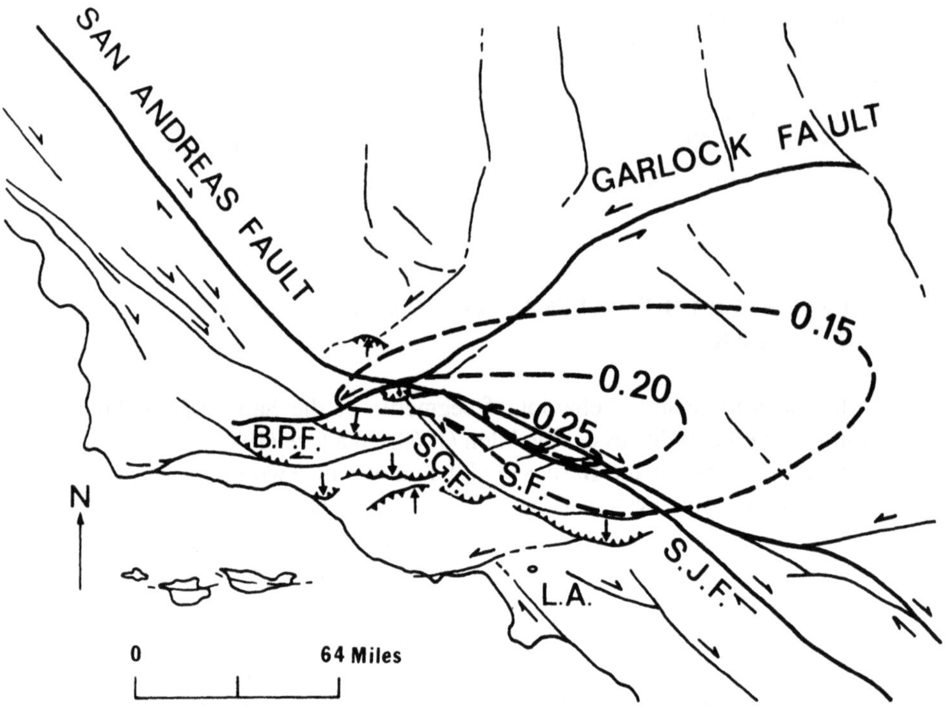

Fig. 6. Note the big bend of the San Andreas fault system near Los Angeles (L.A.). S.F.: epicentral area of 1971 San Fernando earthquake. *Broken lines:* contours of recent crustal uplift during the period 1959 to 1974 which developed around the bend of the main San Andreas fault (after Castle et al., 1976). Note also, the formation of many thrust faults in this area (after Hill, 1954). S.G.F.: San Gabriel fault; B.P.F.: Big Pine fault; S.J.F.: San Jacinto fault

indicates that upliftment should occur in the areas of compressive stress accumulation.

The Gulf of California rift–Salton trough zone is a seismically active area. The geothermal anomaly and presence of recent volcanic rocks indicate active igneous instrusion(s) and probable hydrothermal convection over these magmatic bodies (Fig. 1; Elders et al., 1972; Lawver et al., 1975). Since the Gulf of California rift system is an active zone of mantle intrusion and rifting, wedging of deep-seated diapiric intrusion(s) would produce rifting, fracturing, subsidence and extensional basins on the crustal surface in this area (Bhattacharji and Koide, 1975; Koide and Bhattacharji, 1975a, 1975b). Another seismically active zone in central California is along the San Andreas fault system in the extensional zone on the periphery of the reverse step-like pattern (Fig. 1). Metamorphic fluid (Irwin and Barnes, 1975) may reduce the effective pressure significantly causing seismic activity in this region (Table 1).

6 REVERSE STEP-LIKE PATTERNS OF EARTHQUAKE FAULTS IN JAPAN

Earthquake faults which were formed in the area around the epicenter of the Matsushiro earthquake swarm in Japan (Fig. 7) are very similar to the earthquake faults of the Parkfield-Cholame area cited earlier. At Matsushiro, the development of earthquake faults was observed during the earthquake (Nakamura and Tsuneishi, 1966; 1967). First, tension fractures appeared on the ground surface creating several zones of en echelon tension fractures (Fig. 7B). Each zone of en echelon tension fracture formed a left-lateral shear fissure zone. These shear fissures also arranged en echelon formed a large left lateral shear zone (Fig. 7). The upliftment of the ground surface indicates compression in the gaps of en echelon fissure zones. As these fracture geometries are very similar to those produced by the Riedel shear experiments described earlier, the existence of an underlying left-lateral major fault is expected (Koide, 1971). The formation of conjugate faults in the periphery of the major shear zone is similar to the left-lateral Garlock, White Wolf, and Big Pine faults in the flank of the right-lateral San Andreas fault system (Figs. 1, 6, 7).

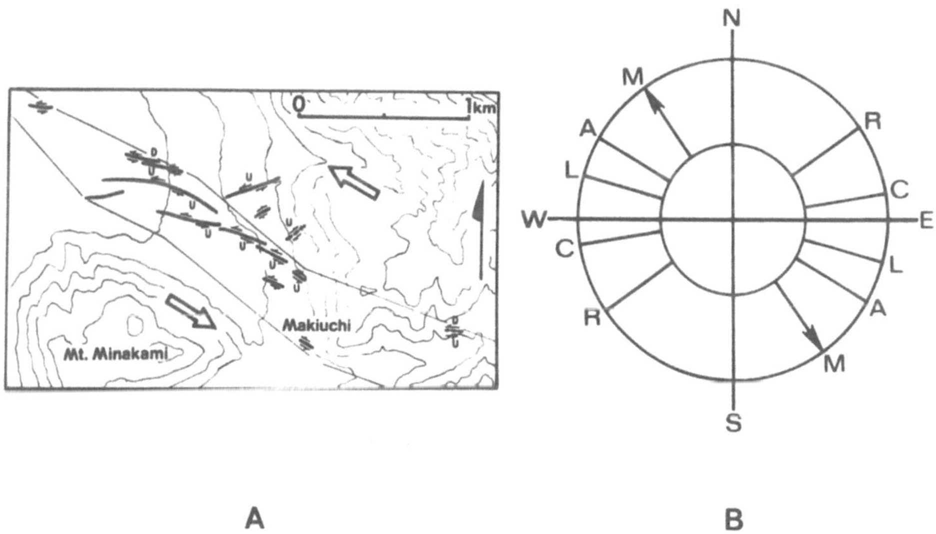

Fig. 7A, B. (A) En echelons fissure (fracture) zones formed by the 1966 Matushiro earthquake swarm, central Japan (after Nakamura and Tsuneishi, 1967). Each fracture zone also consists of an array of short tension fractures. Note the relative upliftment *U* and subsidence *D* along the fissure zones; (B) Relationship among orientations of short tension fractures *C–C,* left-lateral fissure zones *L–L,* right-lateral fissure zones *R–R,* probable underlying fault *A–A* and measured direction of movement of the left-lateral fissure zones *M–M* (after Nakamura and Tsuneishi, 1966; Koide, 1971)

The Hukozu–Yokosuka fault system which was produced by the 1945 Mikawa earthquake ($M = 6.9$) in Japan also shows a reverse step-like pattern (Fig. 8). The ground level change and displacement of triple points with the earthquake indicate the development of thrust fault characteristics in conjunction with the right-lateral strike-slip movement (Ando, 1974). Thrust faultng in the compressional zone in the reverse step-like bend may have been the cause for the Mikawa earthquake.

The Neo Valley fault system in Japan is a major left-lateral strike-slip fault zone (Fig. 9) along which a series of fault scarps were formed in the

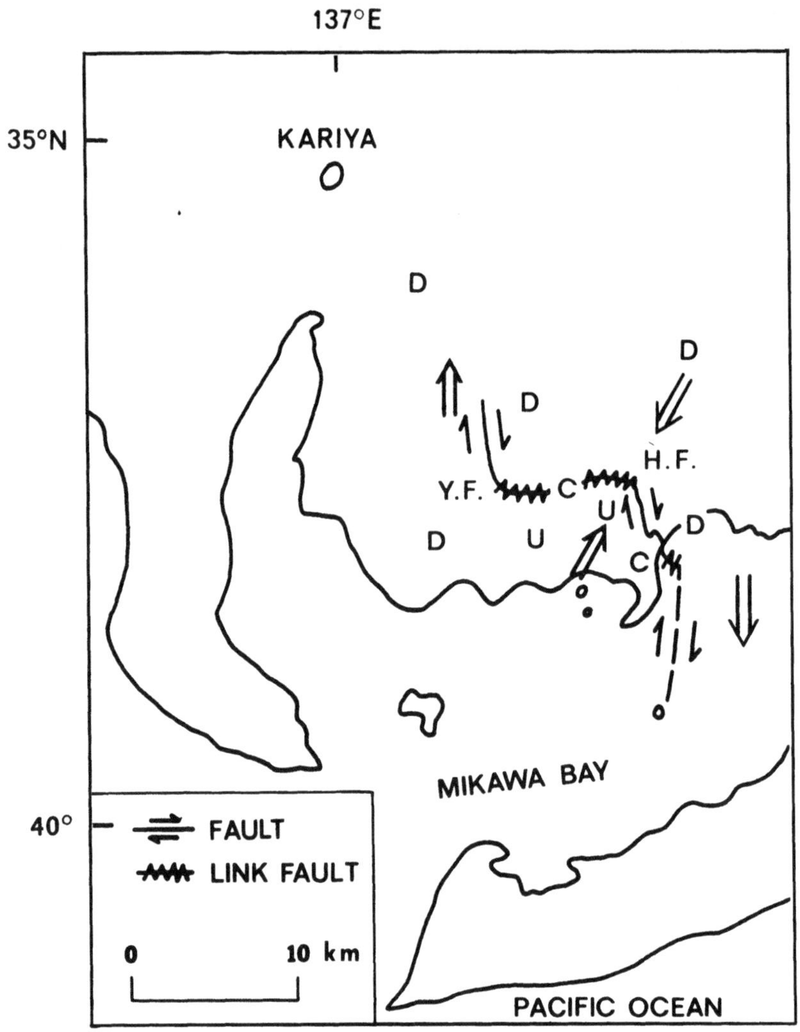

Fig. 8. Hukozu-Yokosuka fault system formed by the 1945 Mikawa earthquake, Japan and the measured ground movement (modified from Ando, 1974). Note the lateral displacement *(arrows),* relative ground upliftment *U,* subsidence *D,* and probable compression zones *C.* Y.F.: Yokosuka fault; H.F.: Hukozu fault

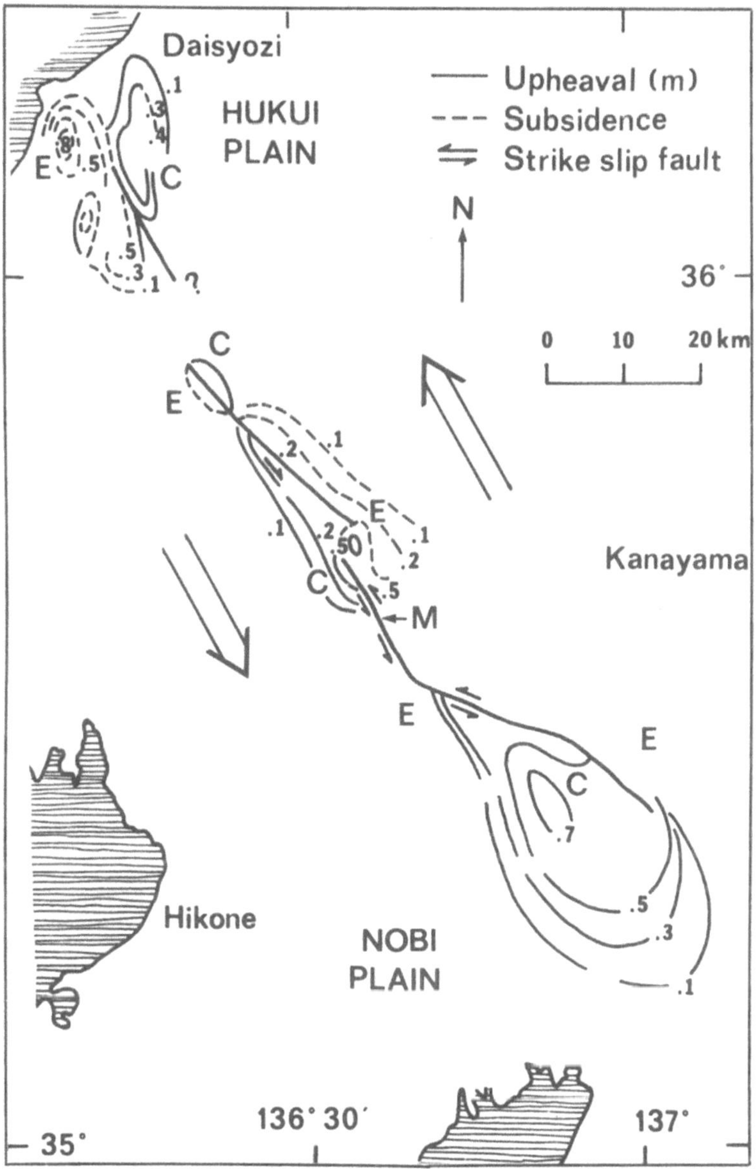

Fig. 9. Neo Valley fault system and vertical displacement caused by the 1891 Mino-Owari earthquake (M = 8.0) and the 1948 Hukui earthquake (M = 7.2) (modified from Muramatu et al., 1964; and Sugimura and Ueda, 1973). Zones of compression *C* and extension *E* are suggested from measured vertical movement. *M:* location of the Midori fault scarp formed from the Mino-Owari earthquake

1891 Mino–Owari earthquake (M = 8.0). The Midori fault scarp which was formed from this earthquake shows an extraordinarily large vertical displacement (5.5 m) compared to the other parts of the Neo Valley fault system (Sugimura and Ueda, 1973). We suggest that the Midori fault scarp was formed in the compressional zone of the reverse step-like bend of the Neo Valley fault system (Fig. 9).

A recent submarine geological survey shows that the Sagami Bay fault system which triggered the 1923 major Kanto earthquake (M = 7.9) in Japan is also of the reverse step-like pattern with en echelon offset (Fig. 10). Kanamori's (1971, 1974) estimation of right-lateral slip (2 m) suggests that secondary faulting with a thrust component in compressional gap of the fault of reverse step-like pattern may have been the cause of the Kanto earthquake. Such an explanation seems more fitting in view of the present observation of extraordinarily low seismic activity in the focal region of the Kanto earthquake (Shimazaki, 1971).

Sugimura and Ueda (1973) have drawn attention to the fact that almost all primary earthquake faults in Japan are either strike-slip or reverse (thrust) faults; virtually none are normal faults. Recently, Matsuda

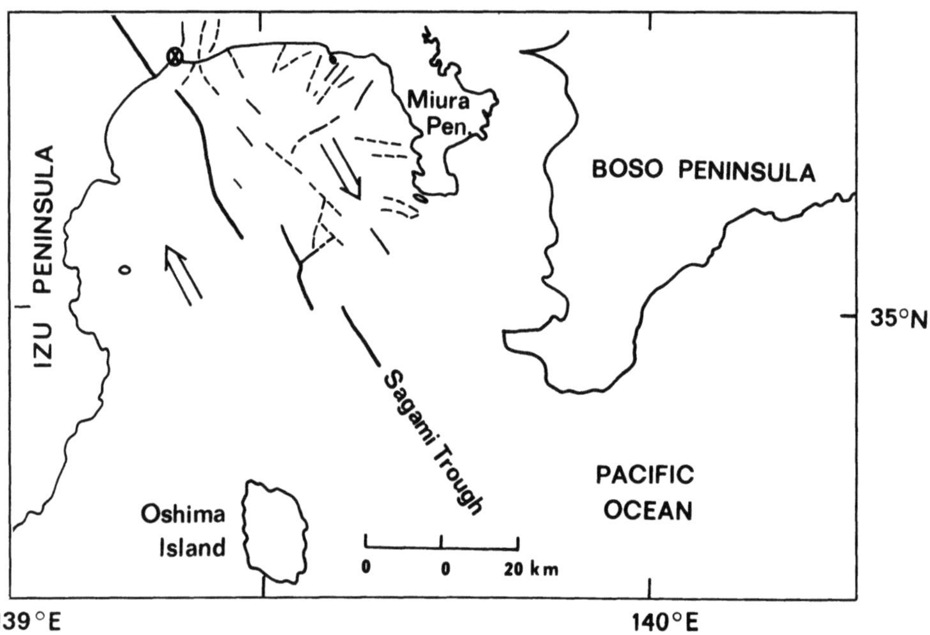

Fig. 10. En echelon pattern of the Sagami Bay fault system (after Kimura, 1976; and pers. comm.). Development of large-scale fault of reverse step-like pattern at northern and southern ends is also suggested

(1967, 1969) has also noted that a large, active strike-slip fault usually tends to bend at its ends toward the region of compression (Fig. 2), while smaller faults tend to curve toward the regions of extension. Anderson's "splay fault" (1951) is a type of branching of a strike-slip fault into a region of compression. This bending or branching of faults into a compressive region may appear paradoxical since formation of fractures in rocks should be facilitated in a region of extension and under tensile stress. However, if this bending or branching represents the partial appearance of a reverse step-like pattern in a fault, the development of this bend or branching toward a region of compression is to be expected under the "Riedel shear" condition of a strike-slip fault. All these observations suggest that the majority of large earthquake faults in Japan, like the San Andreas fault system in California, are compressional and characterized by reverse step-like pattern.

7 CONCLUSION

Many large earthquakes are generated by thrusting or shearing in a bend or gap of a fault of reverse step-like pattern. The faults of such a pattern are generally en echelon shear faults characterized by compressional bend(s) and upliftment. They are produced under relatively ductile (transitional between brittle and ductile behavior) conditions at moderate effective pressure. Strike-slip faults of a reverse step-like pattern can occur in the periphery of active volcanic areas, with transform faults, intraplate fault zones or in subduction and plate collision boundaries.

On the other hand, strike-slip faults of a normal step-like pattern accompany the regions of active rifting, high volcanicity, and geothermal activity. These are characterized by en echelon tension fractures, extensional bends, basins, and areas of subsidence. Brittle condition of rocks, high interstitial fluid pressure, or low effective pressure favor their formation. Frequent small to moderate earthquakes are associated with faults of such a pattern. Recognition of these characteristic geometric patterns of active strike-slip faults together with measurements of ground upliftment or subsidence, accumulation of compressive strain, etc., may serve as guide for establishing the vulnerability of an area in the fault zone to more frequent but less dangerous earthquakes or to infrequent large earthquakes. Strain release by hydrofracturing along main shear faults or along extensional zones can build up compressive stress in the bends or gaps between fault segments showing a reverse step-like pattern and could even accelerate large earthquakes. However, careful strain release by hydrofracturing within compression bends or gaps may reduce the risk of large earthquakes. Here again, fault geometry can aid in deciphering compressional zones.

ACKNOWLEDGMENT

This research has been supported by a CUNY faculty research grant to the junior author (S. B.). The authors are thankful to Lee C. Bhattacharji for helping with this manuscript and to Florence Letts for typing it.

REFERENCES

Allen, C. R.: The tectonic environments of seismically active and inactive areas along the San Andreas fault system. In: Dickinson, W. R., Grantz, A., (eds.) Proc. Conf. Geol. Problems of San Andreas Fault System. Stanford Univ. Publ. Geol. Sci., *11,* 70–82 (1968)

Anderson, D. L.: The San Andreas fault, In: Wilson, J. T., (ed.) Continents Adrift. San Francisco: W. H. Freeman and Co. 1971, pp. 143–157

Anderson, E. M.: The Dynamics of Faulting, 2nd ed. Edinburgh: Oliver and Boyd, 1951, p. 206

Ando, M.: Faulting in the Mikawa earthquake of 1945, Tectonophysics *22,* 173–186 (1974)

Bhattacharji, S., Koide, H.: Mechanistic model for triple junction fracture geometry, Nature (London) *255,* 21–24 (1975)

Brace, W. F., Bombolakis, E. G.: A note on brittle crack growth in compression. J. Geophys. Res. *68,* 3709–3713 (1963)

Brown, R. D., Jr., Vedder, J. G., Wallace, R. E., Roth, E. F., Yerkes, R. F., Castle, R. O., Waananen, A. O., Page, R. W., Eaton, J. P.: The Parkfield-Cholame, California, earthquakes of June–August 1966-Surface geologic effects, water-resources aspects, and preliminary seismic data. U.S. Geol. Surv Prof. Paper *579,* 66 (1967)

Castle, R. O., Church, J. P., Elliott, M. R.: A seismic uplift in southern California, Science *192,* 251–253 (1976)

Castle, R. O., Church, J. P., Elliott, M. R., Morrison, N. L.: Vertical crustal movements preceding and accompanying the San Fernando earthquake of February 9, 1971: Summary in Tectonophysics *29,* 127–140 (1975)

Cloos, E.: Experimental analysis of fracture patterns. Bull. Geol. Soc. Am. *66,* 241–256 (1955)

Cook, K. L.: Active rift system in the Basin and Range province. Tectonophysics *8,* 469–511 (1969)

Crowel, J. C.: Movement histories of faults in the transverse ranges and speculations on the tectonic history of California. In: Proc. Conf. Geol. Problems of San Andreas Fault System. Stanford Univ. Publ. Geol. Sci. *11,* 323–341 (1968)

Dickinson, W. R., Grantz, A.: Proc. Conf. Geol. Problems of San Andreas Fault System. Stanford Univ. Publ. Geol. Sci. *11,* (1968)

Elders, W. A., Rex, R. W., Meidav T., Robinson, R. T., Biehler, S.: Crustal spreading in Southern California. Science *178,* 15–24 (1972)

Harrison, J. C., Mathur, S. P.: Gravity anomalies in Gulf of California in Marine Geology of the Gulf of California. Am. Assoc. Petroleum Geol. Memoirs *3,* 76–89 (1964)

Hill, M. L.: Tectonics of faulting in Southern California. In: Jahns, R. H., (ed.) Geol. Southern California, Calif. Div. Mines Bull., *170* (IV), 5–13 (1954)

Hoshino, K., Koide, H.: Process of deformation of the sedimentary rocks. In: Proc. 2nd Intern. Congr. Soc. Rock Mech., 2–13, Beograd, I.S.R.M. 1970.
Imai, H., Koide, H.: Structural analysis of the fractures in the earth's crust. J. Min. Met. Inst. Japan, *79*, 717–724 (1963) (in Japanese with English abstract)
Irwin, W. P., Barnes, I.: Effect of geologic structure and metamorphic fluids on seismic behavior of the San Andreas fault system in central and northern California, Geology *3*, 713–716 (1975)
Kanamori, H.: Faulting of the great Kanto earthquake of 1923 as revealed by seismological data. Bull. Earthquake Res. Inst. Tokyo Univ. *49*, 13–18 (1971)
Kanamori, H.: Long-period ground motion in the epicentral area of major earthquakes. Tectonophysics *21*, 341–356 (1974)
Kimura, M., Submarine geologic structure around the southern Kanto District, In: Nasu, N. (ed.) Kaiyochishitsugaku (Submarine Geology), Univ. Tokyo Press, 1976, pp. 155–176 (in Japanese)
Koide, H.: A tectonophysical study of the initiation and development of fracture in rocks. Ph.D. thesis, Univ. Tokyo (1968)
Koide, H.: Fractures aligned en echelon and fracture patterns. Soc. Mining Geol. Japan, Spec. Issue *3*, (Proc. IMA-IAGOD Meetings '70 IAGOD Vol.) 107–114 (1971)
Koide, H., Bhattacharji, S.: Formation of fractures around magmatic intrusions and their role in ore localization. Econ. Geol. *70*, 781–799 (1975a)
Koide, H., Bhattacharji, S.: Mechanistic interpretation of rift valley formation. Science *189*, 791–793 (1975b)
Koide, H., Hoshino, K.: Development of microfractures in experimentally deformed rocks (preliminary report). J. Seisml. Soc. Japan, 2nd Ser. *20*, 85–97, 1967 (in Japanese with English abstract)
Koide, H., Hoshino, K., Inami, K.: Fault models based on an extended Griffith's theory. Bull. Geol Surv. Japan *25*, 89–103 (1974) (in Japanese with English abstract)
Lawver, L. A., Williams, D. L., von Herzen, R. P.: A major geothermal anomaly in the Gulf of California. Nature (London) *257*, 23–28 (1975)
Lomnitz, C., Mooser, F., Allen, C. R., Brune, J. N., Thatcher, W.: Seismicity and tectonics of the northern Gulf of California region-preliminary results. Geofis. Intern., Mexico, *10*, 37–48 (1970)
Matsuda, T.: Seismogeology; Zisin. J. Seismol. Soc. Japan, *20*, 230–235 (1967) (in Japanese)
Matsuda, T.: Active faults and great earthquakes. Kagaku (Nat. Sci.) *39*, 398–407 (1969) (in Japanese)
McKinstry, H. E.: Mining Geology. Englewood Cliffs, N.J.: Prentice Hall, 1948
Mogi, K.: On the pressure dependence of strength of rocks and the Coulomb fracture criterion, Tectonophysics *21*, 273–285 (1974)
Muramatu, I., Kazita, S., Suzuki, S., Sugimura, A.: Structure of the Midori fault, a part of the Neo Valley fault in central Japan. Sci. Rept., Fac. Lib. Arts Educ., Gifu Univ. (Nat. Sci.) *3*, 308–317 (1964) (in Japanese with English Abstract)
Nakamura, K., Tsuneishi, Y.: Ground cracks at Matsushiro probably of underlying strike-slip fault origin, I—preliminary report. Bull. Earthq. Res. Inst. *44*, 1371–1384 (1966)
Nakamura, K., Tsuneishi, Y.: Ground cracks at Matsushiro probably of underlying strike-slip fault origin II—The Matsushiro Earthquake Fault. Bull. Earthquake Res. Ins. Tokyo Univ. *45*, 417–471 (1967)

Nason, R. D.: Fault creep and earthquakes on the San Andreas fault. In: Proc Conf. Geol. Problems of the San Andreas Fault System: Stanford, Calif., Stanford Univ. Publ. Geol. Sci. *13,* 275–285 (1973)

Nason, R. D., Tocher, D.: Measurement of movement on the San Andreas fault. In: Earthquake Displacement Fields and the Rotation of the Earth. Mansinha, L., Smylie, D. E. and Beck, A. E. eds., Netherlands: Dordrecht, 1970, pp. 246–254

Newhouse, W. H.: Ore Deposits as Related to Structural Features. Princeton: Princeton Univ. 1942, 280 p

Pakiser, L. C.: Transcurrent faulting and volcanism in Owens Valley, California. Bull. Geol. Soc. Am. *71,* 153–160 (1960)

Riedel, W.: Zur Mechanik geologischer Brucherscheinungen. Center. Bl. Min. Nsw., Abt. *B* 354–368 (1929)

Rusnak, G. A., Fisher, R. L., Sheppard, F. P.: Bathymetry and faults of Gulf of California. In: van Andel, T. H., Shor, G. G., Jr. (eds.) Marine Geology of the Gulf of California. Am. Assoc. Petrol. Geol. Mem. *3,* 59–75 (1964)

Scholz, C. H.: A comparison of the San Andreas fault with the Alpine fault. In: Kovach, R. L., Nur, S. (eds) Proc. Conf. Tecto. Problems of the San Andreas Fault System. Stanford Univ. Publ. Geol. Sci., *13,* 186–191 (1973)

Shepard, F. P.: Submarine topography of the Gulf of California. (Pt. 3 of the 1940 E. W. Scripps cruise to the Gulf of California) Geol. Soc. Am. Mem. *43,* 32 (1950)

Shimazaki, K.: Unusually low seismic activity in the focal region of the great Kanto earthquake of 1923. Tectonophysics *11,* 305–312 (1971)

Sugimura, A., Ueda, S.: Island Arcs, Development in Geotectonics 3. Amsterdam; Elsevier, 1973, 247 p

Walker, G. P. L.: Excess spreading axes and spreading rate in Iceland. Nature (London) *255,* 468–471 (1975)

Wallace, R. E.: Earthquake recurrence intervals on the San Andreas fault. Geol. Soc. Am. Bull. *81,* 2875–2890 (1970)

5

S. M. SCHMID AND M. S. PATERSON

Abstract An oolitic limestone has been experimentally deformed in a wide range of temperatures (500 to 900°C) and strain rates (10^{-3} to 10^{-5}/s) involving flow stresses ranging from 30 to 2500 bars. Strain analysis on the deformed specimens reveals that, in the low stress regime (higher temperatures, lower strain rates), the relative shortening in the ooids is higher than the known shortening strain in the bulk rock, whereas at the higher stresses the ooid strain is identical with the bulk strain. The strain discrepancy at the low stresses is attributed to differences in flow stress arising mainly from differences in grain size between matrix and ooids, the fine-grained ooids being weaker in the low-stress regime. The latter effect is consistent with previous observations on marbles and Solenhofen limestone. It is argued that the low-stress regime in the experiments is representative of many geological situations in terms of the processes of deformation and hence that such strain differences between ooids and matrix could occur naturally in oolites, leading to the possibility of errors in strain analysis based on ooid shape change. The trends in ooid strain versus bulk strain are also compared with theoretical predictions based on viscous inclusion models and the significance of discrepancies discussed.

STRAIN ANALYSIS IN AN EXPERIMENTALLY DEFORMED OOLITIC LIMESTONE

1 INTRODUCTION

For the analysis of finite strain in deformed rocks, oolites are an attractive choice. Although oolites are not as widespread as might be desired, they have properties that would appear to make them nearly ideal for strain analysis. In particular, the original shape of the ooids is often nearly spherical, and many oolites are monomineralic rocks, essentially uniform in composition. This use of oolites is well illustrated in the strain analysis of folded layers, as in the pioneering work of Cloos (1947).

At first sight, the monomineralic character of oolites tends to lead one to expect homogeneous deformation of the strain markers and matrix, and this homogeneity is usually assumed. However, recent laboratory experiments have revealed that the grain size can have a quite dramatic influence on the flow stress of calcite rock, especially at low, geologically realistic stresses. This conclusion follows from comparison of the rheological behavior of the fine-grained Solenhofen limestone with that of coarser-grained Carrara marble (Schmid, 1976 and further results to be published) as well as with that of Yule marble (Heard and Raleigh, 1972). The emphasis on making the rheological comparisons at the stress levels of tens or hundreds of bars thought to occur geologically is based, first, on the experience that the mechanisms of deformation appear to be determined primarily by the stress level and, second, on the principle that meaningful comparison between laboratory and geology can only be made if the same deformation mechanism is inolved; strain rates measurable in the laboratory under such low stresses can only be achieved by working at temperatures substantially above those expected in nature (Paterson, 1976).

Given the emphasis on grain size as an important material parameter in rock deformation and the fact that ooids often have a different grain size from their matrix, doubts then arise about the validity of the assumption of homogeneous deformation in the markers and matrix. Therefore, experimental deformation of such an oolite was undertaken in order to compare the known finite strain in the bulk rock with the strain inferred from the shape change of the ooids. It is shown that a discrepancy between these quantities can arise and, further, it has been possible to discuss the behavior in terms of the relative flow resistances ("viscosity contrast") of ooids and matrix and to make comparisons with the predictions of the theoretical models of Gay (1968) and Bilby et al. (1975). Finally, relevance to field geology is discussed.

2 SPECIMEN MATERIAL

A single block of undeformed oolite of Jurassic age (Hauptrogenstein) from the Jura mountains (Switzerland) was used. The matrix is sparitic, with a grain size of 80 μm to 600 μm, typically around 200 μm. The ooids occupy on the average 63% of the volume, are spaced fairly uniformly and have diameters of 0.7–1.0 mm (Fig. 1). The ooids show a remnant spherulitic structure indicated by a "dark cross" under crossed nicols (Fig. 5a) but this structure has been largely obliterated by recrystallization to ultrafine equiaxed grains of 1–2-μm diameter. Very rarely radial fibres as such survive. Most ooids also show a weak concentric structure. The mineralogical and chemical composition is very nearly uniform over both ooids and matrix, as shown by microprobe analysis of the very small traces of Mg and Fe present. There are very small, dispersed opaque impurities present in the ooids, however, possibly carbon of organic origin. These impurities are very helpful in delineating the shapes of the ooids, especially after recrystallization and grain growth has occurred at the higher temperatures.

A few ooids have formed around fossil debris. In cases where the debris particle has a high aspect ratio the ooids can deviate substantially from an initially spherical form, but such ooids are easily recognizable and very rare. They were rejected for measurements. The majority of ooids approximate fairly closely to spheres. There is a slight tendency for the longest dimension to lie parallel to bedding, giving a weak initial shape

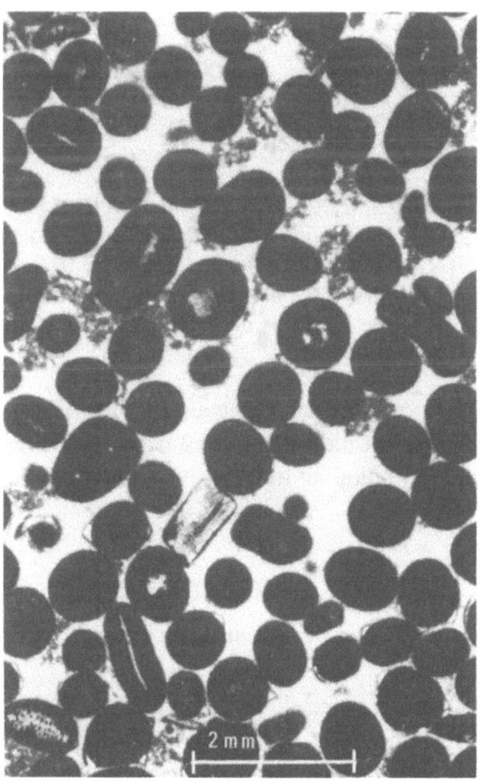

Fig. 1. Thin section of the undeformed oolite. Trace of bedding plane horizontal

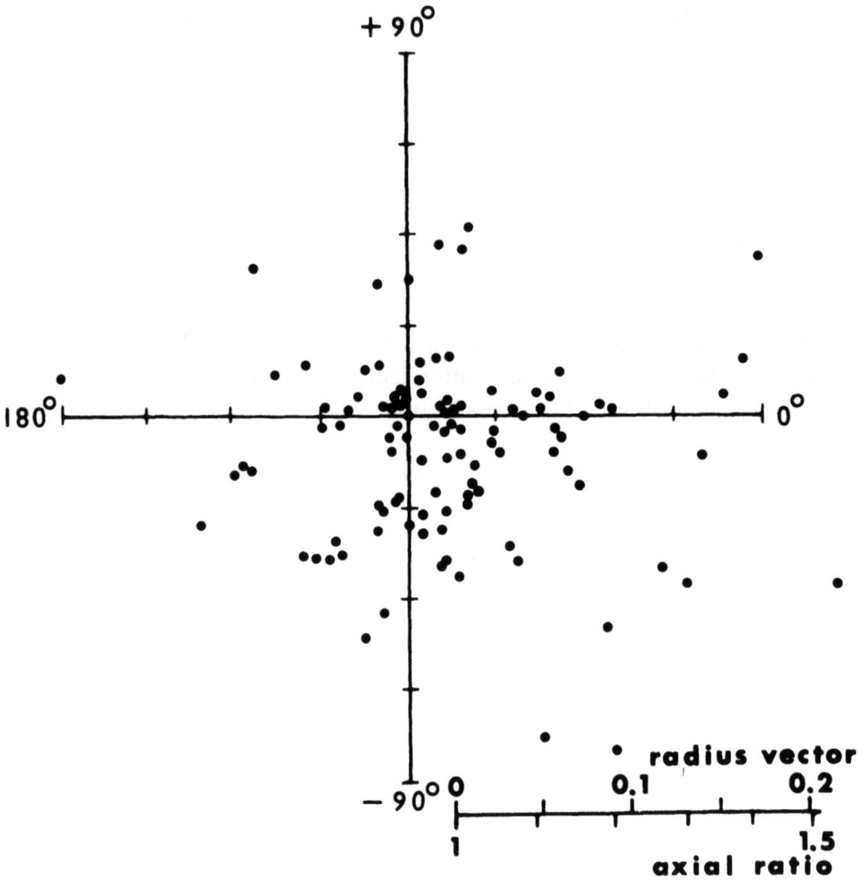

Fig. 2. Elliott plot (Elliott, 1970) of undeformed oolite (100 ooids). Half the natural logarithm of the axial ratio (radius vector scale) and the doubled angle between the long axis and the trace of the bedding plane (reference line) define a point on the graph. The distribution is nearly random and only a very slight preferential alignment of the long axis parallel to bedding is observed

anisotropy (Section 6). These properties of the initial distribution are summarized in the Elliott plot of Figure 2 from which it is seen that this oolite is particularly suitable for an experimental study. Nonoolitic debris is very rare.

3 EXPERIMENTAL DEFORMATION PROCEDURE

The specimens were, in general, deformed at constant strain rates in a high-pressure high-temperature apparatus built and described by Paterson (1970). In addition, a few stress relaxation runs were performed. It was possible to keep the temperature constant within ±5°C over the entire length of the specimen. A sensitive internal load cell allows the detection of differential stresses as low as 10 bars.

Cores of 20-mm length and 10-mm diameter were drilled perpendicular to bedding. The specimens were sealed during deformation with 0.25-mm walled copper jackets. The temperature range of 500 to 900°C was covered, and this made it possible to cover the entire geologically interesting range of flow stresses from about 30 to 2500 bars. A standard confining pressure of 3 kbar was applied.

The stress-strain curves were derived by making the usual corrections for apparatus distortion, copper strength, and cross-sectional increase.

The amount of shortening in the specimen was obtained by measuring initial and final length with a micrometer, rather than from the load-displacement record. Thus, the nominal strain was computed at ambient conditions and is directly comparable with strains obtained by analysis of thin sections. In some of the runs at 500 to 700°C, where relatively high strains were produced at high flow stresses, thin unloading cracks parallel and, more rarely, perpendicular to the compressional axis formed. They introduce negligible errors in the strain determination, compared to other errors discussed later.

4 STRESS STRAIN RESULTS

The experiments were mostly performed at a few standard conditions in order to study the relationships between bulk, ooid and matrix strains as the amount of finite strain is increased (Table 1). Typical stress–strain curves for these conditions are plotted in Figure 3. Runs exceeding 20% shortening generally were interrupted to check the degree of homogeneity in deformation in order to avoid potential damage to the furnace. In all runs, but especially in those at high temperature, some work hardening was observed.

Generally, the degree of work hardening in materials decreases with increasing temperature. However, two reasons for this exceptional case of work hardening at high temperatures can be proposed:

(1) Progressive grain growth in the ooids occurred at high temperatures (Table 2) and increasing grain size effectively increases the flow stress if the predominant controlling mechanism is diffusional creep and/or grain boundary sliding (Schmid, 1976).
(2) It will be shown that as the strain in the specimen is increased the ratio of the strain in the matrix to the strain in the ooids increases (Fig. 9); then since the matrix is more flow resistant at high temperatures

Table 1.

Test no.	Temperature (°C)	Strain rate (per s)	Nominal bulk strain $\bar{e}_b$ (% shortening)*	Mean ooid strain $\bar{e}_{oo}$ (% shortening)*	Mean matrix strain $\bar{e}_m$ (% shortening)*
2810	500	10^{-4}	12.9	9.6	18.5
2741	500	10^{-3}	19.0	17.0	22.4
2738	500	10^{-4}	20.9	21.3	20.2
2740	500	10^{-4}	33.0	33.4	32.3
2728	600	10^{-3}	9.1	10.8	6.2
2724	600	10^{-3}	23.9	24.2	23.4
2730	600	10^{-3}	30.0	31.0	28.3
2737	700	10^{-3}	11.4	12.4	9.7
2739	700	10^{-3}	11.9	13.7	8.8
2681	700	10^{-3}	19.0	21.5	14.7
2722	700	10^{-3}	32.3	35.1	27.5
2675	800	10^{-3}	43.0	46.6	36.9
2755	800	10^{-5}	6.2	8.3	2.6
2771	800	10^{-5}	13.3	16.6	7.7
2815	800	10^{-5}	18.5	24.4	8.5
2767	800	10^{-5}	24.1	27.6	18.1
2812	800	10^{-5}	33.8	38.1	26.5
2699	800	2×10^{-5}	43.1	49.5	32.2
2756	800	10^{-5}	45.7	51.9	35.1
2817	900	10^{-5}	6.2	9.6	0.8
2779	900	10^{-5}	10.9	15.9	2.4
2765	900	10^{-5}	20.4	27.3	8.7
2789	900	10^{-5}	31.0	39.3	16.9
2799	900	10^{-5}	41.9	48.6	30.5

*Note that throughout the text and figures the engineering sign convention for strain is used, under which shortening strains are negative

because of coarser grain size, hardening is to be expected. It is not possible at this stage to evaluate the relative importance of the two factors.

In order to get additional information on the rheological behavior of the bulk rock, stress relaxation runs were performed at 600°, 700°, and 900°C. The experimental procedure and the conversion of stress rates into strain rates have been described earlier (Schmid, 1976). Figure 4 is a combined plot of constant strain rate and relaxation experiments. Stresses were determined at 10% strain in the specimen. Assuming a strain rate vs. stress relationship of the form $\dot{\varepsilon} \propto \sigma^n$ a stress exponent of around $n = 5$ holds at high flow stresses, decreasing to $n < 3$ at low stresses. In terms of this stress dependence the similarities to Solenhofen limestone (Schmid, 1976) are remarkable. Marbles (Yule and Carrara) show a constant slope corresponding to $n \approx 8$ over this same region of temperature and strain rate (Heard and Raleigh, 1972). The transition into low n value ($n < 3$) observed in Solenhofen limestone has been attributed to a change in deformation mechanism (Schmid, 1976) and there is good structural evidence (Section 5) that this is also the case in this oolite (strictly speaking in the ooids only).

A comparison of strength at various conditions between oolite, Carrara marble and Solenhofen limestone is presented in Table 2. The ratio of flow stresses at a given strain rate, corresponding to the "effective

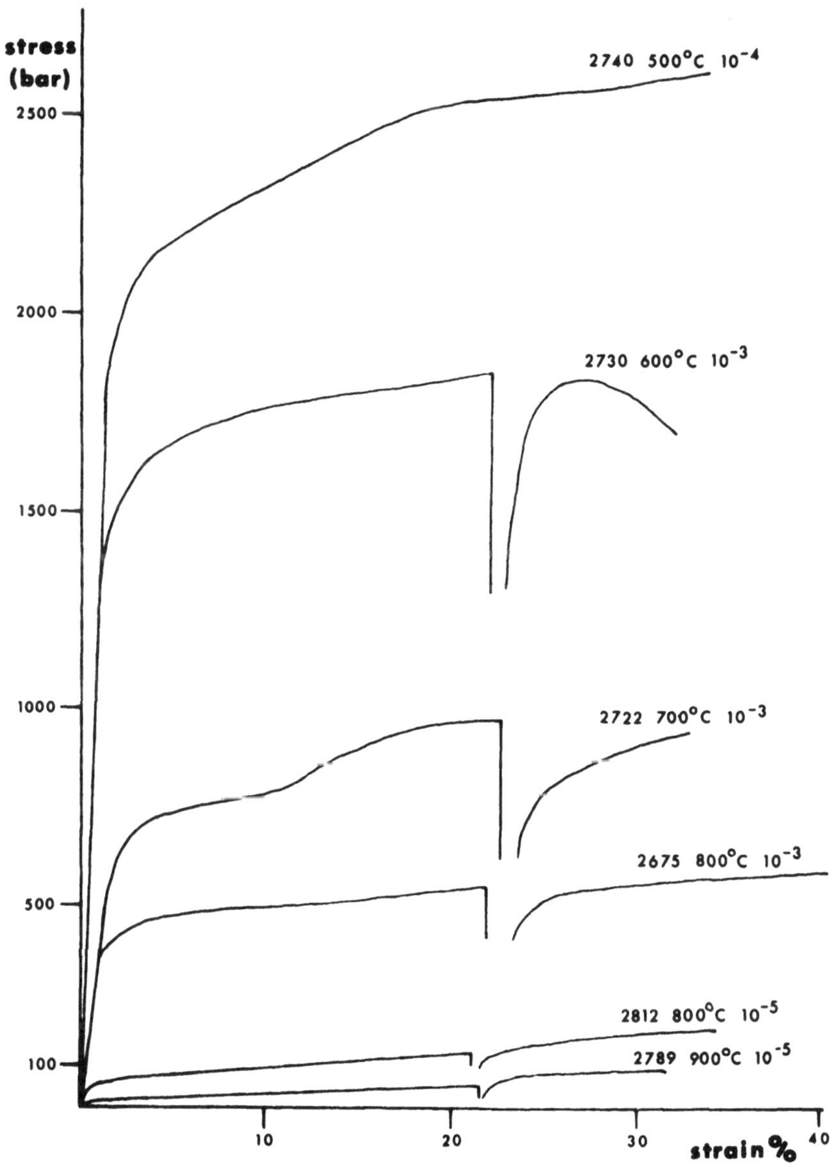

Fig. 3. Stress-strain curves for oolite

Table 2. Comparative strength and "effective viscosity" ratios in three different calcite rocks

Test conditions		Flow stress at 10% strain (bars) (Final grain size in micrometers given in parentheses)			"Effective viscosity" ratios	
Temperature (°C)	Strain rate (s)	Solenhofen σ_s	Carrara σ_c	Oolite σ_{oo}	σ_s/σ_{oo}	σ_s/σ_c
500	10^{-4}	2825 (4.5)	2060 (200)	2531 (1–2)	1.12	1.37
600	10^{-3}	2167 (4.5)	1640 (200)	1762 (2–3)	1.23	1.32
700	10^{-3}	1175 (5)	1148 (200)	860 (3)	1.37	1.02
800	10^{-3}	635 (5)	779 (200)	497 (5)	1.28	0.82
800	10^{-5}	87 (6)	447 (200)	99 (10)	0.88	0.19
900	10^{-5}	10 (7)	296 (200)	29 (20)	0.34	0.03

viscosity" contrast, is also listed in this table. The strength of the oolite is intermediate between Solenhofen limestone and Carrara marble at high stresses, where $\sigma_s/\sigma_c > 1$, as well as at low stresses, where $\sigma_s/\sigma_c \ll 1$. In an intermediate stress range oolite is weaker than both marble and limestone. This observation shows that the oolite behavior cannot simply be modelled as a combination of the properties of a matrix corresponding to Carrara marble and ooids corresponding to Solenhofen limestone. However, at the lowest stresses, where the contrast in flow stresses is very large, σ_s/σ_c can probably be taken as a sufficient first approximation to an "effective viscosity" contrast of ooids vs. matrix.

It is interesting to note that at the low stresses the oolite as a whole is much closer to Solenhofen limestone in terms of flow stress than to Carrara marble. This suggests that under these conditions, where the ooids are believed to deform superplastically (see Section 5), the flow behavior of the oolite in terms both of flow stress and of parameter n is almost entirely controlled by the major constituent in the rock, that is, the ooids. This suggestion is supported by the fact that during the first 10% shortening in the bulk rock the matrix deforms, on average, only by an insignificant amount (Table 1 and Fig. 9). The observation is of general interest because it shows that the rate of deformation in the bulk rock is obviously not governed by the rate of deformation of its strongest constituent.

5 MICROSCOPIC OBSERVATIONS AND COMMENTS ON THE DEFORMATION MECHANISMS

Substantial grain growth inside the ooids has been observed at the highest temperatures (Table 2). It is not clear how other variables affect the amount of grain growth, but the duration of time for which a

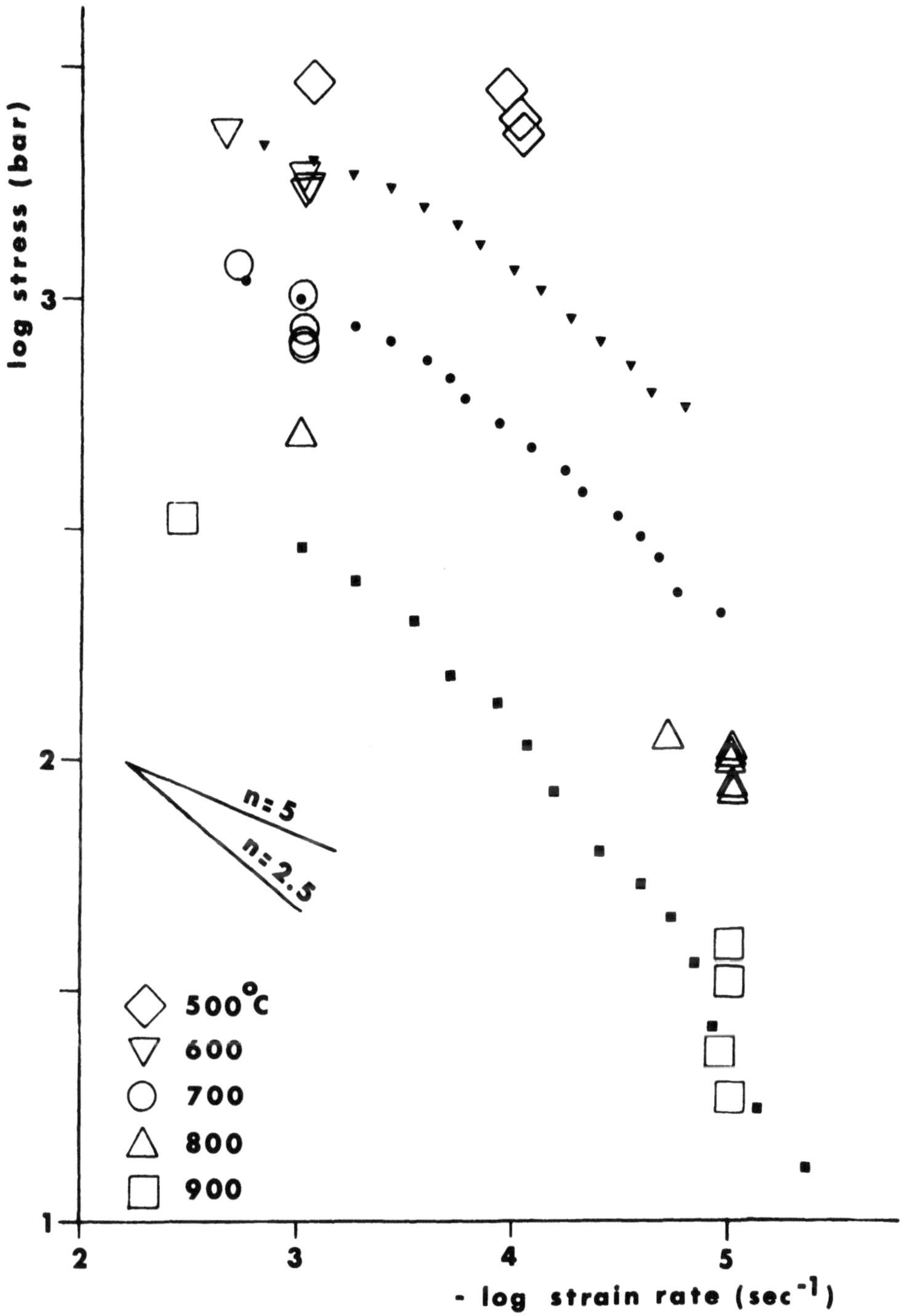

Fig. 4. Log–log plot of strain rate versus flow stress at 10% strain. *Large open symbols:* constant strain rate tests; *small closed symbols:* relaxation tests. The relaxation tests were started at 10% strain and the stress rates observed were converted into strain rates, using the known apparatus compliance

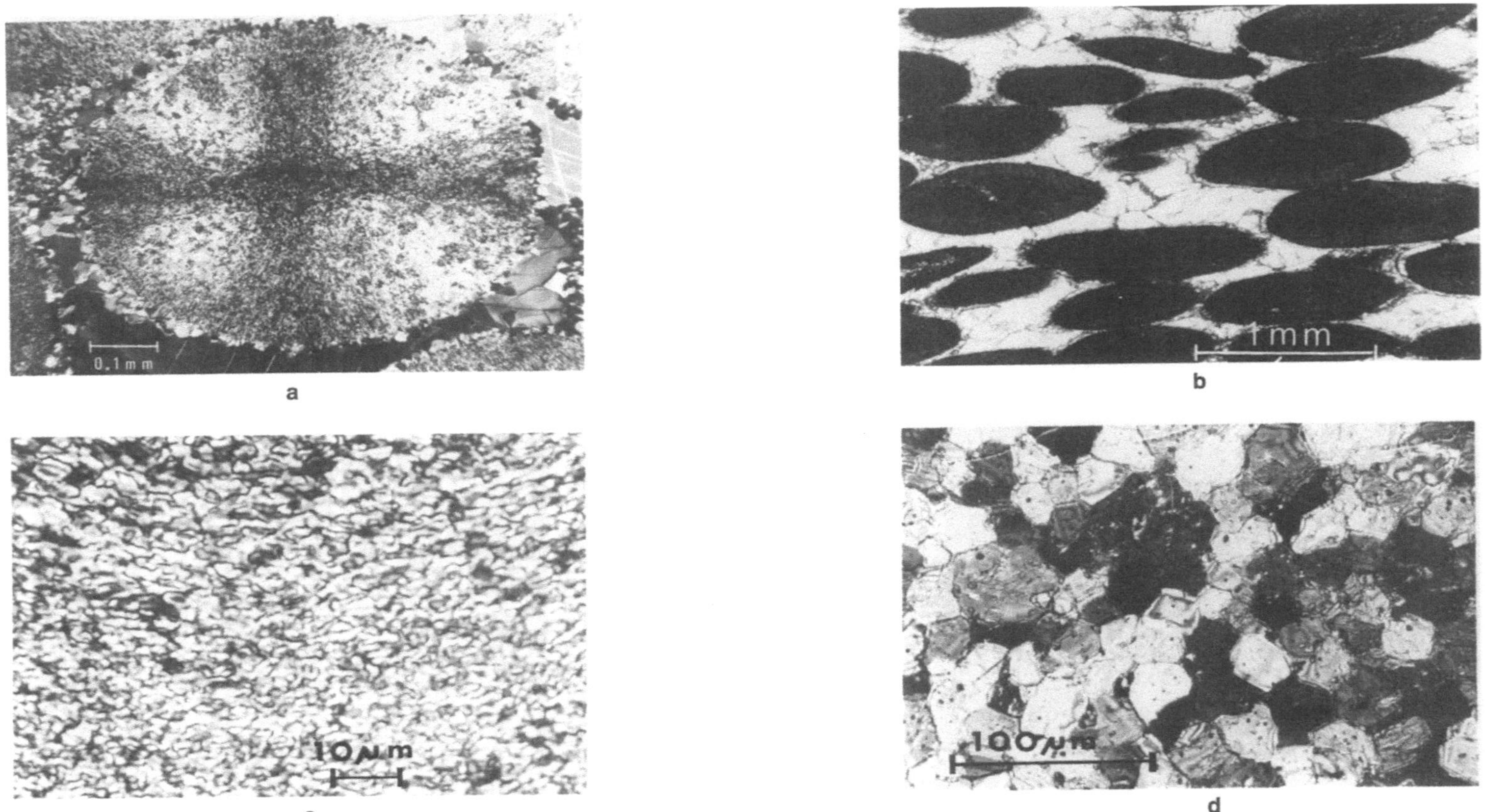

Fig. 5a–d. (a) Ooid, showing "dark cross" under crossed nicols, specimen 2728 (600°C, 10^{-3}/s strain rate, 9.1% shortening). (b) Heavily deformed oolite, specimen 2675 (800°C, 10^{-3}/s strain rate, 43% shortening). (c) Moderate grain growth and flattening of grains within an ooid of specimen 2730 (600°C, 10^{-3}/s, 30% shortening). Crossed nicols. (d) Grain growth in ooid of specimen 2799 (900°C, 10^{-5}/s, 41.9% shortening). Note absence of grain flattening in spite of the high bulk strain. Crossed nicols. Compressional stress applied vertically in all figures

specimen has been at a given temperature is certainly a major factor. Differences in grain size however persist up to the highest temperatures and strains and they make it possible always to define the outlines of the deformed ooids unambiguously. These outlines are accentuated by the higher content of opaque impurities in the ooids. It is interesting that there is also a tendency, although less pronounced, for grain size reduction to occur in the matrix at the highest temperatures. This supports the idea of an equilibrium grain size being approached at any given set of conditions, although it is not clear here what parameter determines the size of this equilibrium grain size. Kohlstedt et al. (1976) suggested that a simple relationship exists between the stress and an equilibrium grain size.

The sectional shape of the deformed ooids coincides surprisingly well with that of an ellipse, even at the higher temperatures, where substantial recrystallization occurred and where a considerable effective viscosity contrast in ooids and matrix is inferred. No systematic departure from an ideally elliptical form could be found to correlate with this viscosity contrast.* This is unfortunate because the field geologist needs such an indication to decide on the reliability of his strain analysis. An alternative method of detecting a viscosity contrast is described in Section 6.

Only in specimen 2799 (900°C, 41.9% nominal strain) is the effective viscosity contrast apparently expressed structurally (Fig. 6c), and this effect in turn probably explains the strong deviation of the mean ooid strain of this sample from the trend in the remainder of the 900°C runs (Fig. 8). Figure 6c shows how virtually undeformed matrix grains are pinched into the weaker ooids. Many matrix grains in this sample show textural characteristics of porphyroclasts in a mylonite. In places boudinage of two matrix grains with "intrusion" of fine-grained oolitc aggregate has been observed (Fig. 6c). Figure 6d presents evidence suggestive for grain boundary sliding between coarse matrix grains.

Intense twinning, undulatory extinction and kink band formation in the matrix grains are typical for highly strained specimens at 500° to 700°C where the matrix obviously deforms by a substantial amount (Fig. 6a). This strain is also directly recorded in substantial flattening of the matrix grains.

*Tan (1974) showed that viscosity contrast between a nucleus and its mantle within an ooid can give rise to departures from ideally ellipsoidal shape. However, the ooids in the rock used in the present study do not contain obvious nuclei and a uniform effective viscosity can be assumed within the domain of the ooid.

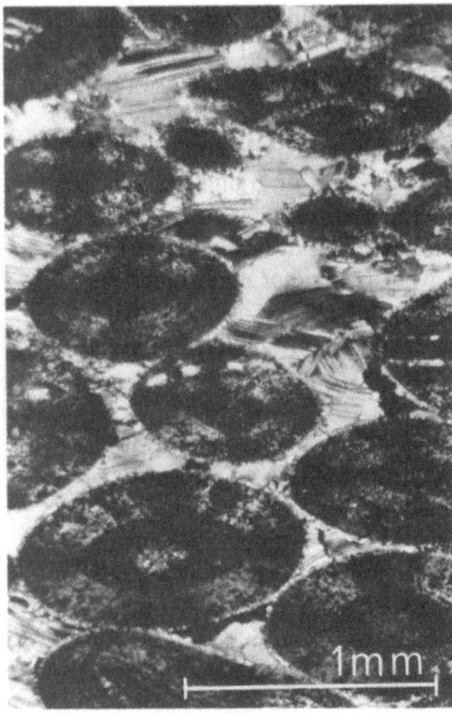

a

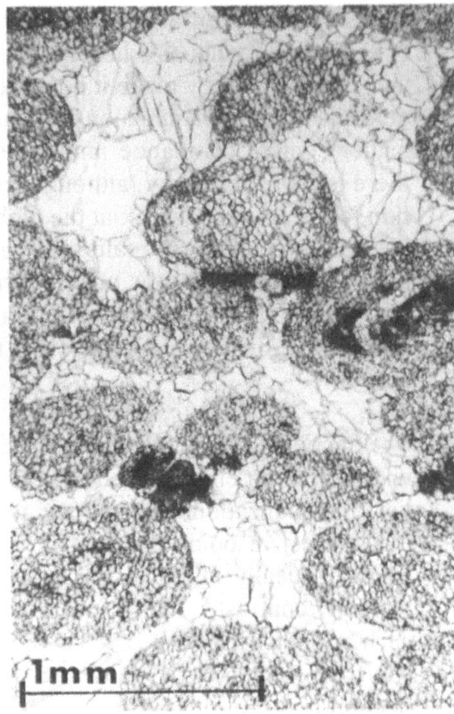

b

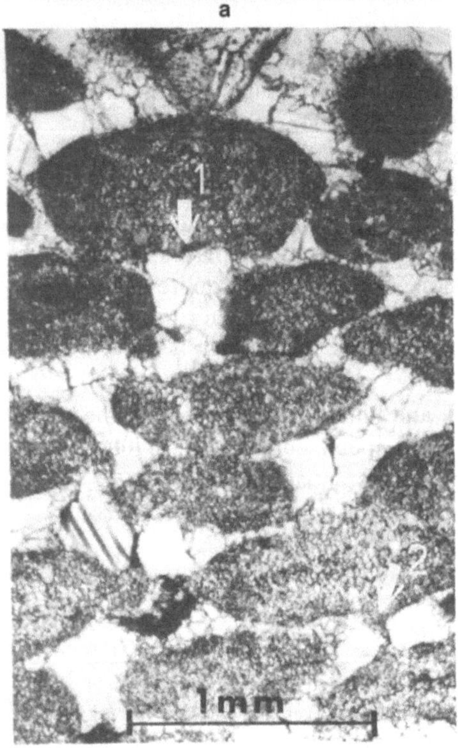

c

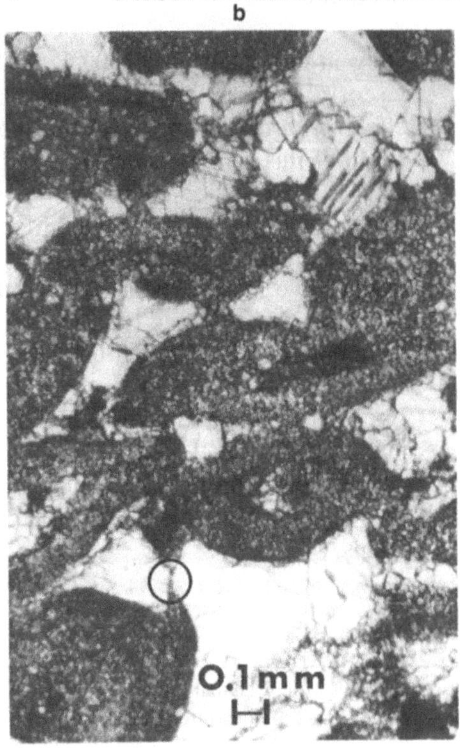

d

Clear signs of grain flattening were also found within the ooids at 500° to 700°C (Fig. 5c) but were absent at 800° and 900°C (Fig. 5d & 6b). The latter observation suggests that the predominant deformation mechanism within the ooids at the highest temperatures is grain boundary sliding, as has been proposed for Solenhofen limestone under similar conditions (Schmid, 1976, and new evidence to be published).

These microstructural observations on the matrix and ooids are consistent with similar observations on Carrara marble and Solenhofen limestone. The change in slope $n = \partial \log \dot{\epsilon} / \partial \log \sigma$ of the logarithmic stress vs. strain rate plot to values less than 3 in Solenhofen limestone and oolite, but not in marble at similar conditions, supports the idea that there is a major change in the deformation mechanism at lower stresses in fine-grained calcite aggregates within the temperature-strain rate regime studied. It is this change within the ooids that allows the substantial viscosity contrasts to build up. Similar changes in n are observed in fine-grained metals which deform superplastically and where grain boundary sliding is generally believed to be the predominant deformation mechanism (Davies et al., 1970; Edington et al., 1976).

6 STRAIN DETERMINATIONS

6.1 Strain Determination Using the Deformed Ooids

Relative shortening of the specimen parallel to its axis is reflected in the aspect ratio of an initially spherical ooid. Since the bulk deformation has rotational symmetry, such an ooid will become an oblate spheroid with axis parallel to the specimen axis. In this case, the aspect ratio as measured in a section parallel to the specimen axis is independent of whether the section is cut through the center of the ooid or not. It is assumed that there is no volume change during deformation of an ooid since bulk measurements on the specimens reveal no significant volume changes.

In this study, where the principal axes of the average bulk strain are known, the measure taken for the aspect ratio is the ratio R of the

Fig. 6a–d. (a) Heavily deformed matrix and ooids in specimen 2730 (600°C, 10^{-3}/s, 30% shortening). Note twinning and slip in matrix grains. Crossed nicols. (b) Recrystallization at 900°C (specimen 2765, 10^{-5}/s, 20.4% shortening). Note grain size reduction in matrix and grain growth in ooids. Also, note absence of flattening of individual matrix and ooid grains. (c) Marked deviation from elliptical form in ooids from run 2799 (900°C, 10^{-5}/s, 41.9% shortening). Note indentation of matrix grains into ooids (region 1) and "intrusion" of ooid aggregate between matrix grains (region 2). (d) Evidence for grain boundary sliding in specimen 2799 (encircled region). Compressional stress applied vertically in all figures diameter of the ooid parallel to the specimen axis to the diameter normal to the axis. This measure will deviate somewhat from the true axial ratio as normally measured in strain analyses on rocks in the field where the principal axes of the bulk strain are unknown. However, for the present purpose of exploring heterogeneity of strain between ooids and matrix this measure is the most relevant quantity. It can be regarded as describing the axial ratio of an imaginary strain ellipse inscribed within the ooid; the axes of this ellipse have not rotated, on average, because of the coaxial nature of the deformation involved in the experiments. The measurements of R were made on enlarged photographs of thin sections of the deformed specimens.

In calculating the ooid strain from the measured aspect ratios account must be taken of a small initial shape anisotropy consisting of a preferential alignment of the short axes of the originally slightly non-spherical ooids perpendicular to bedding, that is, parallel to the axis of subsequent shortening. Using the same procedure as for the deformed material, the initial aspect ratios R_0 were measured on 250 ooids in an undeformed oolite specimen. The arithmetic mean of R_0 was 0.971, with a standard error of 0.008; the standard deviation of the set of individual measurements was correspondingly 0.121.

The engineering axial strain in the ooids is designated e_{oo}. If an ooid is assumed to be initially spherical, with unit diameter, and if the maximum and minimum diameters of the spheroid into which it is deformed are $(1 + e_1)$ and $(1 + e_3)$ respectively, then, from the constant volume condition $(1 + e_3)(1 + e_1)^2 = 1$, it follows that

$$e_{oo} = e_3 = R^{2/3} - 1, \tag{1}$$

where

$$R = \frac{1 + e_3}{1 + e_1}.$$

If the ooid has an initial spheroidal shape with maximum and minimum diameters $(1 + e_1)_0$ and $(1 + e_3)_0$ and if the deformation is coaxial, the constant volume condition is $(1 + e_3)_0(1 + e_1)_0^2 = (1 + e_3)(1 + e_1)^2$ and the axial strain is

$$e_{oo} = \frac{1 + e_3}{(1 + e_3)_0} - 1 = \left(\frac{R}{R_0}\right)^{2/3} - 1, \tag{2}$$

where

$$R_0 = \frac{(1 + e_3)_0}{(1 + e_1)_0}$$

We now wish to relate the strain in the ooids to the bulk strain in the oolite. In doing this in the experimental specimens, the heterogeneity in deformation arising from frictional constraint at the ends must be taken into account. We, therefore, calculate a notional or mean ooid strain $\bar{e}_{oo}$ for the whole specimen under the supposition that the matrix deforms the same as the ooids, in order to be able to compare this quantity with the nominal strain as determined from the micrometer measurements. There is a variation axially in the amount of relative shortening, with a minimum near the ends, and there is some variation radially in both amount and orientation of the principal strains due to "overflow" at the edges of the loading pistons (Fig. 7). Error due to the latter effect was minimized by not making measurements near the lateral boundaries of the specimen (in highly deformed specimens only the central half of the total area available in the thin section was used). The variation in relative shortening along the length of the specimen was allowed for by subdividing the specimen into segments of length l_i, each approximately 1 mm, and treating each segment as having uniform bulk shortening. Using Equation (2), and thus allowing for the initial anisotropy, the notional initial length $(l_i)_0$ that the ith segment would have had if it had undergone a uniform strain equal to that in the ooids within it is given by

$$(l_i)_0 = \frac{l_i}{1 + (e_{oo})_i} = l_i \left(\frac{R_0}{R_i}\right)^{2/3}$$

where the aspect ratio R_i is the arithmetic mean determined within the ith segment (usually containing between 5 and 15 ooids). The mean ooid strain for the whole specimen is given by

$$\bar{e}_{oo} = \frac{L - \sum_i (l_i)_0}{\sum_i (l_i)_0} = \frac{L}{\sum_i l_i \left(\frac{R_0}{R_i}\right)^{2/3}} - 1$$

where L is the final overall length of the specimen. The determined values of $\bar{e}_{oo}$ are listed in Table 1, expressed in engineering strain, and are plotted in Figures 8 and 9 expressed in natral strain, $\epsilon_{oo} = \ln(1 + \bar{e}_{oo})$. A total of 150 to 300 ooids was measured in each specimen.

The accuracy of the strain determination can be illustrated with reference to specimen 2765 in which the bulk deformation was homogeneous over almost the entire length (Fig. 7b). A set of 150 measurements gave an arithmetic mean of the aspect ratios of the ooids, in the homogeneous area, of 0.576, with a standard error of 0.006 and a standard deviation of individual measurements of 0.069. After taking into account the uncertainty in the initial anisotropy, quoted above, the quantity $R/R_0^{2/3}$ has a standard error of 0.006, which leads to a mean ooid strain of 29.4 ±

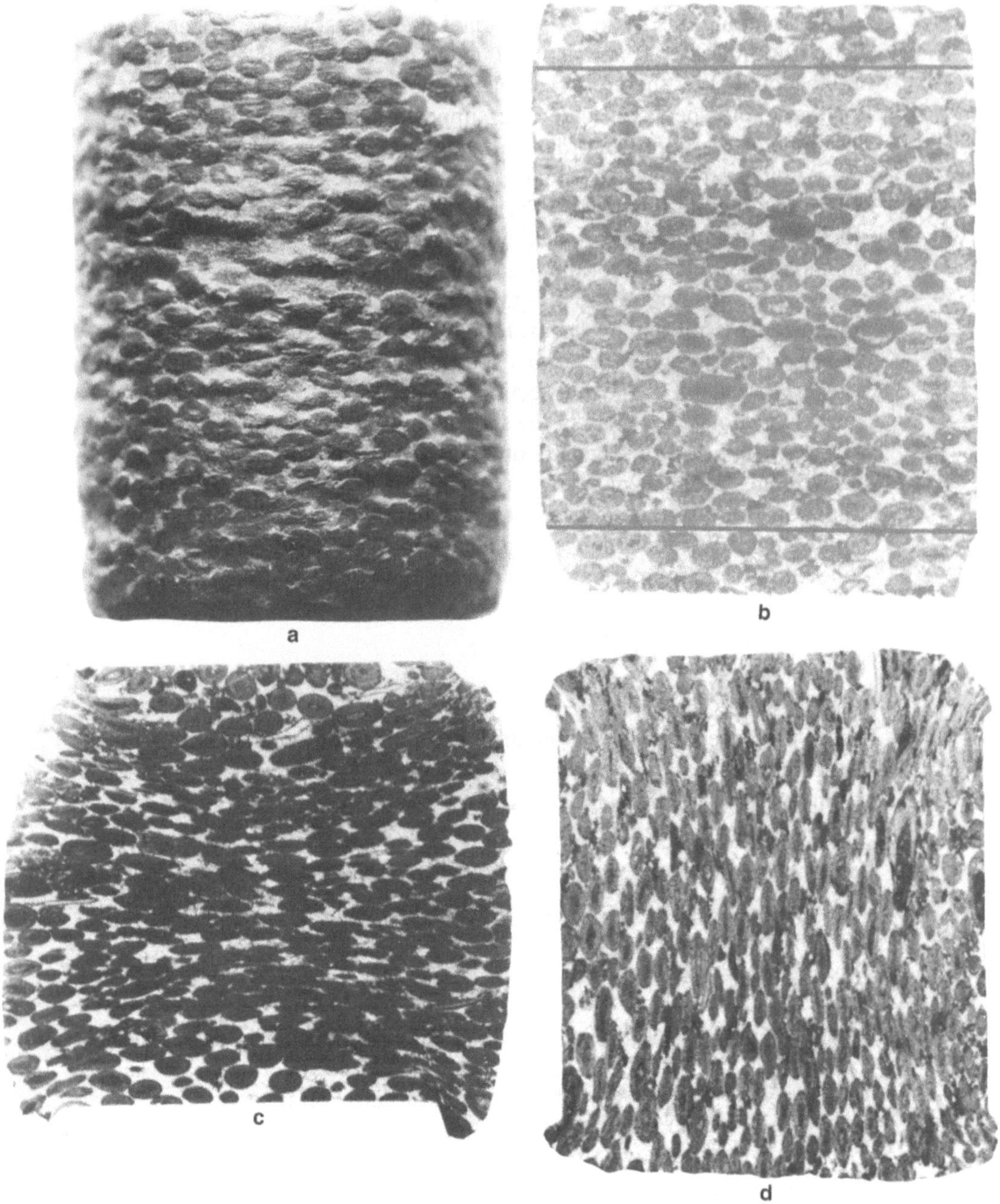

Fig. 7a–d. (a) Cylinder surface of specimen 2767 (800°C, 10^{-5}/s, 24.1% shortening). Note "squeezing" of ooids out of the cylinder surface, directly expressing the contrast in flow properties between ooids and matrix. (b) Specimen 2765 (900°C, 10^{-5}/s, 20.4% shortening), homogeneously deformed region between marker lines was used for the strain analysis reported in Section 6.2 and for statistical evaluations. (c) Very markedly inhomogeneous strain distribution within specimen 2675 (800°C, 10^{-3}/s, 43% shortening). (d) Relatively homogeneous strain distribution in specimen 2699 (800°C, 10^{-5}/s, 41.9% shortening)

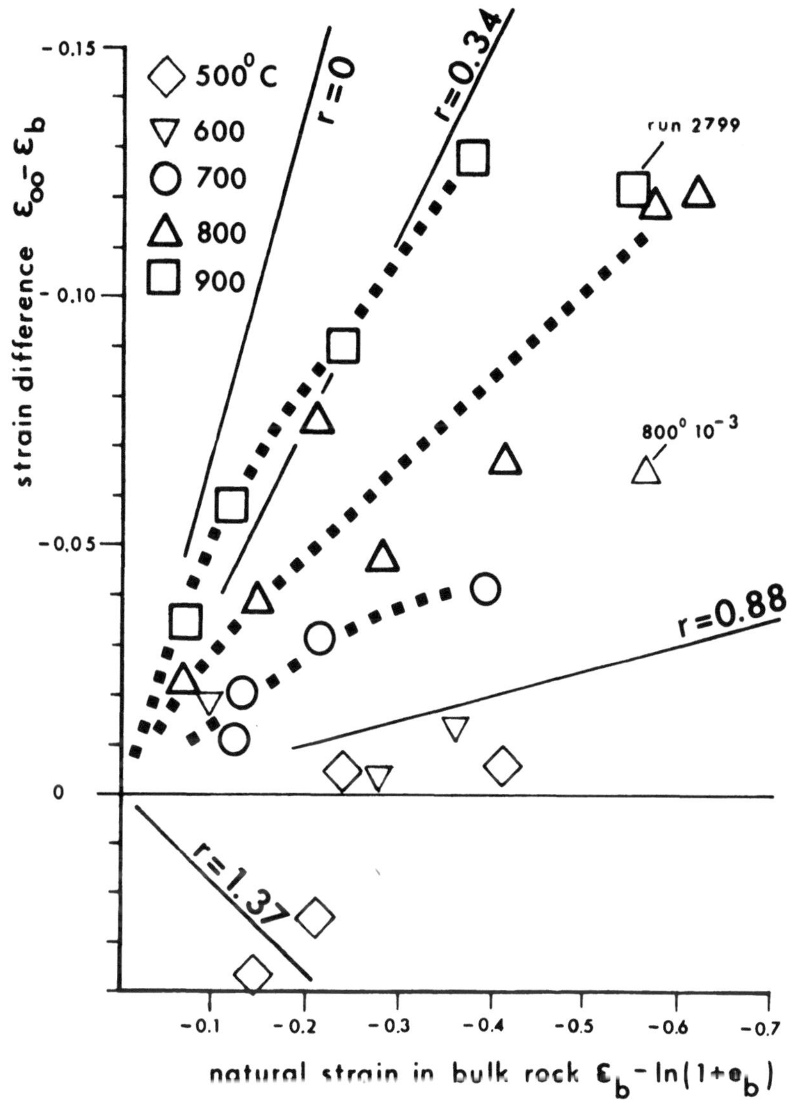

Fig. 8. Plot of the nominal bulk strain in the specimen versus the difference between the mean ooid strain and the nominal bulk strain. This strain difference corresponds to the error arising in a strain analysis based on the ooid shape, assuming that the deformation is homogeneous between ooids and matrix. Strains are plotted as natural strains $\epsilon = \ln(1 + e)$. *Broken lines:* trends at 700°, 800°, and 900°C (estimated visually); *Full lines:* theoretical predictions of the model of Bilby et al. (1975) for different viscosity contrasts r

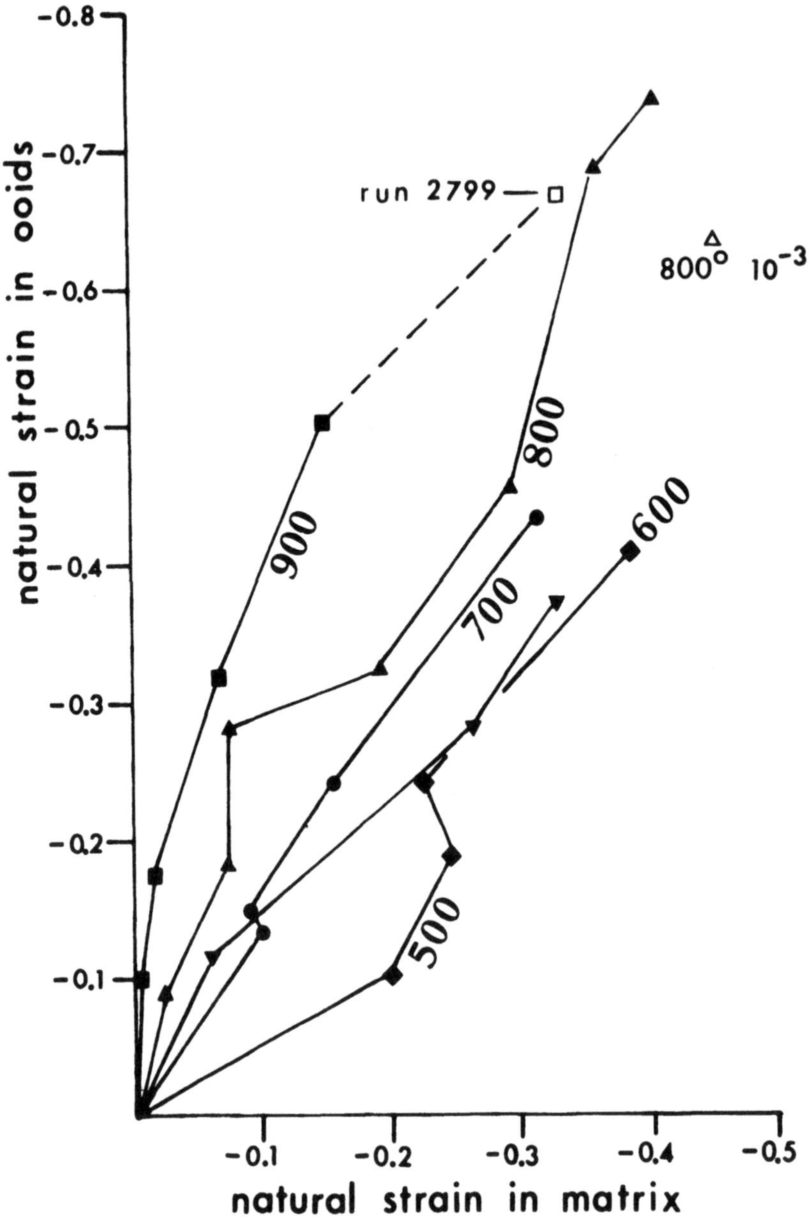

Fig. 9. Plot of the mean ooid strain versus the mean matrix strain, units of natural strain

0.6%. This precision only applies, of course, to this particular case, but it can be expected to apply approximately for all the measurements and not be greatly affected by the magnitude of the mean aspect ratio within the range of strains studied nor by the relatively small variation in strain along the specimen axis to which the summation method described above is applied.

Since the mean ooid strain $\bar{e}_{oo}$ is seen in general to be different from the nominal bulk strain $\bar{e}_b$, it is of particular interest to compare the mean ooid strain with the mean strain $\bar{e}_m$ in the matrix, calculating the latter as follows:

$$\bar{e}_m = \frac{\bar{e}_b - f_{oo}\bar{e}_{oo}}{1 - f_{oo}}$$

where f_{oo} is the volume fraction of ooids. There is clearly a variation in the relation between matrix and ooid strains as temperature and bulk strain are increased (see Table 1 and Figs. 8 and 9). Thus, while at 500°C and 600°C there is no significant departure from homogeneity between matrix and ooids, at higher temperature the concentration of deformation in the ooids, especially at relatively small strains, is well illustrated (the different behavior of specimen 2799 from the others deformed at 900°C was discussed above).

It is noteworthy that an inverse departure from homogeneity of deformation between matrix and ooids is not observed at the lowest temperatures in spite of the fact that an inversion in effective viscosity contrast between Carrara marble and Solenhofen limestone is found experimentally to occur at about 700°C for the strain rates studied here (see Table 2). This presumably is again evidence that the oolite cannot be regarded as being exactly modelled in its behavior by regarding the ooids as equivalent to Solenhofen limestone and the matrix as equivalent to Carrara marble, although such a model has qualitative validity at the higher temperature.

The existence of the large differences between $\bar{e}_{oo}$ and $\bar{e}_m$ poses a problem of strain compatibility between ooids and matrix. Clearly the local strains, especially in the matrix, must be very inhomogeneous and substantial shearing displacements must be accomodated in the matrix. However, this deformation in the matrix is readily achieved by slip and twinning within the large matrix grains.

6.2 Strain Measurement Using the Spatial Distribution of Ooids

This method, described by Ramsay (1967), is suitable for strain analysis in rocks having undergone different amounts of strain in markers and matrix and is exceptional in being, ideally, independent of contrast in flow resistance between matrix and markers. What is needed, however, is a uniform or known spatial distribution of particles (in our case, ooids) through the rock. In case of a uniform distribution, the mean distance between centers of neighboring ooids should be independent of the orientation of the line joining them, while after deformation the distance will vary from maximum to minimum values parallel respectively to the maximum and minimum principal strains. Thus, treating the problem in two dimensions for the case of rotational symmetry, the mean distance d between centers will vary with direction according to

$$d^2 = (1 + e_1)^2 \cos^2 \alpha + (1 + e_3)^2 \sin^2 \alpha \tag{3}$$

where α is the angle between the direction of maximum principal strain, and the line joining the centers of neighboring ooids. This equation is simply a description of the axial section of the bulk strain ellipsoid.

What this describes physically is the result of bodily displacement of ooids that initially lay on a circle around a common neighbor. This bodily displacement is governed by the flow in both matrix and ooids and ideally should give a strain measure for the total rock, provided constant volume is still assumed. The bulk axial engineering strain e_b in an experimental specimen analyzed in this way is, therefore, given immediately by e_3 in Equation (3).

The practical application of this method presents two problems:

(1) Neighbors have to be defined with care, especially after high amounts of strain. Here ooids were counted as neighbors when the line connecting them does not cut through another ooid. Selection according to this criterion is, strictly speaking, influenced by the amount of inhomogeneous strain between matrix and ooids, but only to a very minor and negligible degree.
(2) The accuracy of this method is bound to be low because a very considerable scatter in d at any single α can be expected. However, if the directions of principal strain axes are known beforehand, the measurements of d can be restricted to directions parallel to those axes and an axial ratio is directly obtained. In this way, a great number of measurements can be obtained in a relatively short amount of time.

In an oolite formed by the sedimentation of spherical particles in a gravity field, the distance d between ooids touching each other in three dimensions (as is the case in many oolites including the one studied here) is likely to be large parallel to bedding and to decrease statistically towards a direction perpendicular to bedding. In order to test for such an initial anisotropy 800 measurements of d were made on an undeformed sample and plotted as a function of α, defined as the angle between the direction of d and the bedding plane in a cut perpendicular to bedding (Fig. 10). Because of the large scatter the measurements were grouped in intervals of 5° in α. The arithmetric means of these do not obviously conform closely to a relation of Equation (3) and a statistical fit has not been carried out. However, if only the 0–5° and 85–90° intervals of α are considered the initial anisotropy corresponds to a mean ratio $(1 + e_3) : (1 + e_1)$ of 0.843 with a standard error of 0.051 for the total of 90 measurements (half in each interval), or to an initial equivalent strain of $10.1 \pm 3.6\%$ using Equation (1). This value may be a useful approximation for the initial anisotropy of other oolites also, provided their sedimentary environment of formation was similar.

Because of the rather low inherent accuracy, only one deformed specimen (2765) was analyzed using this method. As noted earlier, this specimen shows homogeneous deformation over almost its entire length and it also shows a large strain difference $\bar{e}_{oo} - \bar{e}_b$. The analysis was done in order to demonstrate that the method does indeed lead to a correct measurment of strain in the bulk rock and also to assess the final accuracy of the result. This information is essential for the field geologist who wishes to estimate effective viscosity contrasts in naturally deformed oolites.

The following results apply within the region of homogeneous bulk deformation of specimen 2765 (Fig. 7b):

(1) The bulk oolite strain e'_b obtained from the spatial distribution of the ooids is 19.1 ± 5.0% (Fig. 10).
(2) The ooid strain $\bar{e}_{oo}$ obtained from analysis of the aspect ratios of the ooids is 29.4 ± 0.6%.
(3) The known bulk oolite strain $\bar{e}_b$ between the marker lines in Figure 7b is 21.9% as evaluated from micrometer measurements of the whole specimen corrected to this gauge length using the strain distibution indicated by the variation of e_{oo} along the specimen axis (this correction is small in the present specimen).

Thus, within the accuracy of the measurements, the bulk oolite strain determined from the spatial distribution of the ooids agrees with the known bulk strain, whereas a bulk strain inferred from the ooid strain under an assumption of homogeneous deformation of ooid and matrix is substantially greater. The accuracy of the former determination could be improved if more ooids were considered; in the above case, 80 measurements were available in the 0–5° interval and 35 in the 85–90° interval, defining an ellipse whose mean axial ratio had a standard error of 0.051; combining this uncertainty with an analogous value for the undeformed material (see above) led to the quoted uncertainty of ±5.0% in the determination of the bulk oolite strain (e_b').

In order to produce an alternative, graphical impression of how well the assumptions underlying this method hold, the initial variation of d as a function of α was analytically "deformed" by applying the known strain of 21.9% (Fig, 10). The "deformed" distribution is seen to agree reasonably well, within the scatter, with the measured variation of d with α in the deformed material.

7 INHOMOGENEOUS STRAIN DISTRIBUTION WITHIN A SPECIMEN

The variation in strain within the specimen is shown for two examples in Figures 7c and 7d. It is seen that the degree of inhomogeneity in bulk strain (as approximated by the ooid strain) differs considerably between

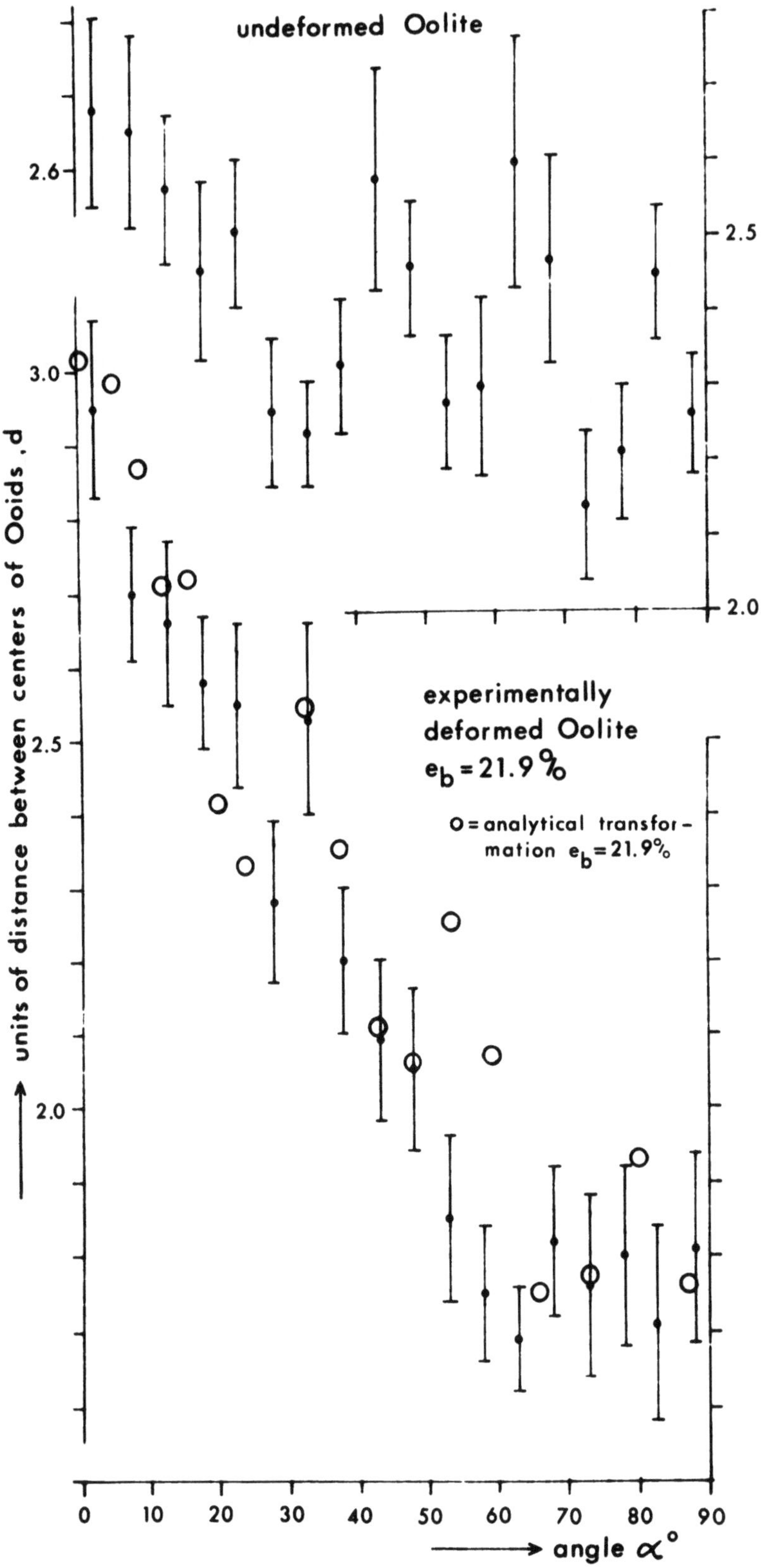

undeformed Oolite
2.6
2.5
3.0
2.0
experimentally
deformed Oolite
$e_b = 21.9\%$
o = analytical transformation $e_b = 21.9\%$
2.5
2.0
units of distance between centers of Ooids, d
0
10
20
30
40
50
60
70
80
90
angle α°

the two samples deformed under different conditions. In general, the degree of inhomogeneity is greater when the deformation involves higher stresses. The same effect has been observed in Solenhofen limestone, where barreling is absent over most of the specimen length in low stress experiments (that is, with differential stresses of a few hundreds or tens of bars).

The inhomogeneity of deformation expressed in the barreling of a compression specimen can be attributed to the influence of the constraint imposed on deformation near the ends of the specimen due to friction between the loading piston and the specimen. This constraint leads to there being adjacent to the endface a roughly conically shaped region in which the deformation is relatively small. The barreling then arises from the distribution of lateral displacement during shortening as these conical regions approach each other. It is proposed here that the degree of barreling in specimens deformed under different conditions is related to the different degree of stress dependence of the strain rate, as measured by the parameter n if the power law holds.

Under conditions where the flow stress is relatively high, n is high (Fig. 4) and, therefore, the strain rate is relatively strongly sensitive to differences in stress. Such differences in stress within the specimen are imposed by the frictional constraints exerted by the pistons on the endfaces of the specimen. In the center of the specimen the deviatoric stress causing flow is given by the stress difference between axial stress and confining pressure. Near the pistons, the additional stress components arising from the frictional constraint have the qualitative effect of increasing the confining pressure and so reducing the deviatoric stress near the piston. Taking the amount of change in deviatoric stress between the end and the center of the specimen as being always the same, it can then be seen that the total variation in strain rate between end and center will be greater the more stress sensitive is the strain rate, that is, the higher is the value of n. Thus, the material having higher n will develop a greater degree of inhomogeneity of strain, expressed as the difference in relative shortening in the axial direction between center and end of the specimen, and hence in the barreling.

Fig. 10. Plot of the length d of the line joining neighboring ooids versus α, its angle to the bedding plane *(top diagram)* or the direction normal to the specimen axis *(bottom diagram)*. *Top diagram* applies to a typical specimen before deformation; *Bottom diagram* applies to specimen 2765 within the region of homogeneous bulk strain. *Closed symbols* and *bars:* arithmetic mean and standard error of groups of measurements within 5° intervals in α. The variation after an analytical "deformation" corresponding to the known 21.9% shortening is shown by the open symbols

8 COMPARISON WITH THEORY

Recently, Bilby et al. (1975) analytically treated the change of shape of a viscous oblate spheroid in a matrix of different viscosity, modifying and extending an earlier study by Gay (1968). Their Equation (17) (Bilby et al., 1975) describes the development of strain in the spheroid parallel to its short axis as a function of the strain remote from the inclusion and of the viscosity contrast, r, defined as the ratio of the viscosity in the inclusion to the viscosity in the surrounding material. Newtonian viscosity is assumed.

The model of Bilby et al. (1975) clearly cannot be used to represent in all respects the oolite studied here because the model considers only a single inclusion within a surrounding material of a homogeneous but different viscosity. In the case of the oolite, the ooids form a multiplicity of inclusions which are closely spaced and interact, that is, the perturbation in an otherwise homogeneous strain field due to the presence of a given ooid overlaps the perturbations due to neighboring ooids. The flow resistance of the coarser-grained matrix, which itself might be represented by that of Carrara marble, is relevant only immediately adjacent to the ooid boundary, while in respect of interaction over longer distances the ooid will behave as if it were immersed in a medium having the properties of the bulk oolite. Therefore, in selecting data for comparing the model of Bilby et al. (1975) with the oolite, the flow stress of the bulk oolite is taken as determining the viscosity of the surrounding medium. To obtain a viscosity for the inclusion itself, we have tentatively used the flow stress of Solenhofen limestone on account of its fine grain size and characteristic flow behavior described above. Another difficulty in applying the model to the present situation lies in the non-Newtonian behavior of the materials of the oolite but we at first overlook this difficulty and return to it later.

The behavior of the model of Bilby et al. has therefore been calculated for viscosity ratios r equal to the ratios of the flow stresses of Solenhofen limestone and the bulk oolite at the relevant temperatures and strain rates (Table 2) and plotted on Figure 8 for comparison with our results. At 900°C, 10^{-5}/s strain rate ($r = 0.34$) excellent agreement is shown. However, at 800°C, 10^{-5}/s ($r = 0.88$), the ooids undergo relatively more strain than would be predicted by the theory; and at 700°C, 10^{-3}/s ($r = 1.37$), the ooids are still undergoing more strain than the average rock even though the theory now predicts less inclusion strain because the viscosity ratio is now greater than unity. Since it seems unlikely that taking into proper account the influence of the non-Newtonian nature of the effective viscosities would change the qualitative significance of a change of viscosity ratio from greater than unity to less than unity, we conclude that the discrepancy between observed and the predicted relative strains lies in the choice of Solenhofen limestone as a model material for the ooids. This conclusion is reinforced by the observation above that at 700°C the oolite is, in fact, weaker than both Carrara marble and Solenhofen limestone (Table 2). Thus, in the 700° and 800°C cases the ooid material must be assumed to be substantially weaker than Solenhofen limestone in spite of the similarity in grainsize, suggesting that differences in impurity content may be significantly affecting the relative strengths. The question then arises as to why the 900°C behavior does not show a corresponding discrepancy with the theory. Possibly the biasing factors affecting the 700° and 800°C behavior are counterbalanced by the effect of the reversed grain size differential that is now more marked due to the extra grain-growth in the ooids at 900°C (Table 2). But it must also be noted that the comparison with theory is less sensitive to variations in viscosity ratio as the latter approaches zero, and it is also possible that the departure from Newtonian viscosity becomes more significant then, too. Unfortunately, it is not possible to assess from the present work the role played by the non-Newtonian nature of the viscosity, except to speculate that it gives rise to what appears to be a somewhat greater curvature in the observed trends in Figure 8 than is predicted by the Newtonian theory; however, this curvature may have another explanation.

9 GEOLOGICAL APPLICATIONS

The scope and accuracy of strain determinations using deformed, originally quasi-spherical objects has been greatly improved by taking into account both the axial ratios and the orientations of the originally ellipsoidal strain markers (Elliott, 1970). However, apart from theoretical studies (Gay, 1968; Bilby et al., 1975), not much progress has been made in evaluating the other main source of errors in the strain determination in rocks, namely the contrast in flow behavior between strain marker and matrix. The present study concerns this second source of error only. Because of the inhomogeneous strain distribution within the test specimens a direct test on more sophisticated methods of strain determination, such as the one described by Elliott (1970) and primarily concerned with eliminating errors arising from the initial shape and orientation of strain markers, could not be made.

Our results indicate that the determination of bulk strain in naturally deformed oolites from the shape of the ooids may involve substantial error if the conditions of deformation are equivalent to those in the higher temperature experiments, whereas such a strain analysis may be reliable if the conditions are equivalent to those in the lower temperature experiments. Which circumstances applies most generally in practice can only be determined directly by the study of situations in the field in which an independent determination of the bulk strain can be made on other bases. (In Sect. 6.2 such an alternative strain analysis was demonstrated to indicate bulk strain in the oolite.) However, there are some pointers to the existence of a strong grain-size dependence in monomineralic calcite rocks and therefore to the likelihood of heterogeneity of deformation between coarse and fine-grained parts of rocks, for example, in oolites. Such fossils as belemnites (coarse grained) and crinoid stems (single crystals), embedded in a matrix of fine-grained limestone, are often disrupted and internally hardly deformed at all (Ramsay, 1967) while foliation or the flattening of ooids indicates high amounts of strain. Such observations would indicate that the high temperature runs are more representative for the geological environment and that viscosity contrasts would occur in an oolite such as the one listed in this study.

Such a conclusion can also be drawn from a more theoretical approach. It has been pointed out earlier (Sect. 5) that the effective viscosity contrast at high temperatures in this oolite probably results from a transition to a deformation mechanism of diffusional flow and/or grain boundary sliding in the finegrained ooids. These deformation mechanisms are grain size dependent, a small grain size favoring faster strain rates at any given stress. A relationship of the form $\epsilon \propto 1/L^a$, where L is the grain size and where $1 < a < 3$, fits most materials behaving superplastically (Edington et al., 1976) and is predicted by models of diffusional flow (Stocker and Ashby, 1973). So the question of the likelihood of such viscosity contrasts in oolites is reduced to the question whether grain size sensitive deformation mechanisms such as the ones mentioned above occur in nature or not. When fields of deformation mechanisms are plotted in the form of deformation mechanism maps (Stocker and Ashby, 1973) in stress–temperature space, it becomes evident that the transition from one deformation mechanism into another is primarily a function of differential stress. In this light it seems likely that the high temperature runs, where "geologically realistic" flow stresses are believed to have been achieved, are the ones representative of many crustal environments.

Viscosity contrasts inferred to have given rise to strain heterogeneity between marker inclusions and the bulk rock can be conversely used as evidence revealing the stress regime of the natural deformation, and hence its temperature–strain rate regime, if the strain heterogeneity can be independently established and if the factors determining the rheological behavior of the component materials of the rock have been well established. To exploit such a possibility is perhaps an ambitious and long-term aim but it is illustrative of one way in which coordination between field and laboratory studies may eventually further our understanding of the geological record.

ACKNOWLEDGMENTS

We wish to thank J. N. Boland, R. S. Coe and I. van der Molen (ANU) as well as W. Müller and O. A. Pfiffner (ETH Zürich) for many discussions and suggestions. W. Müller kindly provided the test material. The technical assistance by T. White (apparatus) and P. Willis (thin sections) is also acknowledged.

REFERENCES

Bilby, B. A., Eshelby, J. D., Kundu, A. K.: The change of shape of a viscous ellipsoidal region embedded in a slowly deforming matrix having a different viscosity. Tectonophysics *28*, 265–274 (1975)

Cloos, E.: Oolite deformation in the South Mountain fold, Maryland. Geol. Soc. Am. Bull. *68*, 843–918 (1947)

Davies, G. J., Edington, J. W., Cutler, C. P., Padmanabhan, K. A.: Superplasticity: A Review. J. Mat. Sci. *5*, 1091–1102 (1970)

Edington, J.W., Melton, K. N., Cutler, C. P.: Superplasticity. Progr. Mat. Sci. *21*, 61–158 (1976)

Elliott, D.: Determination of finite strain and initial shape from deformed elliptical objects. Geol. Soc. Am. Bull. *81*, 2221–2236 (1970)

Gay, N. C.: Pure-shear and simple-shear deformation of inhomogeneous viscous fluids. 1. Theory. Tectonophysics *5*, 211–234 (1968)

Heard, H. C., Raleigh, C. B.: Steady-state flow in marble at 500–800°C. Geol. Soc. Am. Bull. *83*, 935–956 (1972)

Kohlstedt, D. L., Goetze, C., Durham, W. B.: Experimental deformation of single crystal olivine with application to flow in the mantle. In: Petrophysics: The Physics and Chemistry of Minerals and Rocks. Runcorn, S. K. (ed.) London: John Wiley and Sons, (in press)

Paterson, M. S.: A high-pressure high-temperature apparatus for rock deformation. Intern. J. Rock Mech. Min. Sci. *7*, 517–526 (1970)

Paterson, M. S.: Some current aspects of experimental rock deformation. Phil. Trans. R. Soc. London Ser. A. 208, 163–171 (1976)

Ramsay, J. G.: Folding and Fracturing of Rocks. New York: McGraw-Hill, 1967, 567 pp.

Schmid, S. M.: Rheological evidence for changes in the deformation mechanism of Solenhofen limestone towards low stresses. Tectonophysics *31*, T21–T28 (1976)

Stocker, R. L., Ashby, M. F.: On the rheology of the upper mantle. Rev. of Geophys. Space Phys. *11*, 391–426 (1973)

Tan, B. K.: Deformation of particles developed around rigid and deformable nuclei. Tectonophysics *24*,243–257 (1974)

6

S. K. GHOSH

Abstract The drag-patterns of planar structures around rigid spherical and ellipsoidal inclusions may be classified on the basis of their symmetry or asymmetry, on relative lengths of segments showing normal and reverse drag and on the basis of whether the planar structures around the inclusion bow in or bow out. If the bulk deformation is by a combination of simultaneous pure shear and simple shear, the pattern of drag will be determined by the initial orientation of the marker planes (parallel to the long axis of inclusion), the total deformation of the matrix and the ratio between the rates of pure shear and simple shear. These physical factors would decide whether the total rotation of the inclusion is greater than or less than or is equal to the total rotation of the planar structure in the matrix. If the total rotation of inclusion is greater, predominantly normal drag would result. If the total rotation of inclusion is less than or equal to that of the marker, dominantly reverse drag or a symmetrical pattern would develop.

DRAG PATTERNS OF PLANAR STRUCTURES AROUND RIGID INCLUSIONS

1 INTRODUCTION

When a rigid inclusion occurs within a nonrigid matrix, deformation of the host material gives rise to a distortion of the planar structures around the inclusion. The most common examples of such distorted patterns are seen in the drag patterns of bedding or schistosity around porphyroblasts. Similar patterns may also be found in the deformed matrix around pebbles, fossils, or tectonic inclusions. The principles of development of drag patterns around spherical inclusions by either simple shear or pure shear are already known (Ghosh, 1975). In the present discussion these principles are extended to include a more general kind of deformation, viz., a combination of simultaneous simple shear and pure shear, with the compression direction of the pure-shear component acting at a right angle to the direction of simple shear. Again these principles for the development of drag patterns would include both equant and elongate inclusions—to be more specific, inclusions with circular and elliptic sections.

In the following discussion of the two-dimensional case, the direction of simple shear movement is represented by the x-coordinate axis, the line at right angle to x normal to the plane of shear by the y axis, with z normal to the xy plane and parallel to the axis of shear. The direction of maximum compression for the pure-shear component is assumed to be parallel to the y axis and the maximum extension parallel to the x axis. The z axis is a direction of zero strain for both the simple-shear and the pure-shear components (Fig. 1). It is assumed that the passive marker lines in the groundmass were initially parallel to the long axis (the a axis) of the rigid inclusion. If they were not parallel, the possible drag patterns can still be determined by using similar principles. The intermediate axis of the incluson is assumed to coincide with the z-coordinate axis. The initial angle between the y-coordinate axis and the long axis of the inclusion (or the marker line) is designated Φ_0. The final angle between the y axis and the a axis is Φ. The final angle between y axis and the passive marker line is Φ' (Fig. 1). Again, it is assumed that whatever be the ratio between the *rates* of pure-shear to simple shear, this ratio remains constant throughout the deformation. It should be noted that in the earlier publication (Ghosh, 1975) the angles were measured from the x axis, while in the present paper, the corresponding angles are with respect to the y axis.

2 ROTATION OF PASSIVE MARKERS AND INCLUSIONS

For a particular initial orientation (i.e., Φ_0) and for a particular bulk deformation, the nature of drag pattern of the markers around the rigid inclusion would depend on the manner in which the inclusion reorients

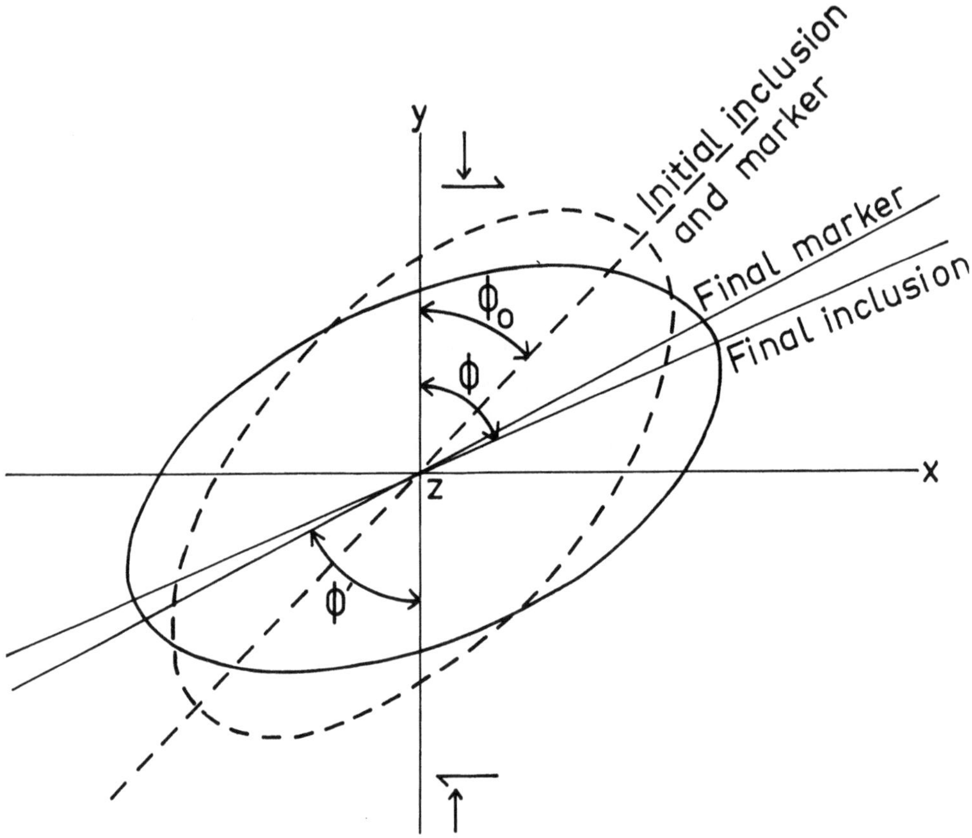

Fig. 1. Initial and final orientations of a passive marker line and an elongate inclusion with reference to the coordinate axes

itself relative to the marker lines. This can be determined with the help of the equations given in Ramberg (1975) and in Ghosh and Ramberg (1976). The final angle (Φ') between the y-coordinate axis and the passive marker line is given by the equation

$$\tan \Phi' = \exp(2\gamma s_r) \tan \Phi_0 + \frac{1}{2s_r}\{\exp(2\gamma s_r) - 1\}, \quad (1)$$

(Ghosh and Ramberg, 1976), which follows from Equation (38) of Ramberg (1975), and where s_r is the ratio ($\dot{\epsilon}_x/\dot{\gamma}$) between the rates of pure shear and simple shear, and γ is the finite shear strain of the simple-shear component. The equation for the finite angle between the y-coordinate axis and the long axis of the inclusion (Ghosh and Ramberg, 1976) depends on certain relations existing between the ratio s_r and the axial ratio R between the long a and the short b axes of the inclusion.

If

$$R > \frac{1 + \sqrt{1 + 4s_r^{2'}}}{2s_r}, \tag{2a}$$

$$\tan \Phi = \frac{P(B + \sqrt{B^2 - AC}) - B + \sqrt{B^2 - AC}}{C(1 - P)}, \tag{2b}$$

where

$$A = \frac{R^2}{R^2 + 1},$$
$$B = \frac{s_r(R^2 - 1)}{R^2 + 1},$$
$$C = \frac{1}{R^2 + 1},$$

and

$$P = \frac{C \tan \Phi_0 + B - \sqrt{B^2 - AC}}{C \tan \Phi_0 + B + \sqrt{B^2 - AC}} \exp (2\gamma\sqrt{B^2 - AC}).$$

If

$$R < \frac{1 + \sqrt{1 + 4s_r^{2'}}}{2s_r}, \tag{3a}$$

$$\tan \Phi = \frac{1}{C} \sqrt{AC - B^2} \tan \left\{ \gamma\sqrt{AC - B^2} + \operatorname{Arc\,tan}\left(\frac{C \tan \Phi_0 + B}{\sqrt{AC - B^{2'}}} \right) \right\} - \frac{B}{C}. \tag{3b}$$

If

$$R = \frac{1 + \sqrt{1 + 4s_r^2}}{2s_r}, \tag{4a}$$

$$\tan \Phi = \frac{1}{C} \left\{ \frac{C \tan \Phi_0 + B}{1 - \gamma(C \tan \Phi_0 + B)} - B \right\}. \tag{4b}$$

In all cases, the rigid rotation, $\omega = \Phi - \Phi_0$.

The physical significance of the differences among the three solutions given by Equations (2b), (3b), and (4b) may be better understood if the rate of rotation of the inclusion is considered. The rate of rotation of a rigid elongate inclusion is given by the equation

$$\dot{\Phi} = \dot{\gamma}(A \cos^2 \Phi + B \sin 2\Phi + C \sin^2 \Phi), \tag{5}$$

(Ghosh and Ramberg, 1976), where $\dot{\Phi}$ is the rate of rotation, $\dot{\gamma}$ is the shear-strain rate for the component of simple shear, and Φ is the angle between the y axis and the a axis at which the rate of rotation is measured. A, B, and C are as given before. When the rates of rotation are plotted for different orientations of the inclusion, the curves obtained can be of three different types (Fig. 2), provided the bulk deformation is by combined pure shear and simple shear (i.e., when s_r is neither zero nor infinite). The conditions under which the three different sets of curves are obtained are the same as those given by Equations (2a), (3a), and (4a). For a particular s_r, if the axial ratio of the inclusion is smaller than a critical value, the rate of rotation of the inclusion will never vanish in any possible orientation (Curve C in Fig. 2). Moreover, the rate of rotation in this case will always be positive, which would mean that the rotation is always in the same sense as the sense of rotation for the simple-shear component. If the axial ratio of the inclusion happens to coincide with the critical value [given by Eq. (4a)], there will be one orientation at which the rate of rotation will be zero (Curve B Fig. 2). At any other orientation the rate of rotation will be positive, i.e., the rotation will be in forward sense. Again, if the axial ratio of the inclusion is larger than the critical value, the rate of rotation can be positive, negative, or zero (Curve A in Fig. 2), depending on the orientation of the inclusion. To be more specific, there will be a large range of orientation at which the inclusion will move in a forward sense, a relatively small range of orientation (between Φ_5 and Φ_6 in Fig. 2) in which the inclusion will move *backward* (i.e., in a sense opposite to that given by the simple-shear component) and two orientations (Φ_5 and Φ_6 in Fig. 2) at which the rate of rotation is zero. In terms of the finite angle of rotation [Eq. (2b), (3b), and (4b)], this would mean that in combined pure shear and simple shear, a relatively short inclusion can rotate in a forward sense through an unlimited angle. On the other hand a relatively long inclusion will rotate forward or backward or will remain stationary depending on its initial orientation, but in any case its total angle of rotation will be limited. With progressive deformation, it will tend to approach a particular orientation which is fixed by a certain relation existing between s_r and R. This critical orientation Φ_5 is given by the equation

$$\tan \Phi_5 = -s_r(R^2 - 1) - \sqrt{s_r^2(R^2 - 1)^2 - R^2} \tag{6}$$

(Ghosh and Ramberg, 1976). The rate of rotation of a passive marker line coincides with that of a rigid inclusion when R tends to infinity.

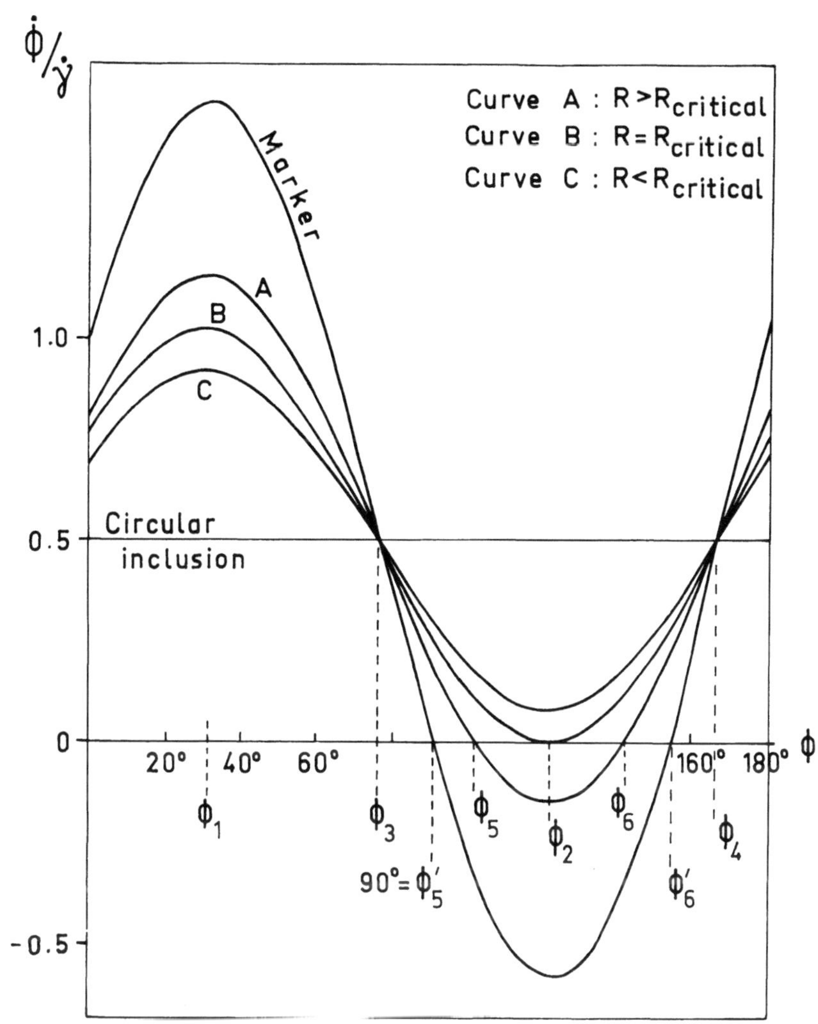

Fig. 2. Variation in the rate of rotation of rigid inclusions and a passive marker line with change in orientation for the case of $\Phi_0 = 0$ and $s_r = 1$. Curve *A* is for *R* = 2, curve *B* for *R* = 1.62 and the curve *C* for *R* = 1.5. Φ_1 and Φ_2 are orientations for maximum and minimum, Φ_3 and Φ_4 for points of inflection, Φ_5 and Φ_6 for vanishing rate of rotation. ϕ_5' and ϕ_6' are orientations at which the passive marker does not rotate

3 MEASURE OF DRAG ALONG PERIPHERY

If the origin of the *xy*-coordinate axes (as defined earlier) is chosen to coincide with the center of an inclusion, the position of a point *P* on the periphery of the inclusion can be expressed in terms of its distance *r* from the origin and the angle ζ_0 which the radius vector makes with the *y* axis. Let us assume that a passive marker line making an angle of Φ_0 with the

y axis met the inclusion at its border at P (Fig. 3). Because of rigid rotation of the inclusion the point P on the inclusion would move to the point R with the same distance r from the origin but with the angle ROY as ζ. The rigid rotation will of course be $\zeta - \zeta_0$. Away from the inclusion the marker line in its deformed state will have an orientation which makes an angle of Φ' with the y axis. The deformed marker line may be curved near the contact of the inclusion but will remain essentially straight at a sufficient distance from the inclusion. If the straight portion

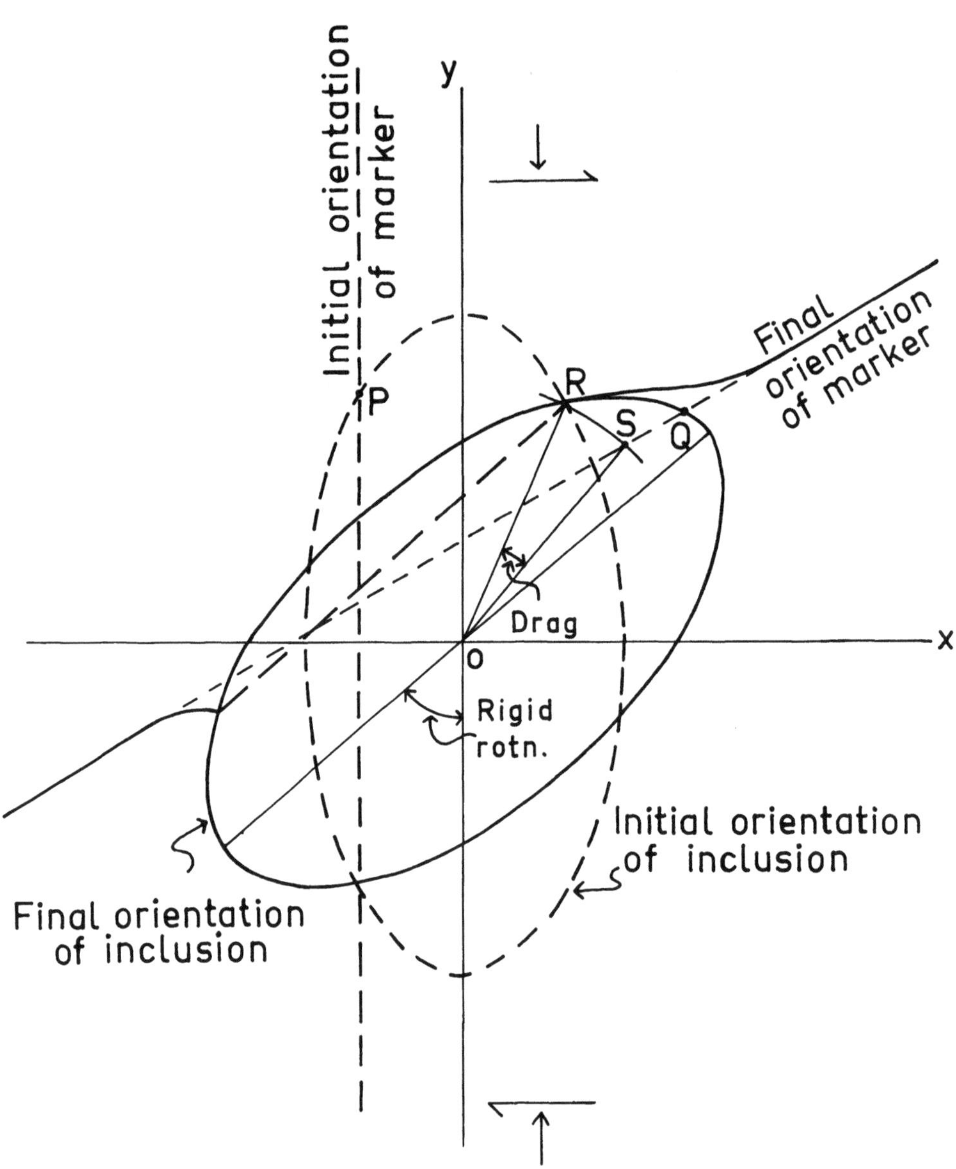

Fig. 3. Definition of drag for an elliptic inclusion

of the marker is produced towards the inclusion, it will intersect the periphery at some point Q and an arc of radius r at some point S (Fig. 3). The drag will be defined as the angle ROS. Although this method of defining the drag was found to be convenient for the present purpose, it is conceivable that the drag might be defined through other parameters for other purposes. If the points Q and R (Fig. 3) coincide, the drag will of course be zero. Again, for circular inclusions, the points S and Q would be the same, and the drag may be measured by the angle ROQ as had been shown in Figure 1 of Ghosh (1975). In Figure 3 the point R is lagging behind the point Q and hence the drag of the marker line on the periphery is opposite to the sense of rotation given by the simple-shear component. This kind of drag may be described as negative drag or reverse drag. If the point R had moved ahead of the point Q, the drag would be in the same sense as the sense of rotation for the simple-shear component. This has been described as positive or normal drag.

It may be noted that the present method of defining drag has some disadvantages. For instance, in case of circular inclusions, the straight portion of a deformed marker line, after being produced, may not intersect the boundary of the inclusion. For those marker lines the drag would remain undefined even though their curved segments meet the boundary of the inclusion.

4 PATTERNS OF DRAG AROUND CIRCULAR INCLUSIONS

The principles of development of drag patterns around circular inclusions have been described in a previous publication (Ghosh, 1975). In the following discussion, it will be seen how the evolution of the patterns is modified when the bulk deformation is by combined pure-shear and simple-shear acting simultaneously. The controlling factors in this case are the ratio s_r between the rates of pure shear and simple shear which is assumed to remain constant during the deformation, the total bulk deformation which can be expressed through the finite strain of either the pure-shear component ϵ_x or the simple-shear component γ or through the axial ratio X/Y of the finite strain ellipse for the combined deformation, and lastly the initial orientaton Φ_0 of the marker lines. The drag at a particular point on the border of the inclusion can be either calculated or geometrically constructed with the help of Equation (1). Since the inclusion is circular, its rigid rotation will be independent of the component of pure shear and will be equal to half the value of the finite simple-shear component (Rosenfeld, 1970; Ghosh and Ramberg, 1976).

If the initial orientation of the marker line is at a high angle to or at a right angle to the direction of simple-shear movement (i.e., if $\Phi_0 \approx 0$), it was found (Ghosh, 1975) that for $s_r = 0$, i.e., for the case of simple shear, the initial drag pattern is strongly asymmetrical, with reverse drag predomi-

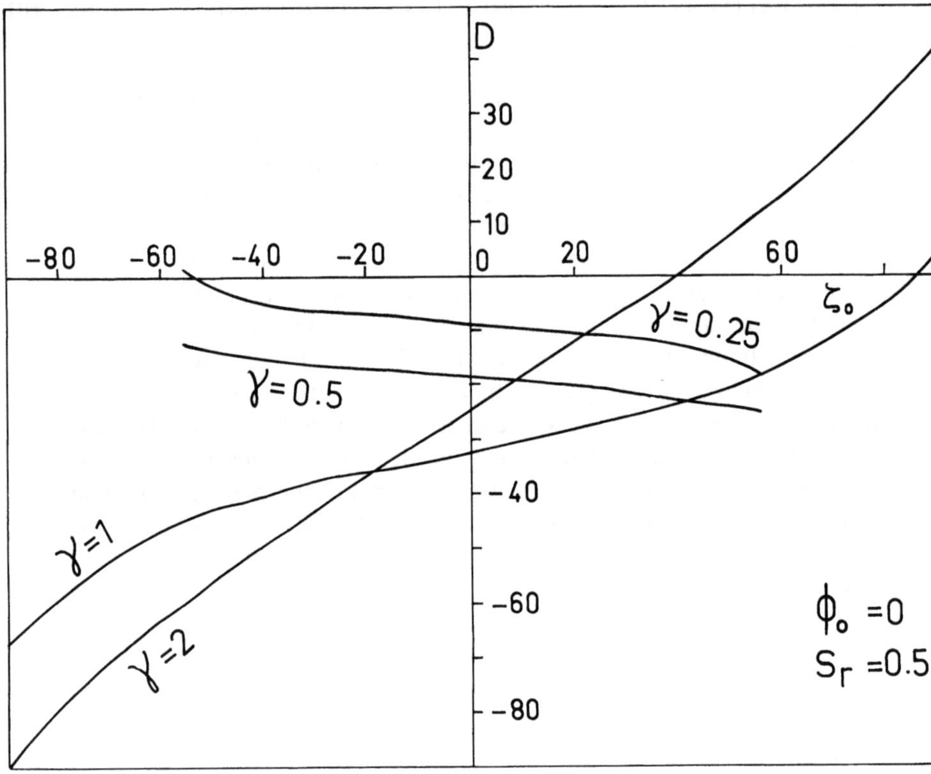
D
30
20
10
0
-80
-60
-40
-20
20
60
ζ₀
γ=0.25
γ=0.5
-40
γ=1
-60
γ=2
-80
ϕ₀ =0
Sr =0.5

a

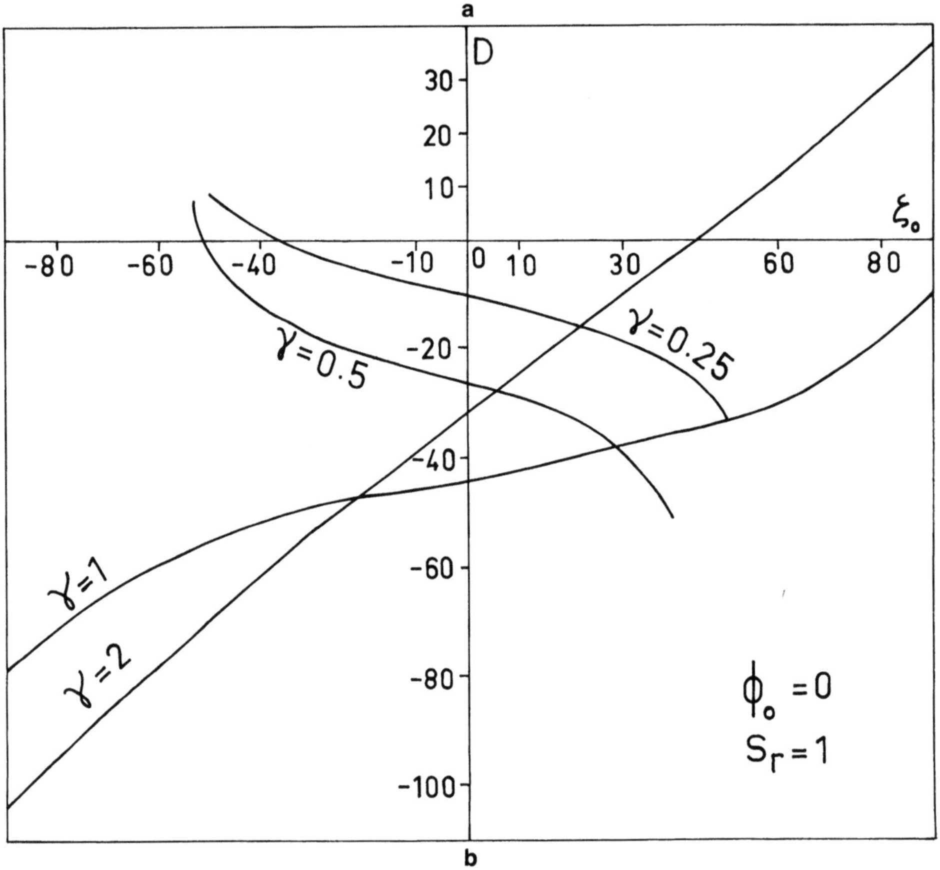
D
30
20
10
ξ₀
-80
-60
-40
-10
0
10
30
60
80
γ=0.25
-20
γ=0.5
-40
γ=1
-60
γ=2
-80
ϕ₀ =0
Sr =1
-100

b

nating over the periphery of the inclusion. With increasing bulk deformation the reverse drag changes over to normal drag over larger and larger segments so that at a moderately large deformation a more or less symmetrical pattern is achieved. With further deformation, the pattern becomes increasingly asymmetric with normal drag predominating over the periphery. With increase in s_r, i.e., with increase in the rate of pure-shear component, this mode of evolution of the drag pattern changes in at least two ways. First, the drag pattern does not become symmetrical unless the bulk deformation is extremely large (Fig. 4 and Fig. 5). Second, in certain situations (for instance in the case of $s_r = 1$), the drag may be entirely reverse all over the periphery (Fig. 4). Thus, in combined pure shear and simple shear a dominantly reverse pattern of asymmetric drag is expected up to large values of bulk deformation, if the initial orientation of the marker was at high angles to the direction of simple-shear movement.

If the marker line is initially oriented at an angle of approximately 45° (i.e., $\Phi_0 = 45°$), then, for $s_r = 0$, the drag pattern is initially symmetrical with roughly equal segments of normal and reverse drag. With progressive deformation the pattern becomes more and more asymmetrical with normal drag predominating (Figs. 5 to 7 and Fig. 20 of Ghosh, 1975). If $s_r \neq 0$, the initial pattern of drag is asymmetrical with reverse drag predominating (Fig. 7a and Fig. 6). With progressive deformation reverse drag changes over to normal drag over longer segments of the periphery till a more or less symmetrical pattern is achieved (Fig. 7b and Fig. 6). With continued deformation the pattern again becomes asymmetrical with normal drag predominating (Fig. 7c and Fig. 6). The total bulk deformation (i.e., the axial ratio of the strain ellipse in the matrix) at which a more or less symmetrical pattern occurs is larger for larger values of s_r.

If the marker line is initially oriented parallel to the direction of simple shear (i.e., if $\Phi_0 = 90°$), then, if $s_r = 0$, the drag-pattern remains strongly asymetric throughout the deformation (Fig. 14, Ghosh, 1975). The drag in this case is always normal and the amount of drag remains the same at every point on the periphery (Fig. 14 of Ghosh, 1975). This pattern is characteristic for simple-shear deformation parallel to the marker planes (Fig. 8a) and does not appear in any other situations. If $s_r \neq 0$, this pattern changes essentially in two ways. First, the amount of drag varies along the periphery. Second, although the drag remains predominantly normal,

Fig. 4a, b. Variation of drag D (in degrees) along the periphery of a circular inclusion, for (a) $\Phi_0 = 0$, $s_r = 0.5$ and (b) $\Phi_0 = 0$, $s_r = 1$. ζ_0 is the *initial* angle between the y-axis and a line joining the center of the inclusion to a point on the periphery. Note that there is no symmetrical pattern of drag, since none of the curves show equal segments of positive and negative drag

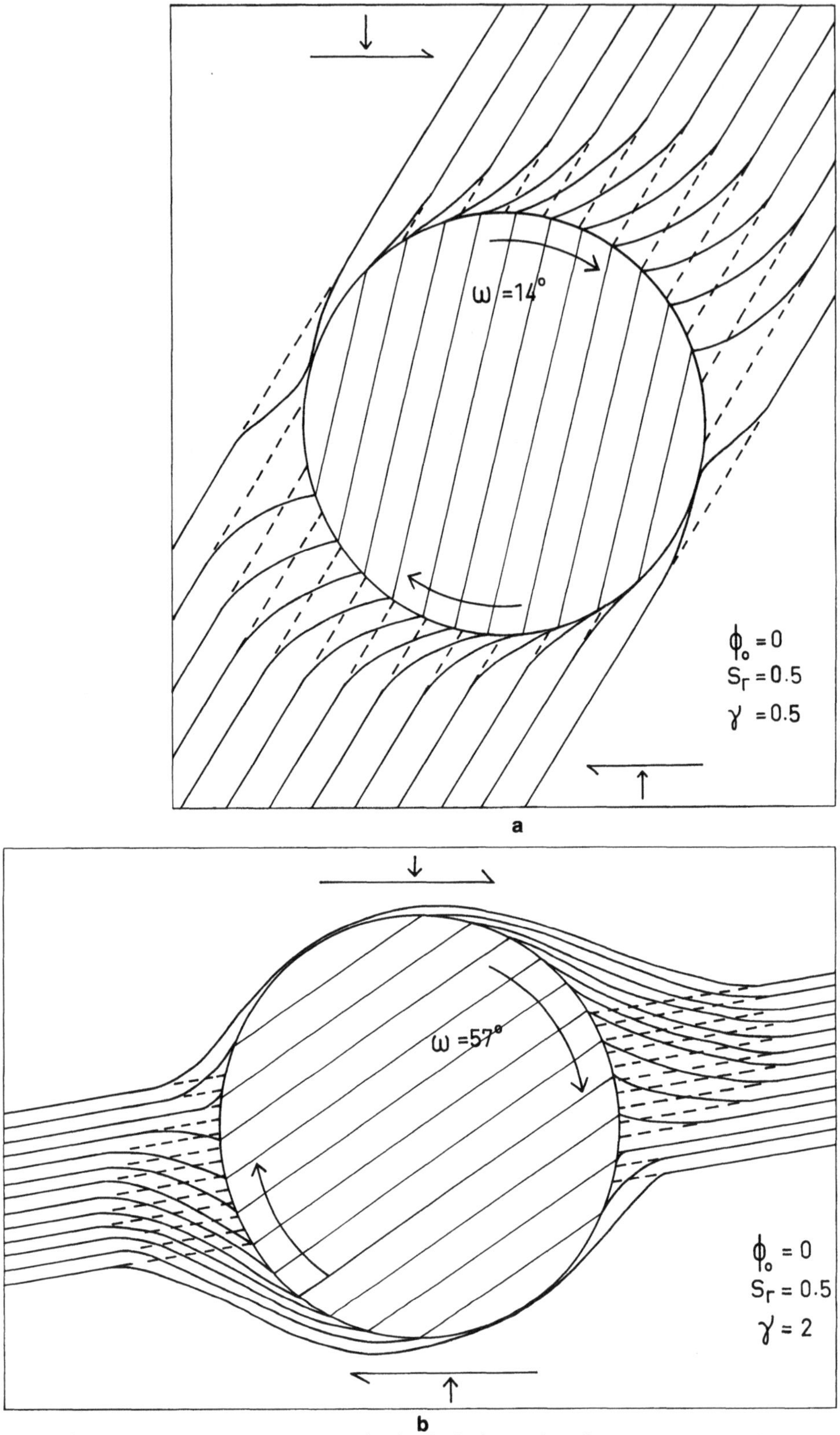

Fig. 5a, b. Drag patterns around circular inclusions when $\Phi_0 = 0$, and $s_r = 0.5$ at two stages of deformation

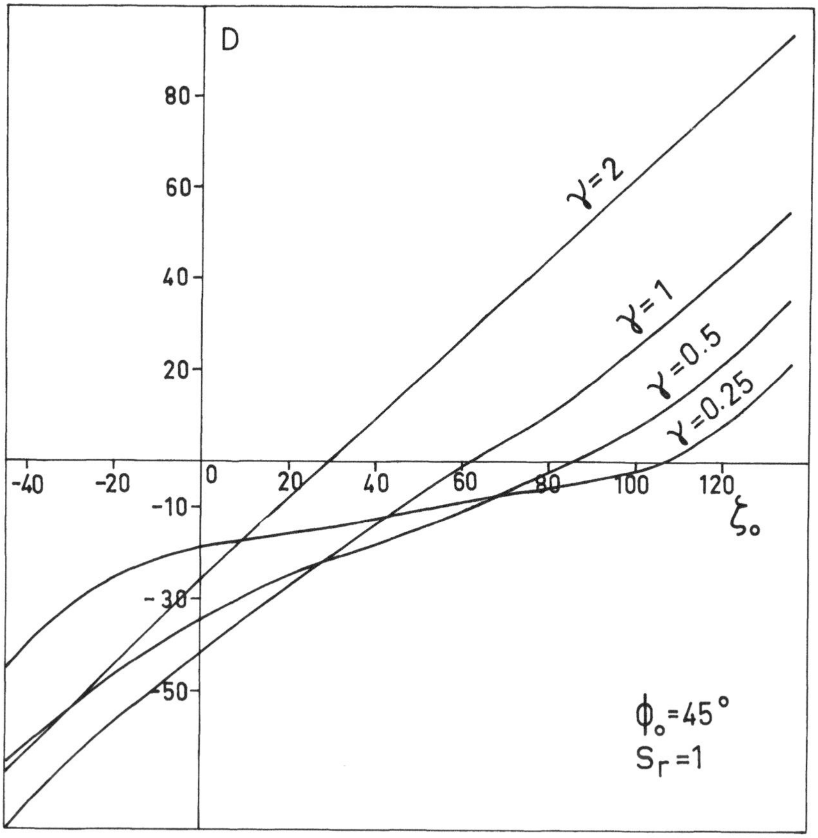

Fig. 6. Variation of drag D (in degrees) along the periphery of a circular inclusion for $\Phi_0 = 45°$ and $s_r = 1$. Note that there are longer segments of positive drag with increase in γ. A symmetrical pattern with equal segments of positive and negative drag would be obtained somewhere between $\gamma = 1$ and $\gamma = 2$

there is a significant segment of reverse drag up to large values of bulk deformation (Fig. 8b and c and Fig. 9). If the ratio between the rates of pure shear and simple shear is not very high, a symmetrical pattern does not develop in this situation.

If the initial orientation of the marker line is at an angle of about 135° to the direction of simple shear, for the case of $s_r \neq 0$ it was found that, initially, a more or less symmetrical pattern of drag develops with a characteristic *bowing in* of the marker lines toward the inclusion (Ghosh, 1975, Fig. 12a). The pattern becomes progressively asymmetrical with increase in bulk deformation. If $s_r \neq 0$, the relative movement of the marker planes and the rigid inclusion may become more complex. While the rigid circular inclusion will always rotate in the forward sense (i.e., in the same sense as the sense of rotation for the simple-shear component), the marker plane may rotate forward or backward or may not rotate at all. The critical initial orientation Φ_6' of the marker (Ghosh

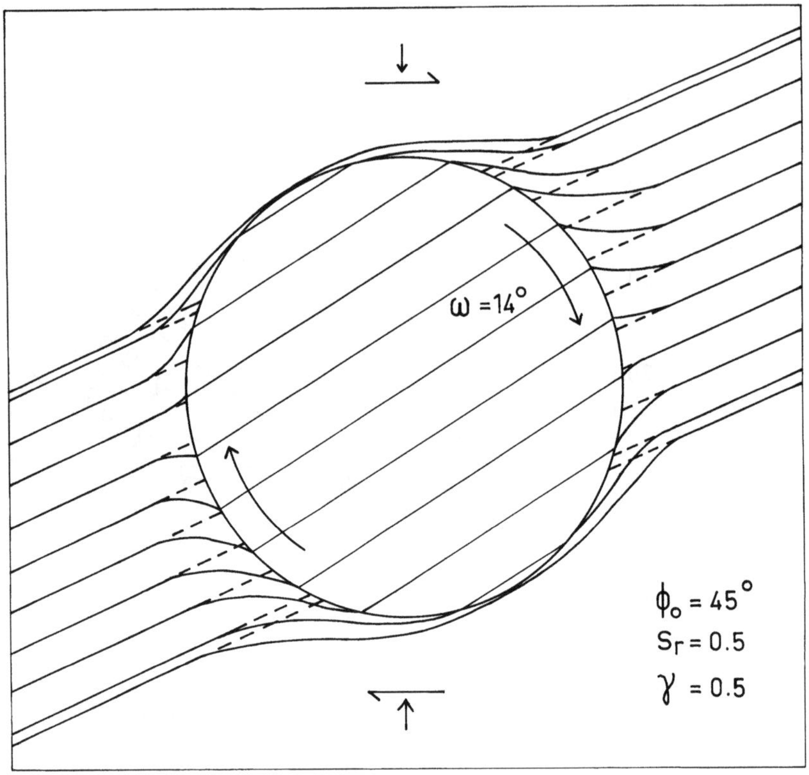

a

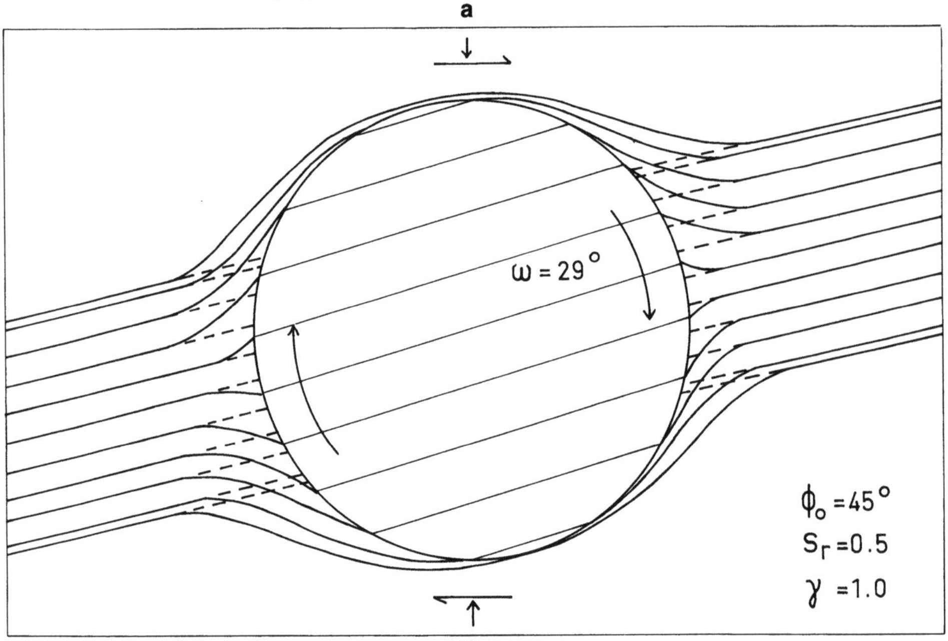

b

Fig. 7a–c. Drag-patterns around circular inclusion for $\Phi_0 = 45°$ and $s_r = 0.5$ at three stages of deformation

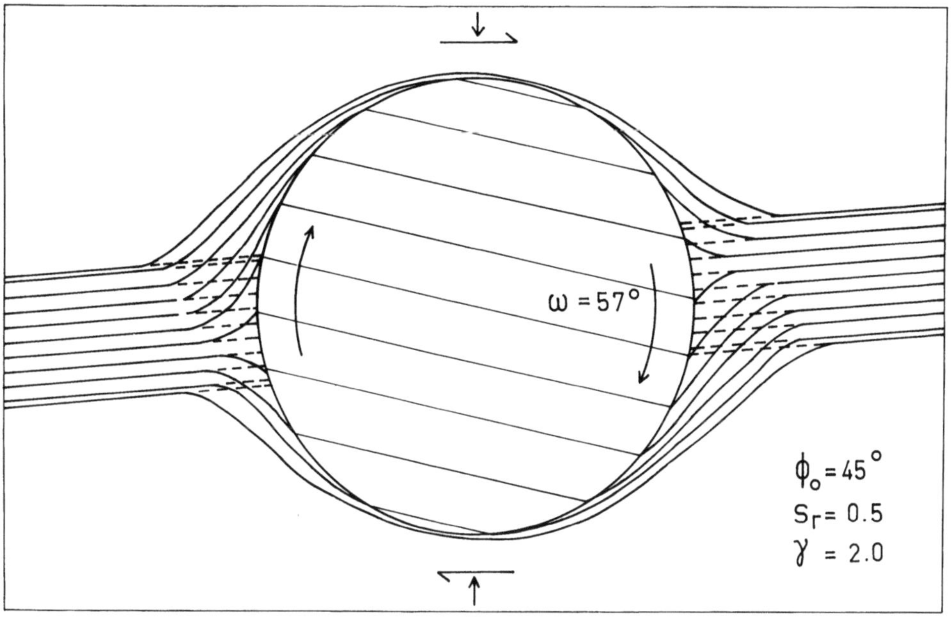

c

and Ramberg, 1976) at which there is no rotation (Fig. 2) is given by the equation

$$\tan \Phi_6' = -\frac{1}{2s_r}.$$

If $\Phi_6' = 135°$, the critical value of s_r will be 0.5. Thus, if s_r is less than a critical value, the marker will rotate forward. If s_r equals the critical value the marker will not rotate (Fig. 11). Lastly, if s_r is larger than the critical value the marker will rotate in a sense opposite to that of the circular inclusion. It is unfortunate that these unlike types of relative movements between the inclusion and the marker cannot be readily distinguished from the corresponding drag patterns. If $s_r \neq 0$, the drag patterns in all cases are asymmetrical (Fig. 10). Up to moderate values of bulk deformation an asymmetric *bowing in* of the markers can be seen so long as s_r does not exceed the critical value (Fig. 11). The drag is dominantly normal. If s_r exceeds the critical value the drag is for most cases entirely normal along the periphery. There is no bowing in of the markers in this case.

5 PATTERNS OF DRAG AROUND ELLIPTICAL INCLUSIONS

A simple method of qualitatively determining the drag pattern of an elongate inclusion is to plot the angle of divergence Φ–Φ' between the major axis of the inclusion and marker line with increasing bulk deforma-

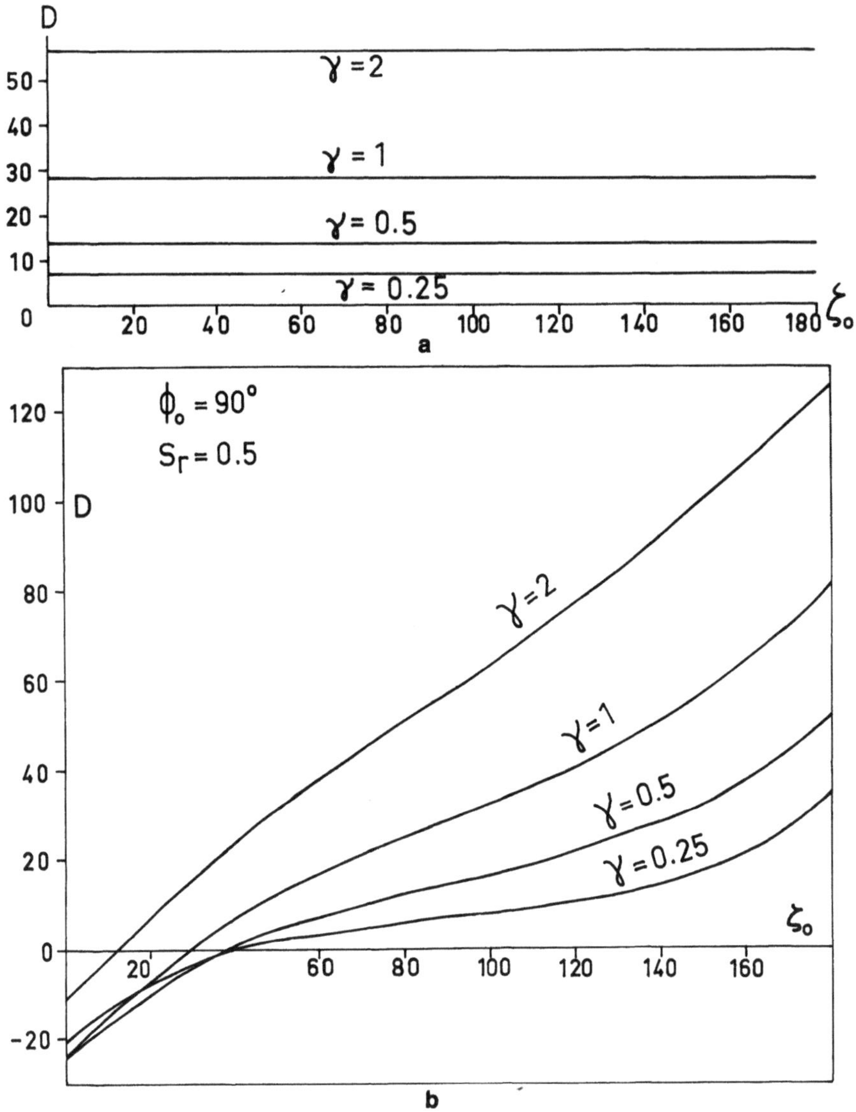

Fig. 8a–c. Variation of drag D (in degrees) along the periphery of a circular inclusion for $\Phi_0 = 90°$ when (a) $s_r = 0$, (b) $s_r = 0.5$ and (c) $s_r = 1$

tion (Fig. 13). If Φ–Φ' is strongly negative, the drag pattern will be asymmetrical, with reverse drag predominating. If Φ–Φ' has a fairly large positive value, the asymmetric pattern will show dominantly normal drag. If Φ–Φ' is near about zero, the drag pattern will be roughly symmetrical.

Numerical calculations of the angles of rotation of rigid inclusions and a marker line show that, if the axial ratio of the inclusion is sufficiently large

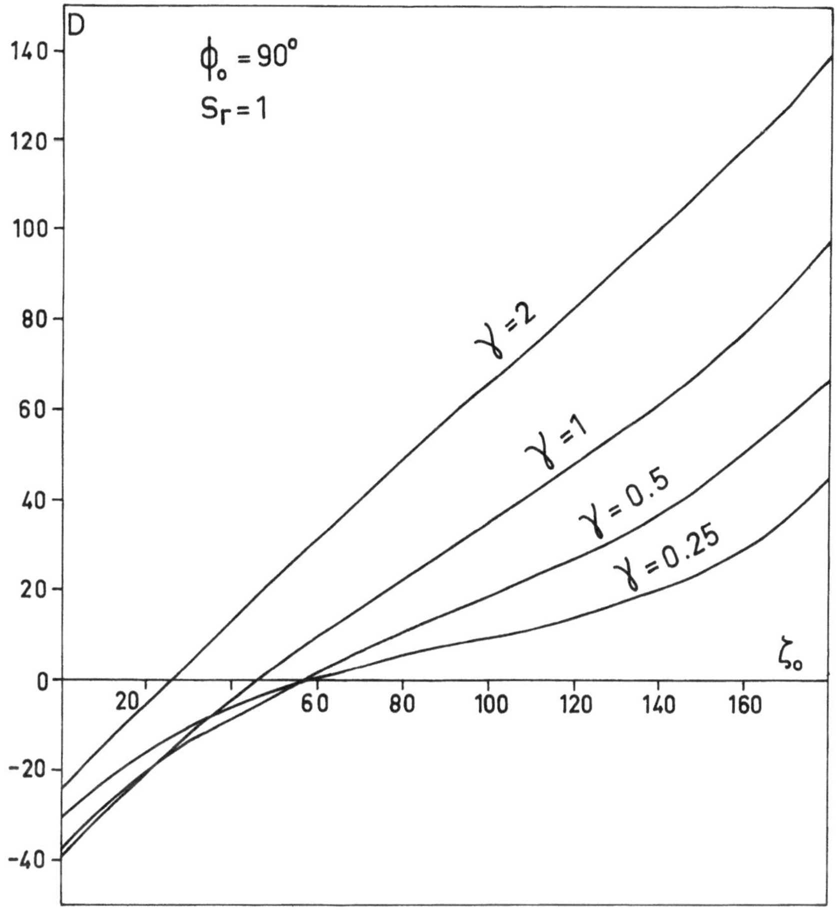

c

(roughly $R > 10$), the rotation of the inclusion virtually coincides with that of the passive marker. Figure 12 shows that, even for moderately long inclusions, the maximum possible divergence in orientation between the marker line and long axis of the inclusion remains rather small. Thus, if the initial orientation of the marker and the inclusion was at a right angle to the direction of simple shear (i.e., $\Phi_0 = 0$), the marker at first moves ahead of the inclusion (i.e., Φ–Φ' is negative). For $s_r = 1$ and $R = 5$, the maximum possible divergence Φ–Φ' for this stage is about $-1.4°$ (Fig. 12). For $R = 4$ and $R = 3$, the corresponding values are $-3.8°$ and $-6.6°$, respectively. With increasing deformation the inclusion moves ahead (giving a positive value of Φ–Φ') but the divergence remains small. For $R = 5$, the maximum divergence for this stage is 1.2° even if the bulk deformation is indefinitely large (Fig. 12). It should be noted in connection with Figure 12 that the axial ratio X/Y of the strain ellipse in the matrix increases very rapidly with increase in γ. For $\gamma = 1$, 1.5, 2, 3, 4, and 5, the corresponding values of X/Y are 8.8, 24.4, 67.8,

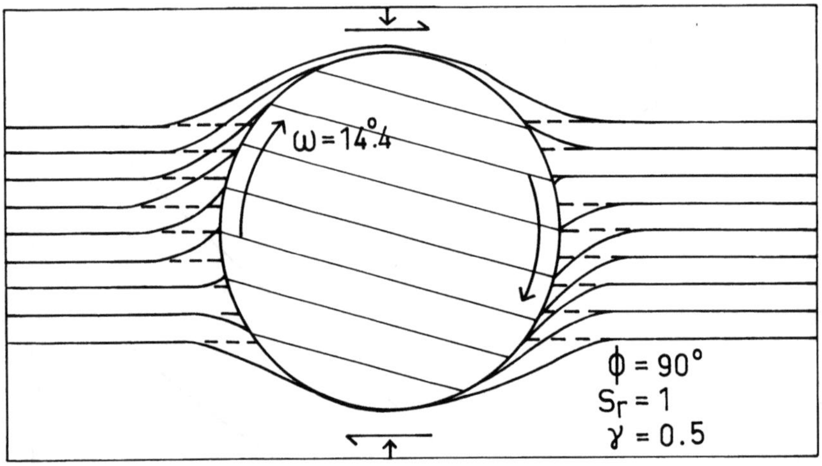

Fig. 9. Drag-pattern around circular inclusion for $\Phi_0 = 90°$, $s_r = 1$, and $\gamma = 0.5$

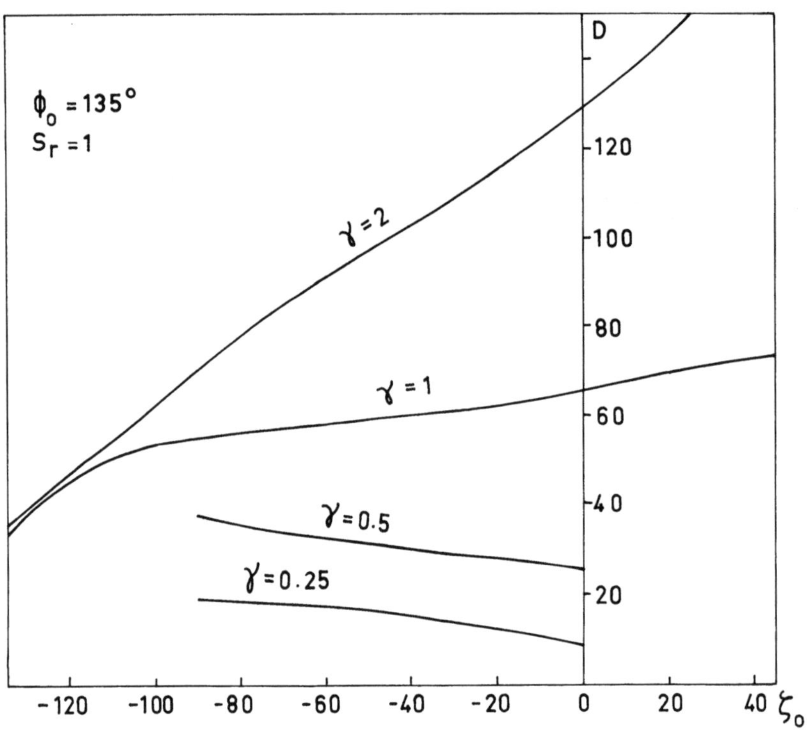

Fig. 10. Variation of drag D along the periphery of a circular inclusion for $\Phi_0 = 135°$ and $s_r = 1$

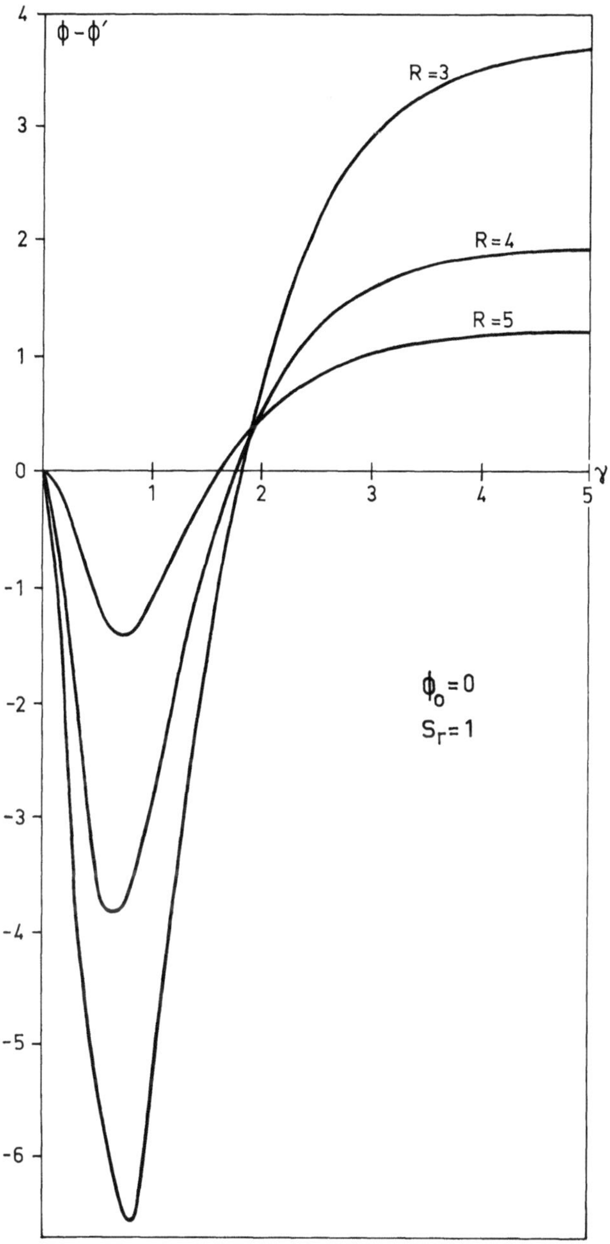

Fig. 11. Drag pattern around circular inclusion showing asymmetric bowing in of markers for $\Phi_0 = 135°$, $s_r = 0.5$, and $\gamma = 1$. *Marker lines away from inclusion* have not rotated

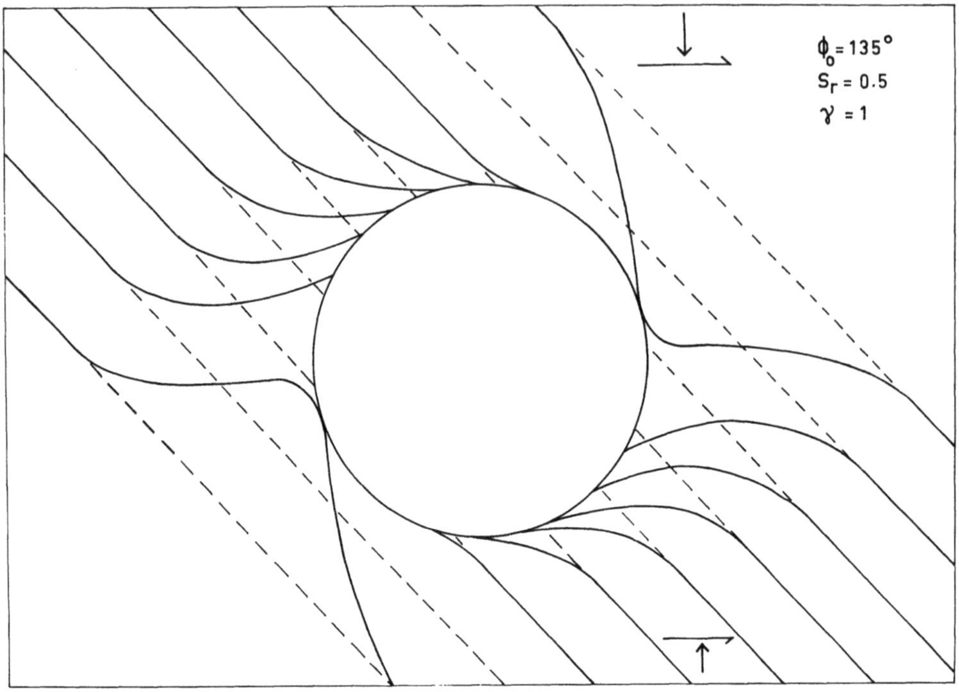

Fig. 12. Variation in Φ–Φ' (in degrees) with γ, for $\Phi_0 = 0$ and $s_r = 1$ for different values of R. Note that Φ–Φ' is very small when $R = 5$

504, 3745, and 25 000. Thus, for reasonable values of geologic strain (say, in this case, $\gamma < 3$) the positive value of Φ–Φ' will be quite small for moderately long inclusions (Fig. 12). A consequence of the small angle between the marker lines and a long or moderately long inclusion is that the drag pattern of the markers around the inclusion will be roughly symmetrical.

For relatively short inclusions, the drag patterns were often found to be more or less similar to those described for the circular inclusions. If Φ_0 is near 0°, and if $s_r = 0$ or small, the pattern is initially asymmetrical with reverse drag predominating (Fig. 14a). This changes over to a more or less symmetrical pattern at moderate bulk deformation and then becomes progressively asymmetrical with normal drag predominating. If the pure-shear component is significantly large, $s_r >> 0$, a symmetrical pattern will not develop for reasonable values of bulk strain X/Y. The drag in this case will be dominantly reverse (Figs. 13a, 14b). For instance in Figure 13a, it can be seen that a symmetrical pattern would develop for $s_r = 0.5$ at $X/Y \approx 16$. For larger s_r, the curves will intersect the line Φ–$\Phi' = 0$ at much larger values of X/Y.

For $\Phi_0 = 45°$ and $\phi_0 = 90°$, the course of evolution of the drag-pattern can be followed from Figures 13b and c. Figures 13b and c are for an inclusion with $R = 2$. The course of evolution remains essentially similar if R is less than 2. Some of the drag patterns are shown in Figures 14c, d, e, and f.

The rotational history of a marker and a rigid elongate inclusion may be quite complex if Φ_0 is a large obtuse angle. There may be several cases, viz., (1) both inclusion and marker rotate forward (Fig. 15a), the marker moving faster ($\Phi_0 < \Phi_3$ or $> \Phi_4$, in Fig. 2), (2) both inclusion and marker rotate forward, the inclusion moving faster ($\Phi_0 > \Phi_3$ but <90° in curve A of Fig. 2; $\Phi_0 > \Phi_6'$ but $<\Phi_4$ in curve A; $\Phi_0 > \Phi_3$ and $<\Phi_4$ in curve C), (3) both inclusion and marker rotate backward (Fig. 15b), the absolute amount of rotation of marker being larger (Φ_0 between Φ_5 and Φ_6 in

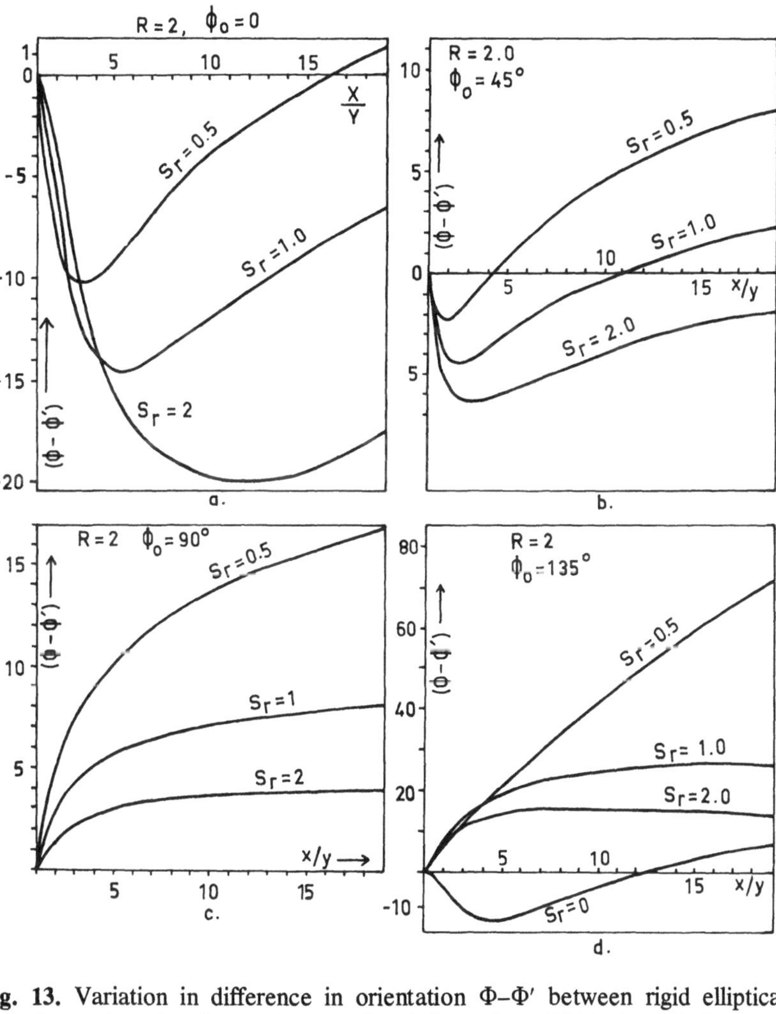

Fig. 13. Variation in difference in orientation Φ–Φ' between rigid elliptical inclusion and marker line with increasing deformation. X/Y is the ratio between the axes of the strain ellipse for the bulk deformation. The origin of each diagram is at $X/Y = 1$, Φ–$\Phi' = 0$

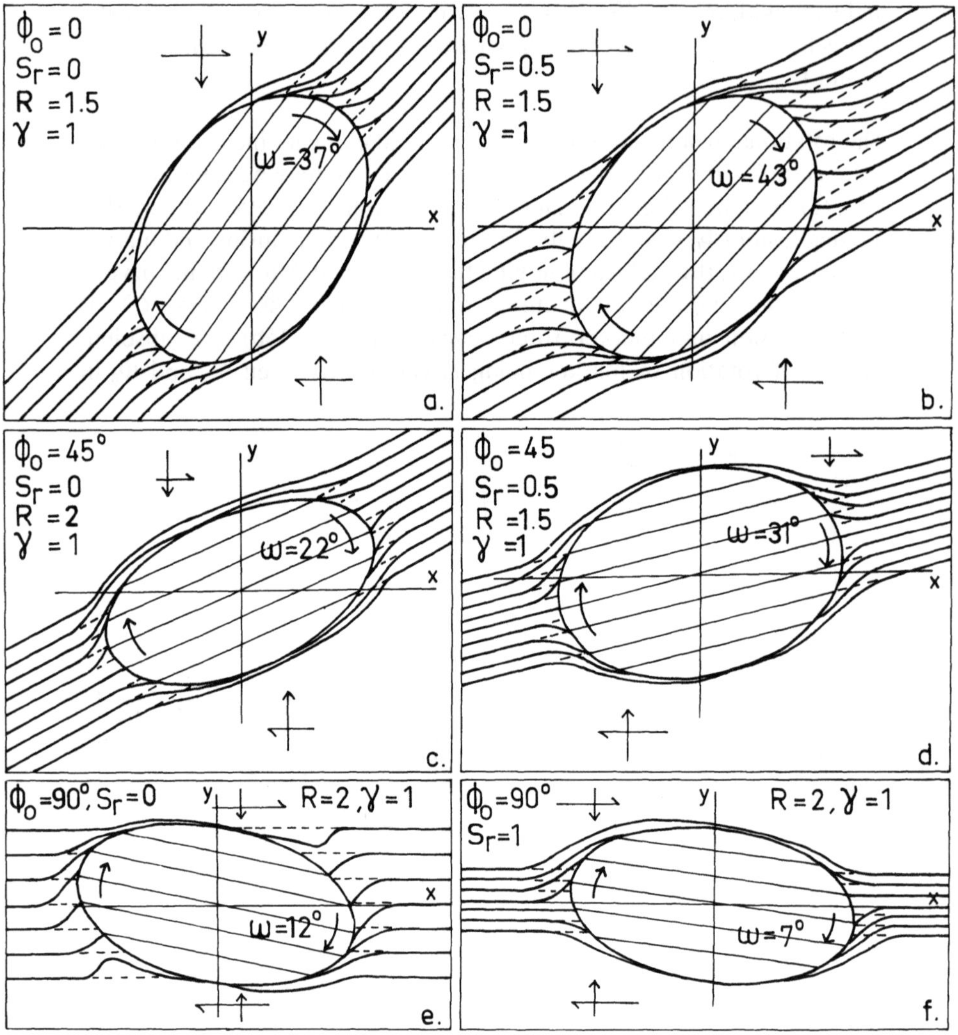

Fig. 14. Drag patterns around elliptical inclusions

curve *A* of Fig. 2), (4) the inclusion rotates forward but the marker rotates backward (Fig. 15c), (for curve *C*, Φ_0 between 90° and Φ_6'; for curve *A*, Φ_0 between Φ_5' and Φ_5 or between ϕ_6 and ϕ_6'), (5) the inclusion does not rotate (Fig. 15d), while the marker rotates backward ($\Phi_0 = \Phi_5$ or Φ_6 in curve *A* of Fig. 2), and (6) the inclusion rotates forward but the marker does not rotate (Fig. 15e) ($\Phi_0 = 90°$ or $\Phi_0 = \Phi'_6$). The patterns of drag for these cases (excepting the second case), are shown in Figure 15. It will be seen that excepting the first case (Fig. 15a), all the other patterns (Figs. 15b, c, d, e) are roughly similar. This is because, excepting the first case, all the other cases give a *relative* forward movement of the inclusion with respect to the marker. Hence in these cases the asymmetric patterns show predominantly normal drag. Only in the first case, where there is a

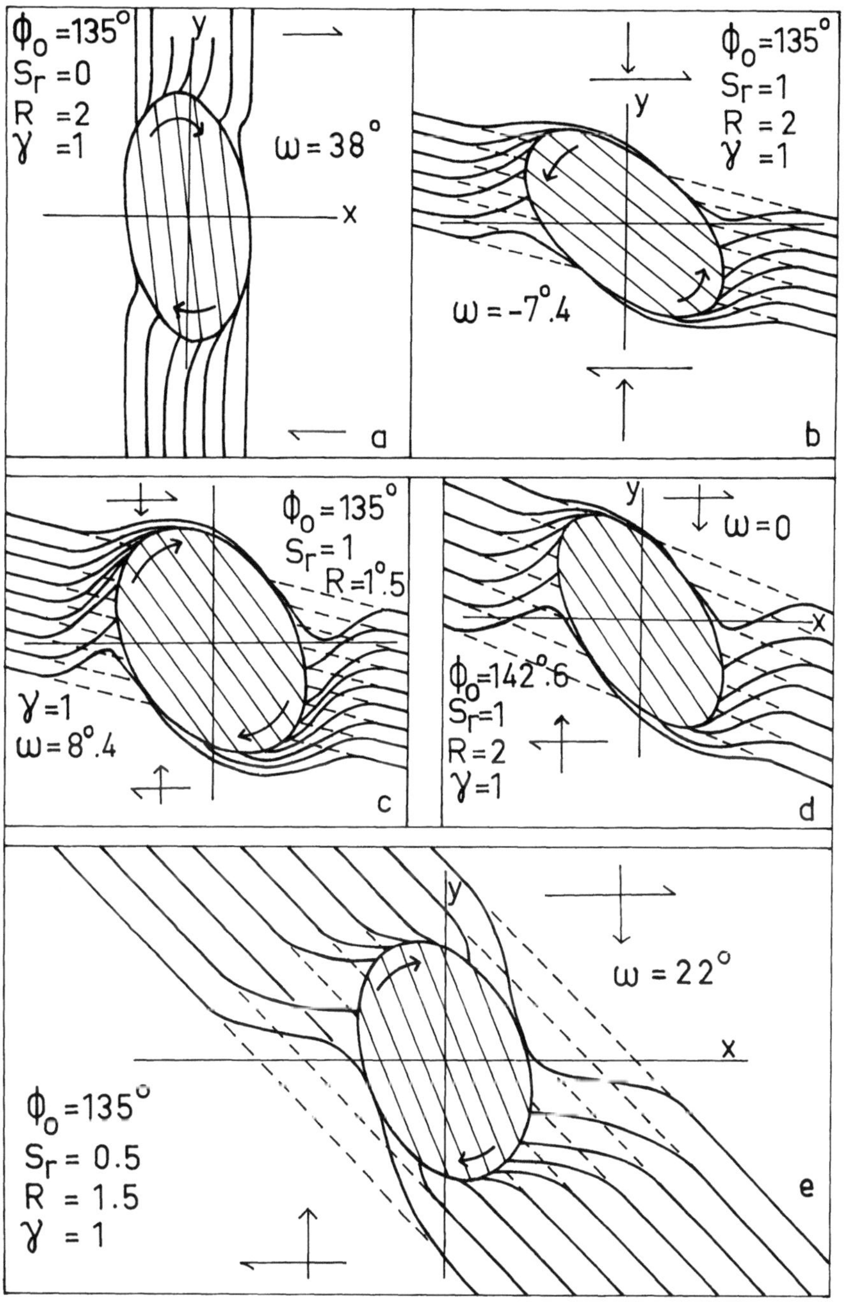

Fig. 15. Drag patterns around elliptical inclusions initially at obtuse angles with *y* axis

relative backward movement of the inclusion, the pattern is predominantly reverse.

When Φ_0 is a large obtuse angle, an unusual drag pattern with a "bowing in" of the markers may develop in certain situations (Fig. 15e). The bowing in of the markers can be seen when both the conditions mentioned below are satisfied: (1) the drag is zero at some point on the periphery of the inclusion, and (2) there is an extension at a right angle to the markers.

6 DISCUSSION

The theoretically determined patterns of drag may be classified on the basis of their symmetry or asymmetry, on relative lengths of segments showing normal and reverse drag and on the basis of whether the markers around the inclusion bow out or bow in. On these bases, the following categories of drag patterns may be distinguished:

(1) Symmetrical bowing out;
(2) Asymmetrical pattern with only normal drag, and with constant amount of drag along the periphery;
(3) Asymmetrical pattern with normal drag only, but with variable amounts of drag along the periphery;
(4) Asymmetrical pattern with both normal and reverse drag, but with normal drag predominant;
(5) Asymmetrical pattern with reverse drag only;
(6) Asymmetrical pattern with both normal and reverse drag, but with predominant reverse drag;
(7) Symmetrical bowing in;
(8) Asymmetrical bowing in.

Excepting the second pattern listed above (Fig. 8a), each of the other patterns can form in a wide range of situations. The second pattern develops only if the bulk deformation is by simple shear (i.e., $s_r = 0$), and if the marker lines are parallel to the direction of shear. Even if a small amount of pure-shear strain-rate is added to the rate of simple shear, the amount of drag will show a strong variation along the periphery (compare Fig. 8a with Fig 8b). Hence the presence or absence of constant amounts of drag of an *S* surface can be regarded as a criterion to decide whether or not the bulk deformation was only by simple-shear parallel to the *S* surface.

The pattern with an approximately symmetrical bowing out of the markers is shown in the following situations: (a) if the bulk defomation is nonrotational and the markers were at right angles to the direction of maximum shortening, (b) if the rates of rotation of the inclusion and the markers were the same during the initiation of deformation (i.e., if $\Phi_0 = \Phi_3$ in Fig. 2) and if the total deformation is very small, and (c) if Φ_0 is somewhere between Φ_1 and Φ_3. If, in addition, it is known whether the bulk deformation was small, moderate or large, a rough idea may be obtained about the initial orientation of the marker. A small bulk deformation would indicate that the initial orientation was close to Φ_3, while a large bulk deformation would indicate that the marker was initially at a fairly large angle with Φ_3 (and hence at a high angle to the x axis).

Similarly, a symmetrical *bowing in* of the markers may develop in the following situations: (a) if the deformation is nonrotational and if the markers are at a right angle to the direction of maximum extension, (b) if the rates of rotation of marker and inclusion are initially coincident (i.e., if $\Phi_0 = \Phi_4$) and (c) if Φ_0 was between Φ_6' and Φ_4 (or a slightly larger angle). A small bulk deformation would indicate that Φ_0 was close to Φ_4 while a large bulk deformation would indicate that Φ_0 was closer to Φ_6'. At large or moderate bulk deformations, an asymmetrical pattern with bowing in of the markers would indicate that Φ_0 is a large obtuse angle ($\Phi_0 = \Phi_6'$ or somewhat larger).

It should be remembered that for a sufficiently long inclusion, the marker and the inclusion would be essentially parallel and the drag-pattern will be symmetrical in all situations. If there is an extension across the markers (with an obtuse angle for Φ_0), the pattern will show a symmetrical bowing in. Otherwise there will be a symmetrical bowing out.

The other types of patterns (i.e., patterns 2 to 6) will not be so informative unless some independent criterion is available to distinguish between normal and reverse drags in natural structures. Normal and reverse drags can be distinguished only if the sense of rotation for the simple-shear component is known. (Incidentally, determination of the sense of paracrystalline rotation from S-shaped si in equant porphyroblasts will also be subject to the same uncertainty if the total rotation is not large. An S-shaped si would give an anticlockwise rotation and an si with mirror-image of S shape would show a clockwise rotation only in those situations (Φ between Φ_3 and Φ_4) where the inclusion was rotating faster than the marker lines. In situations where the marker line was rotating faster ($\Phi < \Phi_3$ or $> \Phi_4$), a paratectonic crystal would show S-shaped si for a clockwise rotation and the mirror image of S shape for anticlockwise rotation).

However, the uncertainty in distinguishing between normal and reverse drag remains only if the angular divergence between the marker and the long axis of the inclusion is not large. This is because the maximum angular divergence which can result from a faster rotation of the marker (with respect to the inclusion) is generally small. It increases with increasing values of s_r. For instance, if $\Phi_0 = 0$, $R = 2$, and $s_r = 0.5$, the maximum divergence is about $-10°$. For $s_r = 1$ and $s_r = 2$ the corresponding values are $-14.5°$ and $-20°$ (Fig. 13a). Hence, if the angular divergence is considerably large (say, near 90° or more), it will be justified to assume that the pattern is showing predominantly positive or normal drag.

Again, in certain instances (for instance where the sense of movement of an overthrust sheet is known), the sense of rotation of the simple-shear component may be independently known and may be utilized to distinguish between normal and reverse drags. In those cases, a pattern of predominantly reverse drag (type 5 and type 6) would indicate that Φ_0 was less than Φ_3 or greater than Φ_4. Again, if the pattern shows predominant normal drag (type 3 and type 4), this would usually indicate that Φ_0 was not much less than Φ_3 or not much greater than Φ_4. If the total deformation was large, Φ_0 would be between Φ_3 and Φ_4. Unless the total deformation was extremely large a predominantly normal drag pattern would not develop if Φ_0 was more or less at a right angle to the x axis.

The pattern of drag showing an inward bowing of the planar structures develop when there is a component of extension across them. If the planar structure is schistosity, the crenulations developed on it may then obscure the inward bowing pattern. The pattern might be recognized if the crenulations are much smaller than the rigid inclusion or if the planar structure is represented by a color-banding without significant contrasts in competence.

In the above discussion it had been assumed that the planar structures could be at any angle to the x- or y-coordinate axis. For a schistosity this would be relevant mostly for the case of superposed deformations. If it is assumed that the drag pattern of a schistosity has developed in continuation of the same movement which had earlier given rise to the schistosity itself (parallel to the major axis of the strain ellipse), it would mean that during the initiation of the drag pattern Φ_0 was somewhere between 45° and 90°. A point of considerable interest is that the critical orientation ϕ_3 (Fig. 2) is not only the orientation at which all inclusions and the passive marker rotate at the same rate (viz., $\dot{\gamma}/2$), and that it separates orientations for faster and slower rotations of the inclusion (with respect to the marker), it is also the orientation for the major axis of the *infinitesimal* strain ellipse for the bulk deformation. This crucial orientation is given by the equation

$$\Phi_3 = 90° + \frac{1}{2}\tan^{-1}\left(\frac{-1}{2s_r}\right)$$

(Ghosh and Ramberg, 1976). The orientation of the major axis of the *finite* strain ellipse will be somewhere between $\Phi = \Phi_3$ and $\Phi = 90°$.

Hence, an inclusion which was initially parallel to the major axis of the bulk strain ellipse will always rotate faster than the major-axis of the ellipse or a marker parallel to it. The distorted pattern of the schistosity in this case will always show predominantly normal drag with variation in the amount of drag along the periphery of the inclusion. Unless the bulk deformation after the development of the schistosity is very small, the drag pattern will be asymmetrical. The drag patterns in this case will not show a bowing in of the passive markers.

ACKNOWLEDGMENT

The work was made possible by grant No. GDP 608-21-04 of the Swedish Geodynamics Project-A, Swedish Natural Science Research Council.

APPENDIX

Explanation of Symbols

Φ_0	Initial angle between y-coordinate axis and the long axis of inclusion (or marker line in the groundmass).
Φ	Final angle between y axis and the long axis of the inclusion.
Φ'	Final angle between y axis and the marker line away from the inclusion.
R	Ratio between long and short axes of elliptical inclusion.
ζ_0 and ζ	Indicate initial and final positions of a point on the periphery of the inclusion and are measured by the angles between y axis and the lines joining the center of the inclusion with the initial and final positions of the points.
$\dot{\epsilon}$	Rate of longitudinal strain for the pure-shear part.
$\dot{\gamma}$	Rate of shear strain for the simple-shear part.
s_r	The ratio $\dot{\epsilon}/\dot{\gamma}$.
A	$\dfrac{R^2}{R^2+1}$.
B	$\dfrac{s_r(R^2-1)}{R^2+1}$.
C	$\dfrac{1}{R^2+1}$.
ω	Rigid rotation of inclusion $\Phi-\Phi_0$.
X/Y	Axial ratio of strain ellipse for combined pure shear and simple shear.

REFERENCES

Ghosh, S. K.: Distortion of planar structures around rigid spherical bodies. Tectonophysics *28*, 185–208 (1975)

Ghosh, S. K., Ramberg, H.: Reorientation of inclusions by combination of pure shear and simple shear. Tectonophysics *34*, 1–70 (1976)

Ramberg, H.: Particle paths, displacement and progressive strain applicable to rocks. Tectonophysics *28*, 1–37 (1975)

Rosenfeld, J. L.: Rotated garnets in metamorphic rocks. Geol. Soc. Am. Spec. Paper *129*, 1–105 (1970)

7

P. J. HUDLESTON

Abstract Discrete, ductile zones of differential shear are common in rocks, also occur in glacial ice, and are a response of crystalline materials to deformation at high temperatures. Such zones imply strain softening under shear deformation. They provide us with excellent opportunities for examining the nature of changes in fabric and texture with changes in state of strain.

Superimposed ice at the margin of the Barnes Ice Cap, Baffin Island, Canada, has been overridden and deformed by the main glacier. Deformation within the overridden ice is of variable intensity, and is characterized by zones of shear which are a few centimeters thick and which lie almost parallel to the bed rock and to the measured direction of maximum shear strain rate. The geometry of the shear zones, as delineated by air bubbles in the ice, indicates that deformation is very nearly one of simple shear. Changes from margin to center across a shear zone include an increase in grain size, the development of a weakly sutured texture, an increase in the aspect ratio of air bubbles, a decrease in the acute angle between the long axes of the bubbles and the shear zone boundaries, and an increase in the degree of preferred orientation of *c* axes. In a selected example, the fabric changes from a fairly uniform distribution of *c* axes outside the shear zone, through a double maximum fabric (not always well defined) of the kind observed in experiments, to an intense

PROGRESSIVE DEFORMATION AND DEVELOPMENT OF FABRIC ACROSS ZONES OF SHEAR IN GLACIAL ICE

deformation and that it is related to a state of finite strain rather than to the stress or strain rate.

Quantitative measures of degree and direction of preferred orientations of *c* axes can be found using the Bingham probability distribution. It may thus be shown that a strong correlation exists between bubble orientation and the degree of preferred orientation of *c* axes, expressed as a concentration statistic. Bubble shape and inclination is related to the amount of shear strain undergone by the ice, and allows estimates of total shear strain across the shear zones to be made. There is considerable uncertainty involved in this procedure, but it appears that the maximum shear strain in the center of the selected shear zone (γ in the range 5 to 8) is about an order of magnitude greater than the maximum value of γ attained in previous experimental work. Considerable strains may be required to attain steady state conditions; such conditions may not obtain even in the centers of the shear zones. The results of this study provide some reconciliation between previous work on experimentally deformed ice at low strains and on glacial ice that has undergone extremely high strains.

1 INTRODUCTION

Inhomogeneous simple shear appears to be a rather common form of deformation in crystalline rocks (Ramsay and Graham, 1970). The ductile shear zones that result from this type of deformation may vary from a few millimeters to a few kilometers across, and the displacements along the zones may range from a few millimeters to many kilometers (Ramsay and Graham, 1970; Hara et al., 1973; Coward, 1974; Beach, 1974; Bak et al., 1975; Mitra, 1975). Differential shear in localized zones has also been observed or is inferred to have occurred in glacial ice (e.g., Goldthwait, 1960; Ragan, 1969), and it may be that the development of ductile shear zones with concomitant strain softening is an important mode of response of massive crystalline material to deformation at temperatures near the melting point. Shear zones often develop in seemingly isotropic and homogeneous rock, and are particularly interesting because they are examples of deformation to varying intensities of strain, under a well-defined pattern of displacement. Foliation often occurs in these zones, becoming progressively better developed towards their centers (Ramsay and Graham, 1970). Strain within the zones can sometimes be computed using deformed objects or displaced veins and dikes (Ramsay and Graham, 1970; Beach, 1974). There is thus the potential for relating the appearance and intensity of the foliation to the amount of strain undergone by the material (Ramsay and Graham, 1970). The fact that the strain path (Ramsay, 1967) is noncoaxial should allow a distinction to be made between effects related to infinitesimal and finite states of strain. Thus Ramsay and Graham (1970) were able to show that foliation develops in response to the total strain, and occurs normal to the direction of maximum finite shortening.

In the margin of the Barnes Ice Cap, Baffin Island, Canada, there are structures very similar to those described from crystalline rocks by Ramsay and Graham (1970). These glacial structures are the subject of this paper. Analysis of the shear zones in glacial ice provides some insight into the nature of ice deformation up to moderate strains and the development of preferred crystallographic orientations. The shear zones in ice afford an example of textural and fabric changes in a material in which deformation is accommodated by intracrystalline plastic flow and recrystallization. Before proceeding to a description of these shear zones, however, it is pertinent to recall briefly the evidence that others have adduced for the existence of shear structures in glacial ice.

The question as to whether or not discrete zones of shear of thrust-like nature exist in glaciers has long been a subject of debate in the literature. The transport of rock debris from the base of a glacier to the surface near the terminus has commonly been associated with shear planes separating active overriding ice from more or less stagnant ice below (Bishop, 1957; Goldthwait, 1960). However, attempts to demonstrate displacements along the supposed planes of shear have generally proved unsuccessful (McCall, 1952), and alternative processes of bringing dirt-laden ice to the surface of a glacier have been proposed that do not require the presence of discrete shear planes (Weertman, 1961; Hooke, 1973). Despite the questionable existence of some shear structures, the evidence for others is very strong. For example, in the Gulkana Glacier, Alaska, Ragan (1969) descibed broad zones of moderately inhomogeneous shear associated with fold development, and also observed thin, more or less periodic shear structures, some of which showed rather sharp offsets of other features. Meier (1960) showed a clear example of primary stratification being offset along foliation in the Saskatchewan Glacier, and Goldthwait (1960) demonstrated that broadly irregular shear exists near the margin of part of the Greenland Ice Cap.

One of the problems of understanding deformation and the development of crystallographic fabric in glacial ice is due to the fact that most ice exposed in the ablation zone of a glacier, where it is most readily accessible, has undergone a long and complex history of deformation which is difficult to reconstruct. The total strain is usually extremely large. If the fabric and response of the ice to the current state of stress is dependent on some previous history of the deformation, that history must be understood if we are to interpret the fabric correctly. Budd (1972) has given some thought to this problem, and has suggested that many of the observed fabric patterns may be due to the cumulative development of fabric under the influence of a changing stress field along a flow line. As will be shown below, the problem of extremely high strains and complex deformation paths is largely avoided in this study: the ice in which the shear zones have developed has been subjected to a relatively moderate shear strain.

Kamb (1972) has examined the progressive recrystallization of ice (between 0° and −5°C) and the development of fabric in torsional experiments closely approximating simple shear. He concluded that the degree of preferred orientation of *c* axes is largely a function of the total strain and is fairly insensitive to stress or time over the range of conditions selected. The results of the present study strongly support this conclusion, and suggest that the same relationship holds for much higher strains, smaller stresses and longer times than those obtained in his experiments.

As is often pointed out, glacier flow provides a good analogy with certain types of rock deformation. The flow pattern of a glacier is broadly similar to that of both the internal and external parts of some orogenic belts (Price, 1972), the concept of tectonic regimes has been applied to both rocks and ice (Wegmann, 1963), and structures such as a coarse foliation and recumbent folds, seen in the mobile cores of orogenic belts, are also found in glaciers (Hudleston, 1976). The similarity appears to extend to shear zones, and since the physical conditions of deformation and the boundary conditions are better known for ice than most rocks, the analysis of shear zones in ice may aid our understanding of them in rocks.

2 SHEAR ZONES IN THE BARNES ICE CAP

The shear zones at the margin of the Barnes Ice Cap occur in superimposed ice (Fig. 1). This ice forms on the surface of the glacier by the refreezing of melt water that percolates through the snow cover. During cooler summers, ice formed in this way is not entirely removed by ablation because near the margin of the glacier the snow cover is thick due to drifting. A zone of net accumulation may thus exist in a strip around the margin of the glacier (Hooke, 1973). Layers of superimposed ice were observed to form in this way from 1970 through 1973. This ice is fine grained and unfoliated. It has a nearly uniform distribution of *c* axes, and contains spherical or subspherical air bubbles (Hooke, 1973). There is a distinct layering parallel to the upper surface of the glacier and marked by variations in bubble content: the layering dips at between 5° and 10°.

The superimposed ice, if preserved, may be overridden by an advance of the main glacier: this has occurred in the Barnes Ice Cap (Hooke, 1973). Along the margin of the ice cap there now exists a wedge of deformed superimposed ice, between several hundred and several thousand years old, overlain by Pleistocene ice (Hooke, 1976). The approximate geometry of this wedge is known from borehole data (Hooke, 1973), and is shown in part in Figure 1A. The observed and partly inferred geometrical outline of this wedge of deformed superimposed ice has been reproduced in a flow model of Hudleston (1976, Fig. 8).

The internal structure of the deformed superimposed ice was observed in an 8-m deep vertical shaft located about 50 m from the glacier margin (Fig. 1). The ice is frozen to the bed rock, and near the base of the shaft the temperature is about −10°C, varying somewhat with seasonal changes. Deformation within the superimposed ice is of highly variable intensity, and is characterized by zones of shear which lie subparallel to the bedrock surface, or dip gently up glacier. The shear zones are also nearly parallel to one of the conjugate pair of measured directions of maximum shearing strain rate (Fig. 1B). They are a few centimeters thick and separated by somewhat thicker regions of weakly deformed ice. The shear zones are delineated and made visible by the deformation of air bubbles, which trace out sigmoidal patterns in the north and south walls of the shaft; the linear feature observed is a strong preferred

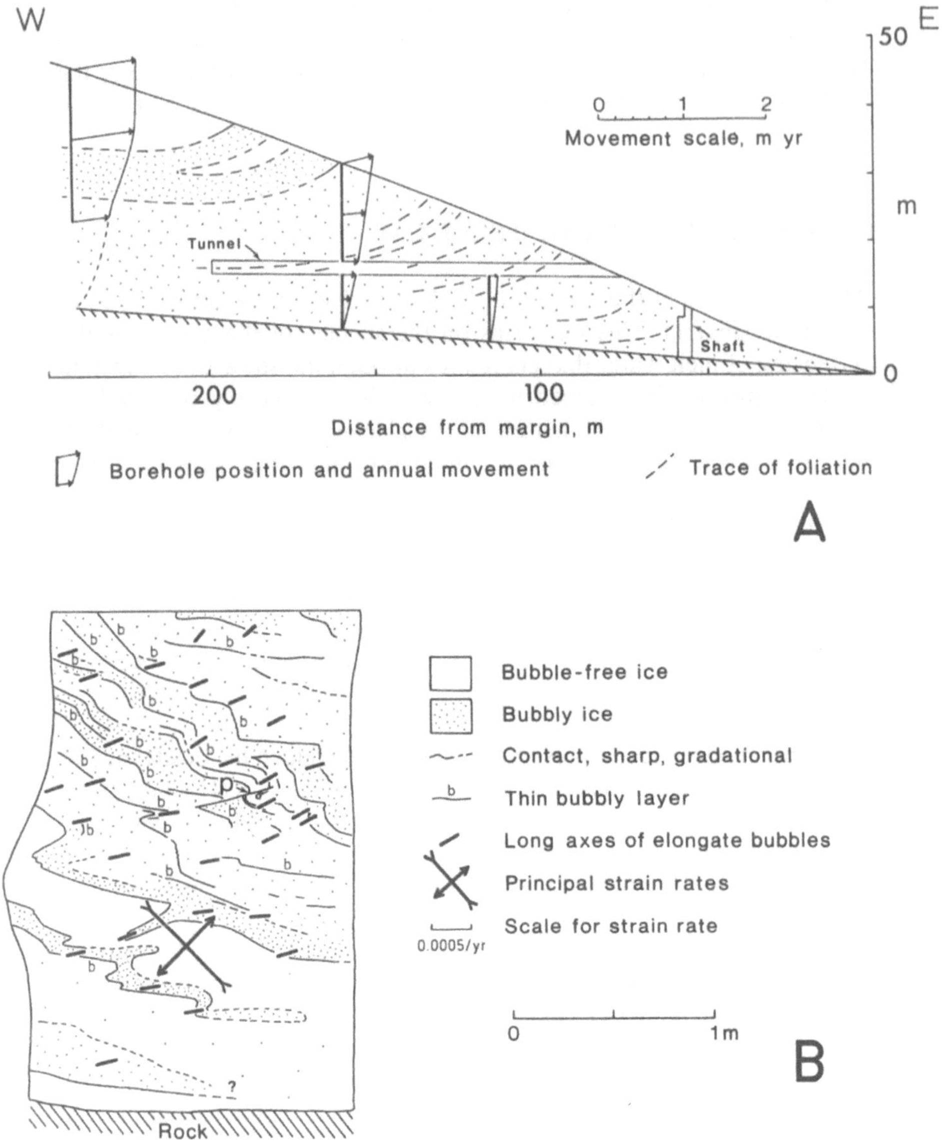

Fig. 1A, B. (A) E–W section perpendicular to the margin of the Barnes Ice Cap (after Hooke, 1973, 1976), showing structure in ice and location of shaft. Deformed superimposed ice is lightly stippled; Pleistocene ice heavily stippled; post-Pleistocene glacial ice unstippled. Vertical exaggeration $2x$. (B) Structural detail at base of shaft in N wall. *Stippling density* denotes bubble concentration. Contacts and individual bubbly layers reflect primary stratification which has been folded and sheared as a result of the inhomogeneous deformation. The prominent shear zone at P is a continuation of the one exposed on the S wall of the shaft which was selected for detailed analysis orientation of elongate bubbles (Fig. 2). The boundaries of the zones defined by the sigmoidal patterns are oblique to the primary stratification in the ice and to the direction of elongation of the bubbles (Figs. 1 and 2).

The sigmoidal patterns delineate zones of shear. This is indicated by offsets of the primary stratification along these zones, by the fact that the bubbles increase in aspect ratio towards the centers of the zones, as their elongation direction becomes inclined at smaller angles to the shear zone boundaries (Fig. 2), and by the observation that the undeformed superimposed ice contains nothing like them. The sigmoidal bubble pattern is quite similar to that which would be expected if the bubbles were simply passive markers in a medium subjected to differential shear.

2.1 Bubble Shape and Orientation: Evidence for Simple Shear

It can be argued from the geometry of the bubble elongation directions that the deformation undergone by the ice is very close to that of simple shear. Ramsay and Graham (1970) describe the conditions that are necessary and sufficient for simple shear. In order to state these, it is convenient to set up a coordinate system as shown in Figure 3A. The x'

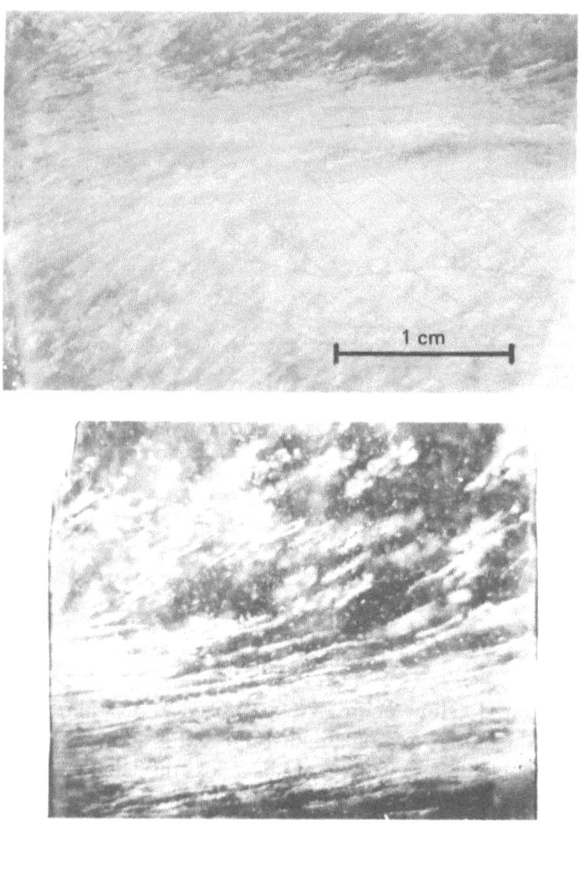

Fig. 2A, B. Photographs of shear zone selected for detailed analysis: (A) As exposed in S wall of shaft (printed in reverse for consistency in sense of shear). Scale bar = 1 cm. (B) In cut slab about 2 cm thick and 6.5 cm across axis is parallel to the edge of the shear zone, and the y' axis normal to the zone boundary. For simple shear deformation, there must be a plane of symmetry ($x'y'$) normal to the shear zone; there must be no volume change; there must be no gradient of the structure parallel to the zone (x' direction); there must be regions of undeformed material outside the zone.

For the shear zones observed in the Barnes Ice Cap, the plane of symmetry $x'y'$ is the vertical plane normal to the glacier margin: in the walls parallel to the margin, the layering is unfolded, with an apparent dip of zero, and the bubbles are more or less equidimensional. Volume changes in the ice will be related to the compression of air bubbles and are likely to be insignificant. The bubbles constitute a small fraction of

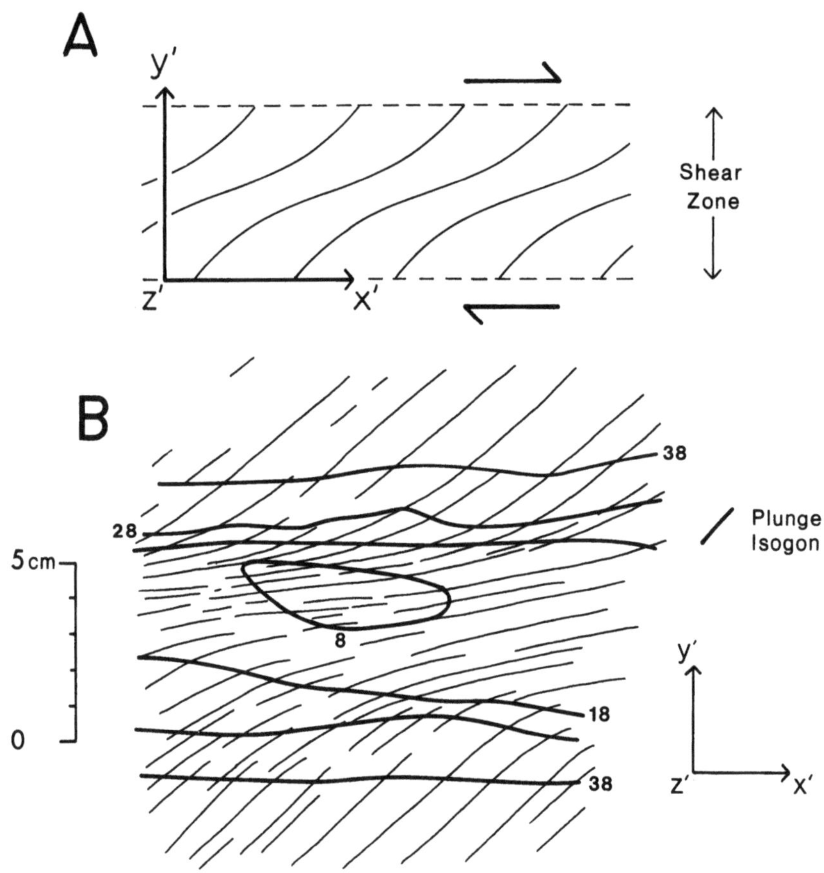

Fig. 3A, B. (A) Idealized shear zone to show coordinate axes used to describe geometry. Sigmoidal lines are trajectories of maximum finite extension. (B) Somewhat smoothed isogons of bubble plunge at 10° intervals across the selected shear zone. Traces of bubble elongation directions are shown as light discontinuous lines the total volume of the ice (less than 5% in all but the most bubbly ice), and in any case, pressure changes across the zones are likely to be small. The extent to which the structure is constant in the x' direction is given by the degree of parallelism of the isogons of bubble plunge (Fig. 3B). The condition that appears to be least satisfied is that of no deformation beyond the borders of the zone. The bubbles are elongate throughout the ice, although they are much more elongate within the shear zones than outside them. Because the ice is frozen to the bed rock, part of the "continuum" is unstrained. The shear zones nearest the base of the glacier, where they are best developed, are subparallel to the bedrock surface, and thus the ice cannot have a large component of strain other than simple shear in this region. If the condition of no deformation beyond the borders of the zones is relaxed, as it may be at increasing distances from the glacier bed, an arbitrary homogeneous strain could, in theory, be superimposed on the ice containing the zones. The bubble shape and orientation both inside and outside the shear zones suggest, however, that the main component of strain is still one of simple shear, even for the zones furthest removed from the bedrock.

Departures from simple shear in the deformed superimposed ice wedge are in the form of a small extensional strain parallel to the ice cap margin, and some horizontal compressional strain perpendicular to the margin (Hooke, 1973; Fig. 3B). The latter is typical of ablation zones of glaciers.

In summary, the geometry of the shear zones, as delineated by the deformed air bubbles, and consideration of the flow of ice which is frozen to its bed, indicate that deformation in the shear zones found in the lower part of the shaft must have been largely accommodated by simple shear.

2.2 Fabric Changes

It is of considerable interest to examine in some detail the changes in texture and crystallographic fabric across the zones of shear. Similar changes were found in all the zones studied, and data will be presented here only for the zone which was most carefully studied. The selected zone is located on the south wall of the shaft, opposite *P* in Figure 1B.

The orientations of ice *c* axes were measured in thin section using a Rigsby universal stage, and fabric diagrams were prepared using standard techniques (Langway, 1958; Kamb, 1962). Figure 4 is a sketch of the thin sections of the selected shear zone, showing the bubble traces and the sampling scheme for fabric analysis. Essentially all grains were measured in these sections, and the orientation data are presented in Figure 5. Fabric data is plotted, across the shear zone, for 1-cm wide strips parallel to the shear zone boundary, as defined by isogons of bubble plunge.

The variation in fabric pattern across the shear zone is striking. Outside the shear zone there is a relatively scattered distribution of *c* axes; towards the middle of the zone the scatter is progressively decreased until a very strong unimodal fabric is developed. The center of this point concentration lies perpendicular to the shear zone boundaries (rows VI to IX, Fig. 5).

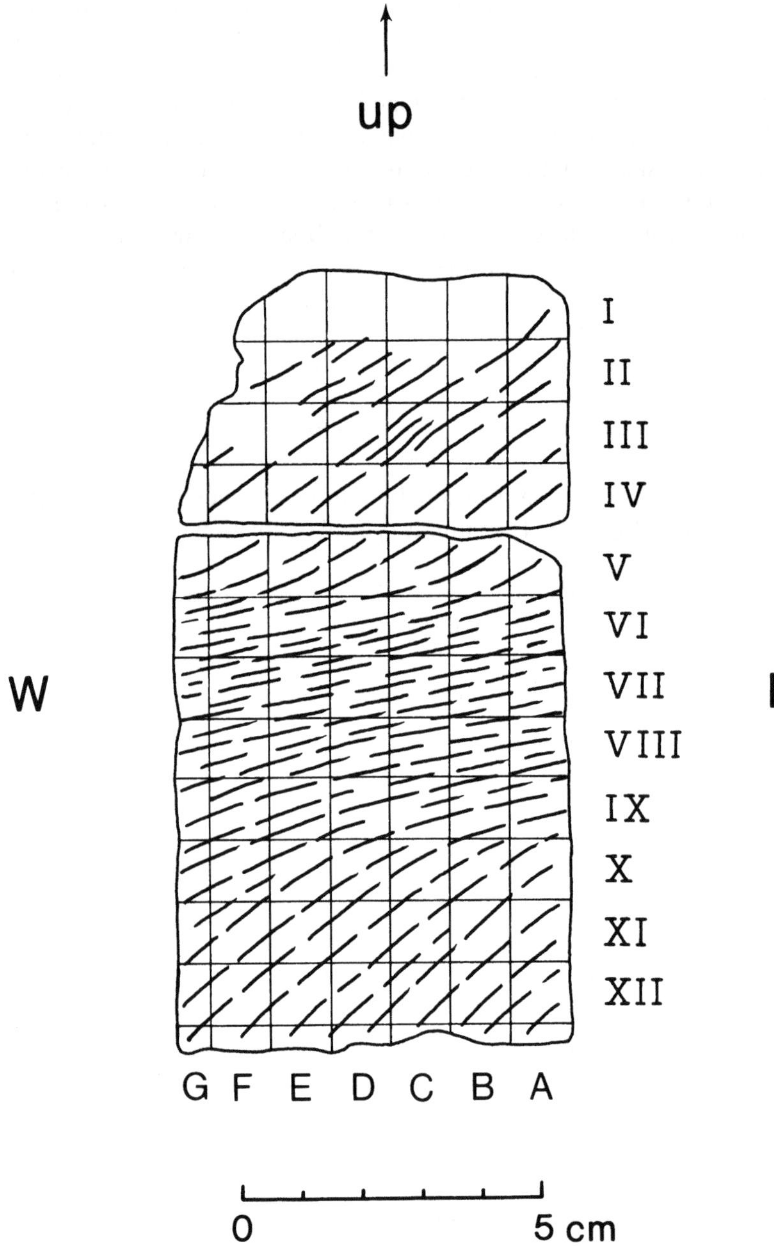

Fig. 4. Accurate sketch of thin section of shear zone showing sampling scheme and trace of bubbles

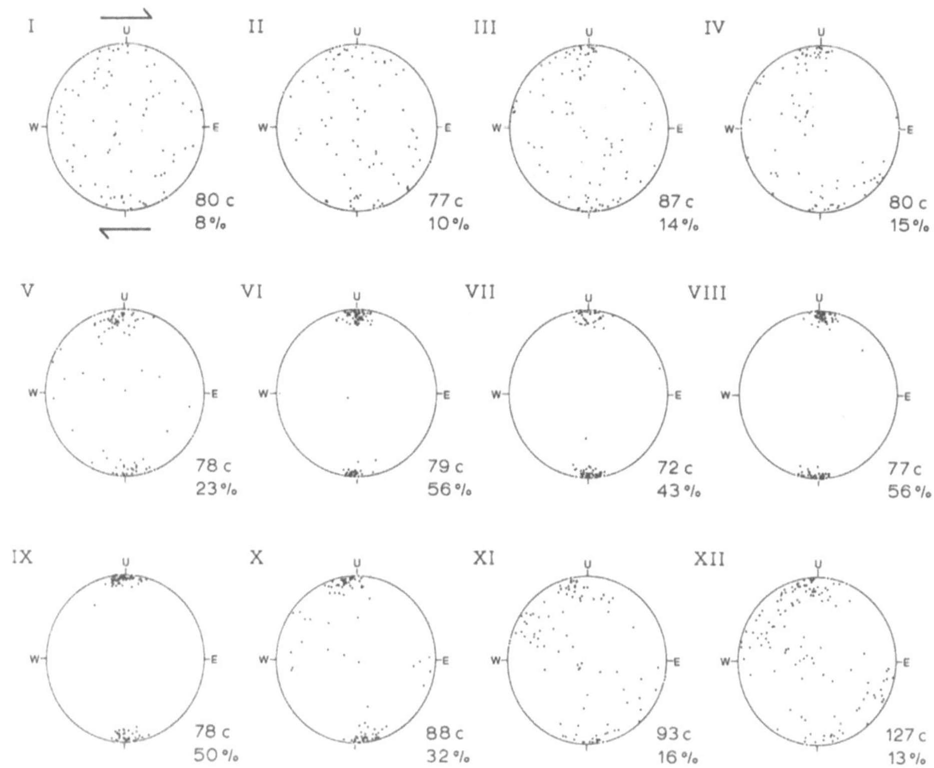

Fig. 5. *c*-axis fabric data for shear zone. *Roman numerals:* rows of Figure 4. Number of crystals measured for each diagram indicated by number followed by *c*. Maximum percentage of points per 1% of area is also given. All equal-area projections (plane of projection is $x'y'$ plane of Figure 3B, x' is *E-W*). *U* is "up," the vertical line

A feature of considerable interest is that away from the center of the shear zone the fabric may be bimodal. In addition to the point concentration normal to the shear zone (at *U* in Fig. 5), a second weaker concentration may be present (between *U* and *W* in Fig. 5) at an angle of about 50°–60° to the main point maximum. This is best displayed in diagrams XI and XII of Figure 5. Also, in the marginal parts of the shear zone, the main point maximum is slightly displaced from *U* toward *W* (diagrams III, IV, X, XI, XII, Fig. 5), by as much as 10°. In any case, there is a distinct absence of points between *U* and *E* in most diagrams.

In the fabric diagrams with the most pronounced preferred orientations (VI, VII, VIII and IX, Fig. 5), there may be one or two points at high angles to the main concentration. These aberrant points are not due to measurement errors, but are valid data from prominent grains which are indeed markedly disoriented with respect to all other grains in the center of the shear zone.

2.3 Textural Changes

There are easily discernible differences in grain size and shape across the shear zone (Fig. 6). Outside the central part of the zone the grains are fairly equant, the grain boundaries straight, and few grains are greater than 3 mm in diameter. In the center of the zone the average grain size is larger—with individual grains up to about 1 cm in size—and there is a weak shape fabric with the longest dimensions of the grains tending to be parallel to the long axes of the bubbles (Fig. 6). Also, the grain bounda-

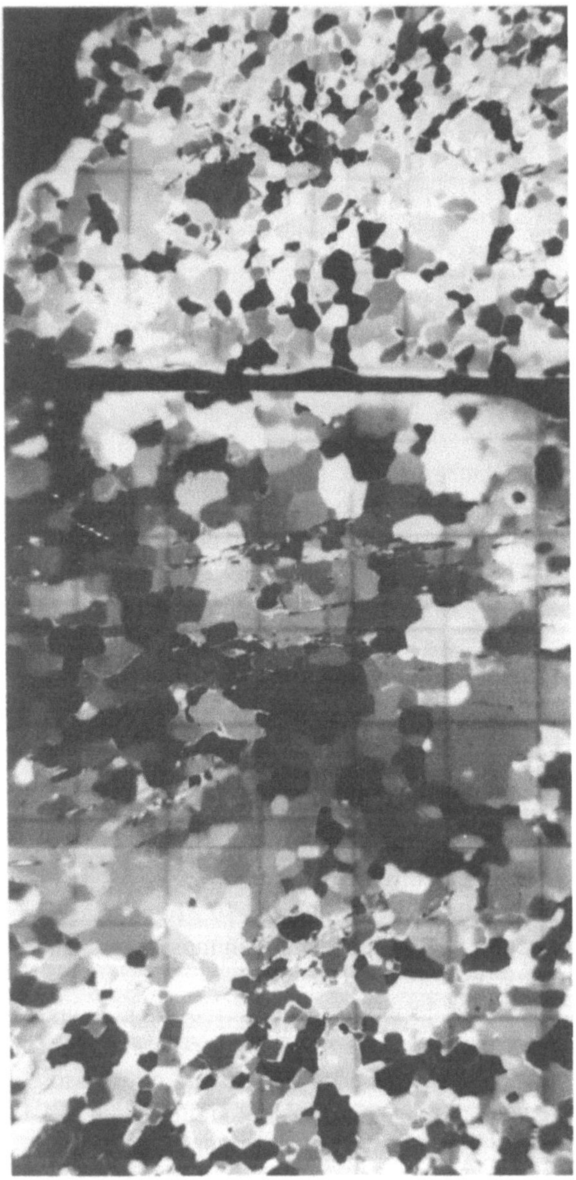

Fig. 6. Thin section of shear zone, photographed through crossed polarisers. The faint grid lines are 1 cm apart. A few elongate bubbles are visible near the center of the section

ries are somewhat sutured and irregular. The similarity in tone of grains at the center of Figure 6 reflects the very strong preferred crystallographic orientation.

3 PROGRESSIVE DEFORMATION IN THE SHEAR ZONES

The changes in crystallographic fabric, texture, and bubble appearance are clearly due to the fact that the ice has been progressively more deformed towards the center of the shear zone. Ice is thus preserved at different total strains. If the deformation is one of differential simple shear, then, at least in terms of strain, the most highly deformed ice has passed through those states of deformation now preserved in the less deformed ice, and all the ice has been deformed along a single deformation path (Ramsay, 1967). If this is also true for the changes in texture and the c-axis fabric, we should be able to relate these changes to the amount of shear undergone. The data from the shear zones would thus be equivalent to those obtained from a series of experiments run to different total shear strains. I will assume that this is the case in interpreting the data presented here, and will attempt to show that there is a quantitative relationship between the degree of preferred orientation of c axes and the amount of shear strain undergone by the ice.

In order to do this, some means of estimating shear strain and quantifying the fabric intensity are required. It is plausible to suppose that, in some way, the shape and orientation of the bubbles are related to the amount of strain undergone by the ice. If these features of the bubbles do not record the total strain in the ice, they at least record some cumulative part of the strain history and do not simply respond to the infinitesimal state of strain at any instant. Bubble orientation is the easiest parameter of the bubble fabric to measure—it constitutes a remarkably regular linear feature (Fig. 2) and is a readily apparent qualitative indicator of deformation: the smaller the angle between the bubbles and the shear zone boundaries, the larger the shear strain. Bubble orientation has been measured for each of the 12 rows in which fabric data were obtained (Fig. 4), and the results are given in the first column of Table 2: they indicate progressive increase in strain towards the center of the zone.

3.1 Estimating Fabric Intensity

The degree of preferred orientation of c axes can be quantitatively assessed in a number of ways. The simplest is to record the maximum concentration of points per unit area of the stereonet (Fig. 5). A second approach would be to utilize the methods outlined by Kamb (1959a), in which some idea of the statistical significance of the concentration would be obtained. The method used here employs the probability distribution of Bingham (1974), that is appropriately applied to undirected axial data of polar, girdle or composite types of preferred orientation. The distribution has orthorhombic symmetry and may be represented by a triaxial ellipsoid. If this type of distribution is applied to the shear zone fabric data, estimates may be made of the principal axes and shape parameters of the distribution (Fig. 7 and Table 1). These statistics provide a quantitative means of describing the shape and orientation of the fabric pattern and of testing for isotropic, polar, or girdle types of distribution (Bingham, 1974).

Applying Bingham's distribution to the data, we conclude that all the fabric diagrams in Figure 5 are significantly nonisotropic, with a high degree of axial symmetry of polar type. The best estimates of the

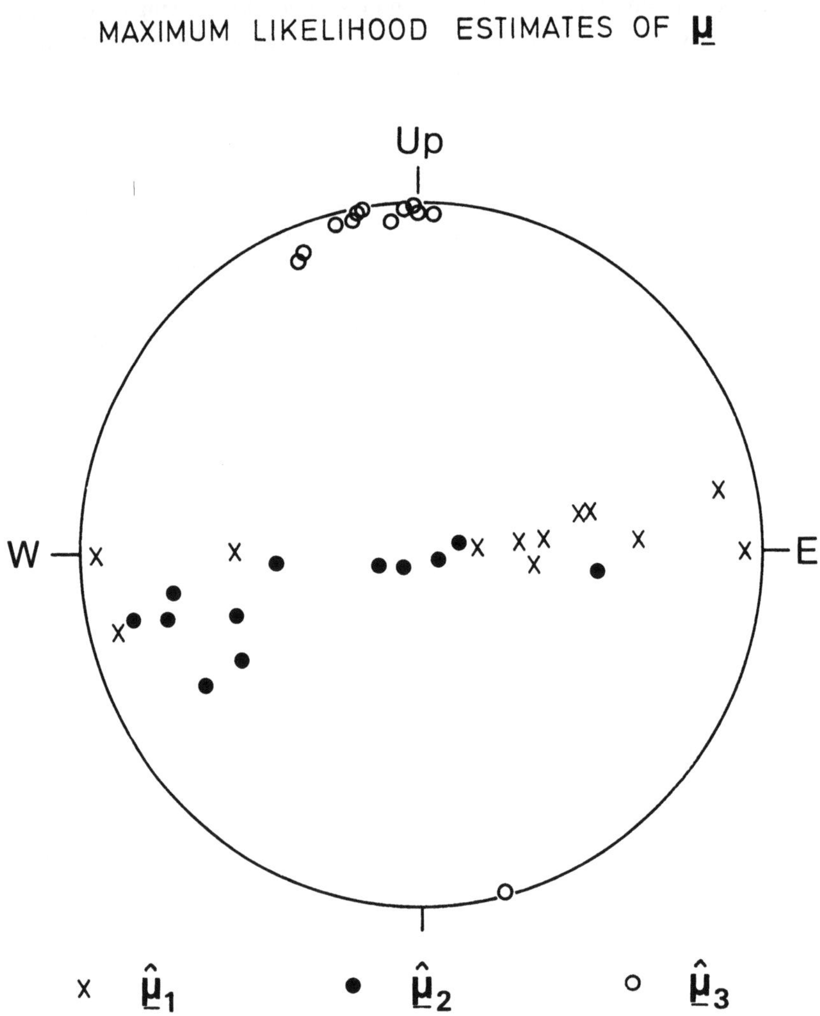

Fig. 7. Maximum likelihood estimates of the orientation parameters of Bingham's distribution for the diagrams of Figure 5

Table 1. Statistics of the Bingham distribution computed for shear zone data, ζ_1, ζ_2, ζ_3, are the maximum likelihood estimates of the shape parameters (making no assumptions about the principal directions of the distribution). $\hat{\zeta}_1 \leq \hat{\zeta}_2 \leq \hat{\zeta}_3$. ζ_3 is arbitrarily set equal to 0. ω_1, ω_2, ω_3, are the eigenvalues of the matrix XX, whose eigenvectors are $\hat{\mu}_1$, $\hat{\mu}_2$, $\hat{\mu}_3$, $X = [x_1, \ldots x_m]$, where x is the unit vector representing each point on the unit sphere. m is the sample size. The values of $\hat{\mu}_1$, $\hat{\mu}_2$, $\hat{\mu}_3$, are not given here, but are plotted in Figure 7. See Bingham (1974) for further details of the distribution

Fabric diagram (Fig. 5)	$\hat{\zeta}_1$	$\hat{\zeta}_2$	$\hat{\zeta}_3$	ω_1/m	ω_2/m	ω_3/m
I	−1.78	−1.21	0	0.235	0.287	0.478
II	−2.46	−1.75	0	0.203	0.258	0.539
III	−3.01	−2.37	0	0.182	0.222	0.596
IV	−4.35	−2.50	0	0.131	0.223	0.646
V	−6.34	−4.94	0	0.090	0.120	0.790
VI	−45.2	−16.9	0	0.011	0.031	0.958
VII	−21.4	−18.3	0	0.024	0.028	0.948
VIII	−39.5	−26.0	0	0.013	0.020	0.968
IX	−37.2	−24.3	0	0.014	0.021	0.965
X	−6.86	−4.90	0	0.082	0.121	0.797
XI	−3.51	−2.62	0	0.161	0.211	0.628
XII	−3.10	−2.43	0	0.168	0.221	0.611

principal axes of the distributions are shown in Figure 7. The axes of maximum density of points $\hat{\mu}_3$ are closely clustered near the normal to the shear zone boundaries. Those $\hat{\mu}_3$ axes that lie furthest from this position (displaced from U towards W in Fig. 7) are associated with the fabric diagrams at the margins of the shear zones (diagrams I to IV, X to XII, Fig. 5). As would be expected where departures from polar circularity are slight, the best estimates of the other two axes of the distribution are not well grouped within the plane normal to $\hat{\mu}_3$. However, the axes of minimum concentration, $\hat{\mu}_1$, do tend to be located between the center of the diagram and E in Figure 7. The fabric diagrams at the margins of the shear zone are responsible for the greatest departures of $\hat{\mu}_1$ and $\hat{\mu}_2$ from the line joining E and W in Figure 7.

In examining the shape of the distribution, a measure of the maximum density of points is the most useful quantity to work with. One appropriate measure is given by the concentration statistic $(\omega_3/m)^{1/2}$ (Bingham, pers. comm.). ω_3 and m are defined in the caption to Table 1. It is of particular interest to examine the relationship between the value of $(\omega_3/m)^{1/2}$ and the bubble orientation, given as plunge with respect to the shear zone boundaries. There is a marked inverse relationship between these two quantities (Figs. 8 and 9). If the bubble plunge is indeed somehow related to cumulative strain, the degree of preferred orientation of c axes is also dependent on cumulative strain in the ice.

In applying the Bingham distribution to the data, I have ignored the complications that arise if the fabric is bimodal, as it appears to be in some cases (e.g., row XI, XII, Fig. 5). Bimodality cannot be reflected in the distribution, which is "blind" to this type of fabric. The statistics, however, are meaningful in describing a best fit distribution that has properties of a triaxial ellipsoid, and in the present case, departures from this are not very great.

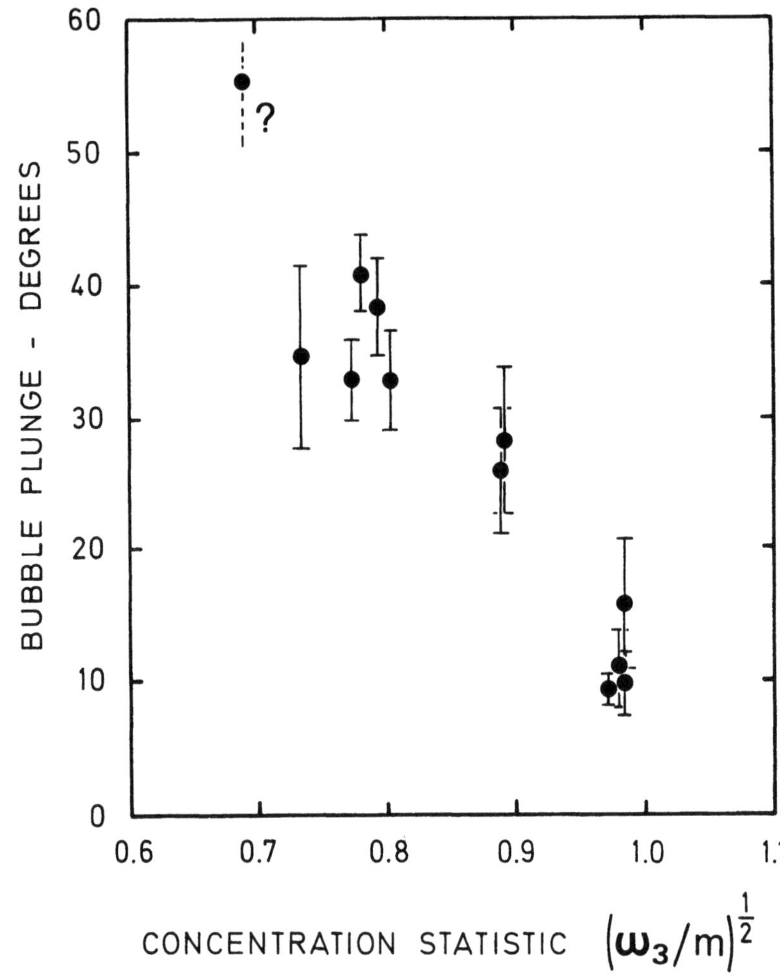

Fig. 8. Bubble plunge plotted against fabric intensity for shear zone data. Concentration statistic $(\omega_3/m)^{1/2}$ is a measure of maximum density of points. Its highest value for perfect point concentration will be 1.0; for no preferred orientation its value will be 0.58. *Error bars:* one standard deviation from mean bubble plunge

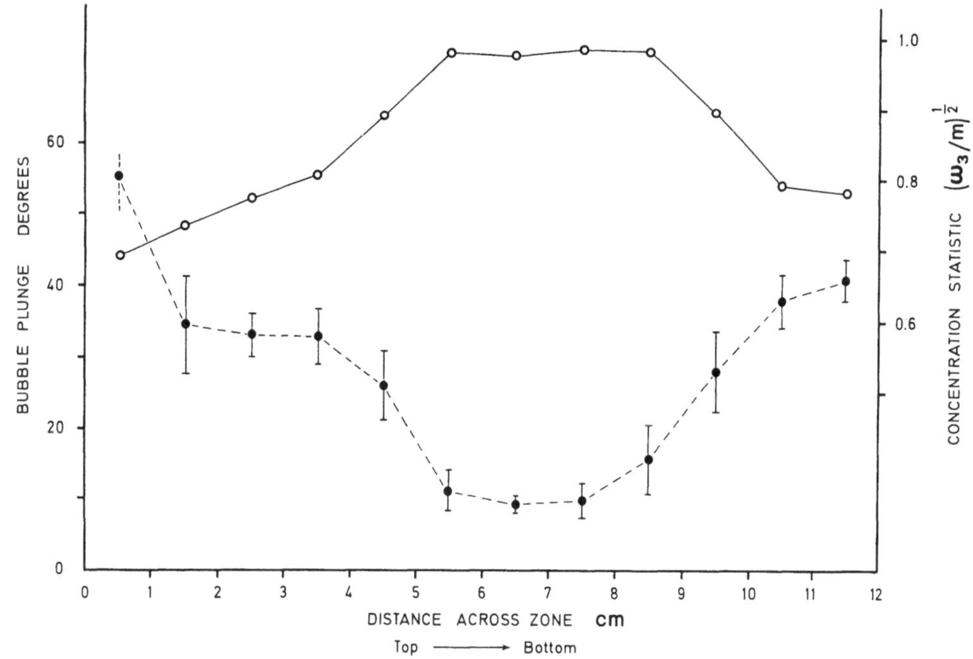

Fig. 9. Data of Figure 8 replotted to show changes at 1 cm intervals across shear zone. *Black dots:* bubble plunge; *open circles:* concentration statistic

3.2 Bubble Behavior

If the behavior of air bubbles in deforming ice were well understood, it should in principle be possible to compute the amount of deformation undergone by the ice by interpreting the present bubble shape and orientation. However, this is not a straightforward task.

If the bubbles are considered to be initially spherical inclusions of zero viscosity in a matrix of higher (Newtonian) viscosity, their changes in shape and orientation under simple shear can be approximately predicted using the theoretical and experimental results of Gay (1968); bearing in mind the fact that only two dimensional (cylindrical) inclusions were considered by him. This work shows that a bubble should become elliptical (in two dimensions), with an axial ratio that depends on the shear strain, and which is greater than the axial ratio of the strain ellipse in the host, well away from the inclusion, at any given amount of strain. Also, the long axis of the bubble makes a greater angle with the shear plane than does the long axis of the strain ellipse at any given strain (Fig. 10). Bubble behavior of this kind may be complicated by several factors. First, ice is anisotropic and probably behaves as a non-Newtonian body, although at the very low stresses existing in the margin of the ice cap, it is probably not markedly non-Newtonian. Greater problems arise when we consider the actual bubble geometry (Figs. 11, 12, 13).

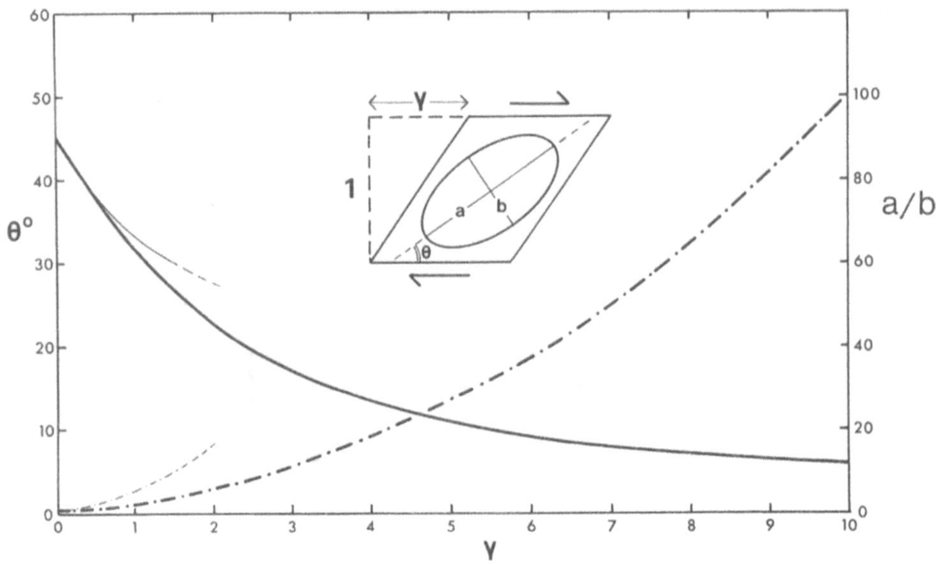

Fig. 10. Finite strain in simple shear deformation. Changes in axial ratio *a/b* of strain ellipse shown by *heavy dot-dashed line,* and orientation of ellipse by *heavy solid line,* for increasing shear strain γ. The equivalent changes for an initially circular inclusion of zero viscosity in a viscous matrix are shown by *light lines* (after Gay, 1968)

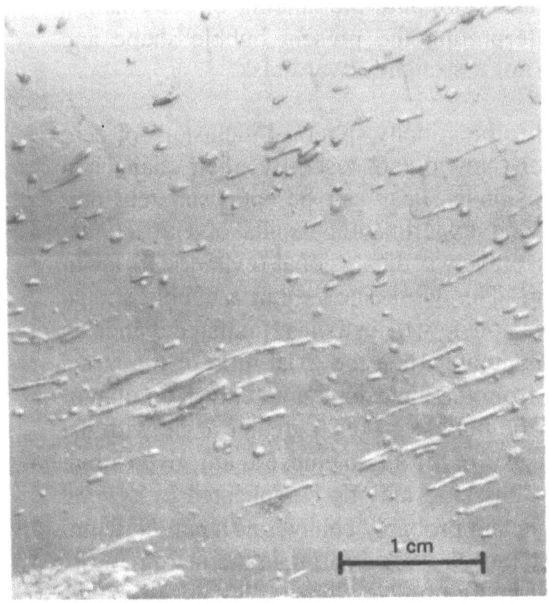

Fig. 11. Details of bubble appearance in a shear zone as photographed in a slab about 0.5 cm thick, cut parallel to symmetry plane of structure ($x'y'$ of Fig. 3). Scale bar = 1 cm

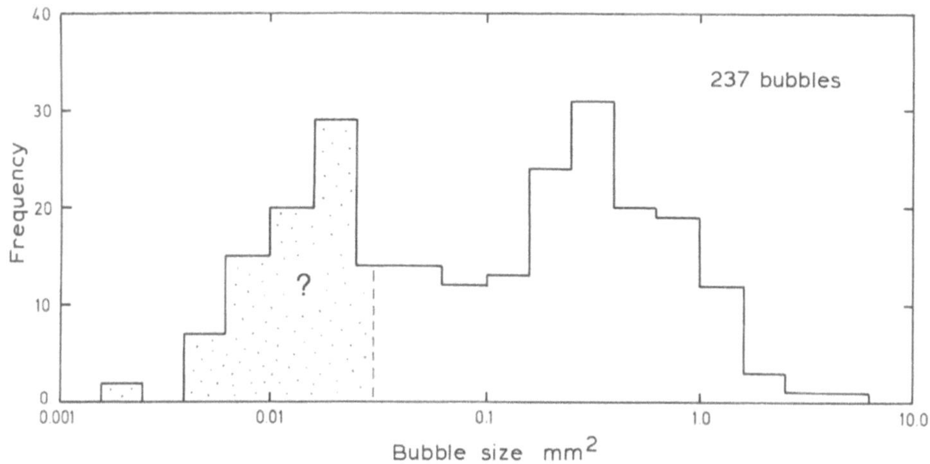

Fig. 12. Frequency histogram of bubble size for bubbles in Figure 11. "Size" is taken as area of bubble measured on the photograph. A circular, elliptical or rectangular outline is assumed and the most complex bubble shapes are ignored. Data are incomplete for bubble sizes below about 0.03 mm², because bubbles become inconspicuous and difficult to measure. Many small bubbles have been missed and their number should therefore be greater than is shown

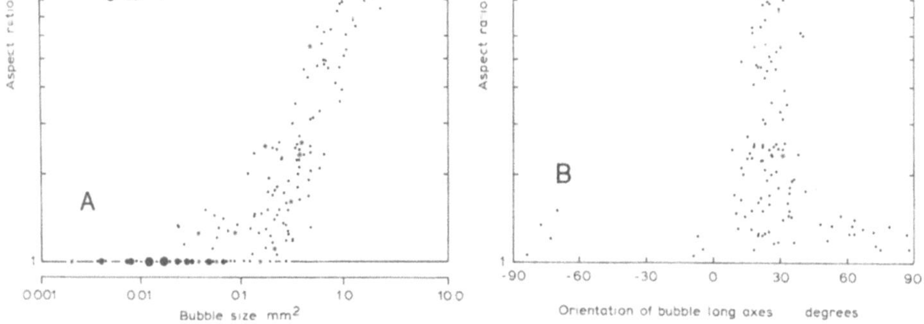

Fig. 13A, B. (A) Aspect ratio of bubbles shown in Figure 11 plotted against size. (B) Aspect ratio of bubbles plotted against orientation of their long axes with respect to some arbitrary direction

The bubble size appears to be bimodally distributed (Fig. 12), and the shape appears to be related to size (Fig. 13A). The smallest bubbles are not ellipsoidal but are almost spherical in shape. This is probably due to the effects of surface tension and diffusion which will tend to restore elongate bubbles to a spherical shape, and thus reduce surface energy. There may be a gradient of water vapor pressure within the bubble that is a functon of the curvature of the interface. In nonspherical bubbles, this will result in diffusion of water vapor from regions of low curvature to those of higher curvature. This will eventually tend to produce spherical bubbles by mass transfer through the bubbles. Comparing bubbles of different size but of the same aspect ratio, the gradient for diffusion will be higher in the smaller ones. Because of this and because the path for diffusion transfer is shorter in the small bubbles, the effects will be strongest for the smallest bubbles. These considerations would suggest that there should be some systematic relationship between bubble size and shape of the kind that is in fact seen (Fig. 13A). There appears to be some critical size—about 0.25-mm diameter in this example—below which almost all bubbles are spherical. This size may well depend on the strain rate.

If the bubbles are being deformed under simple shear to an elongate shape that depends on size, one might expect there to be a relationship between the orientation of the long axis of the bubble and bubble size: the smallest and least eccentric bubbles being stretched at 45° to the shear zone, in the direction of maximum longitudinal strain rate, and the larger bubbles making progressively smaller angles with the shear zone, as they show the effects of finite rotation. If it exists, this effect is not obvious: although there is a fairly broad peak in Figure 13B, bubble orientation seems to be almost independent of shape. This may in part be due to another complication: most of the larger elongate bubbles are not ellipsoidal either; they tend to be more cylindrical in shape with rounded ends, or even dumbbell or peanut shaped, as if the bubbles were being pulled apart (Fig. 11). It seems probable that the rounding of the ends of bubbles is again the result of processes operating to decrease surface energy. Thus the bubbles will not behave like elliptical inclusions and, in fact, once stretched to a cylindrical or dumbbell shape, they will behave more like line elements embedded in the ice. Under simple shear, line elements rotate more rapidly than the axes of the strain ellipse, and their rotation as a function of shear strain γ is easily computed (Ramsay, 1967).

There is another possible way in which the bubbles could have attained their present form. In the undeformed superimposed ice the bubbles are subspherical or irregular in shape and arranged in vertical and partly interconnecting trains. The trains are formed during freezing of the ice. It is possible that these trains impart an initial linear bubble fabric to the ice, the fabric becoming deformed as line elements and rotated from the vertical to form the pattern now seen both inside and outside the shear zones. This hypothesis could help explain why bubbles of equal size have different aspect ratios: there is no reason why the volumes of the initial bubble trains should be proportional to their lengths. It would also help explain why bubbles of different aspect ratios have similar orientations: they would have been similar initially and behaved similarly during flow. However, these features are not inconsistent with other hypotheses, and the vertical trains of bubbles in the undeformed superimposed ice do not define nearly as obvious a linear fabric as the bubbles in the deformed superimposed ice: some means of "improving" the fabric would be required if the bubbles behaved according to the hypothesis of a vertical primary fabric.

From the above discussion it appears that there is no simple way of interpreting the present bubble shapes. Irregularities in the initial bubble shape and inhomogeneities in the strain and strain rates are doubtless responsible for much of the scatter in Figures 12 and 13. Despite this, and despite the theoretical considerations which suggest that there should be a range in bubble orientation associated with the variations in size and shape, the bubble elongation direction is remarkably consistent (Figs. 2, 11, 13B), and changes in a very smooth manner. The most reasonable conclusion is that where the average bubble plunge is constant (as in Fig. 11), this is the result of overall homogeneity of strain. This assumption is made below.

3.3 Estimates of Finite Shear Strain

Finite strain across the selected shear zone has been estimated in three ways (Table 2), all of which assume differential simple shear, but differ in the interpretation of bubble orientation. Bubble shape is not used directly because it is more variable and more difficult both to measure and interpret. The first method assumes that the bubbles are initially spherical, and deform as inclusions of zero viscosity in a linearly viscous matrix until the bubbles attain axial ratios of 2:1, which occurs at $\gamma \approx 0.5$ (Fig. 10). The bubbles are then assumed to become cylindrical, as a result of surface tension and diffusion, and they deform further as line elements (curve *a*, Fig. 14). This scheme is contrived to take the observations on bubble shape (Fig. 11) into account. The second method assumes that the bubbles are initially spherical and become elongated in the direction of maximum finite extension (curve *b*, Fig. 14), an assumption which in general will only be met if the inclusion has the same viscosity as the matrix (Gay, 1968). This method is included primarily to compare the results with those of the first method. The third method assumes a primary linear fabric of vertical bubble trains, which behave as line elements during deformation (curve *c*, Fig. 14).

There is relatively little difference among the three methods in the estimates of γ at the center of the shear zone: they all indicate values for γ of between 5 and 8 (Table 2). Given the measured range in bubble orientations, even if we knew the precise mode of bubble response to simple shear deformation, we would not be able to put closer limits on the estimates of γ. The greatest difference among the methods is in the estimates of γ at the margins of the shear zone. The first method may give the most realistic estimates of γ, because this takes the observed bubble geometry into account, and gives results that are relatively insensitive to the precise stage at which the bubble begins to behave as a line element.

4 DISCUSSION

In the series of experiments run by Kamb (1972), homogeneous shear occurred in initially homogeneous ice, whereas in the glacier zones of differential shear developed. It seems probable that at least some of the shear zones in the Barnes Ice Cap were initiated in ice weakened by a high bubble concentration; many of the zones of most intense shear are in very bubbly ice. Inhomogeneous shear developed in one of Kamb's experiments (Kamb, 1972; Fig. 4), and it is possible that this represented the incipient formation of a shear zone in initially homogeneous (?) ice.

4.1 Comparison with Experimentally-Produced Fabric Patterns

On the basis of his experimental results, Kamb (1972) concluded that the degree of preferred orientation of *c* axes in ice deformed by simple shear

Table 2. Estimates of shear strain γ across the selected shear zone, according to three separate methods *A*, *B*, and *C*, explained in detail in the text. The plus-or-minus figure in the bubble plunge column gives the standard deviation of measured values. The values of γ in parentheses correspond to bubble plunges of one standard deviation from the mean

		γ		
Row	Bubble plunge (degrees)	Bubbles initially spherical, then cylindrical	Bubbles stretched in λ_1 direction	Bubbles form primary vertical fabric
I	56 ?			
II	37 ± 6	0.5 (0.1–0.9)	0.6 (0.1–1.0)	1.3 (1.1–1.7)
III	32 ± 4	0.8 (0.6–1.1)	1.0 (0.7–1.4)	1.6 (1.4–1.9)
IV	33 ± 4	0.8 (0.5–1.0)	0.9 (0.6–1.3)	1.5 (1.3–1.8)
V	27 ± 5	1.2 (0.6–1.7)	1.5 (1.0–2.0)	2.0 (1.6–2.5)
VI	11 ± 3	4.4 (3.2–6.3)	5.0 (3.8–6.9)	5.1 (4.0–7.1)
VII	9 ± 2	5.5 (4.4–7.4)	6.1 (5.0–8.1)	6.3 (5.1–8.1)
VIII	9 ± 2	5.5 (4.4–7.4)	6.1 (5.0–8.1)	6.3 (5.1–8.1)
IX	16 ± 5	2.7 (1.8–4.4)	3.3 (2.2–5.0)	3.5 (2.6–5.1)
X	28 ± 5	1.1 (0.5–1.6)	1.4 (0.9–1.9)	1.9 (1.5–2.4)
XI	39 ± 4	0.5 (0.3–0.7)	0.4 (1.0–0.7)	1.2 (1.1–1.4)
XII	41 ± 3	0.4 (0.3–0.5)	0.3 (0.1–0.5)	1.2 (1.0–1.3)

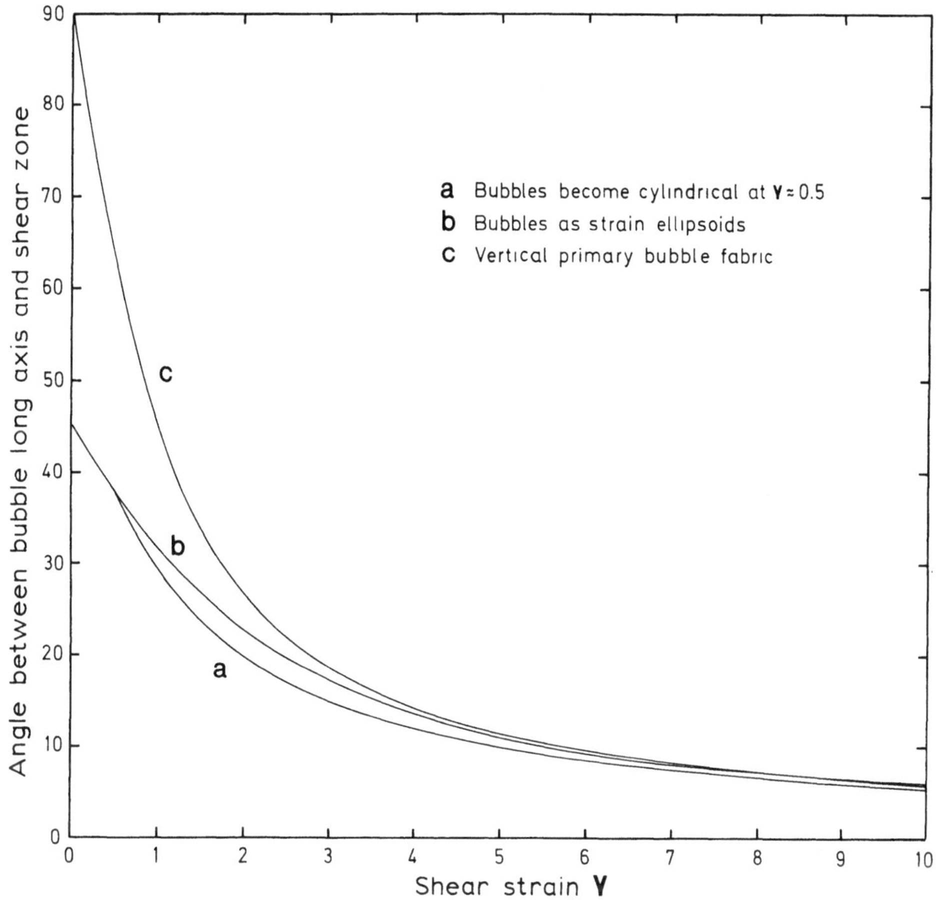

Fig. 14. Three methods of estimating total shear strain using bubble orientation. See text

depends largely on the total strain, and is relatively insensitive to variations in shear stress or time. The results of the present study suggest that this conclusion may also be valid for naturally deformed ice at much higher total strains, and at much lower strain rates ($\dot{\gamma} \approx 10^{-10}$/s on average near the base of the shaft in the Barnes Ice Cap) than those involved in Kamb's experiments ($\dot{\gamma} \approx 10^{-6}$–$10^{-8}$/s).

The maximum shear strain achieved in Kamb's (1972) experiments was $\gamma = 0.4$. At the edges of the Barnes shear zone, the preferred estimates of γ are comparable to this, and the various estimates in the center of the shear zone all indicate γ to be about an order of magnitude greater than $\gamma = 0.4$, bearing in mind the uncertainty involved in the estimating procedure. Kamb (1972) noted the development of two maximum concentrations of c axes in his experiments; one M_1 normal to the shear plane and the other, M_2, somewhat weaker in intensity, at about 70° to the first, in the plane of symmetry of the shear (Fig. 15A). Two similar maxima are found in the fabric diagrams for the margins of our shear zone (notably diagrams XI and XII, Fig. 5), but the angle between them is less than 70°, and M_1 is deflected from the normal to the shear zone towards M_2 by perhaps 10° (Fig. 15B). The present study suggests that with increasing strain (towards the center of the shear zone) M_2 diminishes in intensity and finally disappears, and M_1 becomes normal to the shear zone and intense.

It is of further interest to note the similarity between the maximum concentration of points per unit area in Kamb's fabric diagrams, at the highest strains developed in his experiments, and the maximum concentrations in the fabric diagrams for the margins of the shear zone (Kamb 1972; Fig. 14; Fig. 5, rows II–IV, XI, XII). The values of the associated shear strains are also similar in these cases. One difference between Kamb's experimentaly derived fabric diagrams and the equivalent of the shear zone is that in the latter there is a spread of points between M_2 and R (Figs. 5, 15B), the axis of rotation for the shear, whereas in the former the spread does not reach R (Fig. 15A).

The statistics of the Bingham distribution (Table 1; Fig. 7) indicate that none of the fabric diagrams in Figure 5 depart greatly from axial symmetry about a pole that is normal to the shear zone, or within a few degrees of being so. The vectors, $\hat{\mu}_1$, $\hat{\mu}_2$, $\hat{\mu}_3$, in effect, define the intersections of the planes of symmetry of a "best fit" orthorhombic fabric. Considering

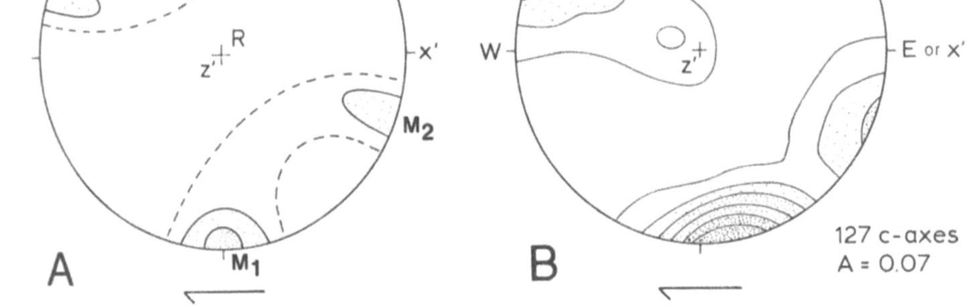

Fig. 15A, B. (A) Synthesis of Kamb's (1972) experimental results. M_1 and M_2 are maxima observed in the experiments. x', y', z' are symmetry axes used to describe kinematics of simple shear. (B) Data of diagram XII, Figure 5, contoured following the method described by Kamb (1959a). Area A of counter is 0.07 of net area; contours at 2σ, 4σ, 6σ, . . . , 14σ

the set of orthogonal axes, x', y', z', used to describe the kinematics of simple shear (Figs. 3, 15), $\hat{\mu}_3$ coincides with y'. Symmetry considerations (Paterson and Weiss, 1961) suggest that $x'y'$ should be a plane of symmetry in the fabric, and thus that one of $\hat{\mu}$ should be parallel to z'. However, because the fabric is of nearly axial type, this is not generally so (Fig. 7); $\hat{\mu}_1$ and $\hat{\mu}_2$ are, therefore, ill-defined and of limited significance. The bimodality of some of the fabric diagrams also lessons the significance of $\hat{\mu}$, particularly $\hat{\mu}_1$ and $\hat{\mu}_2$. It can be argued that with allowance for some statistically acceptable scatter in the distribution of points, the symmetry of the bimodal fabrics is monoclinic, with the plane of symmetry being the plane perpendicular to the shear zone and containing the direction of shear. Kamb (1972) discusses the significance of such symmetry, and the reasons for the lack of compatibility of fabric symmetry with the symmetry and orientation of the states of stress and total strain.

In general terms, the symmetry of the c-axis fabric is monoclinic at low strains, becoming axial and polar at strains of $\gamma \approx 1$–3 or greater.

4.2 Deformation Processes

It is clear that deformation by simple shear favors the development of grains suitably oriented for basal glide in the plane of the shear zone, at the expense of other grains. This accounts for the main concentration of points M_1. The reason for the slight departure of M_1 from y' (Fig. 15B) in the fabric diagrams for the margins of the shear zone is unknown. The secondary maximum M_2, as noted by Kamb (1972), is within 20° of the correct position for grains favorably oriented for slip on the conjugate plane of maximum shear stress and strain rate: because material lines in this position rotate, the grains will be physically rotated in the sense to produce the observed departure of M_2 from its expected position perpendicular to M_1. Kamb (1972) showed, however, that in his experiments, the angular departure of M_2 from x' (Fig. 15) could not in general be caused by physical rotation of the grains, computed from the known total shear, assuming they were initially oriented with their c axes parallel to x'.

That recrystallization and grain boundary migration accompany deformation is indicated by the weakly sutured texture in the shear zone, and by the fact that the grains are more or less equant, or display a weak shape fabric. If intracrystalline glide alone were responsible for deformation, the grains would be highly elongate in the center of the shear zone (with aspect ratios of at least 20:1). Kamb (1972) showed that neither time nor total strain had much influence on the texture and grain size; these stabilized rapidly in his experiments. The recrystallized grains in the shear zones of the Barnes Ice Cap are larger in size, often markedly so, than the grains outside the main zones of shear. This is consistent with the hypothesis that dislocation glide is the main process by which deformation is accommodated, as the larger grain size should enhance the creep rate. If diffusional flow were dominant, the effect on creep rate would be the opposite.

The effect of the bubbles on the deformation is unclear. Presumably the presence of bubbles weakens the ice, and it has already been suggested that the shear zones may have been initiated in more bubbly ice. Kamb (1972) showed, for bubbles of roughly the same size as those in the shear zones studied here, that the bubbles had little effect on recrystallization and grain boundary migration. If they had, the bubbles would have been preferentially located at grain boundaries. Although some of the larger bubbles in the shear zones are located on grain boundaries, there is no marked tendency for this to be so. The regularity of bubble orientation (Fig. 11) suggests a homogeneity of strain on the scale of the grain size (within and between grains) or smoothness of gradient of strain where the bubble plunge changes across a shear zone. This is somewhat surprising, because it seems likely that in a polycrystalline aggregate, while some crystals deform by glide on the basal plane, others, less favorably oriented, will glide less, and some will be merely rotated bodily. Thus, in neighboring grains, one might expect to find variously deformed bubbles stretched in different directions, with grain boundary slip and perhaps some grain boundary migration and recrystallization helping to accommodate the disparate strain within the grains. This is the kind of bubble deformation observed by Nicolas et al. (1973) in experimentally deformed peridotite. They were able to show that the bubbles effectively behaved as passive markers during glide on several slip systems in olivine. Such is unlikely to be the case in ice, because bubble recovery occurs, and because recrystallization and grain boundary migration ensure that bubbles may have several different host crystals during the course of deformation.

To produce homogeneous strain in a polycrystalline aggregate by dislocation glide alone, requires at least five independent glide systems to be operative (Paterson, 1969). At high temperatures, and where cross-slip and grain boundary migration occur, it may be possible to produce homogeneous strain with fewer active slip systems (Hobbs et al., 1976). Other slip systems certainly exist in ice (Fletcher, 1970), and these may be sufficient to allow deformation to be accommodated by glide alone. Basal glide is clearly very important, and it seems that during high temperature creep of the kind that occurs in the shear zones, the processes of glide on the basal plane (and perhaps other glide systems), recrystallization, and grain boundary migration together produce a fairly homogeneous strain, on the scale of the grain size, integrated over the total history of deformation.

The dependence of degree of preferred crystallographic orientation on total strain is predicted by several theoretical models. The mechanical rotation of inequant rigid grains suspended in a matrix undergoing deformation, leads to the progressive development of a fabric, when shape of the grains is crystallographically controlled. Such models have been advanced to explain the origin of slaty cleavage in rocks (Tullis and Wood, 1975). Dislocation glide, on sufficient slip systems, also leads to the progressive development of fabric, whose nature depends on which slip systems are active (Hobbs et al., 1976).

If syntectonic recrystallization becomes important, the situation is less clear. The fabric pattern may differ, in intensity and nature, from that which develops when no recrystallization occurs, or when recrystallization follows deformation (Tullis et al., 1973; Hobbs et al., 1976). There are theoretical predictions that preferred crystallographic orientations will develop when grains recrystallize in a nonhydrostatic stress field (Kamb, 1959b, 1961; Paterson, 1973). In the case of ice, it would appear that failure of such theories to take the effects of plastic glide into account, may be responsible for the general discrepency between theory and observation (Kamb, 1972).

In the present study of fabric in shear zones, preferred orientations are seen to have developed as a function of the total strain, when basal glide, recrystallization, and grain boundary migration were all inferred to have been active.

5 CONCLUSIONS

The estimates of maximum shear strain, γ, in the shear zones developed in the margin of the Barnes Ice Cap, are about an order of magnitude greater than the maximum value of γ attained in Kamb's (1972) experiments. Nature has thus provided us with a way of extending the range of the experimental work, albeit under far less-controlled conditions. The results of this study support and extend Kamb's (1972) conclusion that the degree of preferred orientation of ice c axes is largely a function of the total strain. This conclusion appears to be valid for longer times, larger strains, and smaller rates than those attained in his experiments. The results also support the conclusion that steady-state flow had not been achieved in the experiments: Kamb (1972) suggested that a second mechanism, not operating in his experiments, might occur after higher strains and long term recrystallization. This mechanism would account for the commonly observed four maxima ''diamond'' fabric seen in glacial ice. The present study throws no light on the development of this enigmatic type of fabric, but does show a continuity and compatibility of Kamb's experimental results with a second class of fabric pattern commonly observed in glaciers: the single maximum fabric, with the maximum typically perpendicular to foliation.

The evidence of this study suggests that ice deforming under simple shear, with initially uniform fabric, first develops a bimodal fabric with monoclinic symmetry, as described by Kamb (1972); and at higher strains (perhaps γ in the range of 1 to 3) the pattern becomes unimodal and of axial symmetry with the single maximum perpendicular to the shear plane. The maximum shear strains estimated for the shear zones are small compared with the shear undergone by the ice near the base of the main glacier, which could be several orders of magnitude greater again. Perhaps further changes in fabric pattern will occur at even greater strains, and the single maximum fabric observed in the shear zones does not represent the attainment of steady state. It does seem, however, on the basis of the present study, that the only change in fabric pattern likely to occur with increasing shear strain is further concentration of the single point maximum.

The strain softening observed in Kamb's (1972) experiments and predictable as a result of increasing numbers of grains becoming favorably aligned for basal glide as strain increases, may help explain why shear zones develop. In addition to strain softening, the condition that the direction of incipient shear zone not be one of significant rotation of material lines with respect to the applied stress field, is probably also important.

ACKNOWLEDGMENTS

This work was supported by National Science Foundation grants GA-19310 and GA-42728, and also by the Glaciological Division, Department of the Environment, Canada. The cooperation of G. Holdsworth of the Glaciological Division is particularly appreciated. I am very grateful to R. LeB. Hooke, for arousing my interest in the structures in the Barnes Ice Cap, for valuable discussions on the subject, and for critically reading the manuscript. C. Bingham showed great interest in the problems of statistical analysis posed by the fabric data and very kindly computed the statistics of his distribution.

REFERENCES

Bak, J., Korstgård, J., Sørenson, K.: A major shear zone within the Nagssugtoqidian of West Greenland. Tectonophysics *27,* 191–209 (1975)

Beach, A.: The measurement and significance of displacement on Laxfordian shear zones, north-west Scotland. Proc. Geol. Assoc. *85,* 13–21 (1974)

Bingham, C: An antipodally symmetric distribution on the sphere. Ann. Stat. *2,* 1201–1225 (1974)

Bishop, B. C.: Shear moraines in the Thule area, northwest Greenland. U.S. Army Snow, Ice, and Permafrost Res. Estab., Research Rep. 17, 1957, p. 47.

Budd, W. F.: The development of crystal orientation fabrics in moving ice. Z. Gletscherk. Glazialgeol. *8,* 65–105 (1972)

Coward, M. P.: Flat-lying structures within the Lewisian basement gneiss complex of NW Scotland. Proc. Geol. Assoc. *85,* 459–472 (1974)

Fletcher, N. H.: The Chemical Physics of Ice. London: Cambridge Univ. Press, 1970

Gay, N. C.: Pure shear and simple shear deformation of inhomogeneous viscous fluids. 1. Theory. Tectonophysics *5*, 211–234 (1968)

Goldthwait, R. P.: Study of ice cliff in Nunatarssuaq, Greenland, U.S. Army Snow, Ice, and Permafrost Research Estab. Tech. Rep. *39*, 108 (1960)

Hara, I., Takeda, K., Kimura, T.: Preferred lattice orientation of quartz in shear deformation. Hiroshima Univ. J. Sci., Ser. C *7*, 1–10 (1973)

Hobbs, B. E., Means, W. D., Williams, P. F.: An Outline of Structural Geology. New York-London-Sydney-Toronto: John Wiley and Sons, 1976

Hooke, R. LeB.: Flow near the margin of the Barnes Ice Cap, and the development of ice-cored moraines. Geol. Soc. Am. Bull. *84*, 3929–3948 (1973)

Hooke, R. LeB.: Pleistocene ice at the base of the Barnes Ice Cap, Baffin Island, N.W.T., Canada. J. Glaciol. *17*, 49–60 (1976)

Hudleston, P. J.: Recumbent folding at the base of the Barnes Ice Cap, Baffin Island, Northwest Territories, Canada. Geol. Soc. Am. Bull. *87*, 1684–1692 (1976)

Kamb, B.: Ice petrofabric observations from Blue Glacier, Washington, in relation to theory and experiment. J. Geophys. Res. *64*, 1891–1909 (1959a)

Kamb, B.: Theory of preferred crystal orientation developed by crystallization under stress. J. Geol. *67*, 153–170 (1959b)

Kamb, B: The thermodynamic theory of non-hydrostatically stressed solids. J. Geophys. Res. *66*, 259–271 (1961)

Kamb, B: Refraction corrections for universal stage measurements, 1, Uniaxial crystals. Am. Mineral. *47*, 227–245 (1962)

Kamb, B.: Experimental recrystallization of ice under stress. In: Flow and Fracture of Rocks. Heard, H. C., Borg, I. Y., Carter, N. L., Raleigh, C. B. (eds.). Am. Geophys. Union, Geophys. Monograph *16*, 211–241 (1972)

Langway, C. C.: Ice fabrics and the universal stage. U.S. Army Snow, Ice, and Permafrost Res. Estab., Tech. Rep. *62* 16 (1958)

McCall, J. G.: The internal structure of a cirque glacier: J. Glaciol. *2*, 122–130 (1952)

Meier, M. F.: Mode of flow of Saskatchewan Glacier, Alberta, Canada. U.S. Geol. Survey, Prof. Paper *351*, 68 (1960)

Mitra, G.: Shear zones and mylonites: a look at the mechanical processes involved in the deformation of basement. Geol. Soc. Am. (abstract). Abstr. Progr. *7*, 1202–1203 (1975)

Nicolas, A., Boudier, F., Boullier, A. M.: Mechanisms of flow in naturally and experimentally deformed peridotites. Am. J. Sci. *273*, 853–876 (1973)

Paterson, M. S.: The ductility of rocks. In: Physics of Strength and Elasticity, The Orowan 65th Anniversary Volume. Argon, A. S. (ed.). M.I.T. Press, 1969, pp. 377–392

Paterson, M. S.: Nonhydrostatic thermodynamics and its geological applications. Geophys. Space Phys., Rev. *11*, 355–389 (1973)

Paterson, M. S., Weiss, L. E.: Symmetry concepts in the structural analysis of deformed rocks. Geol. Soc. Am. Bull. *72*, 841–882 (1961)

Price, R. A.: The distinction between displacement and distortion in flow, and the origin of diachronism in tectonic overprinting in orogenic belts. 24th Intern. Geol. Cong. Report, Ser. *3*, 545–551 (1972)

Ragan, D. M.: Structures at the base of an ice fall. J. Geol. *77*, 647–667 (1969)

Ramsay, J. G.: Folding and Fracturing of Rocks. New York: McGraw-Hill, 1967, p. 568

Ramsay, J. G., Graham, R. H.: Strain variation in shear belts. Can. J. Earth Sci. *7,* 786–813 (1970)

Tullis, J., Christie, J. M., Griggs, D. T.: Microstructures and preferred orientations of experimentally deformed quartzites. Geol. Soc. Am. Bull. *84,* 297–314 (1973)

Tullis, T. E., Wood, D. S.: Correlation of finite strain from both reduction bodies and preferred orientation of mica from slate in Wales. Geol. Soc. Am. Bull. *86,* 632–638 (1975)

Weertman, J.: Mechanism for the formation of inner moraines found near the edge of cold ice caps and ice sheets. J. Glaciol. *3,* 965–978 (1961)

Wegmann, C. E.: Tectonic patterns at different levels. Geol. Soc. South Africa Annex to Vol. *66,* 78 (1963)

8

J. L. ROBERTS

Abstract This review provides a short account of the structural relationships developed by metamorphic rocks forming the internal zones of orogenic belts, with particular reference to the methods used to decipher these relationships in terms of a deformation history. It is emphasized that the problems involved in structural correlation must mean that any interpretation of these relationships can only be regarded as a model incorporating all the available evidence into a self-consistent pattern. Discussion is centered on the nature of the structural record and the chronological significance of the deformation history.

THE STRUCTURAL ANALYSIS OF METAMORPHIC ROCKS IN OROGENIC BELTS

1 INTRODUCTION

This paper considers how the structural record within orogenic belts can be established and what interpretations should be placed on the results so obtained. It is a subject which is particularly apposite to the present occasion, honoring the sixtieth birthday of Professor Hans Ramberg, since it forms the observational basis for the sort of theoretical and scale-model studies which he has undertaken with success over many years. However, a word of caution may be in order. It is the purpose of this paper to show how little is really known about the internal structure of orogenic belts, since this is generally so complex that it is difficult to determine with any certainty. Thus, there is a danger that any interpretation concerning the structural evolution of orogenic belts may be based on a view which is oversimplified because a full understanding of the structural relationships is not realized. It is clearly better to proceed by generalization rather than simplification since this allows the salient features of a complex situation to be recognized. This applies with equal force to explanations invoking the concepts of plate tectonics which have had only a superficial influence on the structural intepretation of orogenic belts.

2 THE STRATIGRAPHIC APPROACH

Structural analysis attempts to decipher the complex history of deformation and metamorphism which commonly affects the internal zones of orogenic belts. It is based on the detailed mapping of the outcrop pattern developed by these rocks in relation to the topography. Since most of the metamorphic rocks found within orogenic belts were originally sediments, deposited unconformably on an older basement of igneous and metamorphic rocks, which only appears locally, it was only natural that stratigraphic methods were the first to be applied.

The stratigraphic approach to structural analysis required that sequences of sedimentary formations could be recognized within orogenic belts in their correct stratigraphic order, thus allowing the gross structures to be determined from the present disposition of these formations. This is a method which works best in mountainous terrains like the Alps where the stratigraphy and structure are effectively displayed in cross section by the topography. The stratigraphic relationships in such areas can often be recognized by comparison with the less-deformed rocks of the foreland, where the principle of stratigraphic superposition can be applied, or sufficient palaeontological evidence may be present to allow the stratigraphic order of the various formations to be recognized. Otherwise, the mutual juxtaposition of sedimentary formations within an orogenic belt may enable a stratigraphic sequence to be determined directly from the nature of the outcrop pattern, provided that tectonic contacts can be recognized wherever they occur between these formations. This in itself provides no evidence of stratigraphic order unless some assumption is made concerning the overall structure of the orogenic belt.

It must be admitted that this approach, based essentially on arguments of a stratigraphic nature, does not provide *sufficient* evidence whereby the complex structures of an orogenic belt can be fully understood. The work of Bailey (1922, 1938) on the Dalradian rocks in the Southwest Highlands of Scotland is taken as a case in point. His views were clearly influenced by the discovery that fold nappes and major overthrusts were developed in orogenic belts such as the Alps. Accordingly, Bailey (1922) recognized three nappe complexes within the Southwest Highlands, each formed by a stratigraphic sequence resting on a tectonic slide. These slides, and the recumbent folds affecting the stratigraphic sequences within each nappe complex were regarded as primary structures. These structures were thought to be formed by tectonic movements which were directed toward the southeast throughout the whole of the Southwest Highlands. The level of these primary structures was affected by the subsequent development of secondary structures of a more upright character. These structures were generated by tectonic movements directed away from a central zone of divergence within the Southwest Highlands.

Subsequent work has shown that Bailey's (1922) synthesis was mistaken in many respects. The stratigraphic order of the Lower Dalradian has been reversed, the stratigraphic correlation of the Middle and Upper Dalradian has been changed, major slides have been abandoned, and the primary or secondary designation of particular folds has been altered. Such changes in interpretation have mainly resulted from observations of the structural relationships which can be seen on a minor scale in the field, supplemented by the study of metamorphic fabrics in thin section. Such observations taken in conjunction with the stratigraphic approach to geological mapping are central to the methodology of structural analysis.

3 THE ELEMENTS OF STRUCTURAL ANALYSIS

3.1 Stratigraphic Younging

The use of cross bedding and graded bedding to determine the stratigraphic order of folded rocks was introduced into structural analysis by Leith (1905, 1913) as a result of the experience gained in mapping the Proterozoic rocks around Lake Superior. This general method can also be applied to volcanic rocks if lava pillows are developed, as realised by Wilson (1913) from his work on the Archaen rocks of the Canadian Shield. These structures can be distinguished, by virtue of their common occurrence and effective preservation in metamorphic terrains, from the multitude of other structures which can also be used under favourable circumstances to determine the stratigraphic order of sedimentary and volcanic rocks (Shrock, 1948). All these structures can be used in the field to map major folds as anticlines or synclines, according to whether older or younger rocks occur in the fold cores.

3.2 Structural Relationships

It is a central tenet of structural analysis that each generation of fold structures is represented by folds, developed on different scales but sharing the same axial-planar cleavage or foliation. Although there are early references to this concept, it was Pumpelly (cited by van Hise, 1896) and Leith (1905, 1913) who first established the use of structural relationships in the systematic mapping of major folds. This is based on two principles. First, if the folding is cylindroidal on all scales, then the minor fold axes and the bedding-cleavage intersections define the plunge of any major folds which may be present. A corollary to this principle states that the folding is cylindroidal wherever these structures are parallel to one another throughout a particular region. Second, since folds belonging to the same generation share the same axial-planar cleavage or foliation, a constraint is placed on the vergence of minor folds and the sense of the bedding-cleavage relationships, which are developed on the limbs of the major folds. This allows the major folds to be mapped as antiforms and synforms, simply by observing the nature of these relationships in the field. It may be noted that vergence is best defined by the inclination of the axial-planar cleavage or foliation to the sheet dip of the folded layers (Roberts, 1974), rather than the horizontal (Stille, 1930).

3.3 Structural Facing

Even though the bedding of sedimentary rocks is affected by folding, the structural relationships just mentioned require that the bedding always becomes younger in the same direction, as measured within the axial-planar cleavage at right angles to the fold axes. This direction defines the facing of the fold structures (Shackleton, 1958). It is best observed directly in the field from the relationships shown by the fold structures to the bedding of sedimentary rocks, whose stratigraphic order is known from the evidence of sedimentary structures. However, it may be inferred from a general knowledge of the stratigraphy even if sedimentary structures are not present locally to define the stratigraphic order. It allows antiforms to be distinguished as upward-facing anticlines or downward-facing synclines, and synforms to be distinguished as upward-facing synclines or downward-facing anticlines.

The concept of facing is so important that it is advisable to restrict the term to structural relationships (cf. Shrock, 1948), simply referring to the stratigraphic equivalent as "younging" in the manner originally suggested by Bailey (1934). Facing brings a unity to structural relationships, since a single generation of fold structures will all face in the same direction, provided that they affect a stratigraphic sequence which was neither folded previously nor refolded subsequently. Accordingly, local variations in facing can be used to distinguish different generations of fold-structures from one another. The facing of subsequent fold structures does not have any kinematic meaning, since it depends on the younging shown by stratigraphic sequences which have already been folded. However, the overall facing of the earliest structures is kinematically important since it generally defines the direction of tectonic transport within an orogenic belt as a whole.

3.4 Structural Superposition

The realization that the metamorphic rocks of orogenic belts are typically affected by more than one phase of folding and cleavage formation was originally made by Clough (1897) from a study of the minor structures developed in the Dalradian rocks of the Cowal district in the Southwest Highlands of Scotland. Although the implications of this work were not fully recognised for nearly 60 years, the methods introduced by Clough (1897) to determine the sequential development of fold structures in response to a deformation history of several phases are now regarded as essential to structural analysis.

The earliest phase of deformation produces a slaty cleavage which affects only bedding, while the subsequent phases of deformation result in strain-slip or crenulation cleavages which affect the earlier cleavages as well as bedding. Accordingly, first folds can be recognised by the slaty nature of their axial-planar cleavages, while second and later folds can be identified by the strain-slip or crenulation cleavages parallel to their axial planes. Cleavages can also be dated in relation to one another wherever an early cleavage is transposed and overprinted by a later cleavage. The distinctive character of all these cleavages should be examined in thin section, since it is not always clear in hand specimen.

As Hobbs, Means, and Williams (1976) have emphasized, a slaty cleavage can only be formed in response to the earliest phase in the deformation history if a sedimentary fabric has not previously developed parallel to the bedding of argillaceous rocks. Otherwise, a strain-slip or crenulation cleavage would be produced, affecting the sedimentary fabric in the same way as the later cleavages affect a slaty cleavage (Voll, 1960). However, it is comparatively rare for such a fabric to be developed. This suggests that the consolidation of argillaceous sediment, which is required to produce a sedimentary fabric of this sort, has been insufficient for such a fabric to form, prior to the start of the tectonic deformation. This can be understood since the consolidation of argillaceous sediment requires the expulsion of pore-fluids, which takes place very slowly (Hubbert & Rubey, 1959; Bredehoeft & Hanshaw, 1968). Accordingly, the earliest phase in the deformation history is most likely to affect poorly consolidated sediments containing pore-fluids under anomalous pressures, producing a slaty cleavage by such a dewatering process as advocated by Maxwell (1962).

The use of cleavage fabrics to date fold structures in this way is best undertaken in areas of low-grade metamorphism. The recrystallization and grain growth associated with a rise in metamorphic grade tends to produce coarse-grained schistosities lacking most of the diagnostic features which would allow their age to be determined. Even so, the micas defining a schistosity which developed from a slaty cleavage tend to be distributed rather uniformly throughout the rock, while the micas defining a schistosity which developed from a strain-slip or crenulation cleavage tend to be arranged in discrete zones between which a decussate texture is developed. Such schistosities can be dated as relatively late if they enclose porphyroblasts which contain discordant inclusion trails of an earlier fabric.

The superposition of fold structures belonging to different phases in the deformation history also enables these structures to be dated in relation to one another. Thus, early cleavages will be folded around the hinges of later folds, while late cleavages will cut across the axial planes of earlier folds. Likewise, the development of interference patterns typical of refolding means that one fold set can be recognized as earlier or later than another fold set. All these fold structures can be grouped into coeval sets according to the manner in which particular cleavages or foliations are only developed axial-planar to folds of the same generation.

While these methods allow individual sets of fold structures to be dated relative to one another at each exposure, they do not provide sufficient evidence to erect a deformation history for an entire region. Thus, it is rarely the case that each generation of fold structures is ubiquitous in its development, affecting all the rocks of the region. The only exception to this statement concerns the earliest structures, associated with the formation of a slaty cleavage in rocks previously unaffected by the deformation. These structures can usually be traced from exposure to exposure throughout a region, unless they have been so obscured by deformation and metamorphism that they can no longer be recognized. Even so, it is a reasonable inference that these structures were originally present in all the rocks of the region under consideration, since they are the structures which develop first in response to any deformation that takes place. However, the later generations of fold structures are often restricted to particular zones within an orogenic belt, within which they may be rather sporadic in their development. This means that there are usually chronological gaps in the structural record. Accordingly, a major problem in structural analysis concerns the correlation of fold structures as they are traced from exposure to exposure throughout a particular region.

This problem of structural correlation may be eased if successive fold phases are restricted to areas which become progressively smaller during the course of the deformation history in such a way that these areas overlap one another. This would mean that the first fold structures could be traced into an area affected by the second fold phase, while the second fold structures could be traced from there into an area affected by the third fold phase, and so on. Each set of fold structures can be followed in the field from where they first appear, superposed on a set of earlier fold-structures, to where they are overprinted by a set of later fold-structures, using the axial-planar cleavages as structural markers (Park, 1969).

3.5 Fold Style and the Deformation History

The style of fold-structures can be defined according to a number of parameters (Fleuty, 1964; Ramsay, 1967). It commonly reflects the structural response of the lithological layering to the nature and intensity of the deformation, under a particular set of metamorphic conditions. It may be supposed that the deformation intensity and the metamorphic grade would gradually change during the course of the deformation history, so that successive generations of fold structures would differ in style. However, since these differences depend to some extent on the number of fold phases constituting the deformation history, they would become less obvious as the deformation history became more complex.

The style of fold structures is also controlled by the nature of the deformation which takes place. Since this is partly governed by the structural response of the lithological layering to the deformation, there are some aspects of fold style which reflect more the lithological character of the layers being folded than they indicate the fold phase to which the resulting structures belong. However, even those aspects of fold style which are not controlled by lithological factors cannot be taken as uniquely representative of a particular fold phase in the deformation history. It must be admitted that regional variations in deformation intensity and metamorphic grade could be associated with a single phase of folding and cleavage formation. The structures formed during this phase in the deformation history would then be expected to vary in fold style as they are traced throughout the region concerned. A corollary to this statement implies that structures belonging to different phases in the deformation history could share a common fold style if they were formed under similar conditions in different areas. Accordingly, as Park (1969) has emphasized, fold style is not a satisfactory criterion for structural correlation, since it is controlled by factors which are not unique to particular phases in the deformation history.

3.6 Fold Orientation and the Stretching Direction

The orientation of fold structures is another parameter which has been used for the purposes of structural correlation. However, it is now generally recognized that this was based on an oversimplified view of the structural relationships which develop in metamorphic rocks affected by intense deformation or several phases of folding.

The effect of deformation on the orientation of cylindroidal folds has been studied by Sanderson (1973). His model assumes that fold axes form initially with a Gaussian distribution about a mean fold axis. Such a distribution would be developed wherever the folding was only quasi-cylindroidal from the start of the deformation, with individual folds dying out along their trends.

If subsequent deformation results in stretching within the axial planes, at an angle to the mean fold axis, the initial distribution becomes modified by the relative elongation undergone by individual fold axes as they rotate toward parallelism with the stretching direction. This occurs wherever these fold axes are not strictly at right angles to the stretching direction. The single maximum characteristic of a Gaussian distribution spreads and then splits up to form two maxima, arranged symmetrically about the stretching direction, wherever the mean fold axis was initially at right angles to the stretching direction. Alternatively, the single maximum rotates away from its initial position, while a subsidiary maximum develops at a high angle to the first, wherever the mean fold axis was initially oblique to the stretching direction. Continued deformation then causes all these maxima to rotate toward the stretching direction, so that the individual fold axes ultimately reach a position which is nearly if not quite parallel to the stretching direction.

It should be realized that fold structures undergoing passive rotation and relative elongation in this way preserve their original components of vergence and facing, as measured within the axial planes, parallel to the stretching direction. Accordingly, these components can only be used to establish the nature of the major structures if the orientation of the individual fold axes can be determined relative to the stretching direction. This direction is generally marked by a fibrous mineral lineation which is found associated with the cleavages axial planar to the early folds (Clough, 1897; Voll, 1960). It is thought to represent the direction of maximum extension within the rock since it is parallel to the elongation shown by deformed objects and pressure shadows.

It has long been recognized that the stretching direction within an orogenic belt is very uniform in its attitude, pitching steeply within the cleavage planes at a high angle to the regional trend. Accordingly, while fold axes may form parallel to the regional trend, continued deformation will generally cause these fold axes to rotate into a position at right angles to their original attitude. Such oblique structures should be clearly distinguished from the "cross-folds" which are defined as forming with their fold axes parallel to the stretching direction (Roberts and Sanderson, 1974). It is commonly found that oblique folds of the type described by Sanderson (1973) are characteristic of the early phases in the deformation history, while cross folds are restricted to the zones of intense deformation, such as thrusts and slides, which develop in response to this deformation (Voll, 1960; Roberts, 1974). This means that fold orientation is more a function of the deformation intensity than a factor which would allow particular phases in the deformation history to be recognized.

3.7 Fold Orientation and the Deformation History

The use of fold orientation in structural correlation must also take into account the complex relationships which develop in rocks affected by more than one phase of folding (Ramsay, 1963, 1967). This complexity arises in two ways, assuming that the deformation history only consists of two phases.

Firstly, the two sets of fold structures are usually developed so that they are not coaxial with one another. Accordingly, any cylindroidal character which the earlier folds might have possessed is destroyed by the later folding. This means in general that the earlier folds change in attitude, vergence and facing as they are traced across the later folds.

Secondly, although the later folds are usually associated with an axial-planar cleavage which is rather uniform in its attitude, the later folds may be defined by different surfaces in the rock. This means that the later folds differ in orientation according to whether the fold axes are defined by the cleavage axial planar to the earlier folds or by the layering which is folded by the earlier folds. Moreover, the fold axes defined by the layering will vary in attitude, vergence and facing as they are traced across the earlier folds, provided that these folds are not isoclinal in character.

The two sets of fold structures may differ in scale without affecting the nature of the structural relationships which have just been described. However, this factor does control the form of the structures which are developed as a result. There are three cases to be considered. Firstly, the earlier folds may be smaller in scale than the later folds. This would mean that minor folds would be folded by major folds in a relatively simple manner. Secondly, the two sets of fold structures may be developed on a similar scale. This would result in the formation of interference patterns in the manner described by Ramsay (1962). Three types of interference pattern can be recognized, according to the relative orientation of the two sets of fold structures. Third, the earlier folds may be larger in scale than the later folds. This would cause minor folds to be superposed across major folds, without much effect on the overall structure.

The complexity of these relationships would be increased if the deformation history consisted of more than two phases, and if each phase resulted in folding on different scales. This is a situation which gives rise to certain relationships between the minor fold structures seen in the field and the major folds which are supposed to be present. The major folds can be recognized most easily from their effects on the outcrop pattern of the stratigraphic formations, supplemented as necessary by observations concerning regional variations in the attitude of bedding. The use of sedimentary structures to determine the stratigraphic order of individual beds is particularly important in this context, since it allows the stratigraphic sequence and its order to be defined rather precisely under favorable circumstances. This means that major folds can be mapped in areas where the stratigraphic control is otherwise poor, due to a lack of stratigraphic formations which can be distinguished clearly from one another in lithological grounds. Moreover, sedimentary structures allow the structural position of individual beds to be determined in relation to folding on different scales, thus providing an essential foundation for the interpretation of structural relationships.

Once the presence of major folds has been established in this way, it is possible to consider the minor fold structures seen in the field in relation to each major fold in turn. In general, the minor fold structures associated with a particular major fold can be divided into three groups. One group is directly related to the major fold, changing only in vergence in such a way that they share the same axial-planar cleavage. These fold structures face in the same direction as the major fold, while they correspond in plunge to this fold if the folding is cylindroidal. It is this group which is considered as coeval with the major fold. The remaining fold structures can be divided into two more groups according to whether they are earlier or later than the major fold. Both these groups of fold structures can share some features in common since they may have approximately the same orientation, vergence, and facing wherever they occur in an equivalent position on the limbs of the major fold. However, they can be distinguished from one another with reference to their relationships with the fold structures coeval with the major fold. The earlier fold structures can be recognized as one group because they are overprinted by the fold structures coeval with the major fold, around which they are folded, while the later fold structures can be recognized as another group because they are superposed on the fold structures coeval with the major fold, across which they cut.

It is axiomatic that each group of fold structures should have the same components of vergence and facing, as measured in the direction of tectonic transport, wherever they are developed in an equivalent position on the limbs of the major fold. If not, there would be good reason to suppose that these fold structures were formed in response to more than one phase in the deformation history, unless they were developed as conjugate sets. It should be possible to separate these fold structures into different groups, which can then be dated according to the effects of structural superposition, in such a way that each group is formed by fold structures with the same components of vergence and facing. This must be the case if it is assumed that the folds developed in response to a particular phase in the deformation history must all share the same axial-planar cleavage.

3.8 Structural Synthesis

None of these relationships can be determined unless it is possible to follow different groups of fold structures through the three-dimensional space defined by the major folds. However, there is no absolute and universal method whereby an individual fold structure can be recognized as belonging to a particular phase in the deformation history: fold style and metamorphic fabric are controlled by factors which are not unique to particular phases in the deformation history; the complexity of the structural relationships which arise from the effects of deformation or structural superposition means that fold orientation cannot be used to identify particular phases in the deformation history; and different groups of fold structures can be developed with the same components of vergence and facing wherever they occur in an equivalent position, so that these parameters are not diagnostic of particular phases in the deformation history.

I would argue that this must introduce a subjective element into structural analysis since, if there is no absolute and universal method which would allow individual fold structures to be identified as belonging to a particular phase in the deformation history, the final result of structural correlation can only be a model of the structural relationships which are thought to exist in reality. This model must incorporate all the structural observations which have been made in the field, even although these observations refer to structural features which are not considered diagnostic of particular phases in the deformation history. However, it is this information which constitutes the whole of the available evidence. The construction of such a model takes place on a trial-and-error basis as the investigation proceeds. Thus, a number of different models can be considered at each stage in the investigation, varying in complexity according to the number of fold phases which are supposed to be present. Each model can be tested in an attempt to account for the structural relationships seen in the field. This is done by assigning individual fold structures to particular phases in the deformation history on the subjective basis of style, orientation, and fabric, in such a way that the vergence and facing shown by these fold structures is consistent with an inferred relationship that they are supposed to have with the major folds which are known to be present, while preserving the correct order between fold structures belonging to different fold phases as determined from the effects of structural superposition. The model eventually chosen should offer the simplest account of all the available evidence in a manner which is internally consistent.

4 DISCUSSION AND CONCLUSIONS

The results of structural analysis show that the metamorphic rocks of orogenic belts are typically affected by several phases of deformation and metamorphism, so that the structural relationships developed by these rocks are extremely complex. The methods required to establish this knowledge are time consuming since they involve the detailed mapping of structural features over a very wide area before a regional synthesis can be attempted. Moreover, the subjective nature of any model describing these relationships introduces problems of structural correlation wherever separate areas have been studied by different investigators, particularly if these areas are not contiguous with one another. It is not surprising that a whole generation of structural geologists, using the methods described in the previous section, have only been able to establish the regional structure within local segments of an orogenic belt such as the Caledonides of northwestern Europe.

These difficulties are compounded by the nature of the structural record. Thus, fold structures become increasingly difficult to recognize as the deformation history is traced backwards in time. This is partly the result of structural superposition since it is the earlier structures which are obscured or obliterated by the development of later structures during the deformation history. However, it is also the earlier structures, affected by more phases in the deformation history than the later structures, which are more deformed. This would be the case even if the intensity of the deformation did not decline during the course of the deformation history, as it generally appears to do. Accordingly, it is also the effect of deformation, which generally causes folds to become isoclinal structures and cleavages to rotate into parallelism with the layering, that makes the earlier fold structures not only more difficult to recognise than the later fold structures but also more difficult to distinguish from one another. Metamorphism has a similar effect since it often reaches a climax relatively late during the course of the deformation history. This means that the cleavage fabrics formed during the earlier phases in the deformation history tend to be modified or destroyed by the recrystallisation and grain growth occurring under prograde conditions before this climax is reached. Acting together, the effects of superposition, deformation and metamorphism are sufficient to obscure the structural record so effectively that there are often large areas within orogenic belts where the early structures simply cannot be recognized. The regional structure can only be inferred within these areas by extrapolation from neighbouring areas of lesser complexity where the structural record is preserved in its entirety.

The chronology of the deformation history is a matter for debate. The simplest interpretation would regard each phase in the deformation history to be a distinct and short-lived event, taking place at the same time throughout the whole region and separated from the other phases in the deformation history by longer intervals of geological time. Such a viewpoint implies that fold phases can be used as strict time markers in the study of metamorphic events. It would be favoured by evidence that the fold structures developed in response to different phases in the deformation history are not obviously related to one another. This appears to be the case for the polyphase structures affecting the Moinian rocks in the Highlands of Scotland (Tobisch et al., 1970; Powell, 1974). These structures are defined by interference patterns developed in response to the superposition of major folds belonging to different phases in the deformation history. However, it is commonly found that there is some relationship between the different phases in the deformation history.

The Dalradian rocks in the Southwest Highlands of Scotland can be taken as a case in point. They are affected by a deformation history of eight fold phases (Roberts, 1974). Only the latest fold phase appears to be entirely unrelated to the initiation and evolution of the major folds which now control the structural disposition of the stratigraphic formations. Although these folds started to develop during the first phase in the deformation history, in response to the B_1 deformation, they continued to evolve during the second and third phases, as a result of the polyphase B_{2a} deformation. These major folds are B_1 structures in that a slaty cleavage is developed as the earliest fabric in the rock, parallel to their axial planes. However, the B_{2a} deformation is a "symmetry-constant continuation" of the B_1 deformation in that the B_{2a} structures have the same vergence and facing as the B_1 structures (Voll, 1960). This means that the B_{2a} structures have the same relationship to the major folds as the B_1 structures. The evolution of the major folds continued during the fourth phase, marked by the $B' \perp B_1$ cross folding which affects zones of intense B_1 and B_{2a} deformation associated with the development of slides on the major fold limbs (Voll, 1960). All these fold phases are considered to define the primary deformation (Bailey, 1922). They are closely associated with the penetrative stretching of the whole rock mass which occurred during the initiation and development of the major folds. The subsequent phases in the deformation history constitute the secondary B_{2b} deformation. Although the first phase marks a reversal in the regional stress field which existed at the end of the primary deformation, the first, second and third phases in the secondary deformation continue the process of southeasterly collapse under gravity which affected some of the major folds during the course of the primary deformation.

The close relationship existing between the various phases in this deformation history suggests that the structural evolution of the Southwest Highlands may have occurred in response to a complex history of continuous deformation (Roberts, 1974). This would allow the B_1 structures to form at the start of the primary deformation, while the later structures would develop at various stages during the course of the subsequent deformation. Moreover, although the deformation history generally reflects the sequential development of fold structures belonging to the kinematic stages enumerated in the last paragraph, there are local areas in the Southwest Highlands where certain stages in this sequence are missing. The B_{2a} structures are restricted to those areas where the deformation is relatively intense, while the $B' \perp B_1$ structures only affect slide zones developed in response to the B_1 or B_{2a} deformation. Accordingly, the $B' \perp B_1$ structures may occur in sequence after the B_1 or B_{2a} structures, while the B_{2b} structures may occur in sequence after the B_1, B_{2a} or $B' \perp B_1$ structures. This could imply that either the deformation history was episodic, so that the local gaps in the structural record mark intervals of geological time, or that the kinematic stages in the deformation history did not occur simultaneously throughout the whole region. These statements are not mutually exclusive since, although the deformation history might be discontinuous, the kinematic stages in the deformation history could still be diachronous. However, if it is argued that the deformation history was continuous throughout the whole region, this must imply that the kinematic stages in this deformation history were diachronous.

As Park (1969) has emphasized, there is no direct evidence that fold structures recognized as belonging to a particular fold-phase were ever contemporaneous in their development. This possibility can be tested with reference to the metamorphic history since it implies that the temporal relationships existing between structural and metamorphic events would generally vary throughout the region under consideration (Chadwick, 1968).

These arguments apply with equal force to the development of a slaty cleavage as the earliest fabric of metamorphic origin which is imprinted on a sedimentary rock after its deposition. For example, there is some evidence that the slaty cleavage developed within different zones of the Variscan fold belt in Southwest England differs in age by approximately 40 million years (Roberts and Sanderson, 1971). The junction between these zones shows the earlier slaty cleavage dying out as it is traced to the north and the later slaty cleavage becoming a crenulation cleavage overprinted on this cleavage as it is traced to the south. This interpretation is consistent with the regional synthesis presented by Sanderson and Dearman (1973), which recognizes that the fold belt was affected by mid-Carboniferous movements in the south and late-Carboniferous movements in the north, giving rise to marked differences in the tectonic pattern developed within each region. It is clear that regional differences in the timing of orogenic events must introduce major problems of structural correlation, particularly if the later movements extend into those areas affected by the earlier movements.

This conclusion is reinforced by evidence which has confirmed that the Moinian rocks forming part of the metamorphic Caledonides in the Scottish Highlands were intruded by pegmatites at 730 ± 20 million years and 450 ± 10 million years ago (van Breeman et al., 1974). The Precambrian and Caledonian pegmatites so dated were intruded between different phases in the deformation history of the region, so that there is evidence for Precambrian movements (Powell, 1974). However, these movements must die out as they are traced northwest into the Moine Thrust Zone and southeast toward the Dalradian outcrop, where the earliest movements affect Cambro–Ordovician rocks. These relationships are similar to those developed in the Variscan fold belt of Southwest England in that there is apparently no major unconformity developed between the two groups of rocks affected by orogenic events of different age. This suggests that sedimentation must have continued while the early movements took place at depth, at least in those areas not affected by these movements. However, the stratigraphic record within the area affected by the late movements may well reflect the contemporaneous or subsequent uplift of the area affected by the early movements.

Such problems tend to obscure the contribution made by plate tectonics to the understanding of orogenic belts. However, this theory does provide an insight into their evolution, suggesting that they develop in response to the subduction of oceanic lithosphere below island arcs or cordilleras while recognizing that this process may eventually cause a collision to occur between such a system and a continental margin (Mitchell & Reading, 1969; Dewey & Bird, 1970). It is argued that thermal processes are dominant in orogenic belts formed simply as a result of subduction, while mechanical processes are more important wherever a continental collision takes place. Although the former statement may be doubted, in that mechanical energy is likely to bc generated by the subduction process, the latter is obviously true.

Plate tectonics provides little explanation for the complex history of structural events which affect the metamorphic rocks of orogenic belts. Hobbs, Means, and Williams (1976) have briefly considered this problem. They suggest that such a deformation history may be related to changes in relative displacement or tectonic setting which can occur during the course of subduction. For example, it is possible that each phase in the deformation history might be correlated with the different periods of steady-state subduction which would occur during the evolution of a belt. According to this view, the deformation history would be a continuous process. Alternatively, each phase might be a short-lived event, occurring whenever a change in steady-state subduction took place, so that the deformation history would be episodic in character. A final phase could mark the time of continental collision after a period of steady-state subduction. Such models imply that each phase in the deformation history took place at the same time through the orogenic belt as a whole, unless additional complexities are introduced, while they suggest that different phases in the deformation history are unlikely to be closely related to one another. However, the previous discussion has suggested that deformation phases are unlikely to be synchronous throughout the whole belt, while it is quite clear that the deformation histories of some belts show a consistent pattern of structural evolution, contrary to these views.

I would argue that it is the very intensity of deformation which suggests that other factors are involved in the multiple generation of fold-structures within orogenic belts. The strains are so large that major displacements must take place. It is quite likely that such movements will cause the rocks within an orogenic belt to be affected by different patterns and modes of deformation as they are transported in the direction of flow (Price, 1972). According to this model, the various phases in the deformation history would be generated in turn as these rocks were transported through a deformation field of some sort. This implies a history of continuous deformation, even although the phases constituting this history would be diachronous, generated whenever a particular volume of rock crossed a boundary between different regimes of deformation. This would be the case even if the deformation field did not change with time. However, the gravitational instabilities which are likely to develop as a result of displacements within an orogenic belt, and the constraints which are exerted on these displacements by its boundaries, would probably increase the complexity of this deformation field within the belt during the course of its evolution. The thermal effects associated with regional metamorphism and plutonic activity are another factor to be considered, since changes in temperature have a major influence on the deformation undergone by rocks, while the mechanical effects of plutonic activity must also be taken into account, particularly if the diapiric rise of lighter material is involved (Ramberg, 1967). These effects are likely to be important in further modifying the deformation field within an orogenic belt, since the influence exerted by these factors generally varies in space and time during the course of the deformation history. Accordingly, it may well be argued that no simple explanation is sufficient to account for the complex history of polyphase deformation which is a characteristic feature of metamorphic rocks within orogenic belts.

REFERENCES

Bailey, E. B.: The structure of the Southwest Highlands of Scotland. Quart. J. Geol. Soc. London *78,* 82 (1922)

Bailey, E. B.: West Highland tectonics: Loch Leven to Glen Roy. Quart. J. Geol. Soc. London *90,* 462–525 (1934)

Bailey, E. B.: Regionale Geologie der Erde, Bd. Z: Palaeozische Tafeln und Gebirge, Abt. ii, Northwestern Europe Caledonides. Liepzig (1938)

Bredehoeft, J. D., and Hanshaw, B. B.: On the maintainence of anomalous fluid pressures: I, thick sedimentary sequences. Bull. Geol. Soc. Amer., *79*, 1097–1106 (1968)

Breeman, O. Van, Pidgeon, R. T., Johnson, M. R. W.: Precambrian and Palaeozoic pegmatitics in the Moines of northern Scotland. J. Geol. Soc. London *130*, 493–507 (1974)

Chadwick, B.: Deformation and metamorphism in the Lukmanier region, central Switzerland. Bull. Geol. Soc. Am. *79*, 1123–1150 (1968)

Clough, C. T.: The geology of Cowal. Mem. Geol. Surv. Scotland, 1897

Dewey, J. F., and Bird, J. M.: Mountain belts and the new global tectonics. J. Geophys. Res., *75*, 2625–2647 (1970)

Fleuty, M. J.: The description of folds. Proc. Geol. Assoc. London *75*, 461–492 (1964)

Hise, C. R. Van: Principles of North American Precambrian geology. Ann. Rept. U.S. Geol. Surv., 1894–1895, Pt. *1*, 571–843 (1896)

Hobbs, B. E., Means, W. D., and Williams, P. F.: An outline of structural geology. Wiley, New York 1976

Hubbert, M. K., and Rubey, W. W.: Role of fluid pressure in the mechanics of overthrust faulting. Bull. Geol. Soc. Amer., *70*, 115–206 (1959)

Leith, C. K.: Rock cleavage. Bull. U.S. Geol. Surv., *239*, 1–216 (1905)

Leith, C. K.: Structural Geology, New York: Holt and Co. (1913)

Maxwell, J. C.: Origin of slaty and fracture cleavage in the Delware Water Gap area, Pennsylvania and New Jersey. Mem. Geol. Soc. Amer., *81*, 281–311 (1962)

Mitchell, A. H., and Reading, H.G.: Continental margins, geosynclines and ocean-floor spreading. J. Geol. *77*, 629–646 (1969)

Park, R. G.: Structural correlation in metamorphic belts. Tectonophysics *7*, 323–338 (1969)

Powell, D.: Stratigraphy and structure of the western Moine and the problem of Moine orogenesis. J. Geol. Soc. London *130*, 575–593 (1974)

Price, R. A.: The distinction between displacement and distortion in flow, and the origin of diachronism in tectonic overprinting in orogenic belts. Proc. 24th Intern. Geol. Congr., *3*, 545–551 (1972)

Ramberg, H.: Gravity, deformation and the earth's crust. Academic Press, London (1967)

Ramsay, J. G.: Interference patterns produced by the superposition of folds of similar type. J. Geol. *70*, 466–481 (1962)

Ramsay, J. G.: Structure and metamorphism of the Moine and Lewisian rocks of the Northwest Caledonides. In: The British Caledonides. Johnson, M. R. W., Stewart, F. H., (eds.). Edinburgh: Oliver and Boyd (1963)

Ramsay, J. G.: Folding and Fracturing in Rocks. New York: McGraw-Hill (1967)

Roberts, J. L.: The structure of the Dalradian rocks in the SW Highlands of Scotland. J. Geol. Soc. London *130*, 93–124 (1974)

Roberts, J. L., Sanderson, D. J.: Polyphase development of slaty cleavage and the confrontation of facing directions in North Cornwall. Nature (London) *230*, 87–89 (1971)

Roberts, J. L., Sanderson, D. J.: Oblique fold axes in the Dalradian rocks of the Southwest Highlands. Scott. J. Geol. *9*, 281 (1974)

Sanderson, D. J.: The development of fold axes oblique to the regional trend. Tectonophysics *16*, 55 (1973)

Sanderson, D. J., Dearman, W. R.: Structural zones of the Varsican fold-belt in

SW England, their location and development. J. Geol. Soc. London *129*, 527–536 (1973)

Shackleton, R. M.: Downward-facing structures of the Highland Border. Quart. J. Geol. Soc. London *113*, 361–92 (1958)

Shrock, R. R.: Sequence in Layered Rocks. New York: McGraw-Hill (1948)

Stille, H.: 1930, Uber Einseitigkeiten in der germanotypen Tektonik Nordspaniens und Deutschlands. Nachr. Gessell. Wissensch. Göttingen, Math-Phys. Kl. 379–397 (1930)

Tobisch, P. T., Fleuty, M. J., Merh, S. S., Mukhopadhyay, D., Ramsay J. G.: Deformational and metamorphic history of Moinian and Lewisian rocks between Strathconon and Glen Affric. Scott. J. Geol. *6*, 243–265 (1970)

Voll, G.: New work on petrofabrics. Liverpool. Manchester. Geol. J. *2*, 503 (1960)

Wilson, M. E.: The Kewagama Lake map-area. Mem. Geol. Surv. Can. *39*, 1–51 (1913)

9

P. F. WILLIAMS AND H. J. ZWART

Abstract Structures and metamorphic features are described from the Seve–Köli nappe complex, and an interpretation of their history is presented.

The nappes have undergone intense ductile deformation during their emplacement with the result that earlier foliations are completely transposed into the orientation of the thrust planes, and earlier fold axes have been rotated into approximate parallelism with the direction of thrusting. The latter is parallel to a stretching lineation and to tubular folds and is perpendicular to the trend of the Caledonian Orogen. It is argued that the nappes were emplaced by flow under the influence of gravity, after considerable crustal thickening was achieved by continental collision.

A MODEL FOR THE DEVELOPMENT OF THE SEVE–KÖLI CALEDONIAN NAPPE COMPLEX

1 INTRODUCTION

The Caledonian mountain chain in Scandinavia is characterized by the occurrence of large nappe structures, a fact that was first recognized by Törnebohm (1872). One of such major structures described by him is the so-called Seve–Köli nappe complex. This article deals with some aspects of this complex, namely small- and large-scale structures and metamorphism, and leads to a model for the formation of these large tectonic units.

The Seve–Köli nappe complex stretches for a distance of about 800 km on the eastern side of the Caledonides and outcrops mainly in Sweden. It is overlain by the Rödingsfjäll nappe consisting of undated high grade crystalline rocks and in the south by the Trondheim nappe consisting of lower Paleozoic rocks of variable metamorphic grade. It is underlain by a group of lower grade nappes: the Särv nappe consisting of late Precambrian sediments and some Precambrian crystalline basement, the Offerdal (or granite–mylonite) nappe consisting of deformed Precambrian crystalline rocks, and some parautochthonous nappes (Olden, Blaik) in which lower Paleozoic and late Precambrian sediments occur together with some Precambrian basement. These parautochthonous nappes rest on the crystalline rocks of the Precambrian Baltic shield or on its lower Paleozoic or late Precambrian unconformable cover. Following Törnebohm, it is customary to distinguish two groups of rocks in the Seve–Köli nappe complex: the Seve and the Köli. In a recent publicaton by Gee (1975), these have been called the Seve supergroup and Köli supergroup. We will call these the Seve and Köli nappes or in short the Seve and the Köli. The distinction between the two nappes is made partly on lithology but mainly on the grade of metamorphism. The Köli rocks belong principally to a greenschist facies, and the Seve rocks to the amphibolite or the granulite facies. Our work has shown that they differ also in structural and metamorphic history, and probably in age.

Everywhere the Köli rocks overlie the Seve, and in general the contact is a thrust (Fig. 1). The Seve in turn lies on the lower nappes which have a grade of metamorphism not exceeding greenschist facies, and from the middle of the Seve downward the metamorphic sequence is inverted. The rocks of the Köli nappe are, at least in part Ordovician and Silurian in age as testified by scarce fossils. The absence of fossils makes paleontological dating of the Seve impossible. However, radiometric dating, still in progress, indicates a Precambrian age for at least part of the Seve. (Reymer, personal communication).

For details of the stratigraphy of the Köli, we refer the reader to Zachrisson (1969) and Gee (1975). In the Seve nappe a lithostratigraphy can be recognized (Fig. 1), and the nappe is subdivided into three belts

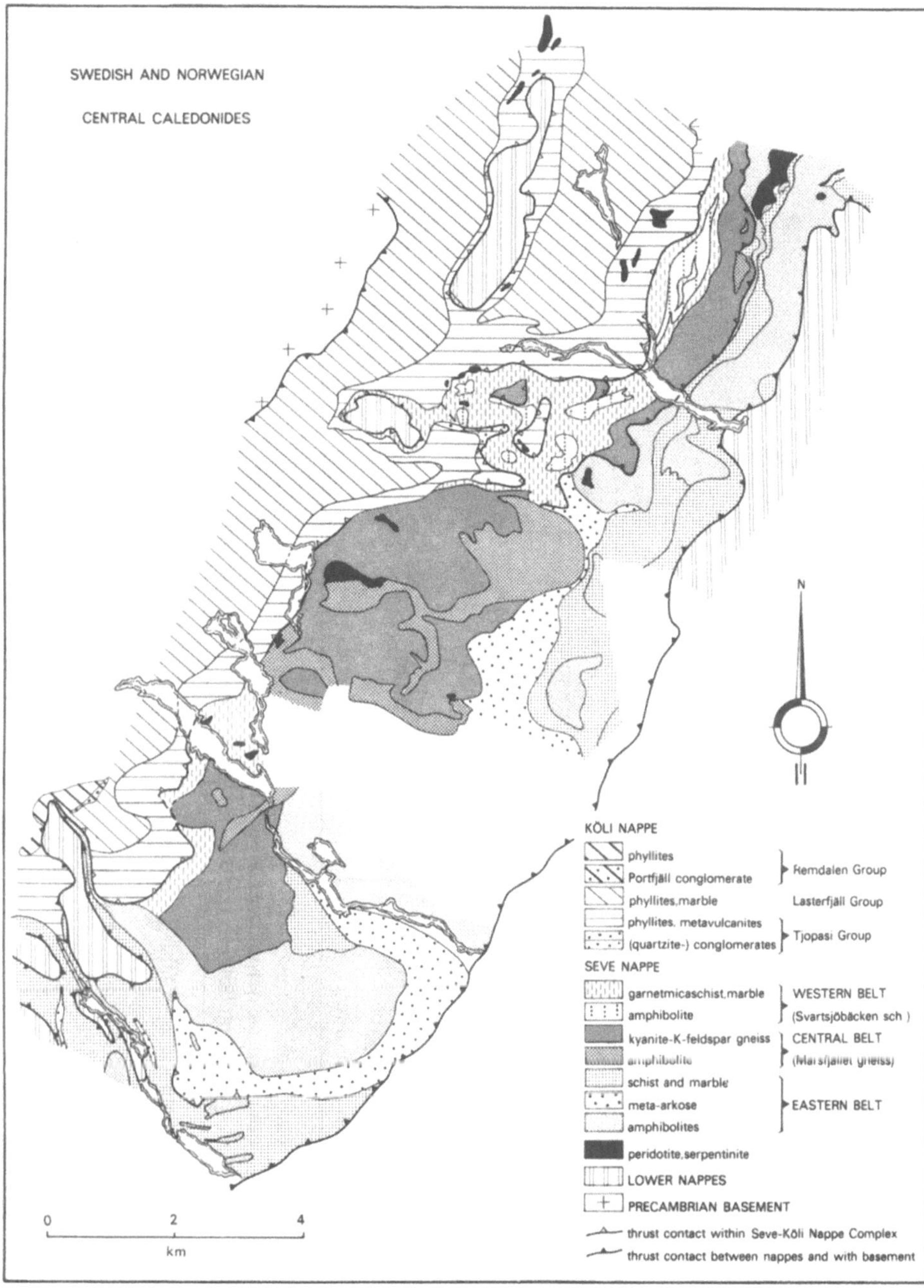

Fig. 1. Geological map of Seve–Köli nappe complex in Jämtland and Västerbotten (Sweden)

Table 1.

Tectonic unit	Rock type	Metamorphic zone	Metamorphic facies
Köli	Conglomerate keratophyre greenschist calcareous phyllite } Lasterfjäll limestone, conglomerate quartz-keratophyre greenschist black phyllite } Tjopasi	Chlorite Biotite Garnet	Greenschist
Western belt	Garnet-micaschist quartzitic garnet-micaschist with amphibolites	Staurolite-kyanite	Amphibolite
Central belt	Migmatic ky-k-feldspar gneiss (with eclogites) quartzo-feldspathic gneiss amphibolite (hbl,opx,cpx,gar) ky-k-feldspar gneiss (+ sill, locally with opx)	Kyanite-potassium-feldspar	Granulite retrogressed to amphibolite

Seve	Garnet-micaschist meta-arkose and quartzite with amphibolite	Staurolite-kyanite	
Eastern belt	garnet-micaschist, marble amphibolite	Garnet	Amphibolite
Särv	Meta-arkose, quartzite muscovite-schist	Biotite	Greenschist
Granite-mylonite (Offerdal)	Augengneisses, mylonites	Biotite	
Lower nappes (Blaik, Olden)	Quartzite, arkose greywacke, shale	Chlorite	
Parautochthonous autochthonous		Unmetamorphosed	

as can be seen from Table 1 which is a compilation of the stratigraphy in Jämtland and Västerbotten. This table also shows the grade of metamorphism associated with the various units.

According to Zachrisson (1969) the major lithological units in the Köli are in normal stratigraphic order, despite the intense deformation. However in the Seve we are certainly not dealing with a normal stratigraphic sequence, but with a structural sequence in which neither the absolute nor relative ages are known. In the Köli nappe there are identifiable sedimentary rocks with clear bedding and it is possible to work out a complete sequence of folding phases. In the Seve it is very rare to find undoubted bedding and a complete structural history is difficult or even impossible to establish. Furthermore, there is evidence that strain in the Seve is, in general, considerably larger than in the Köli. This is consistent with the provisional radiometric dates (see above) which suggest that at least parts of the Seve nappe are likely to have been deformed during both Caledonian and Precambrian orogenies.

In this paper we are concerned mainly with the Seve but we begin by outlining the structural and metamorphic history of the Köli. An understanding of the Köli is useful for unraveling the more complex history of the Seve and is also necessary for developing a model for the emplacement of the Seve–Köli nappe complex.

2 THE KÖLI NAPPE

In the Köli three or locally four generations of folds can be recognized. The first consists of tight to isoclinal folds with an axial plane slaty cleavage. As a rule they are mesoscopic structures but some major folds have been recognized near lake Vesken by Kulling (1933) and on Daunatjåkko by Calon (1973). Second generation structures overprint the earlier structures and are accompanied by a well-developed crenulation cleavage which is commonly differentiated. Very little is known about the orientation of F_1 folds. F_2 foldaxes have a variable orientation; some are parallel to the local Caledonian trend, that is approximately NNE, but others such as the ore-bearing fold from Stekenjokk (Zachrisson, 1971) are perpendicular to it.

The third generation comprises large and small folds with steeply dipping axial planes that have a Caledonian trend. These folds are largely responsible for the large scale outcrop pattern on the map in the Köli. In some hinges there is an axial plane crenulation cleavage; elsewhere, cleavage is lacking. The fourth generation is very local in occurrence and consists of kinks with shallowly dipping axial planes. No large structures belonging to this generation are known.

Thrusts within the Köli sequence have been described by Zachrisson (1969) but at present we are unable to date this thrusting with respect to the fold sequence.

The dating of metamorphism with respect to these fold phases is relatively simple. During F_1, metamorphism was low grade, mainly producing quartz, white mica and chlorite, which define a slaty cleavage. Metamorphism, although not exceeding greenschist facies conditions, increased in grade after F_1 as shown by cross-cutting porphyroblasts of ilmenite, biotite and garnet. The latter two minerals inherited a planar fabric from the matrix, that is defined by inclusions (s_i trails). In many cases porphyroblastesis had ceased before F_2 as demonstrated by the presence of a crenulated matrix around porphyroblasts with planar s_i trails (Trouw, 1973; Zwart, 1974). However the micas and quartz in the F_2 folds are generally recrystallized and strain free, indicating that temperature was still quite high. No new minerals formed during F_3 and F_4 and the strained appearance of quartz and micas affected by these structures indicates that temperature had dropped even further than during or immediately after F_2. Thus the peak of metamorphism occurred between F_1 and F_2.

3 THE SEVE NAPPE

3.1 Deformation Structures in the Seve

Several generations of deformation structures, including folds and thrusts can be recognized in the Seve. The folds and related axial plane foliations and lineations can be divided into two groups on the basis of their relationship to the regional foliation. The latter is found in all rock types including the mylonites associated with the thrusts. The first group comprises folds that can be thought of as contributing to the development of the regional foliation in that they rotate earlier surfaces into parallelism with it. These folds are commonly intrafolial. The second group comprises folds that clearly overprint the regional foliation. These folds also overprint the thrusts.

3.1.1 Group 1 Structures. This group is not a single-fold generation but comprises several generations, which contribute to the progressive transposition of earlier structures into a regional foliation. This foliation is defined by both lenticular layering and preferred orientation of such minerals as micas, quartz and amphiboles. Folds belonging to the group show considerable variation in style but they are generally tight to isoclinal. Quite commonly they have an axial plane schistosity or crenulation cleavage but may lack any form of axial plane foliation. There is a lineation defined by mineral elongation and mineral aggregates; it is parallel to the hinges of some group 1 folds but is apparently folded by other folds of group 1. There is a tendency for the axial plane schistosity and hinge-parallel mineral lineation to be associated with the tighter folds but this is not always true. Despite the considerable variation in fold style it does not appear to be possible to divide the folds into useful groups on the basis of style. This is partly because the various styles are gradational with one another but also because there is demonstrable lack of correlation between style and fold generation (see Olesen et al., 1973). For example, in a single fold some layers may show an axial plane schistosity while in other layers in the same fold there is no axial plane foliation but a folded schistosity. In some cases this difference is related to the quartz content of the rock. In the quartz rich layers where the micas are generally not in contact with one another the deformation has produced a schistosity. In the mica rich layers, where mica is the dominant mineral, the same deformation has produced a crenulation cleavage. Then again, a progressive variation in the tightness of adjacent folds (Fig. 2) accompanied by a transition in axial plane foliation from crenulation cleavage in the more open folds to schistosity in the tighter folds can be seen in some individual outcrops. This type of variation is particularly common adjacent to small thrusts where the thrust is closest to the tightest folds and parallel to their axial planes.

In most large outcrops throughout the Seve it is possible by means of overprinting relationships to demonstrate two generations of group 1 folds and in many outcrops there are three. The most recent of the folds are sometimes more open than the earlier ones but elsewhere, including areas where three generations are present, they may all be isoclinal. Thus the Seve is thoroughly transposed and this is further substantiated by the very lenticular nature of all rock types. Some areas of very planar well-foliated rocks appear quite simple at first sight but careful examina-

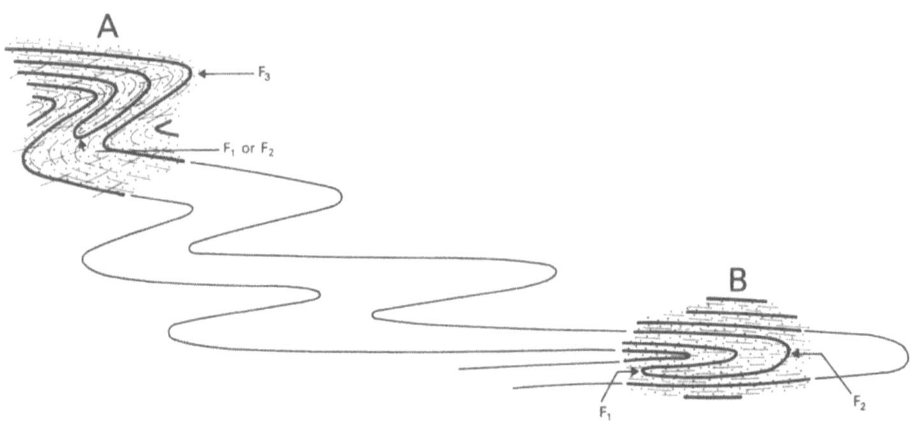

Fig. 2. Diagrammatic representation of variation in fold-style. See text for discussion

tion commonly reveals attenuated isoclinal folds suggesting that the simplicity is due to extreme deformation. Furthermore, in rocks containing boudins, the deformation recorded by folds within the boudins is generally far more complex than that apparent in the country rock. Such structures generally indicate very high strains; for example there are pinch and swell structures which indicate 60% minimum shortening associated with their development and within the swells there are isoclinally refolded isoclinal folds Figure 3. Finite strains in such rocks must be very high. High strains in the Seve are also indicated by rotated garnets. The largest rotation known to us is in excess of 800° for a garnet from the Dikanäs schist (Schoneveld, personal communication). Such a rotation requires a shortening of the diameter of an initial sphere by more than 90%.

On a regional scale, dip directions for the transposition foliation in the Seve are fairly consistently in the north west quadrant. Group 1 folds generally plunge in the same quadrant and typically are reclined and

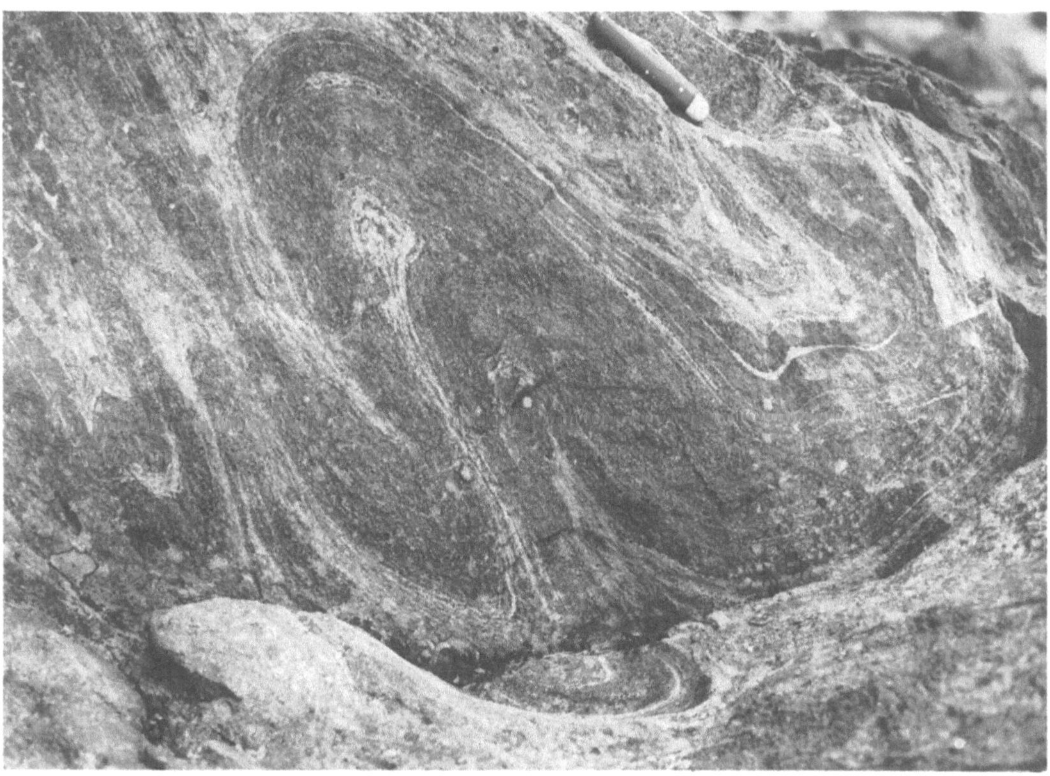

Fig. 3. Isoclinally refolded folds in amphibolite. These folds occur in the swell of a pinch and swell structure

approximately perpendicular to the Caledonian trend. The mineral lineation also plunges in the same direction both in the nappes and in the mylonites between them. Microboudinage of mineral grains associated with this lineation indicates that it is a direction of extension.

Group 1 folds are commonly cylindrical as seen in outcrop. Locally, however, they are markedly noncylindrical and hingelines may turn through angles up to 180° so that the resulting fold resembles a cone or tube closed at one end. This tubular structure may have an approximately circular or more commonly an elliptical cross section. Where elliptical, the aspect ratios can be as high as 25:1. The geometry of such folds can be approximated by a circle or ellipse based cone or cylinder and the orientation of the fold can then be given in terms of the cone or cylinder axis and the long axis of the elliptical base. The cone or cylinder axis of such folds plunges parallel to the mineral lineation (i.e., into the northwest quadrant). This lineation has only been recognized in the portions of the fold hinge that are approximately parallel to it. It has not been recognized in the dome or basin closure of the tube. The long axis of the elliptical section lies in the regional foliation and is approximately parallel to the Caledonian trend.

3.1.2 Group 2 Structures. Group 2 folds are generally gentle to open and some are kink-like. They may have an axial plane crenulation cleavage but commonly lack an axial plane foliation. They vary considerably in the trend of their hinge lines but most of their axial surfaces are steeply dipping. They consistently overprint group 1 structures and produce an open dome and basin structure that folds earlier thrusts and the regional foliation. Locally there is evidence that suggests that the group comprises more than one generation of folds.

3.1.3 Thrusts. Within the Seve several major and minor thrusts have been recognized. A major thrust separates the Köli from the underlying Seve, as testified by the truncation of rock units and a jump in metamorphic grade. Unfortunately, the thrust itself is badly exposed and little is known about it. In the north another important thrust separates the central belt of the Seve from the eastern belt and there is again a jump in metamorphic grade (Trouw, 1973). However, a thrust has yet to be demonstrated between these two belts in the southern part of the area. Several smaller thrusts have been found in the Seve. Like the major thrusts, where visible, they are all characterized by the occurrence of laminated quartz-rich mylonites. These mylonites are generally fine grained, and have a well-developed foliation and lineation. They commonly contain rounded porphyroclasts of garnet, kyanite or feldspar and typically fish-shaped micas. The foliation in the mylonites, defined by the shape of individual minerals or mineral aggregates, has a shallow plunge in the north west quadrant and is parallel to similar lineations in other rocks. Folds occur in the thrust zones and have axial surfaces parallel to the thrust plane and hinge lines parallel to the lineation. Thus the structures in the mylonites belong to group 1.

Locally there are low angle faults that are not accompanied by mylonites, but by strongly retrogressed phyllonitic rocks or sometimes by pseudotachylites. These faults locally cut across the regional foliation and probably represent thrusts of a younger age than the ones described above.

3.1.4 Macroscopic Structure. Group 2 folds occur on a regional scale throughout the Seve and have a very strong influence on the outcrop pattern of various units. They are responsible for a large-scale dome and basin structure in the regional foliation. The thrusts are also important regional structures being traceable for many kilometers and they also influence the outcrop pattern quite strongly. No definite group 1 folds of regional proportions have been recognized in the Seve but units tend to disappear and reappear (Fig. 1) along strike in a manner that is readily interpreted in terms of large-scale reclined folds. Because of the transposition it is impossible to determine the three-dimensional form of such bodies but some of them [e.g., the meta-arkose and amphibolite units of the eastern belt (see Fig. 1)] at least, must either be large lenticular bodies lying in the regional foliation or doubly plunging group 1 folds.

3.1.5 Discussion of the Deformational Structure of the Seve. As stated above, it is not possible to divide the folds in the Seve into generations on a regional scale. Three generations can be demonstrated in many outcrops on the basis of overprinting relationships and in some outcrops there are four generations. However, outcrop is generally not very good and interpretation, therefore, has to rest very heavily on correlation based on style and orientation. This is normally unsatisfactory and is particularly so in the present case since overlap in style and orientation of folds of separate generations can be demonstrated in single outcrops.

The situation represented in Figure 2 is an example of the problems of using style in this area. The figure as a whole depicts a structure that can be seen in some large outcrops. In going from one end of the outcrop to the other, there is a gradual increase in the tightness of the folds. Such a transition may also be accompanied by a change in the axial plane foliation from a crenulation cleavage to a schistosity. So long as the complete transition can be seen, there is no problem, but in practice the evidence is often fragmentary because of the impersistent nature of the outcrop as represented diagrammatically by the heavy lines in Figure 2. Given two such outcrops, if style is used for separating generations the folds will be wrongly identified as in the figure. This is because there are seen to be two style groups but the isoclinal folds in outcrop *B* can be further divided on the basis of overprinting. Thus the folds would be labeled incorrectly as in the figure. This figure also shows the problem of using the concept of regional or main foliation; this concept is commonly used as a basis for dating structures. Thus in outcrop *A* (Fig. 2) the F_3 fold would be said to be younger than the main schistosity whereas in outcrop *B* the F_2 fold would be said to be of the same age as the main schistosity; the folds, however, belong to the same generation. The foliation on the other hand is not of the same generation. In outcrop *A* it is folded and distinguishable from the F_3 axial plane foliation. In outcrop *B* it is transposed and indistinguishable from the F_2 axial plane foliation. In terms of the labels used for outcrop *B* where the "main foliation" is S_2, the "main foliation" in outcrop *A* should be labeled S_1 and the "F_3" ($= F_2$) axial plane foliation should be labeled S_2.

Despite the problem of ascribing the folds to generations, examination of good outcrops where both groups of folds are present reveals a consistent sequence of events. Group 2 consistently overprints group 1 structures. The thrusts also are overprinted by group 2 structures and could be post or syn group 1 folds. However, the fact that they commonly have a mineral lineation that is parallel to the prominent lineation in the country rock, suggests that they mostly developed during one of the later group I fold generations. Any model proposed for the structural development of the Seve must be compatible with these relationships.

Various movement pictures could be proposed for the reclined folds but the tubular folds can only be explained by relative movement of component elements parallel to the tube axis and thus parallel to the prominent lineation. It has been suggested (Johnson, 1965) that reclined folds associated with a similar lineation in the Moine thrust zone have developed by rotation of preexisting folds toward the λ_1, mean strain axis, as a result of shear deformation associated with the thrusting. This model is illustrated in Hobbs et al. (1976, Fig. 6.15). The same model is capable of explaining the tubular folds as shown in Figure 4 and fits well in the

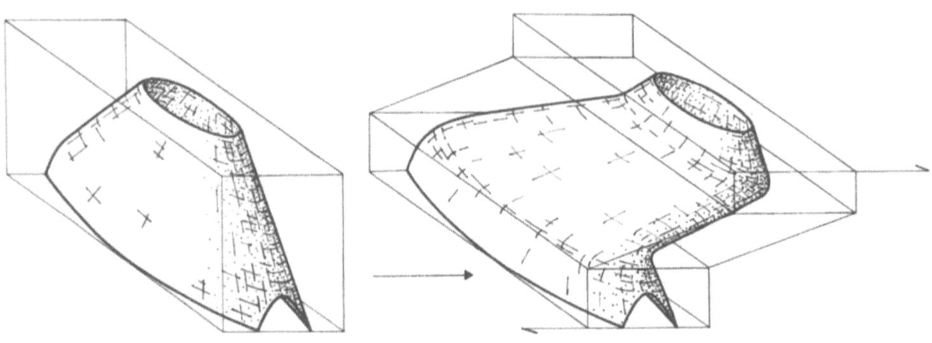

Fig. 4. Development of reclined tubular folds by modification of earlier doubly plunging fold. See text for discussion

general picture of thrusting of rocks from the orogenic zone onto the contnental margin to the southeast. This deformation is represented in Figure 4 by a progressive simple shear, because we believe that the deformation had a large shear component. This model is no doubt an oversimplification but it is believed to represent the essential points. In detail the strain would obviously be very heterogeneous and every type of strain and strain path is to be expected in various parts of the belt. Large strains are of course required to rotate the folds into approximate parallelism with the movement direction but then large strains are necessary anyway to explain observed garnet rotations and such structures as the isoclinally refolded folds preserved in the swells of pinch and swell structure (Fig. 3).

An important aspect of this model is that instead of representing the Seve as a number of comparatively rigid sheets moving on thrust planes it represents the whole sequence as a south easterly flowing rock mass. The thrusts are here interpreted as local highs in strain magnitude which may or may not have been accompanied by a loss of continuity (see Zwart, 1974), or as the surface onto which a given nappe has flowed.

In the Seve it is possible that the group 1 folds were all much earlier than the proposed southeasterly flow but it is perhaps more reasonable to suggest a progressive deformation in which not only earlier folds (possibly belonging to a much earlier deformation) but also new folds rotated as they formed and in so doing developed group 1 characteristics.

In the final stages of the deformation, continued shortening produced the group 2 folds which overprint the earlier structures.

3.2 Metamorphism in the Seve

Metamorphism in the Seve is more complex than in the Köli and this may be partly due to the high grade (granulite or amphibolite facies), Caledonian metamorphism having been superposed on an earlier metamorphic event. A problem requiring explanation is the inversion of the metamorphic zones (see Table 1).

The central belt of the Seve complex is the most problematic. In this belt granulite facies rocks have been recognized from as far south as Åreskutan to Kittelfjäll in the north (Zwart and Senior, in prep.). The granulite facies assemblages have been found in both felsic and mafic rocks and the following minerals are attributed to this phase of the metamorphism: orthopyroxene, clinopyroxene, brown-green hornblende, sillimanite, garnet, hercynite, and possibly scapolite. Eclogites occurring in the Blåsjon area are also believed to have formed during granulite facies metamorphism.

Granulite facies assemblages and eclogites have everywhere been overprinted by amphibolite facies assemblages, with kyanite, staurolite, garnet and green hornblende as some of the characteristic minerals. The granulite facies assemblages are interpreted here as Precambrian in age and the amphibolite facies assemblages as Caledonian. This interpretation is based on the preliminary radiometric ages (Reymer, personal comment) which have yet to be confirmed.

In the western and eastern belts no relics of high grade assemblages have been found. Both belts contain greenschist and amphibolite parageneses which most probably are of Caledonian age.

With the aid of microstructures it is possible to date the metamorphism relative to the fold phases. It appears that the metamorphic crystallization was mainly confined to the period during which the group 1 folds developed. Folds of group 2 are postcrystalline as is shown by strained micas, quartz and amphiboles in fold hinges.

During all stages in the development of the regional foliation temperature was sufficiently high for recrystallization of quartz. However, large variations exist with respect to other minerals. In the western belt (see Table 1) micas are strongly recrystallized, and there is new biotite. The micas define either a schistosity or a differentiated crenulation cleavage with strainfree quartz and mica. In gneisses of the central belt, white micas and kyanite are strongly deformed by group 1 folds.

The same holds true for the eastern belt where deformed fish-shaped boudinaged micas are especially common. Minerals like kyanite, garnet, and feldspar are commonly strained by the last of the group 1 folds, and are rarely recovered or recrystallized. This evidence suggests that in the last stages of group 1 folding, metamorphic conditions were already waning, and that the peak of metamorphism was reached earlier. This is confirmed by many paracrystalline garnets which show considerable rotation, but in which the S_i trails are not continuous with the external schistosity, where the latter is the regional foliation. Again in this case garnets grew and rotated before the end of the group 1 deformation.

Much the same relationships are found in the mylonites. The mylonitic foliation curves strongly around porphyroclasts of such minerals as garnet, kyanite, feldspar, and scapolite which have an almost spherical shape. On the other hand the quartz in the mylonites is always recrystallized and strainfree. This evidence suggests that metamorphic conditions still prevailed during the mylonitization, but that the peak of metamorphism occurred earlier. Thus the history of the mylonite and that of the schists, gneisses, and amphibolites above or below it is similar and this again indicates that the foliation in the mylonites is the same as the regional foliation in the other rocks. The only difference is that the mylonite zones have undergone a much larger strain.

4 CORRELATION OF THE SEVE AND THE KÖLI

It is clear from the described structural and metamorphic histories that the Seve and Köli rocks have at least a part of their histories in common. Group 2 folds in the Seve can be correlated with F_3 and F_4 in the Köli. F_2 in the Köli can be compared with the youngest of group 1 folds in the Seve. F_1 folds in the Köli can be correlated with one of the earlier generations of the Seve. However, there is no doubt that group 1 folds of the Seve are more complex than $F_1 + F_2$ in the Köli. The metamorphic history fits well with this correlation. Both F_3–F_4 and group 2 folds are postcrystalline. F_2 folds in the Köli and the youngest folds of group 1 formed when temperatures were already falling, and the peak of metamorphism is therefore older than this deformation. However, metamorphism in the Seve is more complex than in the Köli which lacks the Precaledonian metamorphism. Much more work is needed to understand the complex metamorphic history of the Seve.

5 TECTONIC MODEL

It has been suggested (e.g., Gee, 1975) that the various nappes were thrust onto the Baltic shield as a series of thin sheets.

In the model presented above, the Seve is believed to have undergone considerable ductile deformation during its emplacement. Strain magnitudes in the Köli are believed to have been smaller, but nevertheless, we suggest that much of its fabric developed by deformation during emplacement. The combined thickness of the Seve–Köli is of the order of 15 km and the width of the complex is at least 80 km. It is inconceivable that such thin, sheet-like bodies could undergo such large internal ductile deformation and at the same time be emplaced by forces pushing on the edge of the sheets as required by a rigid plate, thrusting model. Thus a different tectonic model is required and the model should be consistent not only with the deformation model proposed above, but also with the following features of the Seve–Köli.

(1) The metamorphic sequence in the lower part of the Seve–Köli is inverted and underlying metamorphic rocks are of even lower metamorphic grade.
(2) Metamorphism associated with thrusting is retrograde.
(3) Strains within the thrust slices associated with thrusting are very high, particularly in the Seve, and are much lower in the underlying rocks.

Two models are consistent with the interpretation of the deformation and the formation of the nappes. The first interprets the Seve–Köli as a fossilized Benioff zone exposed by isostasy and erosion (see Ernst,

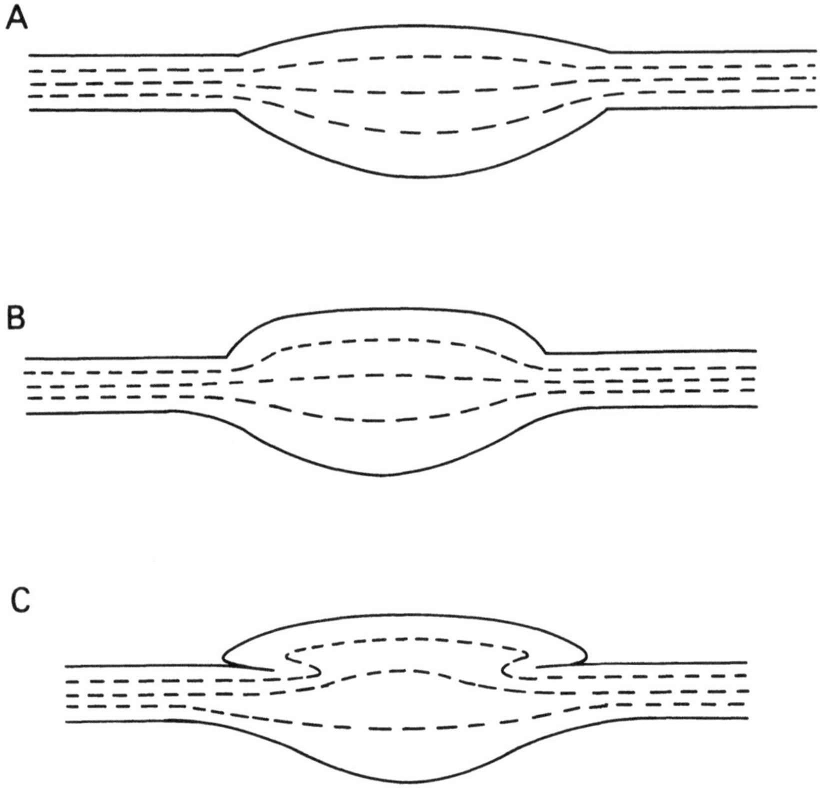

1970). The thrusts are then interpreted as local highs in strain magnitude in a zone of general heterogeneous shear strain as suggested above. The second model interprets the nappes as structures which formed under the influence of body forces due to gravity. This interpretation envisages a body of rock flowing away from a topographic high and the thrusts are again interpreted as local highs in strain magnitude or as the surface onto which the material flowed.

The problem with the first of these models is that it requires that the Baltic shield consisting mainly of granitic rocks be subducted under the denser Seve–Köli for a minimum distance of 80 km. Furthermore, if Gee's interpretation (Gee, 1975) in which the Seve–Köli complex roots west of the present Norwegian coast is correct, as seems reasonable, then the subduction would need to be in excess of 200 km.

The gravity model seems more promising and has received much attention in recent years (Crosby, 1973; de Jong and Scholten, 1973). Many variations on this type of model exist in the literature. For example Price (1973) has interpreted the development of Nappes in the Canadian Rockies in terms of up-slope gravity flow away from a topographic high,

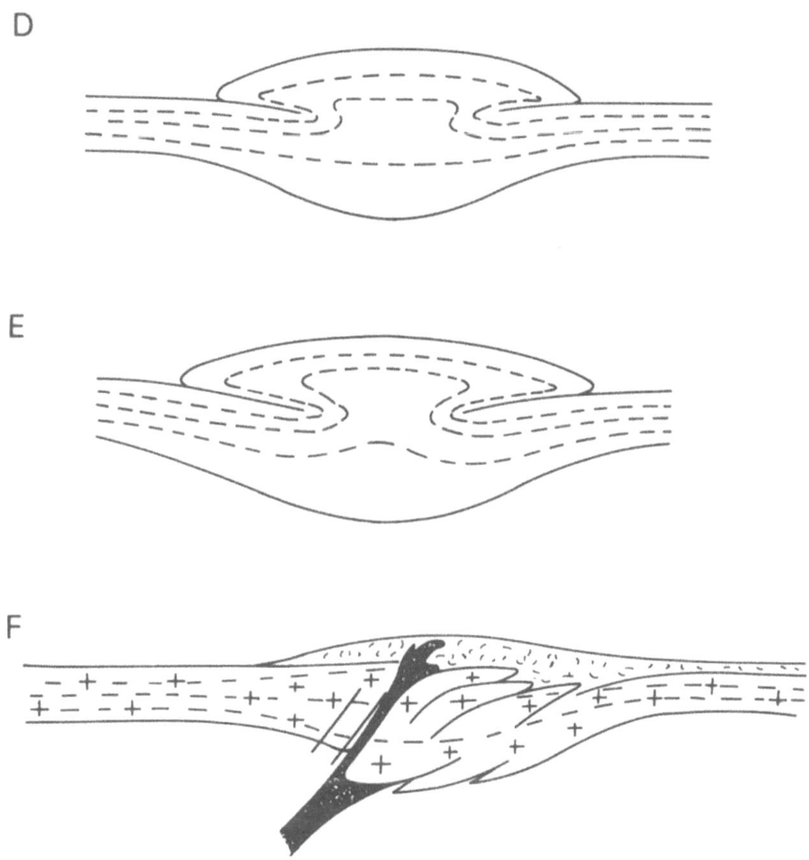

Fig. 5a–f. Diagrammatic representation of the suggested tectonic model (a–e) shows the way in which a homogeneous crust, locally weakened, might deform in response to shortening and uplift. *Broken lines* act as markers to record the deformation; they can be thought of as isograds, though like the rest of the model they are greatly oversimplified. Model is shown here as symmetrical since homogeneity of crust is assumed but departures from homogeneity would obviously result in a more complex deformation. (f) depicts one possible geological situation that could approximate the simplified crustal structure represented in (a). It represents a continent–continent collision and is modified from Dewey and Bird (1970).

but there the internal deformation in the nappe sheets is very weak compared to that in the Seve–Köli. Ramberg (1967) has modeled nappe formation in viscous materials and the model has recently been explored further by Talbot (1974). These models, however, are concerned with emplacing nappes of a given average density onto a substratum of lower density whereas in the case of the Seve–Köli the nappes are more dense than the substratum. However, this does not necessarily exclude density contrast as the driving force and we suggest the following as a possible model for the emplacement of the Seve–Köli nappe complex.

The initial stage of the deformational metamorphic history was a continental collision. This gave rise to an elongate belt of thickened crust which in its lower levels, due to the depth of burial, underwent high pressure and high temperature metamorphism. At least some of the group 1 folds probably formed at this stage. A further consequence of the crustal thickening was a ridge of high ground from which material tended to flow onto the low ground flanking it. Locally this flow may have been uphill (see de Jong and Scholten, 1973, p. 12) but its overall effect was to lower the ridge and thereby reduce the instability. The mountain range, however, continued to rise due to isostatic adjustment as well as to possible continued shortening of the crust in response to plate movements. In this way the high grade metamorphic rocks in the axial zone were brought closer and closer to the surface into the region of gravity flow where nappes were developing. There the environment was one of flow with very large noncoaxial strains predominating. Thus, the early formed structures were modified and new structures such as boudinage and later group 1 folds developed. During this deformation the rocks were being brought gradually closer to the surface so that both temperature and pressure were falling and thus the lower grade metamorphic assemblages were developing.

Figure 5 represents this process very diagrammatically, it makes no attempt to show details of the initial structure since due to the intense deformation associated with nappe emplacement we can only speculate as to its initial appearance. However, it shows in principle how the high pressure, high temperature rocks could be brought to the surface and emplaced on low grade rocks. The initial thickening of the continental crust could well have been due to the collision of two continental plates either by flattening of one against the other or by one overriding the other. Details of this type would also decide the configuration of the isograds. Here we have assumed a very simple set of isograds to demonstrate how they could be inverted and attenuated by the deformation.

Emplacement of the nappes resulted in a density inversion which may have been responsible for the development of the late stage (group 2) dome and basin structure (see e.g. Ramberg, 1967; Fletcher, 1972; Talbot, 1974). This process may or may not have been accompanied by continued shortening.

ACKNOWLEDGMENTS

We are grateful to Dr. P. L. Hancock and Dr. A. Senior for reading the manuscript. The drawings were made by J. Bult and K. A. Hakkert; the typing was done by Miss T. W. Terpstra and Miss L. van der Wijk. To them we express our sincere thanks.

REFERENCES

Calon, T. J.: De geologie van het Kittelfjäll-Daunatjåkko gebied. Internal Rep. Geol. Inst., Leiden (1973)

Crosby, G. W.: The mechanical significance of deformation within overthrust plates. Brigham Young Univ. Geol. Stud. *20* (4), 117–136 (1973)

De Jong, K. A. and Scholten, R. (eds.): Gravity and Tectonics, Wiley, New York, 502 pp. (1973)

Dewey, J. F. and Bird, J. M.: Lithosphere plate-continental margin tectonics and the evolution of the Applachian Orogen. Geol. Soc. Amer. Bull. 81, 103–106 (1973)

Ernst, W. G.: Tectonic contact between the Franciscan Mélange and the Great Valley Sequence, crustal expression of a Late Mesozoic Benioff Zone. J. Geophys. Res., *75* (5), 886–901 (1970)

Fletcher, R. C.: Application of a mathematical model to the emplacement of mantled gneiss domes. Am. J. Sci. *272,* 197–216 (1972)

Gee, D. G.: A tectonic model for the Central Part of the Scandinavian Caledonides. Am. J. Sci. *275*A, 468–515 (1975)

Hobbs, B. E., Means, W. D., Williams, P. F.: An Outline of Structural Geology. John Wiley and Sons, New York, 571 pp. (1976)

Johnson, M. R. W.: The Moine thrust: a discussion. J. Geol., *73* (4), 672–676 (1965)

Kulling, O.: Bergbyggnaden inom Björkvattnet-Virisenområdet i Västerbottensfjällens centrala del. Geol. För. Stockholm Förhandl. *55,* 167–422 (1933)

Olesen, N. Ø.; Hansen, E. S., Kristensen, L. H., and Thyrsted, T.: A preliminary account of the geology of the Selbu-Tydal Area, the Trondheim Region, Central Norwegian Caledonides. Leidse Geol. Meded. *49,* 259–276 (1973)

Price, R. A.: Large-scale gravitational flow of supracrustal rocks, Southern Canadian Rockies. In: Gravity and Tectonics. De Jong, K. A., Scholten, R.: (eds.). John Wiley and Sons, New York, pp. 491–502 (1973)

Ramberg, H.: Gravity, Deformation and the Earth's Crust as Studied by Centrifuged Models. London-New York: Academic Press, 1967

Ramberg, H.: Model studies of gravity-controlled tectonics by the centrifuge technique. In: Gravity and Tectonics. De Jong, K. A., Scholten, R., (eds.). John Wiley and Sons, New York, pp. 49–66 (1973)

Talbot, C. J.: Fold nappes as asymmetric mantled gneiss domes and ensialic orogeny. Tectonophysics *24*, 259–276 (1974)

Törnebohm, A. E.: En geognostisk profil över den skandinaviska fjällryggen mellan Östrsund och Levanger. Sveriges Geol. Undersökn. Arsbok, Ser. *C 6* (1872)

Trouw, R.: Structural geology of the Marsfjällen area, Caledonides of Västerbotten, Sweden. Sveriges Geol. Undersokn. Arsbok, Ser. *C 689,* 1–115 (1973)

Zachrisson, E.: Caledonian geology of Northern Jämtland-Southern Västerbotten. Sveriges Geol. Undersokr. Arsbok, Ser. *C 644,* 1–33 (1969)

Zachrisson, E.: The structural setting of the Stekenjokk ore bodies, Central Swedish Caledonides. Econ. Geol., *66,* 641–652 (1971)

Zwart, H. J.: Structure and metamorphism in the Seve–Köli Nappe Complex (Scandinavian Caledonides) and its implications concerning the formation of metamorphic nappes. Centenaire Soc. Géol. Belgique Géol. Dom. Crist., Liège, 129–144 (1974)

Zwart, H. J., Senior, A.: Granulite facies rocks in the Seve nappe. (In preparation)

10

A. MIYASHIRO

Abstract Although it is often said that typical ophiolites show a definite pseudostratigraphic sequence of rocks and occur in ancient subduction zones, most ophiolites showing a typical sequence occur outside subduction zones, and most ophiolites in subduction zones do not show a regular sequence. So, in this paper, large-scale mafic and ultramafic rock associations in subduction zones and other parts of orogenic belts are called ophiolites, regardless of the presence or absence of a regular sequence. Ophiolites in this sense occur not only in ancient subduction zones but also in ancient island arcs.

Subduction-zone ophiolites and island-arc ophiolites occur in two parallel belts in Hokkaido and are associated with the Kamuikotan high-pressure and the Hidaka low-pressure metamorphic belt, respectively. They were emplaced originally in the subduction zone and the island arc, respectively, formed along an early Tertiary plate boundary.

Three belts of ophiolites occur in northern Greece. Two of them may be regarded as representing subduction-zone and island-arc ophiolites that were originally emplaced along a Mesozoic plate boundary.

SUBDUCTION-ZONE OPHIOLITES AND ISLAND-ARC OPHIOLITES

Island-arc ophiolites form in immature arcs as well as in arcs transitional between the immature and mature stages but do not form in typical mature arcs. Not only in subduction-zone ophiolites but also in island-arc ophiolites, a majority of peridotites may be fragments of the upper mantle pushed up in the solid state. Mafic rocks in subduction-zone ophiolites differ in chemical composition from those in island-arc ophiolites.

The ophiolites in subduction zones, in obduction zones and in island arcs represent three major classes in a classification of ophiolites according to the mode and place of their emplacement. Each of the three probably includes rocks originally created in two or more different tectonic settings.

1 INTRODUCTION

Many authors have claimed recently that the ophiolites are fragments of oceanic crust created in midoceanic ridges (e.g., Dietz, 1963; Moores and Vine, 1971; Gass and Smewing, 1973). This view was opposed by Miyashiro (1973c; 1975a, b), who claimed that ophiolites are created not only in midoceanic ridges but also in some other tectonic settings including island arcs in particular, and that ophiolites of different origins show different chemical charateristics. On the basis of chemical evidence the present author suggests that the Troodos complex in Cyprus and the Vourinos complex in Greece are island-arc ophiolites. Unfortunately, however, geologic relations of the Cypriote and Greek ophiolites are ambiguous. Therefore, it is important that a belt of island-arc ophiolites whose geologic relations are clear, occur in Hokkaido, Japan. The origin of island-arc ophiolites is closely related to the evolution of island arcs. Low-pressure regional metamorphism commonly occurs in island-arc zones. The island-arc ophiolites in Hokkaido are associated with a low-pressure metamorphic belt.

Convergent plate boundaries usually have an island arc on one side and a subduction zone on the other. It is well known that ophiolites, irrespective of their origin, occur in subduction zones such as the Franciscan terrane in California and the Sanbagawa belt in Japan. As high-pressure (glaucophanitic) metamorphism commonly takes place in subduction zones, ophiolites in such zones are commonly associated with high-pressure metamorphic belts. This is the case with Hokkaido.

It appears to me that the subduction-zone ophiolites and the island-arc ophiolites together with obducted ophiolites represent three major categories of ophiolites on earth. The nature of ophiolite belts in Greece and the problem of the chemical characteristics of ophiolites will be discussed from this viewpoint.

2 CONCEPTS OF OPHIOLITE

The term ophiolite is ambiguous. I am not well acquainted with the history of the term, but apparently it has been widely used in Europe as a field name for mafic and ultramafic rocks (commonly with more emphasis on the latter) in orogenic belts, particularly in the Mediterranean region, regardless of their structure and metamorphism. They were generally ascribed to geosynclinal magmatism (e.g., Tyrrell, 1955). When the hypothesis of sea-floor spreading was introduced, the Mediterranean ophiolites were suggested to represent rock assemblages created in an ancient midoceanic ridge (Hess, 1965). On the other hand, mafic and ultramafic rocks of ancient subduction zones, including the Franciscan terrane of California in particular, were regarded as typical ophiolites which represent fragments of ancient oceanic crust originally created in midoceanic ridges (Dietz, 1963). This idea was incorporated in most geologic interpretations of plate tectonics (Hamilton, 1969; Dewey and Bird, 1970; Dickinson, 1971).

It was known for many years that some such ophiolites show a regular pseudostratigraphic sequence, which commonly is an ultramafic mass → mafic plutonic rocks → mafic parallel dike swarm → basaltic rocks (pillow lavas) from base to top. A dike swarm is absent in some cases. Such sequences came to be widely regarded as representing the structure of the original oceanic crust (Moores and Vine, 1971; Dewey and Bird, 1971; Penrose Conference Participants, 1972). When mafic and ultramafic masses of orogenic belts did not show such a sequence, it was commonly assumed (simply assumed) that they had originally had such a sequence but later were dismembered.

In short, varying ideas of ophiolite are (1) the ophiolites show a pseudostratigraphic ophiolite sequence, at least in their original state, and (2) the most typical ophiolites occur in ancient subduction zones. However, a conflict appears to exist between the two.

Most of the well-investigated typical ophiolite sequences occur in obducted masses, and not in subduction-zone rock complexes; e.g., the Bay of Islands complex in Newfoundland and the Semail complex in the Oman Mountains (Dewey, 1976). The site of emplacement of the ophiolite sequences of the Troodos complex in Cyprus and the Vourinos complex in Greece also do not appear to be subduction zones. On the other hand, it is known that even though some masses showing an ophiolite or ophiolite-like sequence in subduction zones may exist, most of the mafic and ultramafic masses now exposed in ancient subduction zones (or high-pressure metamorphic belts) do not show such a sequence. There is strong evidence indicating that many of them had no regular sequence from the beginning. Mafic volcanic rocks in particular usually form beds conformable with the underlying and overlying sedimentary beds, and thus do not represent fragments of any ophiolite sequences (Miyashiro, 1975b).

Thus, if the presence of a definite sequence is regarded as essential to the concept of ophiolites, most mafic and ultramafic rocks of subduction zones are not ophiolites, and many subduction zones may not have ophiolite at all. On the other hand, if mafic and ultramafic rocks in subduction zones are typical ophiolites, probably the presence of a definite sequence cannot be essential to the concept of ophiolites.

In this paper, just as in my previous ones (Miyashiro, 1975b), I use the term ophiolite for any large-scale mafic and ultramafic rock associations in orogenic belts (including subduction zones) regardless of their structure and metamorphism. It appears that only some of such ophiolites in subduction zones as well as in island arcs show ophiolite sequences. This usage is practically the same as that in many recent tectonic syntheses of various regions in the world (e.g., Smith, 1971). There, belts of ophiolites are delineated without paying any attention to the presence or absence of regular sequences. If only the masses with regular sequences are regarded as ophiolites, many of such belts cannot be called ophiolite belts. I do not intend to claim that the usage accepted in this paper is preferable to any others. The present discussion is only to focus attention on the significance of different usages.

The problem of regular sequences will be discussed again at the end of this paper.

3 EVOLUTION OF ISLAND ARCS AND THE UNDERLYING MANTLES

Island arcs show evolution, that is, a secular progressive change as follows: (1) volcanic rocks are piled up resulting in thickening of the crust; (2) the average composition of volcanic rocks becomes more and more felsic and potassic with the passage of time (Jakeš and White, 1971; Condie, 1973; Miyashiro, 1974); (3) volcanic rocks change from mainly tholeiitic trends to mainly calc-alkalic trends (Jakeš and White, 1971; Miyashiro, 1974) as shown in Figure 1; (4) plutonic rocks ranging from gabbro to granite show compositional changes more or less analogous to those of volcanic rocks as will be discussed later; (5) regional metamorphism makes the arc crust more crystalline. Thus, the crusts of immature arcs such as the Kermadecs are thin and mafic, and in this meaning may be called an oceanic-type crust, whereas the crusts of mature arcs such as Northeast Japan are thick and more felsic on an average, and may be called a continental-type crust.

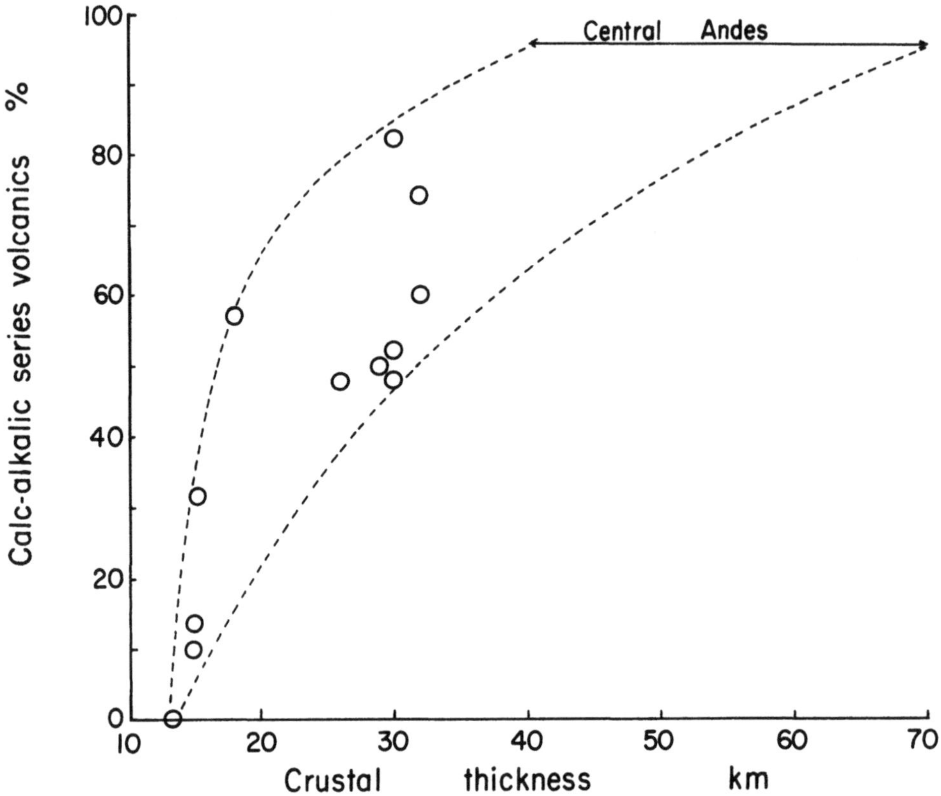

Fig. 1. Relationship between the crustal thickness and the proportion of calc-alkalic series rocks among all the volcanics in island arcs and active continental margins (Miyashiro, 1974)

The present-day arcs with a well-developed continental-type crust have a long history of evolution beginning in Mesozoic or earlier time. Some old arcs such as Japan and New Zealand have Precambrian terranes. So, the arc evolution from an immature to a mature stage takes place over a time span of more than 100 million years.

In the course of such secular evolution, the mantle beneath the arc should also suffer some changes. Island arc magmas may be generated by partial melting of the former oceanic crust at the top of the descending lithospheric slab, or by partial melting in the upper mantle wedge above the slab, or both (e.g., Ringwood, 1974; Miyashiro, 1974). If andesitic magma, generated in the descending slab, rises through the upper mantle wedge, the peridotite in the wedge would be impregnated by felsic materials coming from the rising magma. On the other hand, if basaltic or andesitic magma is generated by partial melting in the upper mantle wedge, the solid residue should become depleted in felsic components. So, it is conceivable that the upper mantle beneath an arc comes to differ in composition from place to place. Moreover, Ringwood (1969, 1974) and Dickinson and Luth (1971) have emphasized that the descending slabs are highly refractory and should be accumulated at depth in the mesosphere in the mantle.

As a result of secular processes, the buoyance of the arc lithosphere increases with resultant general increase in the height of islands above sea level as well as in their areal extent. So, rock disintegration and sedimentation become increasingly more extensive, resulting in a secular increase of the proportion of sedimentary and metasedimentary rocks.

Volcanic arcs on continental margins such as the Andes have a continental crust, similar to or thicker than, the continental-type crust of mature arcs. In this respect they may be regarded as analogous to mature arcs or as representing a more advanced stage of evolution.

Karig (1971) has claimed that marginal seas are created by splitting of continental margins or preexisting arcs. If it is the case, the initial crust of a newly formed island arc is an inheritance from the preexisting structure, independent of igneous and other activity of the new arc. If the volume ratio of such an initial crust to later igneous materials is high, the thickness of the arc crust should not show any regular correlation with later arc evolution. Actually, however, such a correlation is observed (e.g., Fig. 1). So, the possible initial crust of island arcs may be ignored in our discussions as a first approximation.

4 IGNEOUS AND METAMORPHIC BELTS IN CONVERGENT PLATE BOUNDARIES

4.1 Convergent Boundaries with Mature Island Arcs

Paired metamorphic belts, representing ancient convergent plate boundaries, occur in many areas of the circum-Pacific region (Miyashiro, 1961, 1972, 1973a, b; particulary Fig. 1, 1973a). A pair is composed of a high-pressure metamorphic belt in a subduction zone and a low-pressure metamorphic belt in the associated island arc zone. The low-pressure belts in most pairs have a thick continental-type crust. Such pairs may be regarded as representing convergent boundaries with mature island arcs or active continental margins. Typical examples are the Sanbagawa-Ryoke pair in southwest Japan and the Franciscan-Sierra Nevada pair in California.

Some ancient convergent boundaries appear to have only a single (unpaired) metamorphic belt. Such single belts would form, for example, if high pressure metamorphism did not occur in the subduction zone because of too slow a rate of underthrusting of the oceanic plate, or if low-pressure metamorphism did not occur in the island-arc zone because of the scarcity of rising magmas, or if metamorphic recrystallization did not occur either in the subduction zone or in the island-arc zone. An apparently single metamorphic belt may form, if a once-formed high-pressure belt was obliterated by overprinting of a later metamorphic event.

Paired metamorphic belts are usually accompanied by contrasting igneous belts. High-pressure metamorphic belts (ancient subduction zones) usually have ophiolites, whereas low-pressure metamorphic belts (ancient island-arc zones) usually have granitic rocks (Miyashiro 1961, 1972; Zwart, 1967). Mature island arcs are usually capped by volcanic piles made up of basalt, andesite and rhyolite of the tholeiitic and calc-alkalic series (Miyashiro, 1972, 1974). Deeper parts of such piles are commonly weakly altered (or metamorphosed) to become spilite-keratophyre associations (Rocci and Juteau, 1968).

4.2 Convergent Boundaries with Immature or Transitional-Stage Island Arcs

Among the present-day island arcs, immature arcs are relatively common; e.g., the Kermadecs, Tongas and Izu-Bonin. Therefore we have considerable knowledge of the characteristics of volcanic piles above sea level in immature arcs. On the other hand, few examples of immature arcs are known to occur in ancient geologic formations now exposed on land. It appears that immature arcs come to be exposed on land only relatively rarely. This may be because immature arc lithospheres would usually have too little buoyance to keep a major part of them above sea level for denudation. So, our knowledge of the deeper parts of immature arcs is very incomplete and obscure.

The volcanic piles of immature arcs are made up mainly of basaltic rocks of the tholeiitic (not calc-alkalic) series. (In this respect, the Tongas are an exception, being rich in tholeiitic andesites.) Such volcanic piles are probably underlain by gabbroic rocks. Gabbroic inclusions occur rather commonly in such volcanic piles (e. g., Isshiki, 1958). Such volcanic and gabbroic rocks may be underlain by an oceanic crust, because some island arc rock piles are built on preexisting oceanic crusts (Miyashiro, 1975b).

Fisher and Engel (1969) recorded basaltic, gabbroic and peridotitic rocks dredged from the eastern slope of the volcanic piles of the Tongas. Part of the peridotitic rocks in such arc rock assemblages could be cumulates in gabbroic intrusions, but the rest may be fragments of the underlying mantle pushed up into island arc igneous complexes. Granitic rocks do not occur in immature arcs. So, the igneous association characteristic of immature arcs are ophiolitic (i.e., mafic + ultramafic), as was emphasized by Ewart and Bryan (1972) and Miyashiro (1973c, 1975b).

Low-pressure regional metamorphism may or may not occur at depth of immature arcs. Since there are no granitic magmas and associated H_2O, heat transfer would not be so intense as in mature arcs.

We have no direct data on metamorphic and igneous rocks from the subduction zones associated with immature arcs. However, it is likely that high-pressure metamorphism and ophiolite emplacement occur in such subduction zones just as they occur in such zones associated with mature arcs, because there is no reason to suppose that the former differ greatly from the latter. Therefore, such convergent boundaries should be characterized by the occurrence of two parallel belts of ophiolites: one in the subduction zone and the other in the island arc zone.

With advancing evolution of the arc, granitic rocks begin to occur, and the ratio of granitic rocks to ophiolitic rocks increases. At the same time, the ratio of calc-alkalic to tholeiitic series rocks increases in volcanic piles (Fig. 1). Paired metamorphic belts form. Such features of a transitional-stage arc are well observed in the ancient arc of the Hidaka low-pressure metamorphic belt in Hokkaido to be described in the next section. When arcs become well matured, ophiolitic rocks disappear and the ratio of calc-alkalic rocks to tholeiitic series rocks becomes near unity or higher.

5 ANCIENT CONVERGENT BOUNDARY WITH A TRANSITIONAL-STAGE ARC IN HOKKAIDO

5.1 General Features

Figure 2 shows a map of northern Japan, the pertinent part of which is shown in a greater detail in Figure 3. In Hokkaido, paired metamorphic belts, consisting of the Kamuikotan high-pressure metamorphic belt on the west side and the Hidaka low-pressure metamorphic belt on the east side, run in an N–S direction. The two metamorphic belts are separated by an unmetamorphosed (or nearly unmetamorphosed) zone, 10–20 km wide, made up of Mesozoic sedimentary rocks (Cretaceous Yezo group and Jurassic Kamui group). This unmetamorphosed zone is in contact with the Hidaka metamorphic belt on the east through a high-angle thrust dipping to the east.

The Kamuikotan metamorphic belt represents a large anticlinorium. The core of the anticlinorium is made up of high-pressure metamorphic rock masses, which are overlain by Jurassic and Cretaceous volcanic and sedimentary formations that are exposed on both sides of the metamorphic core. Probably high-pressure metamorphic rocks that were produced along the top of the descending slab were broken into blocks, which were then pushed up into the hanging wall to create the anticlinorium. In this interpretation, the trench was not in the Kamuikotan belt (or

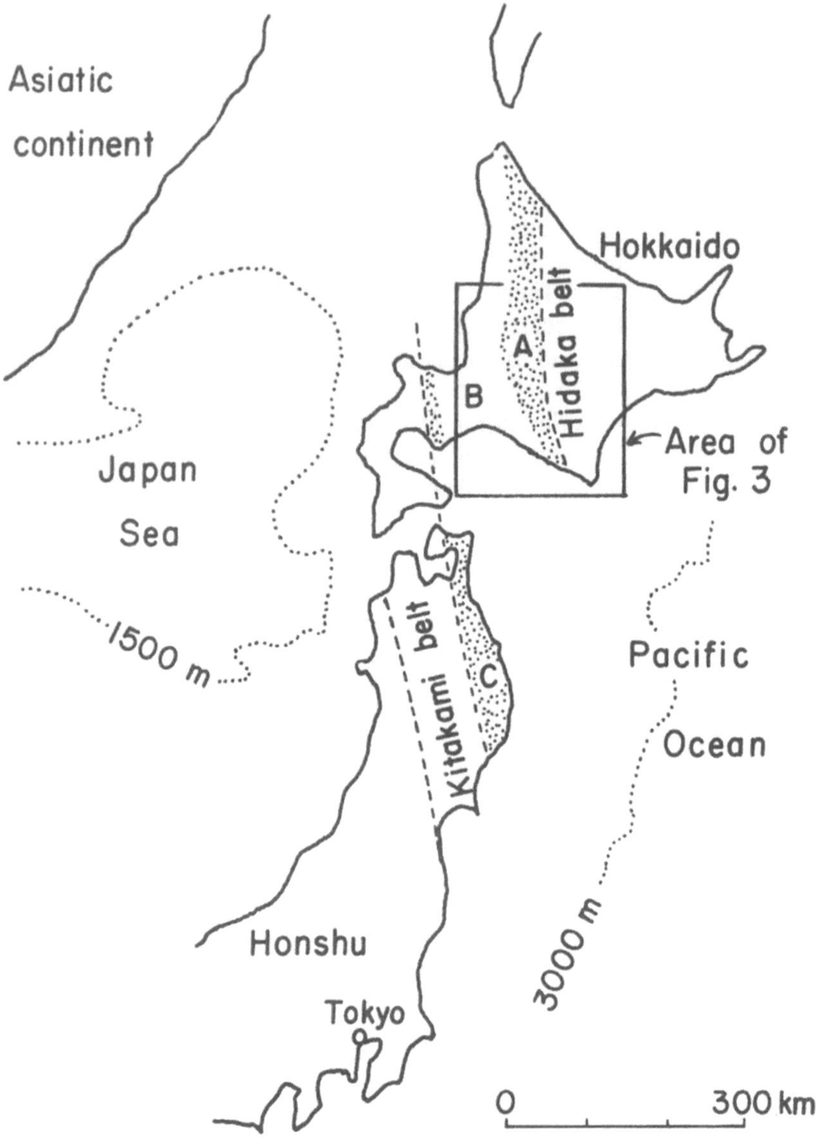

Fig. 2. Hokkaido and northern Honshu, Japan. The region between the Hidaka and Kitakami belts is divided into three zones *A, B,* and *C*. Zone *A:* Kamuikotan metamorphic belt and Mesozoic formations; Zone *B:* Tertiary and Quaternary formations; Zone *C:* characterized by thick Jurassic sediments and Cretaceous granites

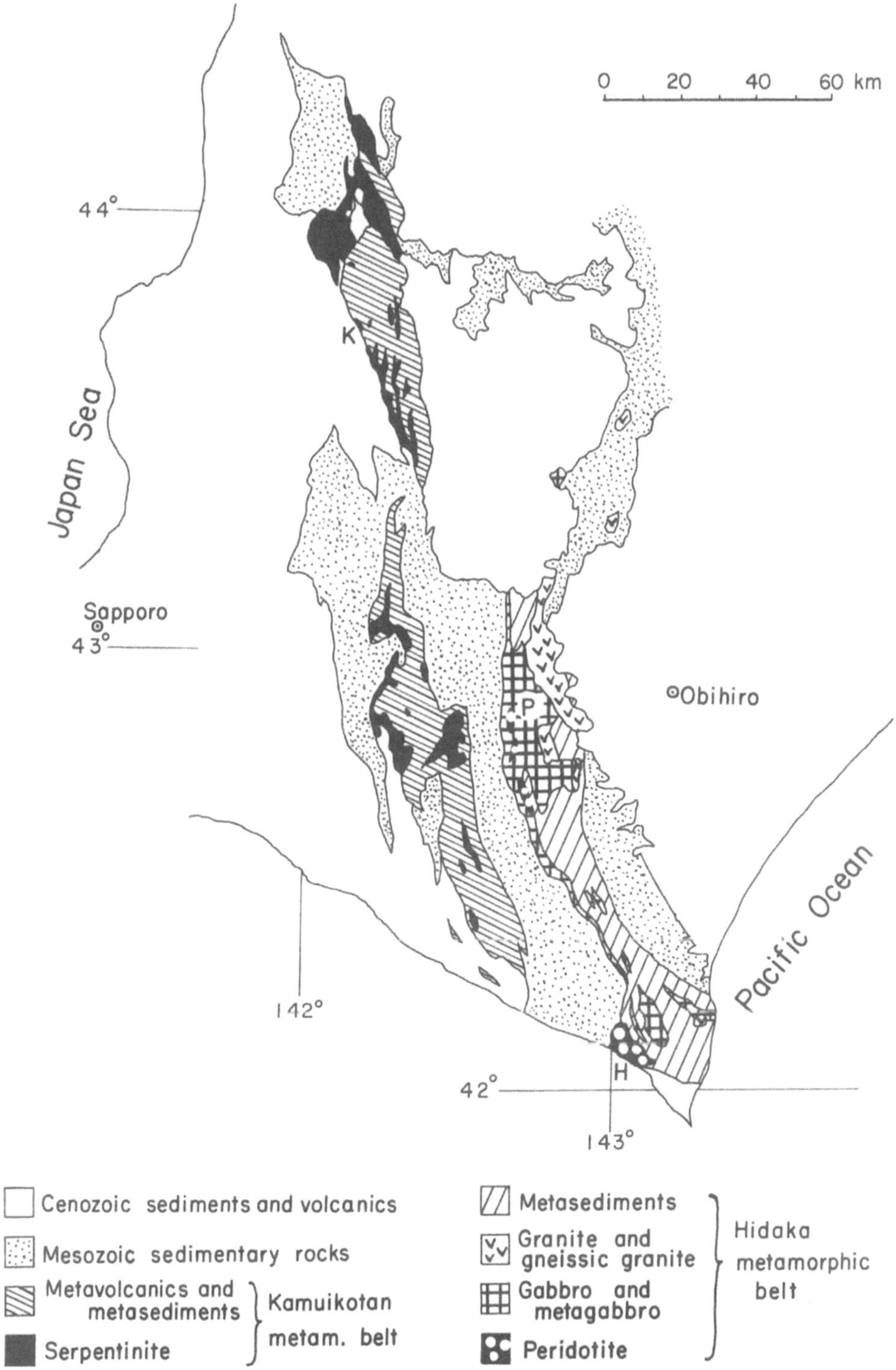

Fig. 3. Kamuikotan high-pressure and the Hidaka low-pressure metamorphic belts in Hokkaido. (The area of this map is shown in Fig. 2.) *K:* Kamuikotan; *P:* Poroshiri-dake; *H:* Horoman. Based on Hashimoto (1975) and others

the anticlinorium) but was at some distance west. In the metamorphic terrane presently exposed, the temperature of metamorphism increases westward in part (Hervé, 1975) and to both the east and west from a central line in another part (Shibakusa, 1974).

The Kamuikotan and Hidaka metamorphic belts are broadly coeval. The majority of metamorphic rocks in both belts were derived probably from Jurassic and early Cretaceous geosynclinal formations by early Tertiary orogeny and metamorphism. In Miocene time, the metamorphic complexes were already widely exposed at the surface as is indicated by metamorphic rock fragments forming Miocene conglomerate formations.

While the Kamuikotan metamorphic belt represents a subduction zone and contains a belt of ophiolite, the Hidaka metamorphic belt representing an island-arc zone, contain not only an ophiolite belt but also granite masses. The Hidaka belt shows positive Bouguer anomalies up to about +50 mgals. This indicates that ophiolitic rocks are abundant not only on the surface but also at considerable depth. Probably this means that the Hidaka metamorphic belt represents an early Tertiary island arc transitional between the immature and mature stages. In this pair the high-pressure belt (subduction zone) is on the west (Asiatic) side of the low-pressure belt (arc zone). In other words, the pair appears to represent a reverse arc, in which the marginal sea floor (or the interarc basin) was underthrust eastward beneath the island arc. This may be the main reason why the evolution of this island arc was halted in the transitional stage. The underthrusting and the growth of the arc were probably of a short duration, because it must have finished when the whole marginal sea floor was subducted.

5.2 Subduction-Zone Ophiolites of the Kamuikotan Belt

The ophiolite belt of the Kamuikotan high-pressure metamorphic belt is a chain of large and small masses of serpentinite together with metabasites (only incipiently metamorphosed basaltic rocks, glaucophane schist, greenschist, and the like). No regular sequences were found. The largest of the serpentinites measures 45 by 7 km in exposure (Suzuki and Suzuki, 1959). Unserpentinized peridotite is rare. An exceptionally large mass of unserpentinized peridotite, measuring 1 by 1.5 km, was found at Iwanai-dake. It is composed of layers of dunite and ortho- and clinopyroxene-bearing dunites. It is not clear whether these unserpentinized peridotites resemble the original rocks from which serpentinites were derived.

5.3 Island-Arc Ophiolites and Granitic Rocks of the Hidaka Belt

The Hidaka low-pressure metamorphic belt shows a thermal axis to which the grade of metamorphism increases from both sides. This axial (median) zone is made up of recrystallized, partly mobilized coarse-grained metasedimentary gneisses commonly containing cordierite.

Hashimoto (1975) demonstrated that the Hidaka belt shows an asymmetric structure in three respects: (1) the metamorphic rocks west of the axis is more strongly schistose than those east, (2) plutonic rocks are progressively more silicic (felsic) eastward, and (3) the time of their emplacement is mostly prior to the end of regional metamorphism in the western and axial parts of the belt and is mostly posterior in the eastern part. Thus, the plutonic rocks in the western part are peridotites and metagabbro (in the epidote amphibolite and amphibolite facies), and those in the axial zone are metagabbros (in the amphibolite and granulite facies), whereas those in the eastern part are gabbros, diorites and granites, mostly unmetamorphosed. In other words, a belt of metamorphosed ophiolites occurs in the western and axial part of the Hidaka belt, whereas a belt of granites accompanied by gabbros and diorites occurs in the eastern part. An ophiolite sequence has not been found.

The asymmetric structure of the Hidaka belt resembles that of the Sierra Nevada in California, which is paired with the Franciscan high-pressure metamorphic terrane on the west. In both Hidaka and the Sierra Nevada, the schistosity of metamorphic rocks is stronger, and the plutonic rocks are more mafic in the subduction-zone side half than in the other half. Thus, such an asymmetric structure may be common in island arc zones.

The ophiolites of the Hidaka belt may be divided into two groups:

(1) Rocks that were created by magmatism in an ancient island arc. This group includes probably most or all gabbros and metagabbros and all diorites.
(2) Rocks that were uplifted in the solid state from below the island arc rock complex, that is, from the mantle or from the marginal sea crust underlying the island arc rock complex if any (as will be discussed below). This group includes some peridotites and possibly some metagabbros. The proportions of peridotites divided between these two groups are not clear.

5.4 Petrography of Peridotites in the Hidaka Belt Ophiolites

The peridotites of the Hidaka belt are mostly fresh, i.e., unserpentinized. They are composed of a large number of layers of dunite, lherzolite and plagioclase lherzolite. It is not clear whether the peridotite masses gave any thermal effect to the surrounding regional metamorphic rocks. Such a thermal effect, if any, must be very small. Some peridotites show high-pressure mineral assemblages stable in the upper mantle though these are partly destroyed by later recrystallization (Komatsu, 1974, 1975). So, they represent fragments of upper mantle peridotites pushed into island arc assemblages.

The largest of the peridotite masses occurs at Horoman (Onuki, 1964; Nagasaki, 1966; Komatsu and Nochi, 1966). It is a sheet, measuring 8 by 10 km in exposure, broadly concordant with the surrounding metamorphic rocks. Its lower part is made up of stacking of cyclic units with an upward sequence: dunite → lherzolite → plagioclase lherzolite → lherzolite. The atomic Fe/(Mg + Fe) ratio of the peridotite varies within a very small range of 0.08–0.11. This ratio increases toward higher horizon only slightly which may not be significant. Nagasaki (1966) observed that the Fe/(Mg + Fe) ratio of orthopyroxene increased upward from about 0.07 to 0.11 within each cyclic unit, but he did not find any upward compositional variation in olivine.

5.5 Petrography of Gabbroic and Metagabbroic Rocks in the Hidaka Belt Ophiolites

Many greenschists, epidote amphibolites and amphibolites show relict textures indicating their gabbroic origin. Postmetamorphic gabbros and diorites show ordinary igneous texture and usual minerals such as olivine, hypersthene and hornblende. They show a wide range of Fe/(Mg + Fe) ratio and show not only rhythmic but also occasional graded layering, suggesting a considerable extent of crystallization differentiation (Nochi and Komatsu, 1967; Miyashita and Hashimoto, 1975). Many gabbros are cumulative with MgO as high as, say, 17%. Their compositions should show a great deviation from their liquid line of descent. So, they are not so good a clue to the tectonic setting of the pertinent magmatism as volcanic rocks. Probably the Hidaka belt was once covered by island arc volcanic piles, but the piles have been eroded away.

Hashimoto (1975, Fig. 45) gives diagrams illustrating the compositional variation of the metamorphosed and unmetamorphosed plutonic rocks, ranging from ultramafic through mafic to dioritic. In the composition range of gabbro and diorite (SI = 40–10), FeO + Fe_2O_3 and TiO_2 tend to increase with advancing fractional crystallization. Since the extent of chemical migration during metamorphism is not clear, only unmetamorphosed gabbros (excluding highly cumulative rocks), diorites and granites are shown in Figure 4. Here, a majority of gabbros and diorite, particularly pyroxene gabbros, show slowly increasing SiO_2 and TiO_2 content with increasing FeO*/MgO. This relation as well as Hashimoto's indicates that the majority of gabbros and diorites belong to the tholeiitic series. Some hornblende gabbros appear to show a tholeiitic trend and others a calc-alkalic trend. The high proportion of tholeiitic series rocks is consistent with the view that the gabbros and diorite were formed by magmatism in a transitional-stage arc, since volcanic piles of such arcs

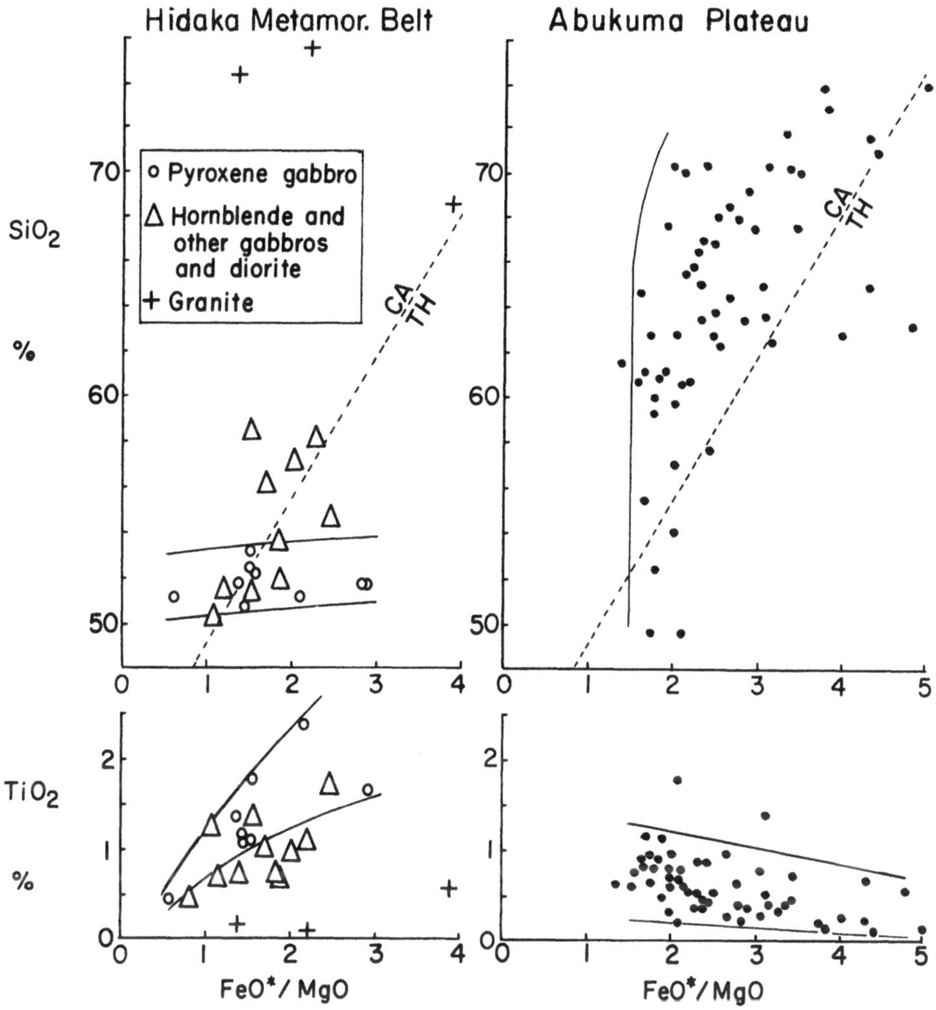

Fig. 4. Compositional variation in gabbroic, dioritic, and granitic rocks in the Hidaka and Abukuma low-pressure metamorphic belts. For FeO*/MgO > 2, dash lines indicate the boundary between the tholeiitic (TH) and calc-alkalic (CA) rocks. Full lines show the tholeiitic trend of pyroxene gabbros in the Hidaka belt and the calc-alkalic trend of most rocks in the Abukuma Plateau. FeO* means total iron as FeO

should show a similarly high proportion of tholeiitic series rocks. Granitic rocks of Hidaka probably belong to the calc-alkalic series.

For comparison, Figure 4 also shows FeO*/MgO ratio of the gabbroic, dioritic and granitic rocks in the Abukuma Plateau, Japan, which is believed to represent an ancient mature arc (Miyashiro, 1973b). SiO_2 increases very rapidly, while TiO_2 tends to decrease, with increasing FeO*/MgO. This means that most rocks belong to the calc-alkalic series. This is analogous to the high proportion of calc-alkalic series rocks in the volcanic piles of mature arcs.

6 ORIGIN OF THE PAIRED METAMORPHIC BELTS AND OPHIOLITES IN HOKKAIDO

6.1 Nature of Events

The K-Ar age of 36–16 million years for granitic rocks of the Hidaka belt (Nozawa, 1975) may be taken to represent an approximate time of intrusion and subsequent cooling. The emplacement of gabbros and regional metamorphism in the belt were probably partly broadly simultaneous with, and partly immediately prior to the granite intrusion.

The paired metamorphic belts are regarded as formed by an eastward underthrusting of a marginal sea floor beneath the island arc of the Hidaka belt, as stated before. The rate of underthrusting must have been high to cause high-pressure metamorphism in the Kamuikotan belt. So, the underthrusting and the consequent orogeny and metamorphism were probably of short duration, because the underthrusting must have come to a stop when the marginal sea floor was consumed. The duration of underthrusting, metamorphism and magmatism was probably not more than 30 million years, even if the time required for the rise of magmas through the upper mantle is taken into consideration. Therefore, we may tentatively conclude that the main phase of orogeny in the paired metamorphic belts was early Tertiary (Paleocene to Oligocene, about 60–30 million years B.P.).

The metavolcanic rocks in the ophiolites of the Kamuikotan belt were probably derived from Jurassic and very early Cretaceous volcanic formations, with which they are associated. So, there exists a large gap of time (about 60 million years long) between their eruption and the early Tertiary orogeny. These two events are presumably independent. If so, what was the tectonic setting of the eruption of the volcanic rocks of the subduction-zone ophiolites?

The volcanic rocks in subduction-zone ophiolites of many terranes including the Kamuikotan, are dominantly basaltic, belonging to both tholeiitic and alkalic series (Miyashiro, 1975b, Fig. 7 and p. 273). Not only in Hokkaido but also in some other regions, such as the Sanbagawa belt, the eruption of such rocks was partly preceded by, and partly coeval with, geosynclinal sedimentation. This suggests the existence of genetic connection between geosynclinal sedimentation and such eruption. It is assumed here for Hokkaido that a lateral extension and thinning of lithosphere occurred in Jurassic time, possibly because of the lateral dragging effect of flows in the asthenosphere caused by a mantle plume, though there is no positive evidence for it. This resulted in a wide submergence and geosynclinal sedimentation. Then in early Cretaceous time, it caused splitting of lithosphere to create a marginal sea. Intense volcanism occurred to cover the marginal sea floor as well as in many parts of adjacent geosynclinal sedimentary terranes. Probably this marginal sea was the one that was subducted during early Tertiary orogeny.

In this interpretation, the ophiolites of the Kamuikotan subduction zone may include fragments of the marginal sea floor as well as mafic and ultramafic rocks intruded into adjacent geosynclinal areas, whereas the ophiolites of the Hidaka island-arc zone were created by magmatism in situ except for peridotites uplifted from the upper mantle.

The island arc of the Hidaka belt was formed probably on, but near the eastern margin of, the marginal sea lithosphere. In other words, the Hidaka belt may be underlain by the marginal sea crust (Fig. 5). It is not impossible that some metabasites exposed in the Hidaka belt represent fragments of the marginal sea crust, broken and pushed up into the overlying island arc rock complex.

So far, I have used the term "island arc" for the Hidaka belt. However, the term magmatic arc may have been more appropriate, because the

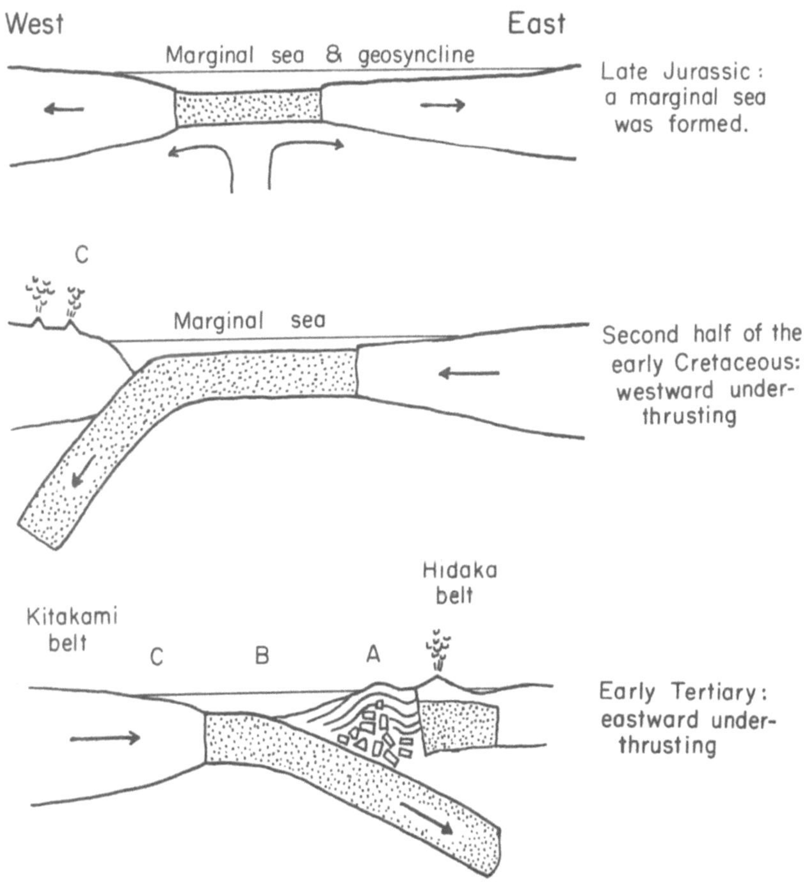

Fig. 5. Schematic diagrams (not to scale) showing the sequence of geologic events including the opening and closing of a marginal sea in western Hokkaido. Zones *A*, *B*, and *C* correspond to those in Figure 2. Zone *A*: Kamuikotan metamorphic belt and associated Mesozoic formations. The lithosphere of the marginal sea floor is *dotted*

existence at that time of a marginal sea on the east side of the Hidaka belt is possible but has not been established.

6.2 Chronological Sequence of Events

The events based on the above interpretation will be summarized below in chronological order (Fig. 5).

The Jurassic and very early Cretaceous time (190–120 million years B.P.), mantle plume activity began below the western part of Hokkaido and adjacent regions. It caused lateral extension, thinning, and then splitting of lithosphere with resultant formation of a geosyncline and then of a marginal sea (or an interarc basin). The geosyncline and marginal sea were limited on the west side by an old continental-type block called the Kitakami belt (Figs. 2 and 5), which had a pre-Silurian crystalline basement covered by Silurian to Permian geosynclinal piles.

In the latter half of early Cretaceous time (120–110 million years B.P.), the western part of the just created marginal sea floor was underthrust westward beneath the Kitakami belt with resultant extensive granitic magmatism not only in the Kitakami belt but also in zone C in Figures 2 and 5.

The marinal sea plate became sufficiently cold by early Tertiary time. Then it was underthrust eastward beneath the Hidaka belt until the whole marginal sea was consumed (60–30 million years B.P.). This underthrusting caused Kamuikotan and Hidaka metamorphism and the volcanism and plutonism in the Hidaka belt, leading to the creation of an island arc and its ophiolites.

The Kamuikotan belt (zone A in Figs. 2 and 5), i.e., the present-day zone of exposure of Kamuikotan metamorphic rocks, represents a large anticlinorium (Fig. 5). Probably the early Tertiary trench was not in this zone but was in zone *B* to the west. At that time, zone *A* was probably in the trench-arc gap. A large number of rock fragments were probably pushed up from the top of the descending slab with resultant formation of a large anticlinorium with disordered masses of metamorphic rocks in its core in the trench-arc gap.

7 OPHIOLITE BELTS OF GREECE

Regional metamorphic belts are not well-developed or not well-exposed in the Mesozoic terrane of Greece. This gives an ambiguity to our interpretation of tectonic environments of Mesozoic ophiolites of this region. Speculation will be made below on the basis of analogy to the two parallel belts of ophiolites in Hokkaido.

Smith (1971) claimed the existence of three distinct parallel belts of ophiolites, trending NW–SE, in the Hellenides (Greece and Albania). I will refer to them as the western, central, and eastern belts (Fig. 6). It appears to me that different belts have different origins, though the available data are very incomplete.

The western belt includes most of well-documented ophiolitic complexes of this region such as the Vourinos and Pindos. Miyashiro (1973c) pointed out the presence of both calc-alkalic and tholeiitic trends in the Vourinos complex, and hence suggested an island arc origin for it. The presence of a calc-alkalic trend in Vourinos was confirmed later with more data by Bebien et al. (1975). Thus, the western belt may be regarded as an island-arc ophiolite belt, although there is a possibility

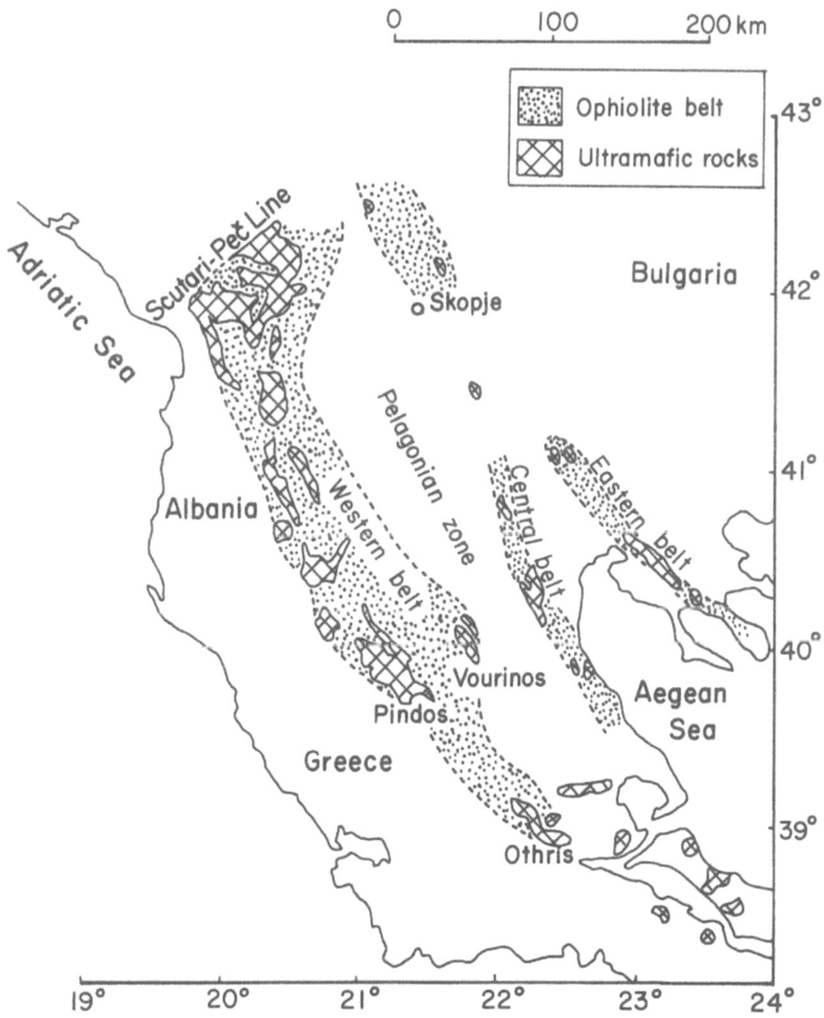

Fig. 6. Ophiolite belts in the Hellenides (Smith, 1971; Zwart, 1973)

that some ophiolite masses may be fragments of a marginal sea floor, as discussed later. The present-day surface may represent too shallow a horizon in the original arc structure to expose the low-presure regional metamorphic terrane, if any exists.

The central belt of ophiolites and its northern extension are apparently associated with a few exposures of glaucophanitic metamorphic rocks. Though the degree of closeness of association is not clear, I presume that the central belt represents subduction-zone ophiolites. Though some glaucophanitic rocks may have been derived from rocks older than the presumed time of subduction (Mesozoic), this is not inconsistent with their possible origin through the subduction.

The western and central belts of ophiolites are about 100 km apart. The intervening zone is made up of a raised block of the Paleozoic (Hercynian) crystaline basement, called the Pelagonian zone. It may be presumed that the western belt and the Pelagonian zone represent the volcanic inner arc and the nonvolcanic outer arc, respectively, of a double arc structure. This double arc and the subduction zone on the east side (now represented by the central belt of ophiolites) were formed probably along a Mesozoic plate boundary.

Probably the sea floor involved in the suduction was an inter-continental ocean basin east of the subduction zone. This ocean basin was conceivably formed by splitting of a preexisting continent. A marginal sea may have formed west of the island arc (double arc). The Vourinos complex lies in the easternmost part of the western belt. It is conceivable that this complex possibly along with a few others in the eastern part of the western belt was formed in an island arc, whereas the complexes in the western part of the western belt represent fragments of the marginal sea floor. Thus, it may be found in future that the Vourinos differs in chemical characteristics from many other ophiolites exposed to the west.

The existence, nature and origin of the eastern belt of ophiolites are outside the scope of this paper.

8 VOLCANIC ROCKS IN SUBDUCTION-ZONE OPHIOLITES AND ISLAND-ARC OPHIOLITES

8.1 Rock Series Classification

Subduction-zone ophiolites contain volcanic rocks of tholeiitic and alkalic series, and belong to class III in Miyashiro's (1975b) classification, whereas island-arc ophiolites usually contain volcanic rocks of tholeiitic and/or calc-alkalic series, and belong to his class I or II.

8.2 Troodos Ophiolitic Complex

Against the majority opinion that the Troodos ophiolitic complex in Cyprus was created in a mid-oceanic ridge (Moores and Vine, 1971; Gass and Smewing, 1973; Moores, 1975), I have proposed the idea that it was created in an island arc (Miyashiro, 1973c; 1975a, b). The evidence for an island arc origin includes (1) the occurrence of calc-alkalic series rocks, (2) the common occurrence of highly felsic (silicic) late differentiates, and (3) its high Sr isotope ratios (0.704–0.707). These are inconsistent with the hypothesis of midoceanic ridge origin.

To explain points (1) and (2), Moores (1975) and Gass et al. (1975) have assumed that intense silica metasomatism occurred in Troodos. However, a more recent study by Pantazis (1977) has demonstrated that the chemical characteristics of Troodos, including (1) and (2), may be regarded as essentially primary features, controlled by crystallization differentiation.

In the course of the above dispute, the existence of chemical difference between the Troodos complex and midoceanic ridge associations has become too clear to be ignored. So, a few recent authors have begun to assume that the Troodos complex represent a marginal sea floor or a mixture of a marginal sea floor and an island arc (Pearce, 1975; Smewing et al., 1975). Prior to these authors, I myself considered a possible relationship of the dike swarm of Troodos with marginal sea spreading (Miyashiro, 1973c, 1975a). In view of the scarcity of petrographic data of marginal sea floors (Miyashiro, 1975b), the marginal sea origin hypothesis is difficult to evaluate. Generally speaking, the existence of some ophiolite masses formed in marginal seas is possible. However, it is unlikely that the whole Troodos complex was formed in a marginal sea, because at least some parts of the complex show too close a similarity to island arc rocks in chemical composition.

8.3 Comparison of Chemical Composition

Table 1 and Figure 7 illustrate that the volcanic rocks in island-arc ophiolites such as in Troodos and Vourinos in the eastern Mediterranean region chemically resemble volcanic piles of the present-day island arcs, and that the volcanic rocks of subduction-zone ophiolites differ chemically from those of island-arc ophiolites. Volcanic rocks in island-arc ophiolites and in the present-day island arcs are relatively low in TiO_2 and show a wide range of FeO*/MgO ratio.

The composition field of volcanic rocks of subduction-zone ophiolites includes that of abyssal tholeiites (midoceanic ridge volcanics), but the former is much wider than the latter. So, part of the subduction-zone ophiolites may have been created in midoceanic ridges, but the rest must have been created in other tectonic settings. At present, the nature of the

Table 1. Ordinary ranges of Sr isotope ratio and TiO_2 content in volcanic rocks

	$^{87}Sr/^{86}Sr$	TiO_2%	FeO*/MgO
Abyssal tholeiites (midoceanic ridges)	0.702–0.704 (mostly 0.702–0.703)	0.6–2.3	0.7–2.2
Oceanic island volcanics	0.702–0.706	1.0–4.0	0.5–4.0
Island arc volcanics	0.703–0.708	0.1–1.8	0.7–7.0
Volcanic rocks in island arc ophiolites (Vourinos, Pindos, and Troodos)	0.704–0.707 (Troodos)	0.1–1.5	0.7–7.0
Volcanic rocks in subduction zone ophiolites (Franciscan and Sanbagawa high-pressure terranes)	0.703–0.708 (Franciscan)	0.1–3.4	0.5–3.5

Note: The Sr isotope data for Troodos are from Peterman et al. (1971) and Coleman and Peterman (1975), while those for the Franciscan are from Sinha and Davis (1971) and Ghent et al. (1973).

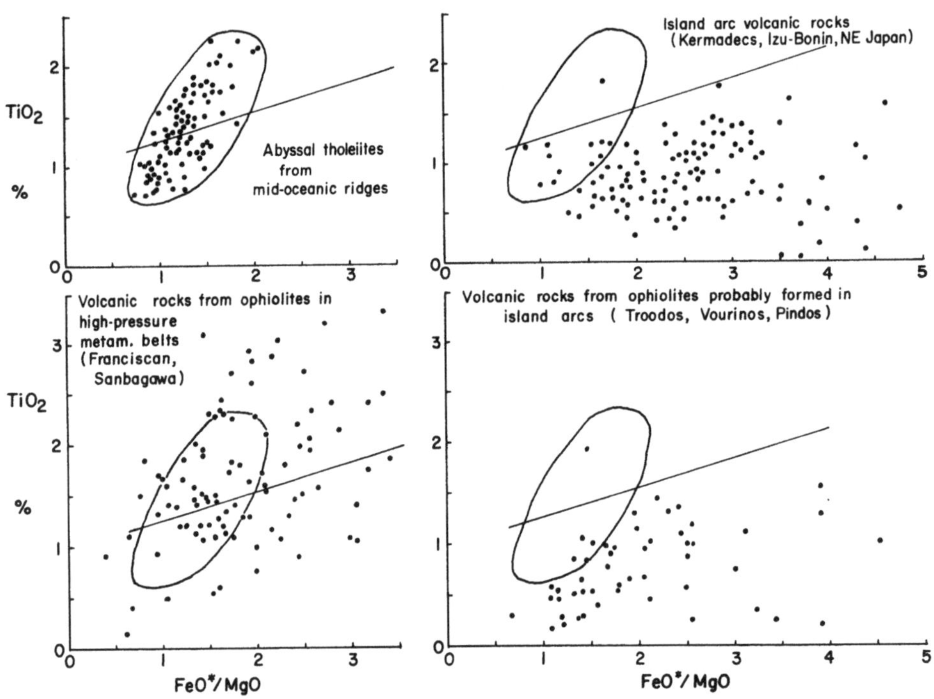

Fig. 7. Midoceanic ridge volcanic rocks (abyssal tholeiites) and present-day island arc volcanics as compared with volcanic rocks in subduction-zone (high-pressure metamorphic-belt) ophiolites and island-arc ophiolites. The outline of the composition field of abyssal tholeiites is shown in all diagrams. *Straight lines:* an upper limit of TiO_2 in ordinary volcanic rocks in island arcs. FeO* means total iron as FeO

latter tectonic settings is not clear. Three possibilities were discussed by Miyashiro (1975b): namely, they may represent (1) oceanic islands and seamounts possibly mixed with normal ocean floors, (2) continental-side walls of trenches, and (3) volcanic formations near the trench-side margins of continents or island arcs, where geosynclinal sedimentation occurs. A fourth possibility (a marginal sea floor origin) for part of the ophiolites of the Kamuikotan belt has already been discussed earlier in this paper. The third possibility means magmatism during geosynclinal sedimentation usually many million years prior to orogeny and metamorphism. We may imagine that the formation of such geosynclines is a result of mantle plume activity, and is usually accompanied by intense basaltic magmatism, and that the activity may lead to the formation of a marginal sea, just as we did in Hokkaido. In such cases, the third possibility is nearly identical to the fourth.

9 REGULAR PSEUDOSTRATIGRAPHIC SEQUENCES IN OPHIOLITES

On the basis of the foregoing discussions, the problem of regular pseudostratigraphic sequences in ophiolites will be reconsidered here.

Not only in ocean floors and marginal seas but also in immature island arcs, the general downward sequence of rocks would be basaltic volcanic piles underlain by gabbroic complexes, which in turn are underlain by mantle peridotite. This agrees generally with what is regarded as a regular ophiolite sequence. So, if fragments of oceanic and marginal sea crust as well as of immature arc crust occur in ophiolites, they may show such a sequence.

As discussed above, the regular squences in the Troodos and Vourinos complexes appear to be of island arc origin. The Troodos complcx has a marked dike swarm, whereas the Vourinos does not. The significance of dike swarm in ophiolites is not clear. The magnetic anomaly stripes in ocean floors were once widely ascribed to such dike swarms, but more recent studies have shown that this view is no longer tenable.

It is important that the regular sequences in the Bay of Islands complex in Newfoundland and in the Oman Mountains were not emplaced in subduction zones but were apparently obducted (Dewey, 1976), whatever the tectonic setting of their original creation may have been. Subduction and obduction (Coleman, 1971) are different processes, though they both may occur occasionally along the same plate boundaries.

It is suggested that some mafic and ultramafic masses in subduction zones have regular sequences. It may be true. However, it is equally important that most mafic and ultramafic masses in subduction zones do not show, and apparently have not had a regular sequence. It is often said that the mafic and ultramafic rocks of subduction zones, showing no regular sequences, lost their original sequences by dismemberment. This may be true in some cases, but it is also known that in many cases it is inconsistent with field evidence as discussed before. Field as well as chemical evidence suggests that the mafic and ultramafic rocks in subduction zones may include a few distinctly different classes having different origins, and that some of them may have had a regular sequence at least originally while others did not.

Some mafic and ultramafic masses in subduction zones and other parts of orogenic belts show what appears to be part of the regular ophiolite sequence, such as volcanic rocks underlain by gabbros, or gabbroic rocks underlain by peridotites. The origin of such partial sequences would be very diverse. The density relation and crystallization differentiation would tend to form such sequences under diverse tectonic conditions.

10 CONCLUSION

The ophiolites may be classified into several classes according to the mode and place of their emplacement. The ophiolites in subduction zones, in obduction zones and in island arcs represent three major classes in this classification.

Each of the three major classes may include rocks originally created in two or more different tectonic settings. For example, a majority of island-arc ophiolites was created by magmatism in situ but some peridotites were pushed up in the solid state from the underlying upper mantle. Part of subduction-zone ophiolites may originally have been created in mid-oceanic ridges, but the rest were formed in some other tectonic settings.

Ophiolites created in marginal sea floors may occur in, or may be associated with, any of the three. The possibility has been suggested that parts of the subduction-zone ophiolites of the Kamuikotan belt as well as parts of the western belt ophiolites, related to an island arc, in Greece may be fragments of marginal sea floors.

Regular pseudostratigraphic sequences occur in some of the island-arc as well as in obducted ophiolites, and may also occur in some subduction-zone ophiolites. It is possible that in some cases, the original sequences were lost by later dismemberment. However, many rocks in subduction-zone- and island-arc ophiolites did not have such sequences from the beginning.

ACKNOWLEDGMENTS

Drs. W. G. Ernst, M. Fonteilles, and J. J. Pamić gave me important advice and criticism. I am solely responsible for the views expressed in this paper.

REFERENCES

Bebien, J., Ohnenstetter, D., Ohnenstetter, M., Paupy, A., Rocci, G.: The role of hypabyssal magmatic rocks in our understanding of ophiolite models. Pétrologie *1*, 157–168 (1975)

Coleman, R. G.: Plate tectonic emplacement of upper mantle peridotites along continental edges. J. Geophys. Res. *76*, 1212–1222 (1971)

Coleman, R. G., Peterman, Z. E.: Oceanic plagioliparite J. Geophys. Res. *80*, 1099–1108 (1975)

Condie, K. C.: Archean magmatism and crustal thickening. Geol. Soc. Am. Bull. *84*, 2981–2992 (1973)

Dewey, J. F.: Ophiolite obduction. Tectonophysics *31*, 93–120 (1976)

Dewey, J. F., Bird, J. M.: Mountain belts and the new global tectonics. J. Geophys. Res. *75*, 2625–2647 (1970)

Dewey, J. F., Bird, J. M.: Origin and emplacement of the ophiolite suite: Appalachian ophiolites in Newfoundland. J. Geophys. Res. *76*, 3179–3206 (1971)

Dickinson, W. R.: Plate tectonics in geologic history. Science *174*, 107–113 (1971)

Dickinson, W. R., Luth, W. C.: A model for plate tectonic evolution of mantle layers. Science *174*, 400–404 (1971)

Dietz, R. S.: Alpine serpentines as oceanic rind fragments. Geol. Soc. Am. Bull. *74*, 947–952 (1963)

Ewart, A., Bryan, W. B.: Petrography and geochemistry of the igneous rocks from Eua, Tongan Islands. Geol. Soc. Am. Bull. *83*, 3281–3298 (1972)

Fisher, R. L., Engel, C. G.: Ultramafic and basaltic rocks dredged from the nearshore flank of the Tonga Trench. Geol. Soc. Am. Bull. 80, 1373–1378 (1969)

Gass, I. G.,Neary, C. R., Plant, J., Robertson, A. H. F., Simonian, K. O., Smewing, J. D., Spooner, E. T. C., Wilson, R. A. M.: Comments on "The Troodos ophiolitic complex was probably formed in an island arc," by A. Miyashiro and subsequent correspondence by A. Hynes and A. Miyashiro. Earth Planet. Sci. Lett. *25*, 236–238 (1975)

Gass, I. G., Smewing, J. D.: Intrusion, extrusion and metamorphism at constructive margins: evidence from the Troodos massif, Cyprus. Nature (London) *242*, 26–29 (1973)

Ghent, E. D., Peterman, Z. E., Coleman, R. G.: Sr^{87}/Sr^{86}, K, Na, Rb, and Sr in some eclogites and associated basalts from California and southwestern Oregon. U.S. Geol. Survey, J. Res. *1*, 643–647 (1973)

Hamilton, W.: Mesozoic California and the underflow of the Pacific mantle. Geol. Soc. Am. Bull. *80*, 2409–2430 (1969)

Hashimoto, S.: The basic plutonic rocks of the Hidaka metamorphic belt, Hokkaido. Part 1. Hokkaido Univ., J. Fac. Sci., Ser. 4, *16*, 367–420 (1975)

Hervé, F.: Petrography of the Kamuikotan metamorphic belt at the Ubun-Orowen cross section, central Hokkaido, Japan. Hokkaido Univ., J. Fac. Sci., Ser. 4, *16,* 453–470 (1975)

Hess, H. H.: Mid-oceanic ridges and tectonics of the sea-floor. In: Submarine Geology and Geophysics. Whittard, W. F., Bradshaw, R. (eds.). London: Butterworths, 1965, pp. 317–332

Isshiki, N.: Petrology of plutonic cognate ejecta from Nishi-yama volcano, Hachijojima, the Seven Izu Islands, Japan. Japan. J. Geol. Geogr. *29,* 55–74 (1958)

Jakeš, P., White, A. J. R.: Composition of island arcs and continental growth. Earth Planet. Sci. Lett. *12,* 224–230 (1971)

Karig, D. F.: Origin and development of marginal basins in the western Pacific. J. Geophys. Res. *76,* 2542–2561 (1971)

Komatsu, M.: Electron microprobe study of unmixing of aluminous pyroxenes from lherzolite and pyroxenite, Uenzaru, Hidaka Mountains, Hokkaido, Japan. Geol. Soc. Japan Mem. *11,* 47–57 (1974)

Komatsu, M.: Recrystallization of the high alumina pyroxene peridotite of the Uenzaru area in Hidaka province, Hokkaido, Japan. J. Geol. Soc. Japan *81,* 11–28 (1975)

Komatsu, M., Nochi, M.: Ultrabasic rocks in the Hidaka metamorphic belt, Hokkaido, Japan, I (in Japanese with English abstract). Chikyu-kagaku *87* 21–29 (1966)

Miyashiro A.: Evolution of metamorphic belts. J. Petrol. *2,* 277–311 (1961)

Miyashiro, A.: Metamorphism and related magmatism in plate tectonics. Am. J. Sci. *272,* 629–656 (1972)

Miyashiro, A.: Paired and unpaired metamorphic belts. Tectonophysics *17,* 241–254 (1973a)

Miyashiro, A.: Metamorphism and Metamorphic Belts. London: George Allen and New York: Halsted Press, 1973b, 492 p

Miyashiro, A.: The Troodos ophiolitic complex was probably formed in an island arc. Earth Planet. Sci. Lett. *19,* 218–224 (1973c)

Miyashiro, A.: Volcanic rock series in island arcs and active continental margins. Am. J. Sci. *274,* 321–355 (1974)

Miyashiro, A.: Origin of the Troodos and other ophiolites: a reply to Moores. Earth Planet. Sci. Lett. *25,* 227–235 (1975a)

Miyashiro, A.: Classification, characteristics, and origin of ophiolites. J. Geol. *83,* 249–281 (1975b)

Miyashita, S., Hashimoto, S.: The layered basic complex of Mt. Poroshiri, Hokkaido, Japan. Hokkaido Univ., J. Fac. Sci., Ser. 4, *16,* 421–452 (1975)

Moores, E. M.: Discussion of "Origin of Troodos and other ophiolites: a reply to Hynes." Earth Planet. Sci. Lett. *25,* 223–226 (1975)

Moores, E. M. Vine, F. J.: The Troodos massif, Cyprus and other ophiolites as oceanic crust: evaluation and implications. Phil. Trans. Roy. Soc. London, Ser. A. *268,* 443–466 (1971)

Nagasaki, H.: A layered ultrabasic complex at Horoman, Hokkaido, Japan. Tokyo Univ., J. Fac. Sci., Sec. 2, *16,* 313–346 (1966)

Nochi, M., Komatsu, M.: Ultrabasic rocks in the Hidaka metamorphic belt, Hokkaido, Japan, II (in Japanese with English abstract). Chikyu-kagaku 21, *5,* 11–26 (1967)

Nozawa, T.: Radiometric age map of Japan: granitic rocks (scale 1:2,000,000). Geol. Survey of Japan, 1975

Onuki, H.: Petrochemical research on the Horoman and Miyamori ultramafic intrusions, northern Japan. Tohoku Univ., Sci. Rept., Ser. 3, *9,* 217–276 (1964)

Pantazis, Th. M.: Major oxide geochemistry and origin of the Troodos ophiolitic complex, Cyprus. Pétrologie (1977 in press)

Pearce, J. A.: Basalt geochemistry used to investigate past tectonic environments on Cyprus. Tectonophysics *25,* 41–67 (1975)

Penrose Conference Participants: Ophiolites. Geotimes, *17,* 24–25 (1972)

Peterman, Z. E., Coleman, R. G., Hildreth, R. A.: Sr^{87}/Sr^{86} in mafic rocks of the Troodos massif, Cyprus. U.S. Geol. Survey Prof. Paper *750-D,* 157–161 (1971)

Ringwood, A. E.: Composition and evolution of the upper mantle. Am. Geophys. Union, Geophys. Monogr. *13,* 1–17 (1969)

Ringwood A. E.: The petrological evolution of island arc system. J. Geol. Soc. (London) *130* 183–204 (1974)

Rocci, G., Juteau, T.: Spilite-keratophyres et ophiolites. Geol. en Mijnbouw 47, 330–339 (1968)

Shibakusa, H.: Glaucophane schists in the Kamuikotan metamorphic belt of the Horokanai area, central Hokkaido (in Japanese with English abstract) J. Geol. Soc. Japan *80,* 341–353 (1974)

Sinha, A. K., Davis, T. E.: Geochemistry of Franciscan volcanic and sedimentary rocks from California. Carnegie Inst. Washington Yearbook *69,* 394–400 (1971)

Smewing, J. D., Simonian, K. O., Gass, I. G.: Metabasalts from the Troodos massif, Cyprus: genetic implication deduced from petrography and trace element geochemistry. Contr. Mineral. Petrol. *51,* 49–64 (1975)

Smith, A. G.: Alpine deformation and the oceanic areas of the Tethys, Mediterranean and Atlantic. Geol. Soc. Am. Bull. *82,* 2039–2070 (1971)

Suzuki, J., Suzuki, Y.: Petrological study of the Kamuikotan metamorphic complex in Hokkaido, Japan. Hokkaido Univ., J. Fac. Sci., Ser. 4, *10,* 349–446 (1959)

Tyrrell, G. W.: Distribution of igneous rocks in space and time. Geol. Soc. Am. Bul. *66,* 405–426 (1955)

Zwart, H. J.: The duality of orogenic belts. Geol. en Mijnbouw *46,* 283–309 (1967)

Zwart, H. J., coordinator: Metamorphic Map of Europe (scale 1:2,500,000), 17 sheets, Paris: UNESCO, 1973

II

ENERGETICS OF GEOCHEMICAL PROCESSES

11

K. S. HEIER AND G. GRØNLIE

Abstract Norwegian heat flow measurements in drill holes and lakes give a mean of 1.02 ± 0.21 heat-flow units (48 values). A zone of possible low mantle heat flow is present east of the Caledonian mountains in Norway.

The heat generation studies of the plutonic rocks of the Oslo Region show that the average surface heat production of the Oslo area is comparable to that of the surface shield crust. The somewhat higher heat production of the Oslo area is related to higher U and K contents in the younger plutonic rocks.

A number of granitic plutons in the south and southeast part of the Precambrian shield have been examined and comparisons and correlations have been attempted.

HEAT FLOW—HEAT GENERATION STUDIES IN NORWAY

1 INTRODUCTION

The Norwegian heat-flow project was started in 1969 and was for the first three years a cooperation project between the Massachusetts Institute of Technology (MIT) and the University of Oslo to measure terrestrial heat flow in diamond drill holes. The fieldwork and temperature gradients were measured at the University of Oslo, while the conductivity measurements were done at MIT. However, in the past three years all the work, including conductivity measurements, has been done at the University of Oslo.

The project planned to make use of existing bore holes which could be found all over Norway, and we are grateful to the Norwegian mining companies which let us use their holes drilled for exploratory purposes. In the past 5 years more than one hundred holes have been located and measured.

Since holes drilled for mining purposes most of the time are drilled in hetereogenous rocks with highly ranging conductivities (Logn and Evensen, 1973), we soon found it necessary to drill our own holes in suitable homogeneous rocks. The Norwegian Research Council for Science and the Humanities (NAVF) has supported a total of 4 holes each ca. 300 m long (Hurum, Hobbøl, Hurdal, Glamsland) and paid the cost of extending two holes ca. 100 m into basement (the two holes at Rena).

The heat-flow project was extended in 1972 when we started a cooperation project with the Niedersächsisches Landesamt für Bodenforschung to measure heat flow in Norwegian lakes. Since Norway is well covered with lakes, we found the method of Hänel (1970) well worth trying in Norway. Till now about 50 lakes have been measured, and 24 heat-flow values have been published.

A total of 48 heat-flow determinations has been published from Norway during the duration of the project. This is regarded internationally as good coverage.

2 GEOLOGY

For the present discussion, it is convenient to consider the geology of Norway in terms of three geological provinces: the Precambrian Baltic Shield, the Paleozoic Caledonian Orogenic Belt, and the Permian Oslo Igneous Province. The following generalized geology includes only the information pertinent to the present study. For a more complete treatment see Holtedahl (1960).

The Baltic shield covers a large portion of Fennoscandia, including southern and west central Norway (Fig. 1). It is Precambrian in age, although it includes different ages of orogenic and intrusive activity. Rock types include metamorphic rocks of varying compositions, and numerous ultramafic to felsic intrusions. The segment of shield in west

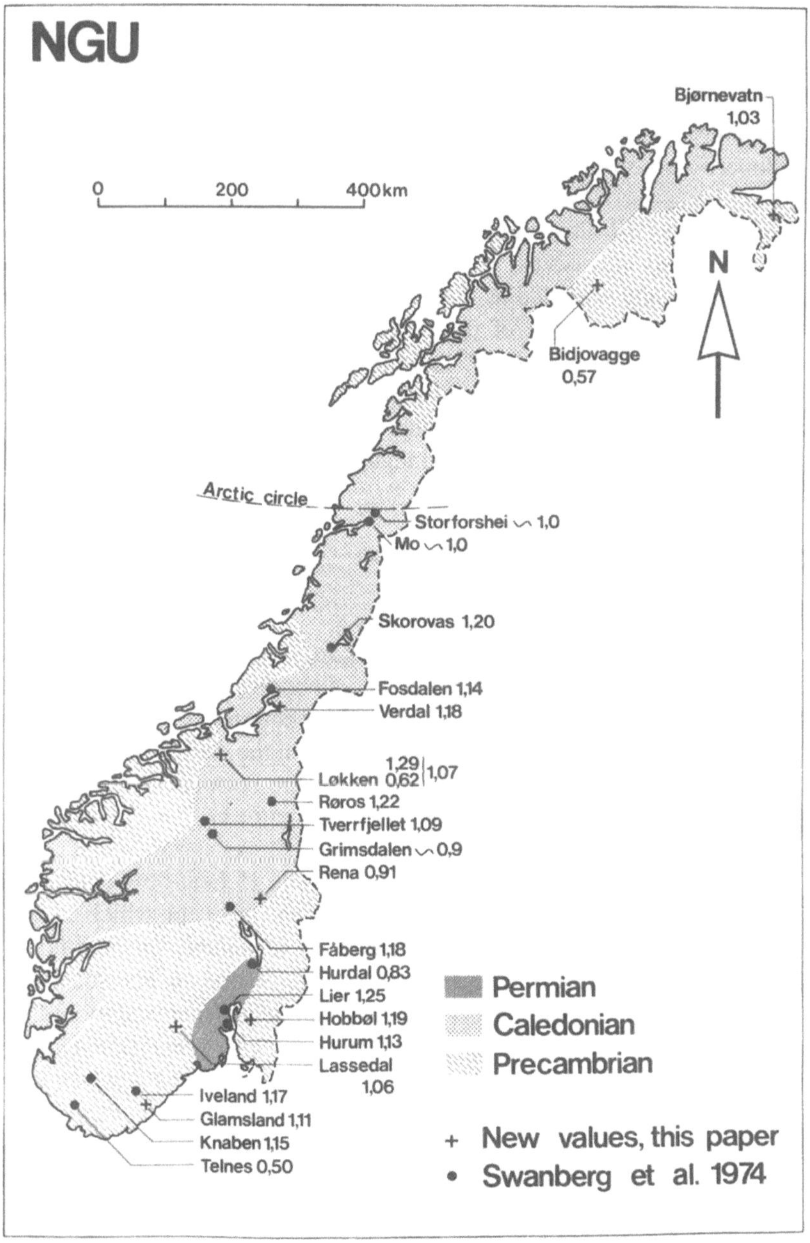

Fig. 1. Simplified geological sketch map of Norway with heat flow values indicated

central Norway is generally similar to the remainder of the shield except that the former, at least in part, was modified during the Caledonian orogeny.

The Paleozoic Caledonian Orogenic Belt extends from south central Norway northeast along the axis of the present-day mountain system to north Norway (Fig. 1) and includes a large portion of the central Scandinavian Peninsula. Rocks range in age from young Precambrian to Early Devonian, and consist of low-grade (greenschist) to moderate-grade (amphibolite) metasediments and metavolcanics. Caledonian intrusives, principally gabbros and trondhjemites, are quite common. Equivalent rocks are also observed on the island of Spitsbergen (80°N).

The Permian Oslo Province, a subvolcanic igneous province of alkaline intrusives and extrusives, is confined to the Oslo graben, a tectonic feature generally regarded as representing a Permian rift system. Crustal structure and the nature of the crust-mantle interface in this area are quite complex (Kanestrøm, 1971; Ramberg and Smithson, 1971), and the thickness of the crust appears to increase from about 32 m in the Oslo area to about 41 km in the interior of central Norway (Sellevoll and Sundvor, 1969; Kanestrøm and Haugland, 1971).

3 HEAT-FLOW MEASUREMENTS

3.1 Measurements in Bore Holes

The results are presented in two publications, Swanberg et al. (1974) and Grønlie et al. (1976). For details of data collection and reduction we refer to Swanberg et al. (1974). The data from the bore holes are presented in two tables; Table 1 gives heat-flow values from the Caledonian Orogenic Belt and Table 2 from the Baltic Shield and the Permian Oslo Province. The heat-flow values are also presented in Figure 1.

Swanberg et al. (1974) found that only 5 of the 15 heat-flow determinations from relatively homogeneous crystalline rocks for which reliable heat-generation values could be obtained, were suitable for heat-flow—heat-generation studies. The remainder are located in heterogeneous Precambrian gneisses or in Caledonian (Paleozoic) metasediments and metavolcanics. A plot of heat flow against heat generation (Fig. 2) confirms the linear relation between heat flow and heat generation as observed by Birch et al. (1968) in North Ameria. The slope of the line ($b = 8.4$ km) is in good agreement with the value of $b = 7.5$ obtained by Roy et al. (1968) for the eastern United States heat-flow province, but the intercept ($Q_0 = 0.48$) is significantly lower than the value of $Q_0 = 0.8$ of Roy et al. (1968) and suggests the presence of a zone of low heat flow and probably low-mantle heat flow in southern Norway. It is suggested that this zone of low heat flow is a real feature of southern Norway, as

Table 1. Heat-flow data from the Caledonian orogenic belt

Location	Hole	Depth range (m)	dT/dZ* (°C/km)	Error	dT/dZ† 10^{-3} cal (cm·s°C)	Error	K 10^{-3} cal (cm·s°C)	Error	Number of samples	Rock type	Heat flow (heat flow units)
Skorovas	97	100–180	13.23	0.12	15.18	0.12	7.84	0.29	22	Greenstone	1.19
Skorovas	98	80–200	12.98	0.09	14.65	0.10	8.26	0.40	13	Greenstone	1.21
Average											1.20
Løkken	888‡	830–910	17.88	0.13	17.90	0.13	6.97	0.20	20	Greenstone	1.25
Løkken	1050§	940–1040	17.84	0.26	17.84	0.26	7.30	0.11	23	Greenstone	1.30
Løkken	1062§	930–1060	18.38	0.13	18.38	0.13	8.21	0.30	20	Greenstone	1.51
Løkken	1071§	930–1060	16.00	0.47	16.00	0.47	7.02	0.10	26	Greenstone	1.12
Løkken	1243	151–452	15.44	2.47	15.51	2.49	6.83	0.20	10	Greenstone	0.63
Løkken	1240	38–414	20.58	0.46	18.78	0.43	6.96	0.35	3	Greenstone	0.60
Average											1.07
Fåberg	1	360–560	16.85	0.09	15.39	0.09	7.69	0.28	20	Sparagmite	1.18
Røros	104	250–320	16.68	0.18	16.28	0.17	7.69	0.72	7	Quartz mica schist	1.25
		370–460	20.36	0.13	19.98	0.12	5.86	0.31	16	Greenstone	1.17
Røros	107	280–380	22.31	0.05	22.39	0.05	5.51	0.20	15	Quartz mica schist	1.23
Average											1.22
Tverrfjellet	60-G**	330–470	13.89	0.06	13.81	0.06	7.11	0.33	14	Greenstone	0.98
		460–560	15.80	0.11	15.66	0.11	6.59	0.34	4	Bio-Gar mica schist	1.03
Tverrfjellet	61-G**	400–450	14.65	0.63	14.55	0.62	7.40	0.94	9	Amphibolite greenst.	1.08
		460–500	17.78	0.00	17.62	0.00	5.84	0.74	5	Bio-Gar mica schist	1.03
		490–560	16.81	0.38	16.66	0.37	7.72	0.31	5	Bio-Gar mica schist	1.29
		400–560	16.95	0.18	16.80	0.17	6.98	0.51	19	Entire hole	1.17
Average											1.09
Fosdalen	H-12	90–180	16.39	0.10	17.24	0.11	6.00	0.33	10	Greenstone	1.03
		170–240	20.31	0.94	21.03	0.95	5.07	0.28	4	Greenstone	1.07
		230–270	22.11	0.30	22.62	0.34	6.36	0.50	9	Greenstone	1.44
		90–270	18.98	0.50	19.76	0.48	5.94	0.24	23	Entire hole	1.17
Fosdalen	H-14	100–230	18.79	0.08	19.71	0.09	5.79	0.20	13	Greenstone	1.14
		300–380	17.58	0.22	18.14	0.23	5.88	0.56	11	Greenstone	1.07
Average											1.14
Verdal	38	90–250	17.94	0.99	17.52	0.96	6.82	0.18	10	Limestone	1.18
Average Caledonian belt											1.18

*Corrected for deviation from the vertical †Corrected for topography and deviation ‡Hole at 480-m level in mine §Holes at 900-m level in mine **Holes at 280-m level in mine

Table 2. Heat-flow data from the Baltic shield and the Permian Oslo province

Location	Hole	Depth range (m)	dT/dZ* (°C/km)	Error	dT/dZ† (°C/km)	Error	K 10^{-3} cal (cm·s°C)	Error	Number of samples	Rock type	Heat flow (heat flow units)
Hurum	1	100–250	13.93	0.06	13.89	0.06	8.14	0.06	19	Drammen granite	1.13
Lier	FUG	30–150	14.78	0.11	15.57	0.11	8.01	0.17	15	Drammen granite	1.25
Hurdal	2	100–300	18.28	0.03	17.21	0.02	4.82	0.05	15	Syenite	0.83
Knaben	7.1	80–260	15.26	0.13	15.30	0.12	7.49	0.19	6	Granite gneiss	1.15
Iveland	5	80–160	15.55	0.19	14.64	0.19	8.01	0.51	6	Amphibolite	1.17
Tellenes	8V-4N	60–160	11.13	0.11	10.53	0.09	4.15	0.08	6	Anorthosite	0.43
Tellenes	8V-4N	160–230	9.93	0.07	9.45	0.07	5.01	0.04	12	Ilmenite	0.47
Tellenes	10V-6L	80–170	13.29	0.07	12.20	0.07	3.96	0.07	3	Anorthosite	0.48
Tellenes	11V-4L	130–220	14.15	0.03	13.19	0.03	4.51	0.27	2	Ilmenite	0.59
Tellenes	8V-1N						5.21		2	Diabase dike	
Tellenes	10V-4N						5.11		3	Diabase dike	
Tellenes	8V-2S	80–200	13.53							Anorthosite	
Tellenes	12V-5V	100–230	14.59							Anorthosite	
Rena/Galå	6	120–280	17.63	2.22	17.25	2.16	7.87	0.89	7	Sandstone, gneiss	1.32
Rena/Slettås	7	80–290	8.64	0.95	8.56	0.93	8.20	0.75	7	Sandstone, gneiss	0.50
Average Rena											0.91
Hobbøl	130–280		15.64	0.93	15.64	0.93	8.10	1.26	10	Gneiss	1.19
Bjørnevatn	116	280–401	12.61	0.48	12.71	0.49	9.49	9.04	4	Gneiss, iron ore	1.09
Bjørnevatn	112	180–313	13.47	0.78	13.53	0.79	8.19	0.77	8	Gneiss, iron ore	0.97
Average Bjørnevatn											1.03
Bidjovagge	S40B	195–226	8.91	0.59	9.03	0.61	7.88	2.37	4	Amphibolite + disseminated ore	0.66
Bidjovagge	S44A	173–225	9.17	0.35	9.30	0.35	6.29	0.20	5	Diabase + disseminated ore	0.48
Average Bidjovagge											0.57
Glamsland		120–250	16.86	0.55	17.20	0.57	6.58	1.24	5	Pegmatite	1.11
Lassedal	14	69–182	11.74	1.46	11.52	1.41	6.99	0.27	6	Gneiss	0.80
Lassedal	12	80–169	13.79	0.43	13.25	0.41	8.26	0.58	5	Gneiss	1.06
Lassedal	10	79–177	13.88	0.97	13.49	0.93	7.46	0.12	6	Gneiss	1.01
Lassedal	8	77–192	14.15	0.44	13.84	0.43	8.88	0.34	7	Gneiss	1.12
Lassedal	6	72–171	15.83	0.70	15.51	0.68	9.14	0.15	6	Gneiss	1.33
Average Lassedal											1.06

*Corrected for deviation from the vertical †Corrected for topography and deviation

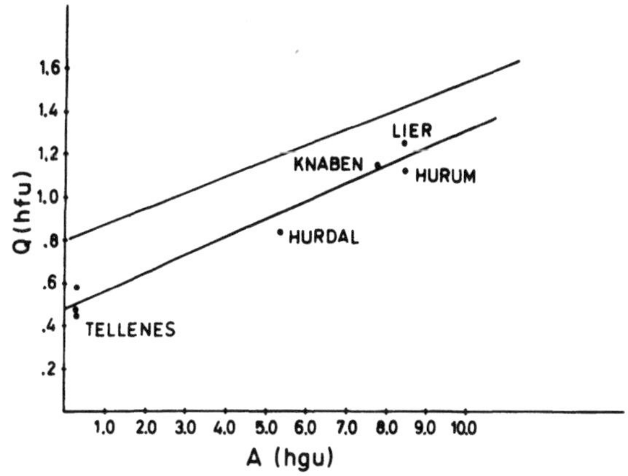

Fig. 2. Heat-flow—heat-generation plot for homogeneous plutons of Precambrian and Permian ages from Southern Norway. *Upper line:* "Eastern United States" heat flow province of Roy et al. (1968) (from Swanberg et al., 1974). Reprinted by permission.

the data were collected from plutons of differing age, composition, and geologic environment, and from widely separated localities in southern Norway.

The areal extent of this low heat-flow zone is difficult to estimate. Although there is no reliable heat-flow—heat-generation data from the Caledonian orogenic belt or from the segment of the Precambrian shield in west central Norway, we believe that the zone of low heat flow does not extend under the Caledonian orogenic belt, as this would require some rather questionable properties of the sub-Caledonian crust.

The eastern extent of this zone of low heat flow is likewise difficult to gauge. Lubimova et al. (1972) report mantle heat-flow values between 0.60 and 0.75 heat flow units for the Baltic shield on the Kola Peninsula, U.S.S.R. Two unpublished heat-flow values from Finland, Puranen et al. (1968) from peridotite and leptite (granulite) where heat generation can be expected to be very low (see Lambert and Heier, 1967; Heier and Thoresen, 1971) average 0.7 heat flow units. Neither of these areas appears to reflect the low heat flow observed in southern Norway. Heat-flow values from areas of mostly basic rocks in Sweden are generally low (D. Malmqvist, pers. comm.), but a detailed analysis of these data has not yet been published.

Swanberg et al. (1974) also found that heat-flow values from the Caledonian orogenic belt ranged from 1.09 to 1.29 heat flow units with an average value of 1.18 heat flow units. These values were similar to model values from other stable continental areas, including areas of Paleozoic orogenic activity (Lee and Uyeda, 1965).

The interpretation of the Caledonian heat-flow values depends upon whether or not the zone of apparent low heat flow observed in southern Norway extends under the Caledonides and beyond to the segment of shield in west Norway. The apparent "normal" nature of Caledonian heat-flow values would seem to suggest a normal (i.e., 0.6–0.8 heat flow unit) mantle heat flow for the Caledonian orogenic belt. Furthermore, if the zone of low heat flow were to extend under the Caledonides, it would imply that the Caledonian heat-flow values are higher than indicated by their numerical values.

Grønlie et al. (1976) presented six low values from the Baltic Shield and the Permian Oslo Province. The mean value of these is 0.94 ± 0.29 heat flow unit which is in good agreement with the results of Swanberg et al. (1974). The low values support the idea of a zone of anomalous low-mantle heat flow to the east of the Caledonian orogenic belt in Norway. Grønlie et al. (1976) also present two heat-flow determinations from the Caledonian mountain belt, and these are consistent with published results from Norway and other Paleozoic orogenic belts.

3.2 Measurements in Lakes

The determination of the terrestrial heat flow in lakes was carried out according to the method described by Hänel (1970) and with a heat-flow probe developed for this method. Measurements were made in 24 lakes, four of which proved to be unsuitable because of an intense water inflow near the measuring point (Årdalsvatn, Engeren, Sandungen, Øyeren). The results from two other lakes (Bygdin, Gjende) could not be interpreted because of water convection. The results of the heat-flow measurements are tabulated in Table 3 and shown in Figure 3 from Hänel et al. (1974). This figure also includes the relevant heat-flow values from Swanberg et al. (1974).

In order to calculate the temperature gradient given in Table 3, consideration was given to the inclination of the heat-flow probe in the sediments, the sedimentation rate, topographic influence and, where necessary, the mean annual temperature wave in the water. The inaccuracy for single values is estimated at ±15% to ±20%.

The heat-flow values are shown in Figure 3 together with values from Swanberg et al. (1974). All the measurements are from lakes which are situated east of the Caledonian mountain belt. Consequently the lake heat-flow values indicate heat flow from the Precambrian (Baltic Shield) in Norway and the Permian Oslo Graben.

Swanberg et al. (1974) concluded that there was anomalous mantle heat flow from the Baltic Shield in Norway, but the basis of their argument was a plot of heat flow versus heat generation with only five values. The 24 new heat-flow values seem to favor the conclusion of Swanberg et al.

Table 3. Heat flow data from 24 lakes in southern Norway

No.	Lake	Latitude (degree-minute)	Longitude (degree-minute)	Water level (m)	Depth of water (m)	Heat conductivity (10^{-3} cal/cm/sec/°C)	Temperature gradient (10^{-3}°C/cm)	Heat flow (μcal/cm²/s)
1	Bandakvatn	59–23.9	8–17.9	72	260	1.89	0.331	0.63
2	Birkelandsvatn	58– 9.0	7–45.4	43	61	1.68	0.648	1.09
3	Blankvatn	60– 1.6	10–40.2	351	52	1.57	0.664	1.04
4	Eikeren	59–41.3	9–54.4	19	154	1.98	0.649	1.28
5	Gjersjøen	59–47.3	10–46.9	42	51	1.88	0.672	1.26
6	Krøderen	60–17.8	9–40.4	132	110	2.54	0.303	0.77
7	Lundevatn	58–25.1	6–35.7	45	202	3.05	0.321	0.98
8	Lundevatn	58–22.2	6–36.6	45	300	1.91	0.527	1.01
9	Mjøsa	60–38.4	11– 7.2	123	428	3.05	0.336	1.12
10	Mjøsa	60–35.0	11–11.2	123	417	3.58	0.205	0.73
11	Nordsjø	59–19.1	9–16.6	15	164	2.24	0.311	0.70
12	Nordsjø	59–15.9	9–23.7	15	168	2.44	0.361	0.88
13	Rørholtfjorden	59– 0.9	9–19.8	60	85	1.65	0.456	0.60
14	Rørholtfjorden	59– 0.9	9–13.8	60	88	1.75	0.302	
15	Sirdalsvatn	58–37.7	6–42.5	170	157	1.72	0.510	0.88
16	Sirdalsvatn	58–34.2	6–42.4	170	159	1.62	0.782	1.27
17	Sperillen	60–32.4	10– 1.6	150	109	1.90	0.323	0.62
18	Sperillen	60–31.1	10– 4.6	150	82	1.91	0.497	0.92
19	Storsjøen	61–38.5	11–11.2	251	267	1.81	0.496	0.90
20	Svinsjøen	59–49.3	10–22.5	180	34	1.85	0.577	
21	Svinsjøen	59–49.3	10–22.5	180	34	1.88	0.601	1.23
22	Svinsjøen	59–49.3	10–22.5	180	35	2.33	0.567	
23	Tinnsjø	59–50.8	8–58.0	191	357	1.85	0.474	0.88
24	Totak	59–43.8	7–51.0	687	309	3.03	0.564	1.14
25	Totak	59–42.4	7–56.1	687	105	1.64	0.683	1.12
26	Tronstadvatn	58–11.6	7–45.7	41	92	1.65	0.711	1.17
27	Tyrifjorden	59–59.1	10–16.6	63	288	2.34	0.411	0.96
28	Tyrifjorden	59–56.5	10–19.0	63	270	2.04	0.421	0.86

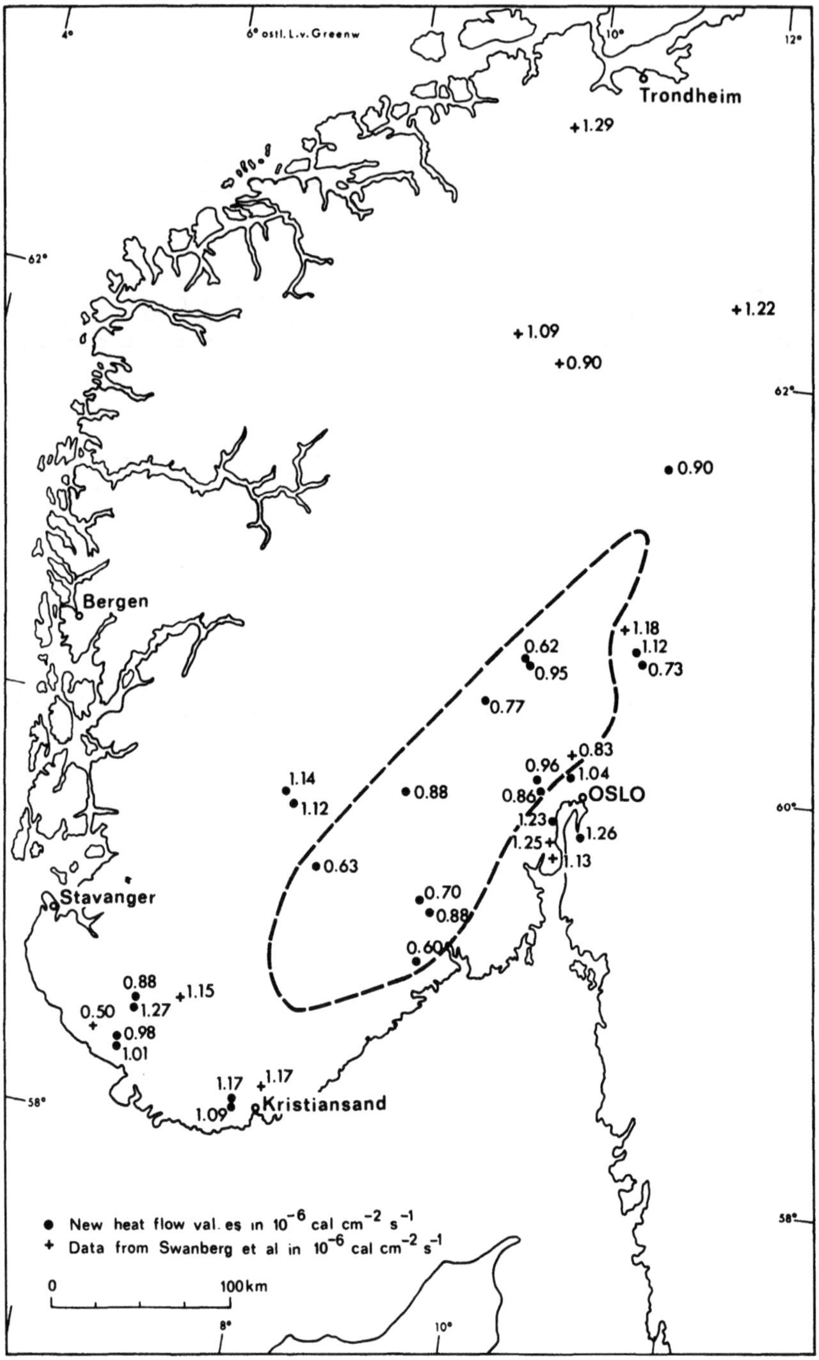

Fig. 3. Heat-flow map of southern Norway (from Hänel et al., 1974). Reprinted by permission.

that there is a zone of low (mantle) heat flow in the Baltic Shield to the east of the Caledonian mountains (Fig. 3, inside the dashed line).

The results of Swanberg et al. (1974) from the Caledonian belt give a mean value of 1.18 heat flow unit and their results from the Precambrian Shield give an average of 0.99 heat flow unit (3 values). Our lake measurements give a mean of 0.96 heat flow unit (standard deviation 0.21 heat flow unit), which compares well with the shield values of Swanberg et al. (1974) and also is within the error limits of their Caledonian mean value. No heat generation study has been made specifically for this project, but a general heat generation study on Precambrian granites in southern Norway has been published (Killeen and Heier, 1974, 1975a,b,c,d,e; Jacobsen, 1975). These studies show that the heat production from the Precambrian in southern Norway is not negligible.

4 HEAT-GENERATION STUDIES

At the same time as the heat-flow project, a study of surface radioactivity and heat production was started.

The analytical technique was described in detail by Raade (1973) and Killeen and Heier (1975a).

Briefly, the experimental apparatus consisted of a 400 channel pulse analyzer, utilizing a 12.7-cm × 12.7-cm (5 × 5-in) NaI (Tl) crystal for a detector, inside a chamber with steel walls 19.5 cm thick. Samples were coarse crushed and placed in cylindrical plastic containers 11.5 cm in diameter and 4.5 cm high, which were placed in contact with the detecting crystal. Sample weight averaged roughly from 700 to 800 g for the rocks in this investigation. The three energy peaks used were 1.46 MeV (K^{40}), 1.76 MeV (Bi^{214}) and 2.62 MeV (Tl^{208}) for K, U, and Th, respectively.

Both "synthetic" standards, U.S. Atomic Energy Standards of uraninite mixed with dunite for U, monazite mixed with dunite for Th, and KBr for K and four USGS standard rocks (GSP-1, AGV-1, BCR-1, and G-2) were used to ensure the apparatus produced accurate results.

The effect of differences in sample weights, sample packing density and background fluctuations were investigated by Raade (1973). The K values are corrected for sample weight variations.

Accuracy, counting errors, and reproducibility was discussed by Raade (1973) and Killen and Heier (1975a).

The studies were concentrated about two areas of particular interest (Fig. 1):

(1) The plutonic rocks of the Oslo region;
(2) Granitic plutons in the south and south-east part of the Baltic shield in Norway.

4.1 The Plutonic Rocks of the Oslo Region

The area was described by Raade (1973). Approximately one thousand samples were collected from an area covering about 5000 km². The major conclusion of Raade are listed in Table 4. It is interesting to note that according to Raade the average surface heat production of the Oslo area is comparable to that of the surface shield crust as described by Lambert and Heier (1968). The somewhat higher heat production of the Oslo area is related to higher U and K contents in these younger plutonic rocks.

4.2 Granitic Plutons in the South and Southeast Part of the Precambrian Baltic Shield in Norway

The results of this study are described in a series of papers by Killeen and Heier (1974, 1975a,b,c,d,e). The granites investigated are shown on Figure 4 and exact sample localities are given by Killeen and Heier (1975a).

A total of 695 samples from the 22 granitic plutons were analyzed to determine their thorium, uranium and potassium concentrations by gammaray spectrometry. The samples are representative of 22 bodies located throughout the Precambrian basement rocks of Southern Norway. Ten of the granite plutons in theTelemark area, which range in size

Table 4. Concentration of radioactive elements and surface heat production in the plutonic rocks of the Oslo region (from Raade, 1973)

Rock type	km²	% of total area	Estimated average ppm U	ppm Th	% K	Surface heat prod*
Akerite	52.2	1.0	3.5	12.6	3.28	5.10
Kjelsåsite/Larvikite	1871.0	36.9	4.2	15.1	3.50	5.98
Lardalite/Foyaite	65.0	1.3	5.1	19.3	4.32	7.45
Normarkite/ grefsensyenite/ pulaskite/ hedrumite	1425.0	28.1	3.0	14.2	4.67	5.35
Ekerite	821.0	16.2	6.3	29.0	4.11	9.76
Bio granite	840.0	16.5	5.0	27.3	4.16	8.71
Total Oslo	5074.2	100	4.3	19.1	4.04	6.88
Surface shield Crust (Lambert and Heier, 1968)			3.0	20.0	2.6	5.86

*Surface heat production: 1 heat production unit = 10^{-13} cal/cm³ s

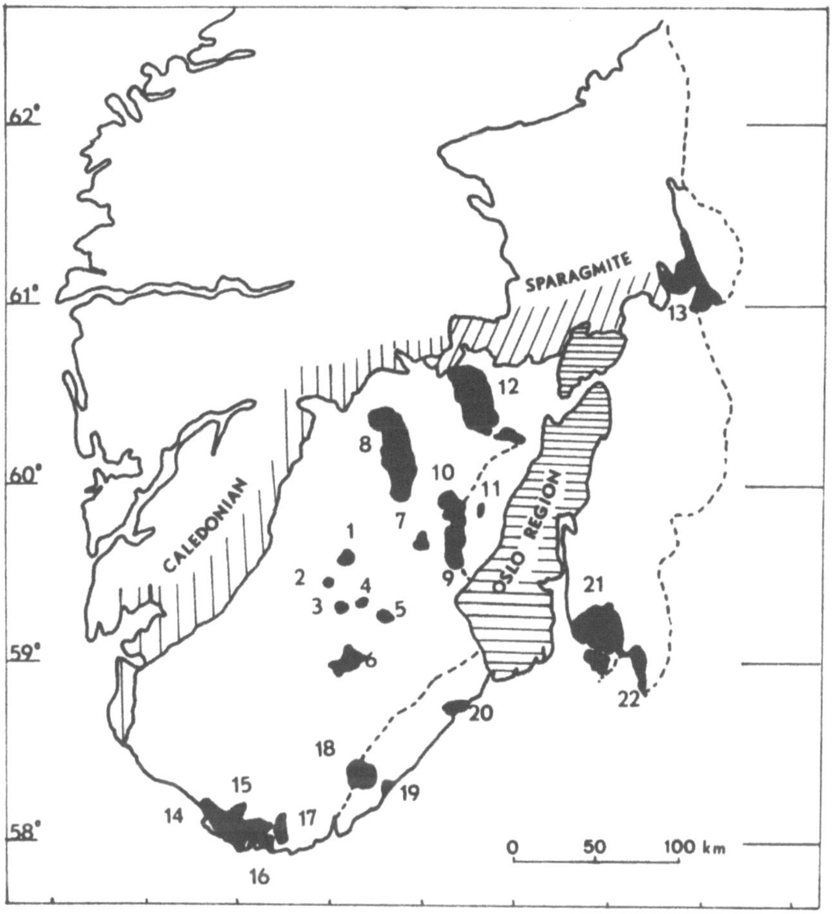

Fig. 4. Locations of granitic bodies in which Th, U, K, and heat production were investigated (from Killeen and Heier, 1975a). Reprinted by permission.

from about 30 km² to 400 km², are presented by 172 samples. A total of 115 samples were obtained from the postorogenic Grimstad and Herefoss granites, and the Levang granite which are all situated in the Bamble area, along the Skagerrak coast. The large postorogenic Flå and Iddefjord granites are represented by 186 samples. An additional five samples were collected from the Bohus granite in Sweden, adjacent to the Iddefjord granite, 53 samples from Trysil granite in East Hedmark, northeast of Oslo, and five samples from the Vatnaas granite, a small body in the Kongsberg area. The granitic rocks of the Farsund area, including the Holum and Snik granites, and the light (normal) and dark (charnockitic) varieties of farsundite are represented by 159 samples.

Some of the preliminary results are summarized by Killeen and Heier (1975a) and listed below:

(1) The Farsundite "normal" or "light" facies (i.e., biotite hornblende quartz monzonite variety) appears to be genetically related to the Holum granite.

(2) The entire area covering the dark charnokitic and normal facies of the Farsundite, the Snik, and the Holum granites appears to be a low-heat prouction province, probably related to the low radioelement content of the Egersund area rocks, and the local high-grade metamorphic facies.
(3) The Herefoss granite appears to have concentric radioelement pattern suggesting that differentiation occurred during cooling.
(4) The Flå, Iddefjord, and Bohus granites appears to be all relatively high in Th and U, adding evidence to support their suspected cogenesis.
(5) The granites of the Telemark area show a large variation in radioelement content. Two in particular (Bessefjell and Fyresdal) appear to be abnormally high in Th and U for granitic rocks.
(6) The Levang granite (or "gneiss dome") as it has been described more recently by O'Nions and Baadsgaard (1971) shows an interesting increase in Th and U to the south. O'Nions and Baadsgaard have dated the main mass of the Levang granite at 1616 million years but they found ages of 1170 million years for the granitic gneisses of the Portør peninsula to the south.

It is possible then that the observed enrichment of Th and U toward the south is related to the Sveconorwegian metamorphic event.

Later studies showed the following: The radioelement concentrations and their distributions in the light (biotite hornblende) Farsundite and dark (charnockitic) Farsundite suggests that they are distinct plutons and not comagmatic. The trend surfaces of radioelement concentrations coupled with available gravity data indicates that the light Farsundite has a closer relationship to the Holum granite than to the dark Farsundite. Average radioelement concentrations for the three plutons suggests they are all part of a low heat production province which may be related to the low radioelement contents in the rocks of the Egersund area to the west (Killeen and Heier, 1975b).

The radioelement concentrations in the ten granite plutons shown on Figures 5 and 6 were determined by gamma-ray spectrometric analysis of 172 samples. There is a wide range in heat production values, the lowest values being found in the Bandak and Rollag granites, 4.6 and 5.5 cal/cm^3/s 10^{13}, respectively, and the highest in the Bessefjell and Fyresdal granites, 19.5 and 14.8 cal/cm^3/s 10^{13}, respectively. Radioelement concentrations in individual granites appear to be related to the degree of granitization shown by the pluton. In addition, the regional variation derived by studying the overall result suggests that there exist parallel, NE–SW trending zones of differing radioelement enrichment, which may be related to orogenic events. The picture becomes obscured as a new orogeny overprints a previous one. The general radioelement distribution pattern suggests the formation of zones of enrichment during both the Svecofennian and the Sveconorwegian orogenies (Killeen and Heier, 1975e).

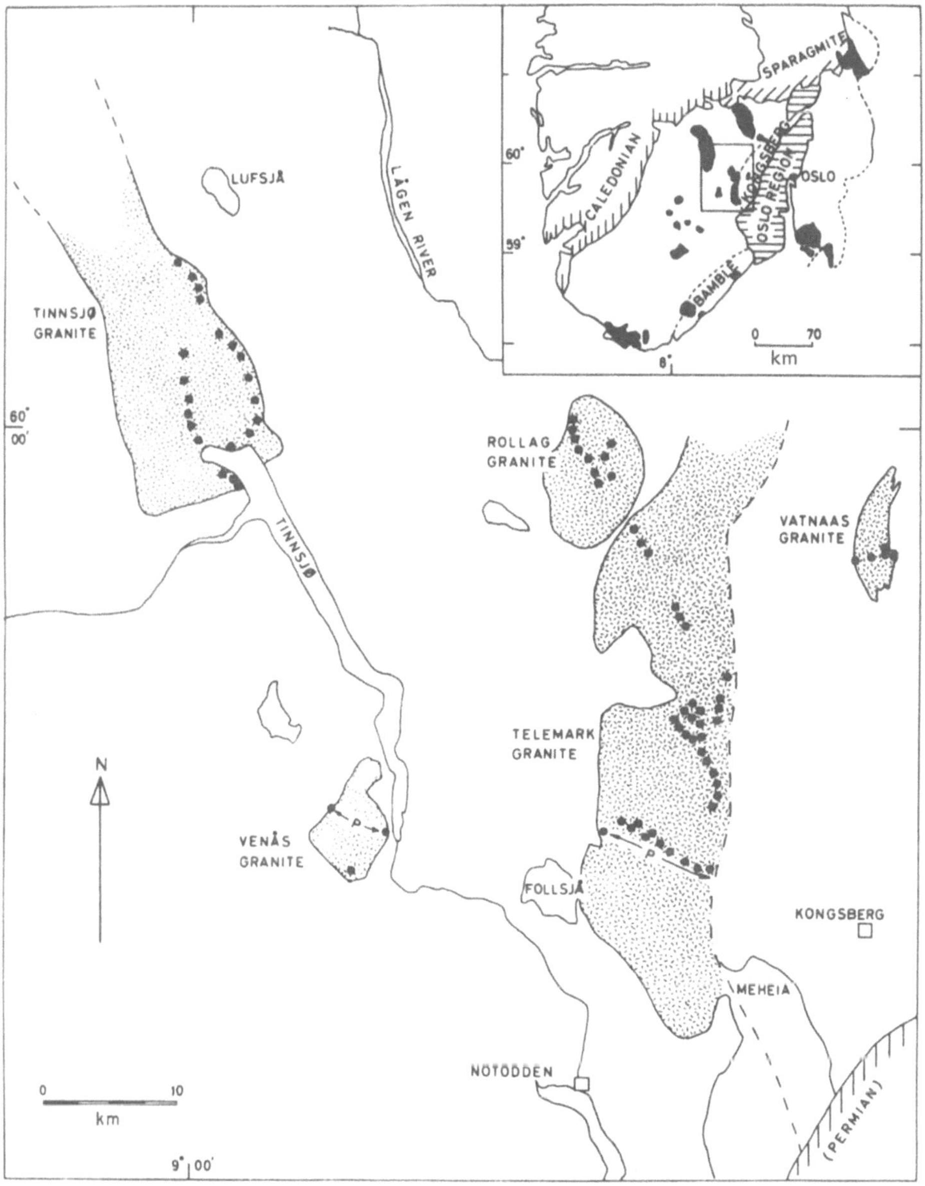

Fig. 5. Locations of the Tinnsjø, Venås, Rollag, and Telemark granites in the Telemark area showing distribution of sample localities. *P:* profiles discussed by Killeen and Heier (1975e). Reprinted by permission.

One rather interesting observation concerns the Precambrian Flå, Iddefjord and Bohus granites described by Killeen and Heier (1975d). These granites (Fig. 7) lie along a line striking roughly north-west which crosses the Permian Oslo Province to the southwest of Oslo. Radioelement investigations in the three bodies show that they all contain abnormally high thorium and uranium concentrations relative to the

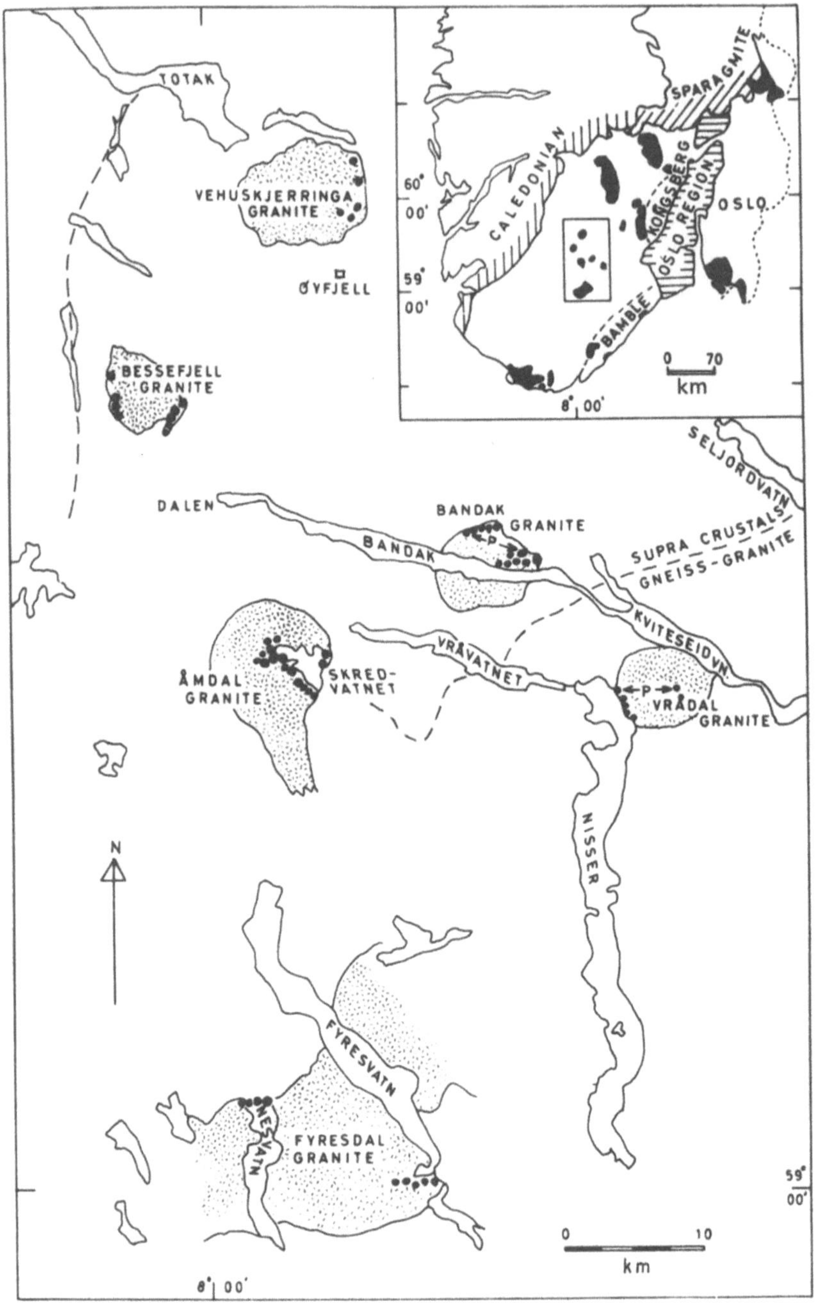

Fig. 6. Location of the Vehuskjerringa, Bessefjell, Bandak, Åmdal, Vrådal, and Fyresdal granites in the Telemark area showing distribution of sample localities. *P:* profiles discussed by Killeen and Heier (1957e). Reprinted by permission.

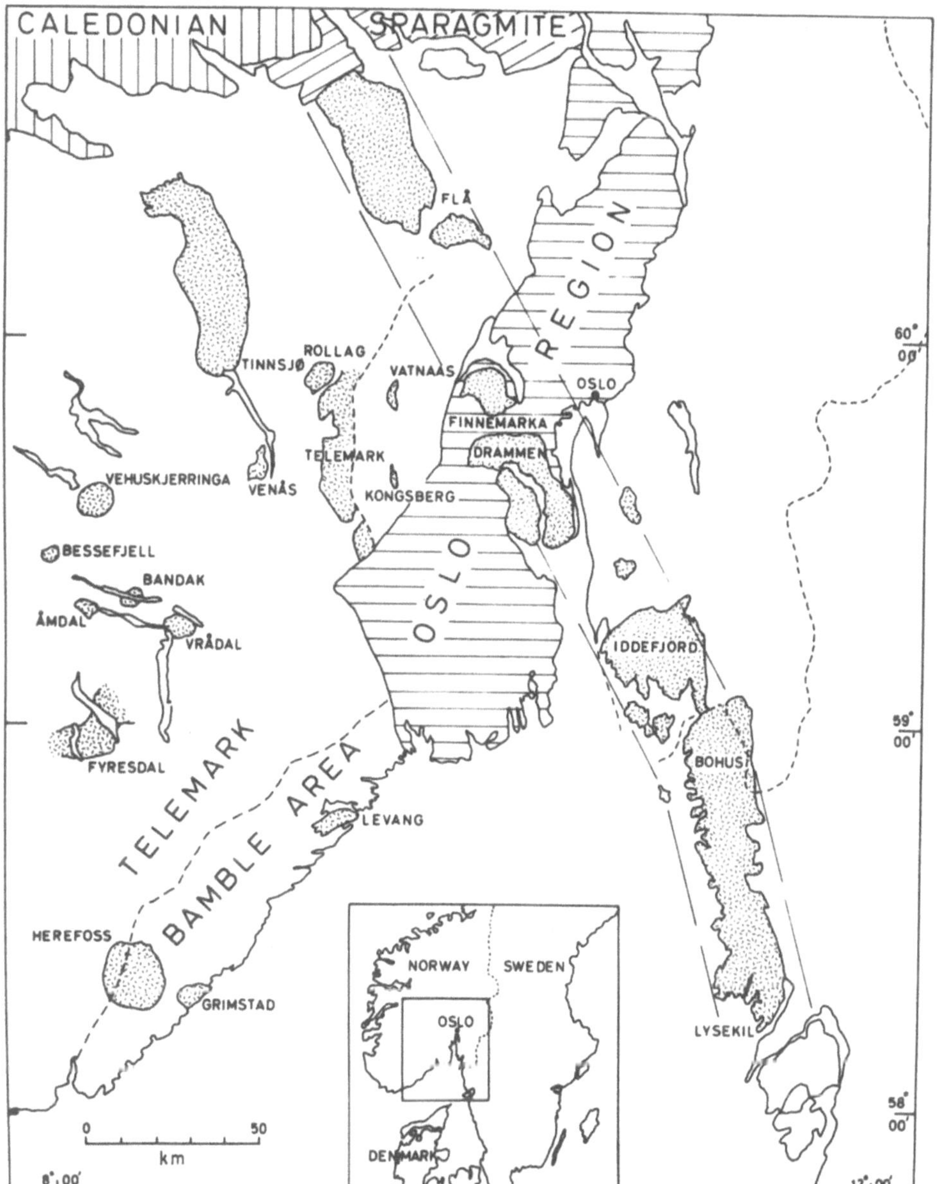

Fig. 7. Locations of some granitic bodies is southeastern Norway showing the Flå-Iddefjord-Bohus granitic belt cutting across the plutonic rocks of the Oslo region (from Killeen and Heier, 1975d). Reprinted by permission.

published literature on average radioelement contents of granitic rocks, Killeen and Heier (1975c). Trend surface analysis of the radioelement distribution in the Iddefjord granite suggests there was relative movement of uranium to the east with respect to thorium, possibly as the result of Permian activity in the adjacent rocks. Geological considerations, radiometric evidence and published gravimetric data suggest that the three granites represent a continuous belt enriched in thorium and uranium during the Sveconorwegian orogeny. A portion of the belt was later involved in the Permian igneous activity which produced the igneous Oslo Province. We present here some evidence that the Permian Drammen and Finnemarka granites represent that part of the belt which was modified in Permian time.

5 SUGGESTIONS FOR FURTHER WORK

In Southeast Norway a comparatively thorough investigation has been carried out with respect to heat flow and heat production measurements in the plutonic rocks of Permian age in the Oslo region and the granitic plutons of the Baltic shield. However, little work has been done on the gneissic rocks which constitute the bulk of the Precambrian in the area. In order to obtain a better impression of the heat-flow—heat-production relationship in this area, a systematic investigation of the gneisses is suggested. This investigation should be supplemented by a small but carefully selected number of heat-flow measurements in drill holes.

A small beginning to the study of radioactive elements in the gneisses in presented by Jacobsen (1975).

REFERENCES

Birch, F., Roy, R. F., Decker, E. R.: Heat flow and thermal history in New England and New York. In: Studies of Appalachian Geology: Northern and Maritime. Zen, E-an, White, W. S., Hadley, J. B., Thompson, J. B., Jr. (eds.) Interscience, New York: 1968, pp. 437–451

Grønlie, G., Heier, K. S., Swanberg, C. A.: Terrestrial heat flow determinations from Norway. Norsk Geol. Tidsskr. (in press 1977)

Heier, K. S., Thoresen, K.: Geochemistry of high-grade metamorphic rocks, Lofoten-Vesterålen, North Norway. Geochim. Cosmochim. Acta *35,* 89–99 (1971)

Holtedahl, O.: Geology of Norway. Norg. Geol. Unders. *208,* 540 p. (1960)

Hänel, R.: A new method for the determination of the heat flow in lakes. Z. Geophys. *36,* 725–742 (1970)

Hänel, R., Grønlie, G., Heier, K. S.: Terrestrial heat flow determinations from lakes in southern Norway. Norsk Geol. Tidsskr. *54,* 423–428 (1974)

Jacobsen, S. B.: Geochronological, geochemical and petrological investigations of Precambrian rocks in the Kongsberg area. Unpubl. Cand. Real. Thesis. Univ. Oslo (1975)

Kanestrøm, R.: A crust-mantle model for the NORSAR area. In: Crustal Structure in Southeastern Norway from Seismic Refraction Measurements. Seismo. Obs., Univ. Bergen, Sc. Rep., *5*(2) 46–71 (1971)

Kanestrøm, R., Haugland, K.: A refraction profile through southeastern Norway. In: Crustal Structure in Southeastern Norway from Seismic Refraction Measurements. Seismol. Obs., Univ. Bergen, Sci. Rep., *5*(1), 1–45 (1971)
Killeen, P. G., Heier, K. S.: Radioelement variation in the Levang granite gneiss, Bamble Region, South Norway. Contrib. Mineral Petrol. *48,* 171–177 (1974)
Killeen, P. G., Heier, K. S.: Radioelement distribution and heat production in Precambrian granitic rocks, Southern Norway. Det Norske Videnskaps-Akademi, I. Mat.-Naturv. Klasse. Skrifter, Ny *55,* pp 32 (1975a)
Killeen, P. G., Heier, K. S.: Trend surface analysis of Th, U and K, and heat production in three related granitic plutons, Farsund area, South Norway. Chem. Geol. *15,* 163–176 (1975b)
Killeen, P. G., Heier, K. S.: Radioelement and heat production measurements in the Trysil granite, east Hedmark, Norway. Norsk Geol. Tidsskr. *55,* 179–184 (1975c)
Kileen, P. G., Heier, K. S.: A. Uranium and thorium enriched province of the Fennoscandian shield in southern Norway. Geochim. Cosmochim. Acta. *39,* 1515–1524 (1975d)
Killeen, P. G., Heier, K. S.: Th, U, K and heat production measurements in ten Precambrian granites of the Telemark area, Norway. Norg. Geol. Unders. *319,* 59–83 (1975e)
Lambert, I. B., Heier, K. S.: The vertical distribution of uranium, thorium and potassium in the continental crust. Geochim. Cosmochim. Acta *31,* 377–390 (1967)
Lambert, I. B., Heier, K. S.: Estimates of the crustal abundances of thorium, uranium and potassium. Chem. Geol. *3,* 233–238 (1968)
Lee, W. H. K., Uyeda, S.: Review of heat-flow data. In: Terrestrial Heat Flow. Lee, W. H. K. (ed.), Geophys. Monogr. Ser. 8 AGU, Washington D.C. p. 87–190 (1965)
Logn, Ø., Evensen E.: Thermal conductivities of some ores and rocks in Norway. Norg. Geol. Unders. *300,* 11–19 (1973)
Lubimova, E. A., Karus, E. V., Firsov, F. V., Starikova, G. N., Vlasov, V. K., Lyusova, L. N., Koperbakh, E. B.: Terrestrial heat flow on Precambrian shields in the U.S.S.R. Geothermics, *1,* 81–89 (1972)
O'Nions, R. K., Baadsgaard, H.: A radiometric study of polymetamorphism in the Bamble region, Norway. Contrib. Mineral. Petrol. *34,* 1–21 (1971)
Puranen, M., Järvimäki, P., Hämäläinen, U., Lehtinen, S.: Terrestrial heat flow in Finland. Geoexploration, *6,* 151–162 (1968)
Ramberg, I. B., Smithson, S. B.: Gravity interpretation of the southern Oslo graben and adjacent Precambrian rocks, Norway. Tectonophysics *11,* 419–531 (1971)
Roy R. F., Blackwell, D. D., Birch, F.: Heat generation of plutonic rocks and continental heat flow provinces. Earth Planet. Sci. Lett., *5,* 1–12 (1968)
Raade, G.: Distribution of radioactive elements in the plutonic rocks of the Oslo region. Unpubl. Cand. Real. Thesis, Univ. Oslo, (1973) pp 162
Sellevoll, M., Sundvor, E.: A P_n timeterm survey in southeastern Norway. Seismol. Obs., Univ. Bergen, Sci. Rep, *2,* 13 p. (1969)
Swanberg, C. A., Chessman, M. D., Simmons, G., Smithson, S. B., Grønlie, G, Heier, K. S.: Heatflow—heat-generations studies in Norway. Tectonophysics *23,* 31–48 (1974)

12

H. K. MAO AND P. M. BELL

Abstract Chemical disproportionation of ferrous iron-bearing phases to an equilibrium mixture of ferric iron-bearing phases plus metallic iron is stabilized at high pressures. The effect is observed in olivine, and is the basis of a model of the earth's mantle and core that is evolved from a primitive earth. The model is consistent with an equilibrium process of formation of the earth, and satisfies known physical properties of materials that are suitable for the mantle.

DISPROPORTIONATION EQUILIBRIUM IN IRON-BEARING SYSTEMS AT PRESSURES ABOVE 100 KBAR WITH APPLICATIONS TO CHEMISTRY OF THE EARTH'S MANTLE

1 INTRODUCTION

Reactions of the chemical disproportionation of ferrous iron in minerals to produce assemblages in which both ferric and metallic iron coexist were discovered to be important at pressures above 100 kbar, corresponding to depths of 300 km or more in the earth (Mao, 1974). In effect, an equilibrium assemblage produced by the chemical disproportionation of ferrous iron is a mixture of oxidized and reduced states of iron. Such a mixture is suggested here for the first time as possible in the earth's mantle. However, the earth contains the mixture on the broad scale of crust-mantle-core, and perhaps the most restrictive constraint used in models of the origin of the earth is the assumption that however the earth formed, disequilibrium between crust, mantle, and core was a fundamental condition of the formational process (Ringwood, 1966). Disproportionation does not include this constraint, and proves the assumption is considered unnecessary. This paper is a discussion of the geochemical consequences of disproportionation that includes by necessity, a model of formation of a hypothetical planet. If disproportionation has occurred in the earth, it must have been initiated when the earth formed, and then progressed and thus contributed to the present geochemical state of the interior of the earth.

2 DISPROPORTIONATION REACTIONS

The first disproportionation reaction demonstrated experimentally at high pressure was that involving the breakdown of spinel (fayalite composition) to a nonstoichiometric wüstite ($Fe_{0.91-00.97}O$) plus metallic iron (plus stishovite) at 250 kbar, 750°C (Mao and Bell, 1971). The results were confirmed in an independent set of experiments by Bassett and Ming (1972). The equilibrium coexistence of metallic iron with ferric iron was well known in metallurgical and ceramic chemistry at low oxygen fugacity (Darken and Gurry, 1945; Johnston, 1964) although no mention of disproportionation reactions has been made in models of the earth.

The reactions involving ferrous iron oxide are, in the case of the earth's mantle, stabilized by high pressures rather than by low oxygen fugacity according to the following relationships (Mao, 1974):

$$FeO = Fe_xO + (1 - x)\,Fe \tag{1}$$
$$FeO = \tfrac{1}{4}(Fe_3O_4 + Fe) \tag{2}$$
$$FeO = \tfrac{1}{3}(Fe_2O_3 + Fe) \tag{3}$$

Using the relationships

$$\Delta G = \int_{P_0}^{P_1} \Delta V \, dP, \tag{4}$$

(where ΔG is the change in free energy of the reaction and ΔV is the change in molar volume of the reaction) and

$$\Delta S/\Delta V = dP/dT, \quad (5)$$

where ΔS is the entropy difference of the reaction, one can determine the equilibrium reaction most likely to occur at a particular depth in the mantle.

Wüstite, nonstoichiometric at 1 atm, has, with magnetite, hematite, and metallic iron a phase diagram that contains $Fe^{2+} - Fe^{3+} - Fe^{0}$ equilibria (Muan and Osborn, 1965). As wüstite is produced by silicate disproportionation reactions at high pressures, it enters a pressure region where it becomes stoichiometric. At high pressures wüstite again becomes nonstoichiometric. This behavior is the result of a delicate balance in molar volume, that is, perhaps, the most crucial factor in determining the character of oxide phase relations of the mantle, and can best be evaluated from existing experimental data on the iron oxide mineral phases and high pressure phases.

3 WÜSTITE

Two apparently contradictory reports of the stoichiometry of wüstite at high pressure resulted from observations of the behavior described above. Katsura et al. (1967) observed that nonstoichiometric wüstite becomes more iron-rich with increasing prssure and forms the stoichiometric compound FeO at 775°C and 35 kbar. In other studies, Mao and Bell (1971) and Bassett and Ming (1972) found that the spinel form of Fe_2SiO_4 disproportionated to nonstoichiometric wüstite Fe_xO with $x =$ 0.91 to 0.95 instead of to FeO at 250 kbar and 800°C. The apparent discrepancy can be reconciled when the change of the volume difference of Equation (1) with pressure is considered as follows:

For Equation (1), when $x \rightarrow 1$,

$$\Delta V = (1 - x)(V_{\text{Fe}} - V_{\text{Fe in wü}}), \quad (6)$$

where V_{Fe} is the molar volume of the metallic iron. The partial molar volume of Fe in wüstite $\bar{V}_{\text{Fe in wü}}$ is defined by

$$\bar{V}_{\text{Fe in wü}} = \frac{\partial V_{\text{wü}}}{\partial x} = 3ma^2_{\text{wü}} \frac{\partial a_{\text{wü}}}{\partial x},$$

where $a_{wü}$ is the lattice parameter of wüstite and m is a constant for converting the unit-cell volume to the molar volume. At 1 bar, using the values $a_{wü}$ = 4.323 Å for FeO (Katsura et al., 1967) and $(\partial a_{wü}/\partial x$ = 0.5826 Å (Jette and Foote, 1933), the quantity $\bar{V}_{Fe\ in\ wü}$ is found to be 4.92 cm³/mol. This value is significantly smaller than V_{Fe} = 7.093 cm³/mol. Thus, ΔV is large and positive. According to Equation (4), wüstite should approach stoichiometry with increasing pressure, as Katsura and coworkers observed.

Nevertheless, nonstoichiometric wüstite, because of its structural vacancy, is expected to be more compressible than FeO. Therefore, the quantity $\bar{V}_{Fe}$ in wü is expected to increase with pressure. On the other hand, the quantity V_{Fe} is expected to decrease with pressure by volume compression and by α-ϵFe transition (130 kbar, 25°C). From Equation 6, ΔV decreases with pressure and becomes negative at high pressures. Thus the reaction is reversed and stable wüstite again becomes nonstoichiometric.

The relative stability of wüstite can be calculated quantitatively. The isothermal bulk modulus of wüstite (K_0) is 1.42 Mbar with x = 0.924 (Mao et al., 1969) and 1.54 Mbar with x = 0.947 (Clendenen and Drickamer, 1966). A linear extrapolation of the two points gives

$$K_0\,(\text{Mbar}) = 5.22x - 3.40 \tag{7}$$

for FeO = 1.82 Mbar and $(\partial K_0/\partial x)$ = 5.22 Mbar. Using a Murnaghan equation with $(\partial K/\partial P) = 4$, the partial molar volume of Fe in wüstite at high pressures can be expressed by:

$$V_{Fe\ in\ wü} = (1 + 2.2P)^{-0.25}\,[4.92 + 19.17/(P^{-1} + 2.2)], \tag{8}$$

where P is the pressure in megabars.

The result of Equations (7) and (8) provides the experimental basis for postulating equilibrium disproportionation of iron oxides in the mantle. Figure 1 shows the values of V_{Fe} and $\bar{V}_{Fe}$ in wü plotted against pressure where a significant reversal in sign of ΔV occurs. At the α–ϵ transition Δ_V changes sign, and pressure destabilizes FeO. Based on Figure 1 and Equation (4), the phase diagram for FeO is constructed in Figure 2. The results are in agreement with the experimental observations mentioned above. Stoichiometric FeO first becomes stable at pressures above 50 kbar and then becomes unstable and disproportionates to nonstoichiometric wüstite and metallic iron at pressures above 250 kbar. This analysis demonstrates that at the pressures and temperatures of the lower mantle, the disproportionation of FeO should occur.

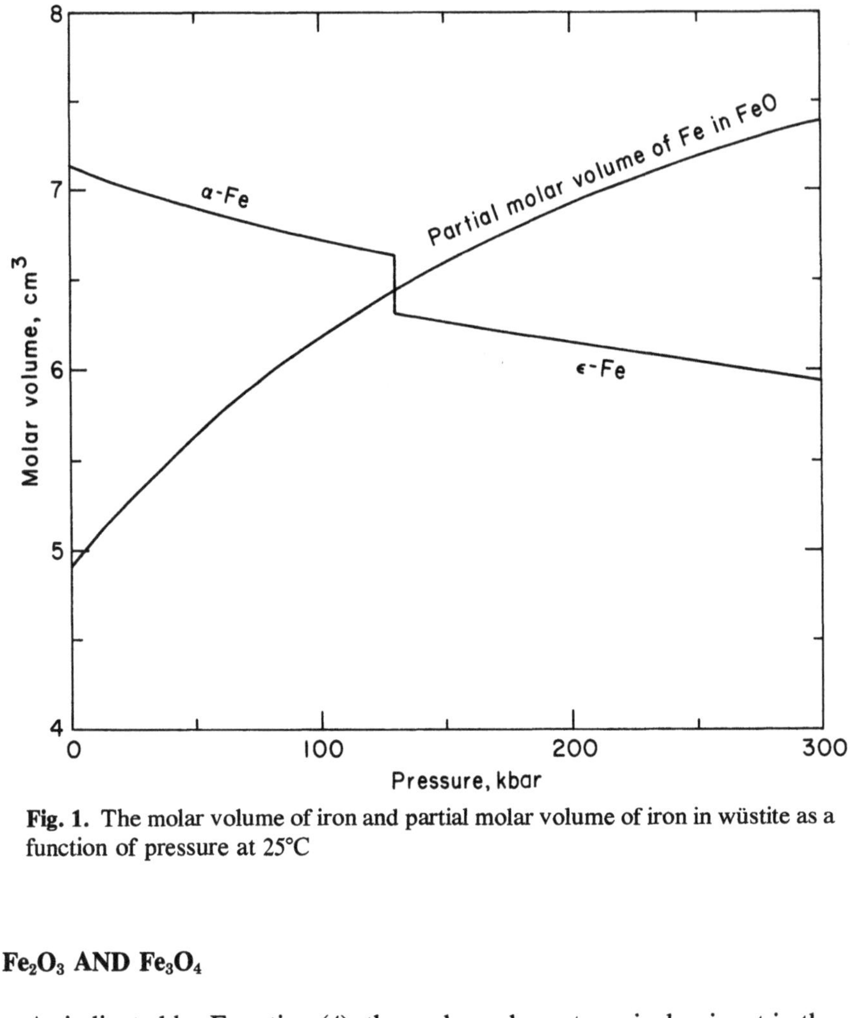

Fig. 1. The molar volume of iron and partial molar volume of iron in wüstite as a function of pressure at 25°C

4 Fe_2O_3 AND Fe_3O_4

As indicated by Equation (4), the molar volume term is dominant in the effect of pressure on reactions 2 and 3. At 1 bar, the volume of magnetite + αFe or hematite + αFe is larger than the volume of FeO wüstite; therefore, pressure tends to stabilize wüstite. At high pressures, several phase transitions occur. At 130 kbar and 25°C, αFe transforms to ϵFe. Phase transitions in magnetite and hematite have been observed in shock-wave studies at pressures less than 600 kbar (McQueen et al., 1967). The pressure-induced phase transition of magnetite is also observed in a diamond cell at 250 kbar and 25°C, and the X-ray pattern of the high-pressure phase can be indexed as being monoclinic (Mao et al., 1974). All these transitions reduce the volume of the right-hand side of Equations (2) and (3), and ΔV becomes negative. As a result, pressure tends to disproportionate FeO into the high-pressure phase of magnetite (hFe_3O_4) + ϵFe or the high-pressure phase of hematite (hFe_2O_3) + ϵFe. Calculated trajectories of the volume-pressure curves for the phases involved in reactions 2 and 3 are presented in Figure 3. The Murnaghan

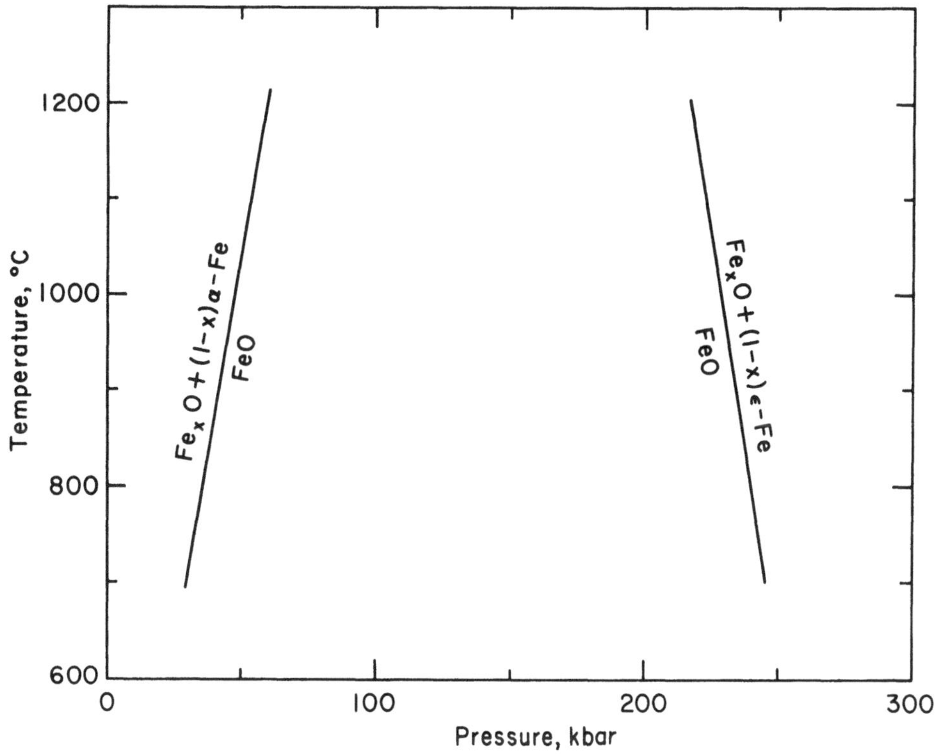

Fig. 2. The stability of stoichiometric and nonstoichiometric wüstite

equation and the bulk modulus and their pressure derivatives of εFe (Mao et al., 1967), FeO [Eq. (7) and Mao et al., 1969], magnetite (Mao et al., 1974), hematite (Liebermann and Schreiber, 1968), hFe_3O_4 and hFe_2O_3 (Ahrens et al., 1969) were used in the calculations. One should note the close proximity of the $\bar{V}$-P curves, and the sometimes slight density differences between them that cause a stabilization of a given phase region.

The volume-pressure trajectories were used in conjunction with available or estimated values of entropy and free energy, to construct a hypothetical P–T phase diagram of the binary Fe–O system in Figure 4. The diagram shows that the FeO phase reacts to form Fe_3O_4 + Fe or Fe_2O_3 + Fe under conditions equivalent to those of the lower mantle.

It should be noted that the phase diagrams of Figures 2 and 4 are only approximations. The diagrams are sensitive to the estimated values and to the extrapolations referred to above used in constructing them. As depicted in Figure 3 the volume difference among the various assemblages is less than 2% at 1000 kbar. Any further phase transition or spin transition in iron oxides at higher pressures would change the volume by approximately 10% and would, therefore, affect ΔV significantly. Solid

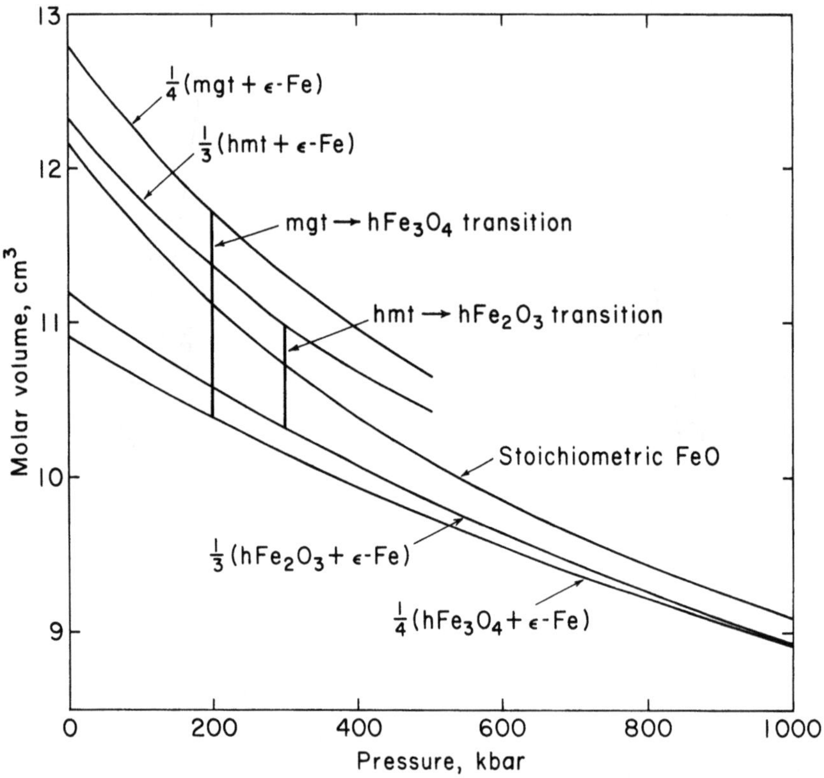

Fig. 3. Volume of FeO and its disproportionating products at pressure and 25°C

solutions of iron with silicon or sulfur, FeO with periclase, $h\mathrm{Fe_2O_3}$ with other sesquioxides, and $h\mathrm{Fe_3O_4}$ with other phases of related structure affect the relative stabilities and may yield a very different phase diagram. Therefore, the main purpose of the two phase diagrams (Figs. 2 and 4) is to show that one can assume the possible existence of iron oxide in the lower mantle as Fe_2O_3, Fe_3O_4, Fe_xO, or other forms more oxidized than FeO, and that such an assumption is, in principle, as valid as those that assume the only oxide to be FeO.

The notion of iron oxides with Fe:O less than 1 in the mantle does not conflict with the concept that ferric iron may be reduced to the ferrous state at high pressures (Drickamer and Frank, 1973). The Drickamer effect is more an internal process of redistributing electrons than an external process of changing oxygen fugacity. For instance, the ferric iron in Fe_2O_3 can be reduced to ferrous iron as $Fe_2^{2+}O_3^{4/3-}$, the Fe:O ratio remaining as 2:3. This reduction is also in keeping with the concept that chemical bonding becomes more covalent at high pressures. By analogy with sulfides, the highly covalent oxide may have more variable Fe:O ratios.

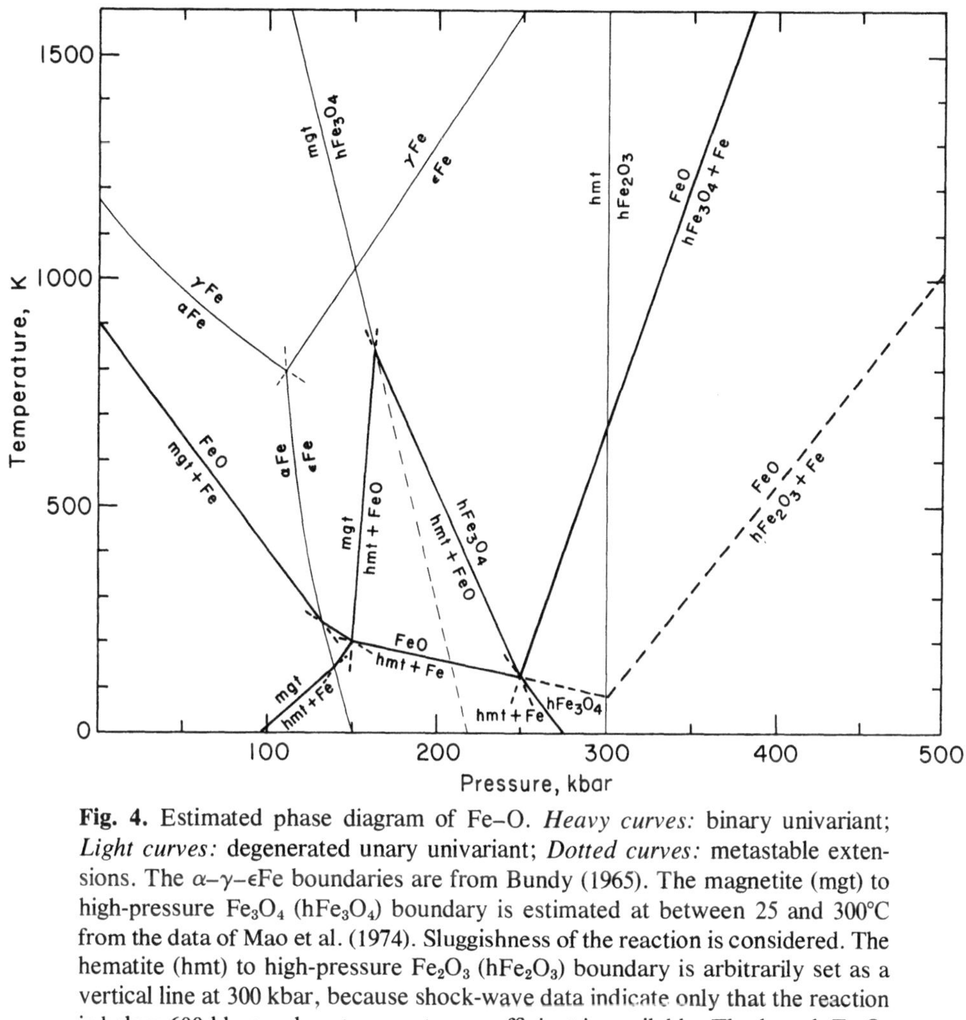

Fig. 4. Estimated phase diagram of Fe–O. *Heavy curves:* binary univariant; *Light curves:* degenerated unary univariant; *Dotted curves:* metastable extensions. The α–γ–ϵFe boundaries are from Bundy (1965). The magnetite (mgt) to high-pressure Fe_3O_4 (hFe_3O_4) boundary is estimated at between 25 and 300°C from the data of Mao et al. (1974). Sluggishness of the reaction is considered. The hematite (hmt) to high-pressure Fe_2O_3 (hFe_2O_3) boundary is arbitrarily set as a vertical line at 300 kbar, because shock-wave data indicate only that the reaction is below 600 kbar and no temperature coefficient is available. The hmt–h Fe_2O_3 transition was not observed in the diamond cell below 300 kbar

Iron oxides with other stoichiometry, such as Fe_2O (Strens, 1974) or nonstoichiometric Fe_2O_3 or Fe_3O_4, may be stable at high pressures. If so, FeO would break down further, thereby lending support to the present hypothesis.

5 THERMAL AND ELECTRICAL CONDUCTIVITY OF THE EARTH'S LOWER MANTLE

The effectiveness of radiative heat transfer depends upon the transparency of the media in the visible and near-infrared region. If FeO disproportionates to Fe and $Fe_{1-x}O$ (which could be Fe_2O_3, Fe_3O_4, or nonstoichiometric wüstite with its mixed oxidation states), $Fe_{1-x}O$ would absorb strongly by charge-transfer processes. An analogous observation is that magnetite and nonstoichiometric wüstite are opaque, but stoichiometric FeO is transparent in the visible region. As demonstrated by Mao (1973) and Bell and Mao (1974) charge-transfer absorption is greatly enhanced at the pressures and temperatures of the lower mantle.

Pressure may reduce trivalent iron to the divalent state (Drickamer and Frank, 1973), thus reducing the absorption, but it may also reduce iron from divalent to monovalent and may increase the covalency of the chemical bonding and enhance metal-ligand charge transfer. The net effect of pressure iron-bearing compounds should, therefore, increase their absorption. Spin-pairing in Fe^{2+} in the lower mantle, could change the crystal-field spectra. However, the crystal-field effects are minor in comparison with charge-transfer effects, that can effectively block radiative heat transfer.

In addition to the increase in optical absorption, the electrical conductivity should increase (Mao and Bell, 1972). Because the charge-transfer effect is basically an extrinsic semiconducting process, rather than one resulting from the close of the energy gap, it is unlikely that the electrical conductivity will approach that of a metal. The pressure-induced increase of electrical conductivity of the lower mantle material will probably decrease and approach that of a semiconductor, such as magnesiowüstite (Mao, 1973). For such a semiconductor with low activation energy, the electrical conductivity should be less dependent on temperature and more independent on iron content and structure.

High-pressure phases of Fe_3O_4, Fe_2O_3, and stishovite all have extremely high bulk moduli. From the relation of bulk modlus and lattice thermal conduction (Mao, 1973), the lower mantle should have very high lattice thermal conductivity, that compensates for the ineffectiveness of radiative heat transfer.

6 APPLICATION TO MODELS OF THE CHEMICAL EVOLUTION OF THE EARTH

It is desirable to establish as many initial and boundary conditions as possible for the origin of the earth, and then to deduce a model satisfying all these constraints (Urey, 1952). One of the most severe constraints employed in the past is the postulated disequilibrium relation between the mantle and the core (Ringwood, 1959, 1966). This constraint was used to reject all previous models of the formation of the earth that were based on differentiation of an originally homogeneous silicate-metal mixture. In an attempt to satisfy this constraint, most subsequent models of the evolution of the earth employ complicated processes such as the inhomogeneous accretion model of Clark et al. (1972) or an "inside out" configuration where the core material was accreted last and then displaced to its present position without reacting with the remainder of the earth (Ringwood, 1966). If the present notion of disproportionation of FeO into an oxide and metallic iron is adopted, it is obvious that the hypothesis of disequilibrium is not a constraint.

On the basis of the present calculations, silicates in the deep interior of a homogenous body react to form simple oxides, and the FeO component of the oxides breaks down further to metallic iron and an iron oxide. The process simultaneously includes oxidation and reduction reactions. A portion of the iron in FeO is reduced to metallic iron as oxidation forms $Fe_{1-x}O$. The oxidation phase could be Fe_2O_3, Fe_3O_4, or nonstoichiometric wüstite, depending on relative stability.

In the process postulated here it is assumed that the initial earth, regardless of bulk composition or state of homogeneity was heated by radioactive elements, and when the temperature reached that of the beginning of melting of iron alloy at the center of the earth, the dispersed metal began to melt and sink to the core, leaving a more oxidized layer of lower bulk density. The oxidized layer could have moved through unmelted mantle to shallower depths by gravitational instability and by convection. At the same time the overlying undifferentiated silicate layer would sink, disproportionate, differentiate, and repeat the process.

Further fractionation of the mantle could occur at the same time, or later, to concentrate certain nonvolatile elements such as Ca, Al, U, Th, Ba, and REE in the crust. Volatile elements, such as K and S, that are depleted in the mantle could either have been dissolved in the metallic core through metal-oxide partitioning (Murthy and Hall, 1972) or have been depleted before accretition.

After most of the metallic iron in the mantle has been removed, residual metallic iron in the oxides ascending from the lower mantle would be insufficient to reduce all the ferric iron to ferrous iron in silicates. Thus, the oxide layer has a low Fe^{2+}/Fe^{3+} ratio and would appear to be relatively oxidized. If iron oxide from the lower mantle has an average Fe:O stoichiometry of 0.95:1, the resulting Fe^{2+}/Fe^{3+} ratio in the mantle would be as low as 6, which is approximately the ratio observed in basaltic magma. The process differentiates metallic iron to the core and simultaneously differentiates oxygen toward the surface. Thus, an oxidized crust and upper mantle and a reduced core could be developed by an equilibrium reaction within an originally intimately mixed material.

The degassing products from the oxidized upper mantle should have high H_2O/H_2 and CO_2/CO ratios (Holland, 1962). The reaction is illustrated by the schematic phase diagrams in Figure 5. Figure 5A shows the

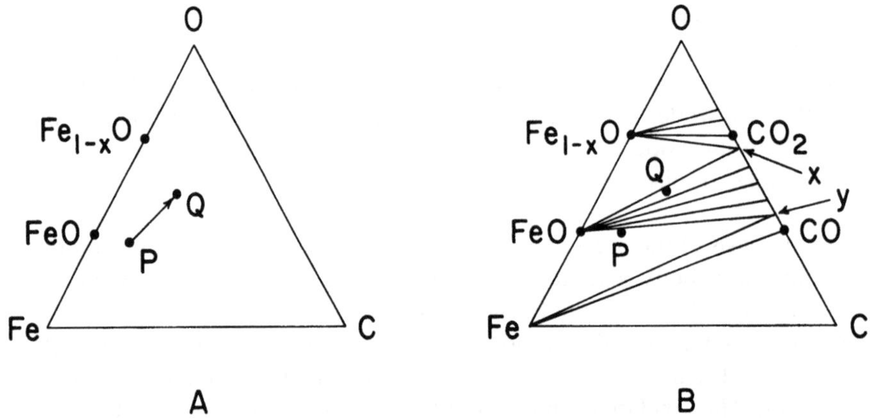

Fig. 5A,B. Schematic phase diagrams of a Fe–C–O system

reaction at high pressure where FeO is unstable relative to Fe and $Fe_{1-x}O$. Because no degassing is expected in the lower mantle, the reaction with carbon is not shown. When iron is removed by fractionation, the composition shifts from P to Q. Figure 5B shows the low-pressure phase diagram. As the composition Q rises to the upper mantle, $Fe_{1-x}O$ reacts with carbon to form FeO or ferromagnesian silicates, projected as FeO, and a gas with high CO_2/CO ratio X, although a low CO_2/CO ratio Y is inferred by the original composition P and by the assemblage Fe and FeO. An analogous argument applies to the reduction by H_2.

The amount of Si existing in the earth's core is an open question. The relatively low density of the core was used as the original argument for the predicted presence of Si, but a low density could also result from significant amounts of S (Murthy and Hall, 1972) and iron oxide that was produced from early disproportionation reactions in the lower mantle. Nevertheless, the existence of a reduced metallic core with high silicon content and an upper mantle with oxidized iron does not imply chemical disequilibrium.

In the present model, the melting and separation of iron from oxides occurs in the lower mantle and core. The amount of Ni, Co, Au, Pt, Ir, and Re in the oxides is thus determined by the partitioning of these siderophile elements between oxides (including silica) and metal at high pressures in the lower mantle. When the oxides rise to the upper mantle and recombine to form silicates, the same siderophile elements are preserved in silicates. The partition coefficient at high pressure has to be known before the content of these siderophile elements in the mantle can be used as a criterion for core-mantle relations. As estimated by Brett (1971), the partition coefficient of Ni between oxides and metal at high pressures is much higher than that between silicates and metal at low pressures, and a high content of these elements in the mantle may imply equilibrium.

Actually, even without the postulated process, an oxidized crust and a metallic core are not necessarily out of equilibrium. As an illustration, an hypothetical planet made of a simplified metal-oxygen system is shown in Figure 6. The pressure in the interior of the planet is caused by gravitation alone. The bulk composition and temperature are assumed to be the same for the three models, *A*, *B*, and *C*. Local equilibrium within each body is assumed. Figure 6A is a homogeneous planet. With its pressure gradient and the different densities of the two phases, obviously it is not in gravitational equilibrium. In Figure 6B, two phases have been differentiated according to their density. Model *B* is closer to chemical equilibrium than *A* but farther from equilibrium than *C*, which has been further differentiated to produce an oxidized crust, even though the oxidized crust is incompatible with the metallic core if they are at the same pressure.

The reason *C* is closer to equilibrium is that the crust and the core are at different pressures (the free energy in a pressure gradient is defined by the integration of *VdP;* (see Mao and Bell, 1971; Bell and Mao, 1972). Changing from model *C* to *B* includes the replacement of high-density metal at high pressure with low-density oxygen at low pressure; a large increase in free energy is involved. Conversely, changing from *B* to *C* involves a decrease in free energy. This is basically the "squeezing" effect suggested by MacDonald and Knopoff (1958). If the equilibrium condition *C* was obtained by the diffusion of oxygen, the rate would be prohibitively slow (Ringwood, 1961), but with the disproportionation of FeO and subsequent differentiation as proposed here, equilibrium conditions of an oxidized upper mantle and a reduced core could easily be achieved.

In general, then, the only requirement for the proposed process is that a reaction involving more than one stoichiometry of the Fe–O system occur in the lower mantle. Aside from the breakdown of FeO discussed previously, the process could also be one of partial melting with selec-

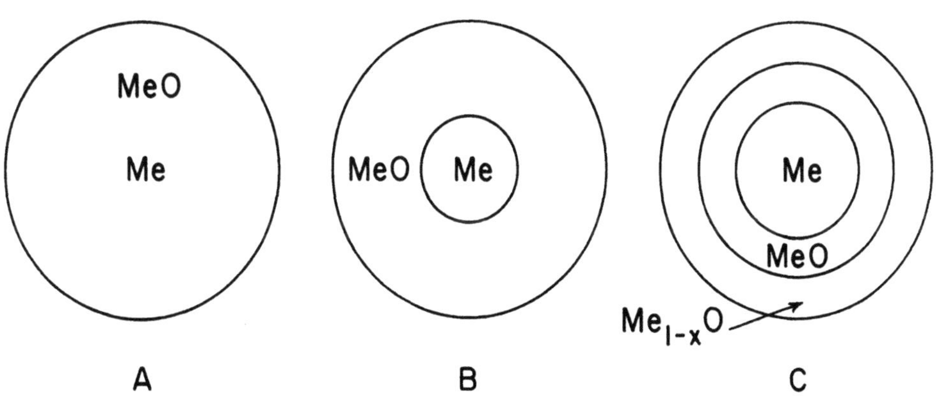

Fig. 6A–C. Hypothetical planet: (A) homogeneous body, (B) differentiated, (C) highly differentiated tive Fe–O partitioning or could involve two immiscible melts. Examples at low pressure might be the incongruent melting of ferrosilite and the immiscible Fe–Fe_xO melts of FeO composition. Most partial melts of silicates show some degree of differentiation of Fe–O, and it is likely that this process occurred in the earth's lower mantle.

REFERENCES

Ahrens, T. J., Anderson, D. L., Ringwood, A. E.: Equations of state and crystal structures of high-pressure phases of shocked silicates and oxides. Rev. Geophys. *7*, 667–707 (1969)

Bassett, W. A., Ming, L. C.: Disproportionation of Fe_2SiO_4 to 2FeO + SiO_2 at pressures up to 250 kbar and temperatures up to 3000°C. Phys. Earth Planet. Interiors *5*, 154–160 (1972)

Bell, P. M., Mao, H. K.: The Hypothesis of Melting at Stress Dislocations in the Earth. Carnegie Inst. Washington, Year Book 71, 416–418, (1972)

Bell, P. M., Mao, H. K.: Pressure Effect on Charge-transfer processes in Minerals. Carnegie Inst. of Wash. Yearbook *73*, 507–510 (1974)

Brett, R.: The earth's core: Speculations on its chemical equilibrium with the mantle. Geochim. Cosmochim. Acta, *35*, 203–221 (1971)

Bundy, F. P.: Pressure-temperature phase diagram of iron to 200 kbar, 900°C. J. Appl. Phys. *36*, 616–610, 1965

Clark, S. P., Jr., Turekian, K. K., Grossman, L.: Model for the early history of the earth. In: The Nature of the Solid Earth. Robertson, E. G. (ed.). New York: McGraw-Hill, 1972, pp. 3–18

Clendenen, R. L., Drickamer, H. G.: Lattice Parameter of nine oxides and sulfides as a function of pressure. J. Chem. Phys. *44*, 4223–4228 (1966)

Darken, L. S., Gurry, R. W.,: The system iron-oxygen, I, The wüstite field and related equilibria. J. Am. Chem. Soc. *67*, 1398–1412 (1945)

Drickamer, H. G., Frank, C. W.: Electronic Transitions and the High Pressure Chemistry and Physics of Solids. London: Chapman and Hall, 1973

Holland, H. D.: Model for the evolution of the earth's atmosphere. In: Petrologic Studies: A Volume in Honor of A. F. Buddington. Engel, A. E. J., James, H. L., Leonard, B. F. (eds.). Boulder, Colorado: Geol. Soc. Am., 1962, pp. 447–478

Jette, E. R., Foote, F.: An X-ray study of the wüstite (FeO) solid solution. J. Chem. Phys. *1*, 29–36, 1933

Johnston, W. D.: Oxidation-reduction equilibria in iron-containing glass. J. Am. Ceram. Soc. *47*, 198–201 (1964)

Katsura, T. B., Iwasaki, Kimura, S.: High-pressure synthesis of the stoichiometric compound FeO. J. Chem. Phys. *47*, 4559–4560 (1967)

Liebermann, R. C., Schreiber, E.: Elastic constants of polcrystalline hematite as a function of pressure to 3 kbar. J. Geophys. Res., *73*, 6585–6590 (1968)

MacDonald, G. J. F., Knopoff, L.: On the chemical composition of the outer core. Geophys. J., *1*, 284–297 (1958)

Mao, H. K.: Electrical and optical properties of the olivine series at high pressure. Carnegie Inst. of Wash. Yearbook *72*, 552–564 (1973)

Mao, H. K.: A discussion of the iron oxides at high pressure with implications for the chemical and thermal evolution of the earth. Carnegie Inst. of Wash. Yearbook *73*, 510–518 (1974a)

Mao, H. K.: A discussion of the iron oxides at high pressures with implications for the chemical and thermal evolution of the earth. Carnegie Inst. of Wash. Yearbook *73,* 510–518 (1974b)

Mao, H. K., Bassett, W. A., Takahashi, T.: Effect of pressure on crystal structure and lattice parameters of iron up to 300 kbar. J. Appl. Phys., *38,* 272–276 (1967)

Mao, H. K., Bell, P. M.: High pressure decomposition of spinel (Fe_2SiO_4). Carnegie Inst. of Wash. Yearbook *70,* 176–178 (1971a)

Mao, H. K., Bell, P. M.: Generation of Magma Along Faults and Stress Dislocation in the Earth. Carnegie Inst. Washington, Yearbook *70,* 229–233, 1971b, p. 13

Mao, H. K., Bell, P. M.: Electrical conductivity and the red shift of absorption in olivine and spinel at high pressure. Scence, *176,* 403–406 (1972)

Mao, H. K., Takahashi, T., Bassett, W. A., Weaver, J. S., Akimoto, S.: Effect of pressure and temperature on the molar volumes of wüstite and of three $(Fe,Mg)_2SiO_4$ spinel solid solutions. J. Geophys. Res., *74,* 1061–1069 (1969)

Mao, H. K., Takahashi, T., Bassett, W. A., Kinsland, G. L., Merrill, L.: Isothermal compression of magnetite to 320 kbar and pressure-induced phase transformation. J. Geophys, Res. *79,* 1165–1170 (1974)

McQueen, R. G., March, S. P., Fritz, J. N.: Huginiot equations of state of twelve rocks. J. Geophys. Res. *72,* 4999–5036 (1967)

Muan, A., Osborn, E. F.: Phase Equilibria among Oxides in Steelmaking. Reading, Mass.: Addison-Wesley Publ. Co. Inc., 1965

Murthy, V. R., Hall, H. T.: The origin and chemical composition of the earth's core. Phys. Earth Planet. Interiors *6,* 123–130 (1972)

Ringwood, A. E.: On the chemical evolution and densities of the planets. Geochim. Cosmochim. Acta *15,* 257–283 (1959)

Ringwood, A. E.: Silicon in the metal phase of enstatite chondrites and some geochemical implications. Geochim. Cosmochim. Acta *25,* 1–13 (1961)

Ringwood, A. E.: Chemical evolution of the terrestrial planets. Geochim. Cosmochim. Acta *30,* 41–104 (1966)

Strens, R. G. J.: The Stability of Fe_2O at High Pressure (abstract). Petrophysics Conference, NATO Advan. Study Inst. Newcastle, England, 1974

Urey, H. C.: The Planets: Their Origin and Development. New Haven, Conn.: Yale Univ. Press (1952)

13

J. GANGULY

Abstract The compositional effects on metamorphic mineral equilibria, as well as the possible buffering of solid-gas reactions in natural environments, have been discussed within the framework of the principles of classical thermodynamics. Small adjacent volumes of rock might be buffered, owing to their initial differences in bulk composition, along different f_{O_2}–T paths during progressive metamorphism leading to the development of contrasting mineralogies at the same P–T conditions. Thermodynamic formulations have been presented that permit evaluation of the effects of compositional variation of at least three component vapor phase on solid-gas equilibria. The effects of the lowering of fluid pressure from the load pressure within open pores of rocks, which might happen in some shallow environments, can be treated in terms of local equilibrium in the immediate neighborhood of the pores. The compositions of the mineral phases register the activities of the reactive fluid components in a metamorphic terrain. The same relations also permit graphical representation of the stability relations of minerals of fixed compositions as functions of the activities of fluid components at constant P and T. The expression of the activity of a component in solid solution in terms of the "ionic solution" model, which commonly yields the simplest activity–composition relations for the silicate and oxide solutions, and the general framework of thermodynamic relations governing the effects of solid solution on mineral equilibria, have been discussed. Small perturbations of free energy due to such factors as nonequilibrium ordering and nonstoichiometry might, in some cases, significantly affect mineral equilibria, especially in the experimental studies. In conclusion, an analysis is presented as to the criteria of recognition, and the dimensions of chemical equilibrium in metamorphism.

COMPOSITIONAL VARIABLES AND CHEMICAL EQUILIBRIUM IN METAMORPHISM

1 INTRODUCTION

The effect of bulk composition on the mineralogical characteristics of rocks has been recognized in the classical concept of metamorphic facies. However, attention was drawn primarily to the effects of very major differences in the bulk composition of the condensed system, such as the distinctive mineralogical characteristics developed in the isofacial metamorphism of basic, pelitic, and carbonate rocks. Since the introduction of eight metamorphic facies by Eskola, many have been added in later years, and some further divided into subfacies, each supposedly representing a restricted range of conditions in the P–T space. Systematic analysis of mineral equilibria, however, leads us to conclude that the P–T regime of formation or equilibration of a mineral assemblage could be quite sensitive to the compositional variations of the fluid and solid phases in the natural environments. The petrogenetic interpretation of metamorphic rocks, thus, requires an adequate recognition of the effects of the compositional variables. Furthermore, chemical compositions of the minerals themselves, when related within an appropriate thermodynamic framework, yield important informations as to the conditions through which the rock has evolved. It is the purpose of this article to discuss some of the interrelationships of pressure, temperature, and compositional variables, as seem appropriate to the study of metamorphic rocks, in terms of the principles of classical thermodynamics. An outline of the basic thermodynamic principles, related to the developments in the text, is provided in the Appendix, since it is assumed that the reader has an elementary knowledge of the subject.

2 CHEMICAL EQUILIBRIUM AND HETEROGENEOUS MINERAL REACTION

The equilibrium state of a system is independent of the history of the system, and is determined completely by the extremum of an appropriate thermodynamic potential at the stipulated boundary conditions. In geological studies we are able to reconstruct the history of a system or rock by examining its various states of quenched equilibrium, that is, the mineralogical characteristics which the rock has attained by equilibration to some early geological conditions, and which have survived through later geological processes.

At a given P, T, and bulk composition, the chemical equilibrium of a system is determined by the minimization of Gibbs free energy G of the system, whence the chemical potential μ of any component i which is *free to move* in all directions must be the same throughout the system. The gradient in chemical potential provides the driving force for the flow of matter in the system. The free energy change of a reaction is given by

$$\Delta G = \Sigma \nu_i \mu_i \tag{1a}$$
$$= \Delta G^* + RT \ln Q \tag{1b}$$
$$= \Delta G^0(1 \text{ atm}, T) + \int_1^P \Delta V_c^0 \, dP + RT \ln Q'$$

where ν_i is the stoichiometric coefficient of the components or formula units in the reaction, being taken as positive for the product and negative for the reactant, ΔG^* is the standard free energy change of the reaction, $\Delta G^0(1 \text{ atm}, T)$ is the free energy change of the reaction when all components are in their respective pure states at 1 atm pressure and the temperature in question, and

$$Q = \prod_i a_i^{\nu_i} \tag{2a}$$
$$Q' = \prod_i (a_i^{\nu_i})_c (f_i)_g^{\nu_i}, \tag{2b}$$

where c and g stand, respectively, for the condensed and gaseous components. Equation (1) can be rewritten as

$$\Delta G = RT \ln (Q/K) \tag{1a)'$$
$$= RT \ln (Q'/K') \tag{1b)'$$

where the Ks are the equilibrium constants that are functions of only P and T, and are defined as

$$K = \exp\left(-\frac{\Delta G^*}{RT}\right) \tag{2c}$$
$$K' = \exp\left(-\frac{\Delta G^0(1,T)}{RT}\right) \exp\left(-\frac{1}{RT}\int_1^P \Delta V_c^0 \, dP\right). \tag{2d}$$

At equilibrium $\Delta G = 0$. Consequently, (3a)

$$Q_{\text{equilibrium}} = K(P,T) \tag{3b}$$
$$Q'_{\text{equilibrium}} = K'(P,T).$$

The displacement of an equilibrium due to compositional effects depends on the relative change of the chemical potentials of the product and reactant assemblages. The lower the chemical potential is, the greater the stability. The qualitative displacement can be readily determined by finding the ratio Q/K, or Q'/K'. If this ratio is greater than unity, then from Equation (1), $\Delta G > 0$ in which case the reaction is driven to the left, causing an expansion of stability field of the reactant assemblage relative to the specified product, and vice versa. The equilibrium constant can be determined from thermochemical data, if available, or by knowing Q or Q' at equilibrium at P and T for the initial bulk composition. To determine the extent of reaction due to compositional variations until equilibrium has been reestablished in the system, we define, according to the law of definite proportions, a parameter ξ (Prigogine and Defay, 1954), such that $n_i - n_i^0 = \nu_i\xi$, where n_i and n_i^0 are, respectively, the final and initial mole numbers of the species i. ξ is called the *extent of reaction*. Thus, the mole fraction of the species i is given by

$$X_i = \frac{n_i}{\sum_i n_i} = \frac{n_i^0 + \nu_i\xi}{n^0 + \nu\xi} \tag{4}$$

where $n^0 = \sum_i n_i^0$ and $\nu = \sum_i \nu_i$. When equilibrium has been reestablished at the same P and T, we have, from Equations (3) and (4)

$$\prod\left(\frac{n_i^0 + \nu_i\xi}{n^0 + \nu\xi}\gamma_i\right)^{\nu_i} = K(P,T) \tag{5}$$

where γ_i is the activity coefficient of the component i. We can solve for ξ knowing K and γ_i. It can be shown (Prigogine and Defay, 1954) that ξ has the largest value when the initial concentration of the product phases are nil, and those of the reactants are proportional to their respective stoichiometric coefficients, the proportionality constant being given by $n^0/(\nu)^R$, where $(\nu)^R$ is the sum of the stoichiometric coefficients in the reactant assemblage.

It is instructive at this stage to discuss the P–T sensitivity of heterogeneous equilibria to the effects of the compositional variables. By the latter we not only mean the effects due to solid solution and gaseous mixtures, but also those due to such other factors as order–disorder, nonstoichiometry, etc. We have for a reaction,

$$\left(\frac{\partial \Delta G}{\partial T}\right)_P = -\Delta S \tag{6a}$$

$$\left(\frac{\partial \Delta G}{\partial P}\right)_T = \Delta V \tag{6b}$$

where ΔS and ΔV are the entropy and volume changes of the reaction under consideration. It follows from these relations that smaller the ΔS and ΔV near the equilibrium condition, the more sensitive are, respectively, the equilibrium temperature and pressure to the perturbations of the free energies of the phases due to the effects of the chemical variables (cf. Greenwood, 1972). This is illustrated graphically in Figure 1.

The number of independent relations such as Equation (3) among the intensive variables that should exist in a given system or a volume of rock in internal equilibrium, and not subject to any independent external

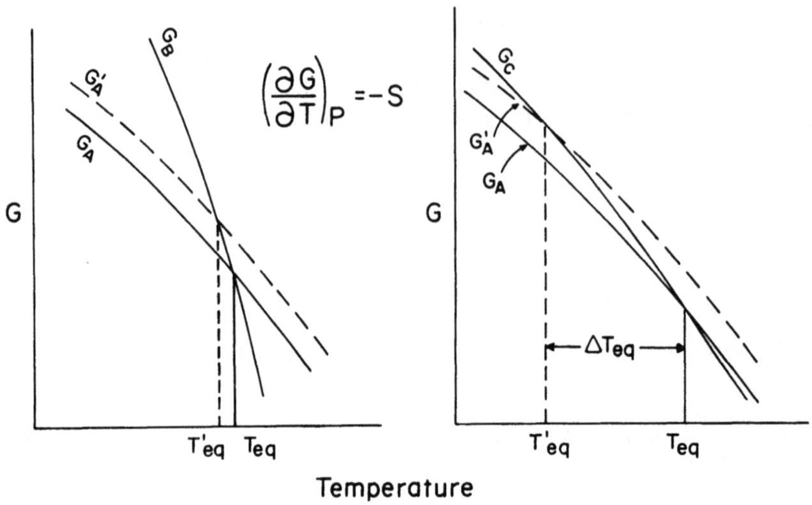

Fig. 1. Schematic illustration of the effect of the perturbation of free energy G on equilibrium temperature (stable or metastable) of reactions with different entropy changes. For the same perturbation of free energy, the effect is greater for the reaction with smaller entropy change. Analogous relation also holds between P_{eq} and ΔV of reaction

control of the chemical potential of the components, can be readily ascertained from the Gibbs phase rule, expressed as

$$F = C - P + 2. \tag{7}$$

Here F is the number of independent intensive variables, or degrees of freedom, and C is the smallest number of chemical species, or "components," required to express the compositions of the P phases in the system. Phase rule essentially expresses the difference between the total number of intensive variables and the number of independent relations among them in a system under equilibrium. Since it is not concerned with the extensive variables, it is applicable to both open systems, which permit transfer of matter to and from the system, as well as closed systems, where the masses of the components remain fixed. If, however, in an open system, the chemical potential of ϕ components are controlled externally, the degrees of freedom of the system decrease by ϕ. The phase rule, then, takes the general form

$$F = (C - \phi) - P + 2. \tag{8}$$

For a *closed* system, that is a system of fixed bulk composition, there are only two independent variables. This is known as Duhem's theorem (cf. Prigogine and Defay, 1954). Thus, the equilibrium state of a closed-system of known initial composition can, in principle, be *completely determined* in terms of all intensive and extensive properties, when any two variables are specified. The number of intensive variables that may be specified is, however, limited by the phase rule consideration. For a natural system of known initial composition, and assumed to have remained closed, the equilibrium phase assemblages including composition and masses of minerals can, therefore, be predicted as functions of temperature and pressure provided the necessary thermochemical data are available (cf. Brown and Skinner, 1975). The comparison of the observed and predicted assemblages would lead to important insights as to the extent to which progressive metamorphism can be treated as isochemical process and the nature of mass transfer involved in the production of the observed mineralogy. An alternative approach to this problem is provided by the analytical method of Greenwood (1967a). It should be obvious that the often sought correlation between the composition of a *single* mineral phase and metamorphic grade has unambiguous interpretation in terms of $P-T$ variation only if the bulk composition of the rock-fluid system has remained constant.

3 SOLID-GAS EQUILIBRIA

It is indeed possible in many cases, by certain means of geothermonetry and geobarometry, to determine the $P-T$ conditions of a rock without any recourse to the solid–gas equilibrium. It is impossible, however, to understand the mineralogical evolution of metamorphic rocks unless we have a thorough appreciation of the interaction between the solid and the vapor phases. The most common mineralogical changes in metamorphism may be represented in terms of equilibria involving gaseous species belonging to the system C–O–H–S; addiional components of interest are F and NaCl. We would here confine our attention to the most predominant ternary subsystem of the vapor system, namely the subsystem C–O–H.

It is now well established that the stabilities of ferromagnesium silicates, the appearance and disappearance of which have often been interpreted in terms of changes of $P-T$ conditions alone such as in the classical concepts of isograds and facies, are also sensitive to changes of f_{O_2} (as shown later in Fig. 4). It is, therefore, important to inquire about the f_{O_2} conditions that are likely to prevail in a C–O–H vapor phase during metamorphism. As calculated by Miyashiro (1964), and recently by Nokleberg (1973), the f_{O_2} generated in the system C–O–H of any initial composition by homogeneous equilibrium (that is, in the absence of a f_{O_2} buffer) at $P-T$ conditions of metamorphism is well within the stability field of hematite (Fig. 2). However, the f_{O_2} conditions compatible with the commonly observed iron oxide and ferromagnesium silicate assemblages in metamorphic rocks lie within the field of magnetite. Consequently, the fluid phase must have undergone reduction by reaction with the solid phases. Miyashiro (1964), thus, drew attention

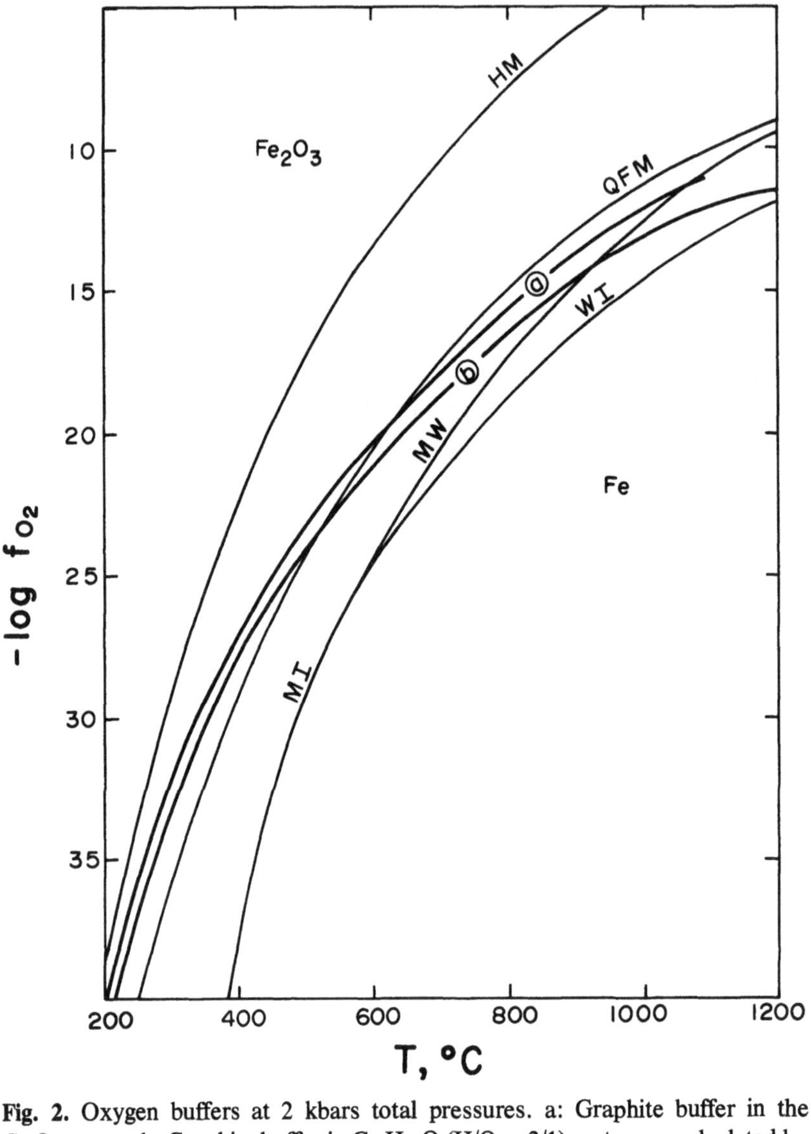

Fig. 2. Oxygen buffers at 2 kbars total pressures. a: Graphite buffer in the C–O system, b: Graphite buffer in C–H–O (H/O = 2/1) system as calculated by French (1966). WI (wustite–iron) and MI (magnetite–iron) data from Eugster and Wones (1962). Other data from sources as summarized in Ganguly (1972)

to the widespread occurrence of graphite in pelitic and semipelitic metamorphic rocks, and its effect on the reduction of the fluid phase. Further, Greenwood (1975) has recently discussed and emphasized the effectiveness of redox reactions of ferromagnesium silicates themselves in buffering of the fluid composition in metamorphism.

According to phase rule, the graphite (C–O–H) gas equilibrium is univariant at a given *P* and *T*. Consequently, in the absence of any other constraint, the f_{O_2} in equilibrium with graphite may be anywhere below the upper stability limit of graphite (Fig. 2). The explicit determination of f_{O_2} has to be made on the basis of the mineralogical characteristics of the rock. It is desirable, nonetheless, to try to gain some idea about the possible magnitudes of f_{O_2} that are likely to prevail in equilibrium with graphite in metamorphic rocks. The amount of oxygen released owing to the reduction of C–O–H fluid has been calculated by Nokleberg (1973) as a function of f_{O_2} both above and below the upper stability limit of graphite, in the P–T regime of 527–827°C and 500–2000 bars. These calculations (Fig. 3) suggest that exceedingly small amount of oxygen is released at f_{O_2} conditions above the upper stability limit of graphite. However, the amount of oxygen released increases very rapidly at f_{O_2} within the stability field of graphite, and virtually all oxygen available in an initial $CO_2 + H_2O$ fluid mixture is given off at f_{O_2} nearly half way between QFM and MW/I buffers. The amount of

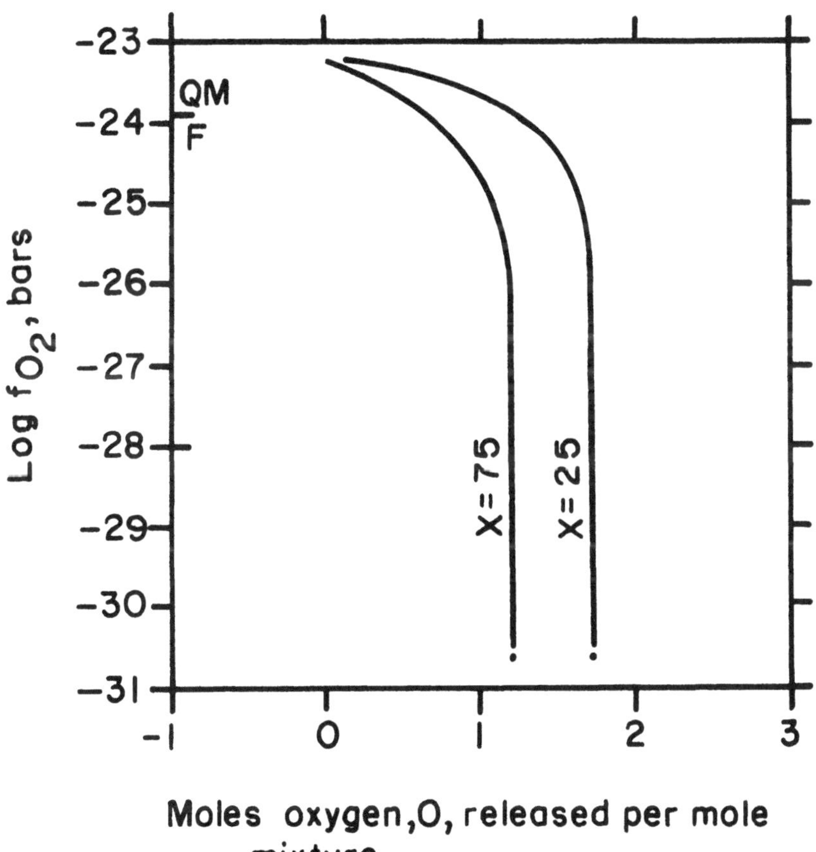

Fig. 3. Amount of oxygen O released during mixing of CO_2 with a $H_2O + H_2$ fluid buffered to oxygen fugacities below the upper stability limit of graphite, 527°C, 500 bars fluid pressure. X: mole percent of $H_2O + H_2$ end member (from Nokleberg, 1973)

oxygen released increases with increasing proportion of CO_2 in the initial fluid mixture owing to homogeneous reactions such as $3CO_2 + 2H_2O \rightarrow 2CO + CH_4 + 3O_2$. It, therefore, seems likely, as noted by Nokleberg, that if the C–O–H fluid phase were buffered initially at f_{O_2} near the lower limit of the magnetite field, the buffering capacity of the mineral assemblages would eventually be exhausted due to the large amounts of O_2 released at such conditions. Also, there are few natural mineral assemblages which can act as an f_{O_2} buffer within the field of stability of graphite. Therefore, f_{O_2} considerably below the upper stability limit of graphite is unlikely to persist during metamorphism, more so during the metamorphism of calc-mica rocks or carbonates owing to the effect of CO_2 in increasing the amount of O_2 released. The f_{O_2} in equilibrium with graphite in natural situations would, however, always be somewhat less than that defined by the upper stability of graphite due to the presence of additional components in the vapor phase. As an example, we have shown the f_{O_2}–T relation that would develop in equilibrium with graphite, as calculated by French (1966), at 2 kbar in the system C–O–H which has an H/O ratio of 2/1. An arbitrary compositional restriction, such as specification of H/O ratio, is required by phase rule to make f_{O_2} a function of only P and T. In the case where pore water is the only source of volatile, an H/O ratio of 2/1 would develop due to the stoichiometry of H_2O, irrespective of the nature of the volatile species, so long as the system remains closed. It can be shown that increase of pressure beyond 2 kbar has very small effect on the position of the graphite buffer. Considering the possible effects of various perturbations, such as the escape of volatiles, release of CO_2 by the decomposition of carbonate minerals, and release of some H_2 from the conversion of hydrocarbons to graphite, it would seem that the f_{O_2} in equilibrium with graphite during metamorphism would be restricted to a narrow band around the H/O = 2/1 buffer curve. In case where graphite is absent, or completely consumed by reaction with the fluid, f_{O_2} would evidently be above the upper stability limit of graphite. The various subsolidus oxidation equilibria of ferromagnesium silicates and oxides, which appear at the upper part of magnetite stability (e.g., Ganguly, 1968, 1972), would, however, often act as highly effective f_{O_2} buffers (cf. Greenwood, 1975). Owing to the very small amount of O_2 released per mole of the fluid phase at such conditions, there has to be very little oxidation of the mineral assemblages to buffer the f_{O_2} except when exposed to relatively large volumes of fluid, or when the rocks have a relatively high oxygen content due to their sedimentary or diagenetic history. In the latter case, the rock itself would release a considerable amount of O_2 if its f_{O_2} condition is higher than that of the buffering equilibria, causing significant oxidative transformation of the ferromagnesium silicates.

As discussed by Mueller (1967), diverse f_{O_2} conditions in small adjacent volumes of rock, once established under relatively oxidizing environment in nature, are extremely difficult to homogenize by the diffusion of the gaseous species during the subsequent evolution of the rock, unless subjected to certain overriding influence such as large influx of H_2O and CO_2. A large body of observational data seem to support this conclusion (e.g., Chinner, 1960; Hounslaw and Moore, 1967; Condie, 1967; Rumble, 1973). These diverse oxidation characteristics, wherever present, are most likely to have been established during sedimentation and/or diagenesis by differences in rock bulk composition. It may be noted that small disequilibrium of f_{O_2} at low temperature conditions could be the reason for large disequilibrium of f_{O_2} between adjacent rock volumes at high temperature! This can be appreciated by considering two adjacent volumes of rock consisting respectively of the hematite-magnetite buffer assemblages and of graphite. Although there would be only a small disequilibrium of f_{O_2} between these volumes at low temperature, they would become routed to diverging f_{O_2}–T paths in progressive metamorphism (Fig. 2).

Let us now consider, with reference to Figure 4, the evolution of a pelitic rock consisting of adjacent graphitic and nongraphitic domains. The figure shows the isobaric stability relations at 5 kbar of Fe–chloritoid, Fe–staurolite, almandine, aluminum silicates, and iron oxides, in the presence of O–H vapor phase, as functions of f_{O_2} and T. Our initial assemblage consists of chloritoid + kyanite + Fe_3O_4 ± graphite. There are many natural examples of such closely occurring graphitic and nongraphitic rocks with contrasting oxidation characteristics. For the sake of simplicity, we are ignoring in the following discussion the effects of additional components such as CO_2 and CH_4, that would exist in the vapor phase in the presence of graphite, on the solid-gas equilibria. Thus, the mineralogical changes that would take place under equilibrium conditions in the graphitic portion of the rock are rapid disappearance of chloritoid and appearance of staurolite around 530°C, and rapid disappearance of staurolite coupled with formation of almandine and kyanite around 690°C. Since f_{O_2} defined by the stoichiometric dissociation of H_2O is in the hematite field, f_{O_2} would rise in the graphite-free domain until it reaches the HM buffer, whence it would be driven to the invariant point I along the buffer. The isobaric invariant reaction at I is $12Chd = 2Fe_3O_4 + 3Fe_2O_3 + 12Ky + 7H_2O + 5H_2$. Thus, as the temperature is raised further, chloritoid would rapidly disappear, and the f_{O_2} would continue to be buffered by the HM buffer. The resultant assemblage would be kyanite, and iron oxides. Thus, at a temperature within the thermal stability limits of staurolite, say 625°C, the graphite-bearing and graphite-free assemblages would consist respectively of staurolite without kyanite, and kyanite without staurolite. Because of the extremely high f_{O_2} buffering capacity of the reaction $2Fe_3O_4 + O_2 \rightarrow 2Fe_2O_3$, the rock would consist of very small amount of Fe_2O_3. Thus, *even though f_{O_2} varied as a function of temperature,* contrasting mineralogies would be observed in the graph-

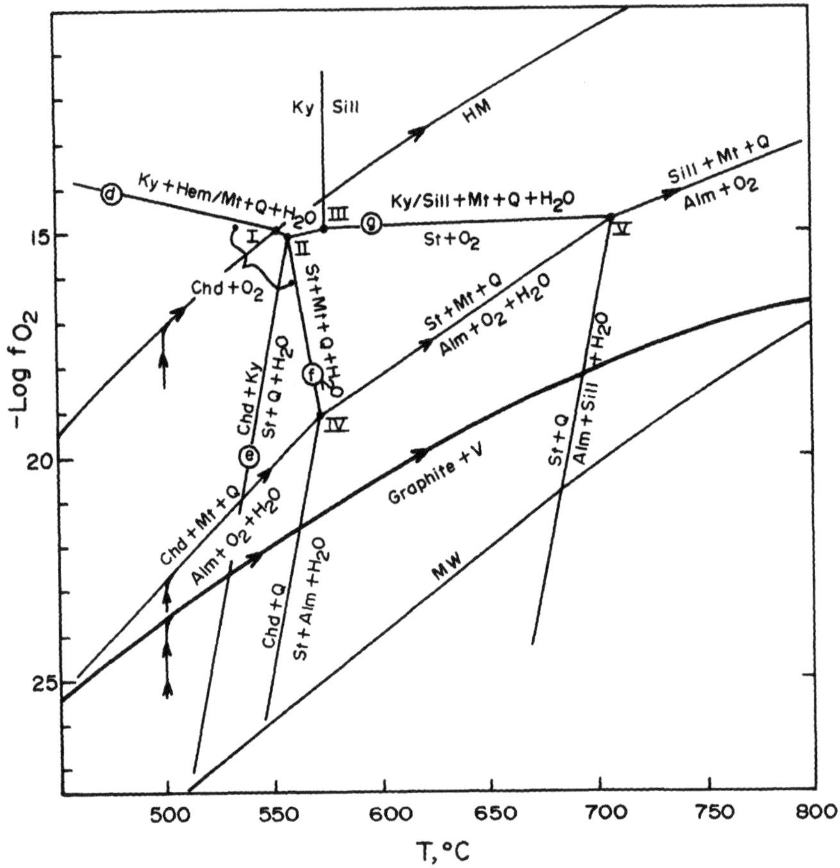

Fig. 4. Approximate stability relations of chloritoid, almandine, aluminum silicate, and iron oxides, along with graphite—C–O–H (H/O = 2/1) buffer at 5 kbar. Silicate stabilities are based on the data of Ganguly (1972). Graphite-Vapor equilibrium is after French (1966). Arrows: paths of the geologically effective buffers

ite-bearing and graphite-free portions of the rock at temperatures within the field of thermal stability of staurolite, similar to what would have been found if f_{O_2} had varied independently of temperature. Natural examples elucidating such variations in the spatial distribution of chloritoid, garnet, staurolite, and aluminum silicates as functions of f_{O_2} have been discussed by Ganguly (1968, 1972), Ganguly and Newton (1968), Rumble 1973, and Harte (1975). Buffering of the fluid composition by graphite and by mineral reactions, however, tends to limit the number of possible paths that the rocks might follow, some of which might be indistinguishable in terms of their effects on the major petrographic characteristics of the rock. This, in turn, introduces certain regularity and repetitive pattern in the evolution of rocks during progressive metamorphism.

The C–H–O gas in equilibrium with graphite is a multicomponent phase consisting essentially of the species H_2O, CO_2, CO, H_2, CH_4, and O_2 (French, 1966). The concentration of these species varies as a function of f_{O_2} T, and P (Figure 5). For the f_{O_2} conditions likely to prevail in natural situations in equilibrium with graphite as discussed earlier, the C–H–O gas phase may, to a good approximation, be treated as a mixture of only three species, CO_2, H_2O, and CH_4. Above the field of graphite, the gas consists essentially of CO_2 and H_2O, depending on their initial proportions. It may be noted from the Fig. 5, as emphasized by several workers before (e.g., Kerrick, 1974), that P_{H_2O} in equilibrium with graphite is significantly less than P_{gas} or P_{total} for a considerable range of metamorphic conditions, and is also a function of f_{O_2} at a given temperature and pressure. Thus, not only the temperature of dehydration equilibria in natural graphitic schist would be lower than that when $P_{H_2O} = P_{total}$, but also would vary with the f_{O_2} condition of the rock. Similar conclusion holds also for decarbonation equilibria involving C–H–O vapor phase in equilibrium with graphite, such as in graphitic–calc–mica schists. Holdaway (1975), for example, has explained the presence and absence of chloritoid in adjacent non-graphitic and graphitic volumes of rocks,

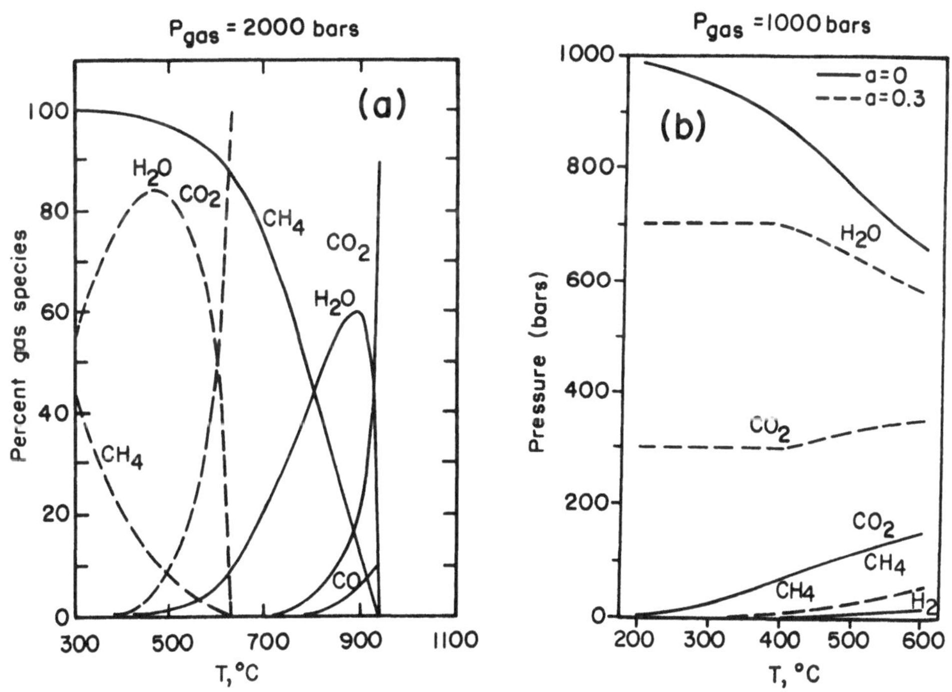

Fig. 5a, b. Composition of a C–H–O vapor phase in equilibrium with graphite as a function of P, T, and f_{O_2}. (a) *dashed line:* f_{O_2} defined by QFM buffer; *solid lines:* f_{O_2} defined by MI/W buffer (from Eugster and Skippen, 1967). (b) $a = CO_2/(H_2O + CO_2)$ of starting fluid which reacted with graphite (from Yui, 1968)

respectively, in the precambrian of New Mexico on the basis of the effect of graphite in depressing the dehydration temperature of chloritoid. Carmichael (1970) has reported that the reaction biotite + calcite + quartz ⇆ Ca-amphibole + CO_2 + H_2O has taken place at anonomously low temperature in a thin graphite- and pyrrhotite-bearing calcareous formation in the Whetstone Lake Area, Ontario. He suggested this to be due to the dilution of CO_2 and H_2O in the vapor phase by equilibration with graphite at relatively low f_{O_2} (Fig. 5a). It is also evident from Fig. 5 that P_{H_2O} or the amount of H_2O in the vapor phase decreases rapidly with increasing temperature from medium to high grade metamorphism, thus providing the relatively H_2O poor environment that seems to be required (such as discussed by Newton, 1974), for the formation of granulite facies rocks. The progressive consumption of graphite with increasing temperature (Yui, 1968) might lead to complete removal of graphite prior to the granulite facies condition. The fluid phase would become depleted in H_2O, nonetheless. Thus, the presence of graphite in the granulite facies is not germane to the validity of the above idea. It may be emphasized here that the carbonaceous matter in metamorphic rocks is not graphitized until probably the upper greenschist to amphibolite facies (Izawa, 1968; Grew, 1974). Thus, at lower grade, the model of fluid composition developed in terms of graphite-gas equilibrium is probably a qualitative representation of that in natural carbonaceous environment, even if the vapor phase is restricted to the system C–H–O.

Greenwood (1976b) has provided the basic theoretical framework for the study of mineral equilibria involving mixed volatile confined to the binary H_2O–CO_2 system. This work has greatly improved our understanding of solid-gas equilibria in metamorphism, especially in calc-mica schists in which both CO_2 and H_2O are liberated in the vapor phase. Reactions involving different stoichiometries of H_2O and CO_2 have different sensitivities to the variation of H_2O/CO_2 of the fluid phase. For an ideal fluid mixture, the isobaric equilibrium temperature is maximum when the composition of the fluid phase is the same as that evolving from the reaction and approaches the temperature axis asymptotically as the difference between the two fluid compositions increases. Three types of equilibria, having qualitatively different responses to the variation of fluid composition, may be distinguished (Fig. 6):

$A \rightleftarrows B + V_1$ (e.g., Cc + Q $\rightleftarrows$ Woll + CO_2)

$A + \nu_1 CO_2 \rightleftarrows B + \nu_2 H_2O$

(e.g., Dol + 4Q + H_2O $\rightleftarrows$ Tc + 3Cc + $3CO_2$)

and

$A \rightleftarrows B + \nu_1 CO_2 + \nu_2 H_2O$

(Trem + 3Cc + 2Q $\rightleftarrows$ 5Diop + $3CO_2$ + H_2O).

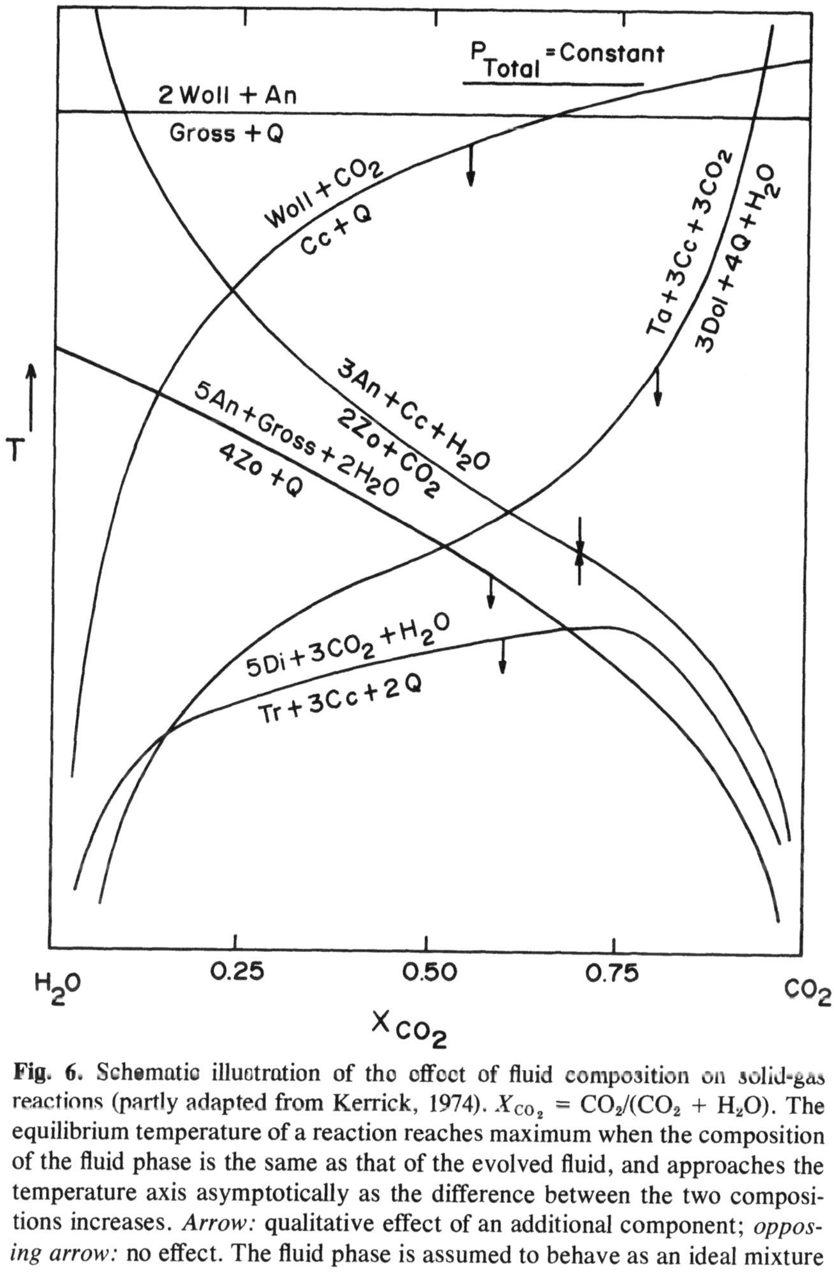

Fig. 6. Schematic illustration of the effect of fluid composition on solid-gas reactions (partly adapted from Kerrick, 1974). $X_{CO_2} = CO_2/(CO_2 + H_2O)$. The equilibrium temperature of a reaction reaches maximum when the composition of the fluid phase is the same as that of the evolved fluid, and approaches the temperature axis asymptotically as the difference between the two compositions increases. *Arrow:* qualitative effect of an additional component; *opposing arrow:* no effect. The fluid phase is assumed to behave as an ideal mixture

The isobaric reaction boundaries of such equilibria might intersect one another, and others, in a complex manner with the variation of H_2O/CO_2 in the fluid phase. This could, in some cases, lead to intersection of reaction boundaries or of the so called "isograds" in the field such as discussed by Carmichael (1970). *In general, great caution must be exercised in relating isotherms to "isograds," especially those defined by solid-gas equilibria.*

Inasmuch as the vapor phase in equilibrium with graphite involves additional components in significant proportions, it is useful here to seek certain extension and reformulation of Greenwood's work appropriate for somewhat more general natural conditions.

For a reaction involving a single vapor species, 1

$$\sum_i r_i R_i = \sum_j p_j P_j + \nu_1 V_1, \tag{a}$$

where r_i, p_j, and ν are the stoichiometric coefficients, R is reactants, and P is solid products, we have, at equilibrium,

$$\Delta G = 0 = \Delta G^* + RT \ln a_{v_1}^{\nu_1}$$

or

$$d\Delta G = 0 = d\Delta G^* + d(RT \ln a_{v_1}^{\nu_1}) \tag{9}$$

it being assumed, for simplicity, that the solid phases are in their respective standard states. Using the relations

$$d\Delta G^* = -\Delta s^* dT + \Delta V^* dP \tag{10}$$

$$\bar{S}_i = -\frac{\partial \mu_i}{\partial T} = S_i^* - R \ln a_i - RT\frac{\partial \ln \gamma_i}{\partial T} \tag{11}$$

and rearranging terms, we have from Equation (9)

$$\left(\frac{\partial T}{\partial X_1}\right)_P = \frac{RT}{\Delta \bar{S} + \dfrac{\nu_1 RT \partial \ln \gamma_i}{\partial T}} \left[\frac{\nu_1}{X_{\nu_1}} + \frac{\nu_1 \partial \ln \gamma_i}{\partial X_1}\right]. \tag{12}$$

Equation (12) is valid for the most general case in which the vapor component in the reaction (a) is a constituent of a multicomponent gaseous mixture. For ideal gaseous mixture, $\gamma_1 = 1$, it reduces to the same equation derived by Greenwood (1967b) for ideal binary gas mixture. In terms of fugacity, Equation (12) can be rewritten as

$$\frac{\partial T}{\partial f_1} = \frac{RT}{\Delta \bar{S} + \nu_1 RT \dfrac{\partial \ln \gamma_1}{\partial T}} \left(\frac{\nu_1 f_1^*}{f_1}\right). \tag{12'}$$

For a reaction

$$\sum_i r_i R_i = \sum_j p_j P_j + \nu_1 V_1 + \nu_2 V_2, \tag{b}$$

involving two reacting volatile species in a multicomponent gas phase, we can derive, following the above procedure

$$\left(\frac{\partial T}{\partial X_2}\right)_P = \frac{RT}{\Delta\bar{S} + RT\dfrac{\partial \ln \gamma_1^{\nu_1}\gamma_2^{\nu_2}}{\partial T}}\left[\frac{\nu_1}{X_1\gamma_1}\left(\frac{\partial a_1}{\partial X_2}\right) + \frac{\nu_2}{X_2} + \frac{\nu_2 \partial \ln \gamma_2}{\partial X_2}\right]. \quad (13)$$

This expression will reduce to that derived by Greenwood (1967b) for ideal binary gas phase for which

$$\frac{\partial a_1}{\partial X_2} = \frac{\partial X_1}{\partial X_2} = -1 \quad (14)$$

[The term expressing temperature dependence of γ is missing from the $\Delta\bar{S}$ term of Greenwood (see Eq. (11) above), so his general $\partial T/\partial X$ expression would differ from that derived above.] For the gas phase involving additional components, however, it is more convenient to deal with the effect of a third component on the binary T–X relation rather than with the expression (13) because of the difficulty in evaluating $\partial a_1/\partial X_2$ for such cases. For the case of ideal gas mixture, we have at equilibrium for the reaction (b)

$$-\Delta\bar{S}\left(\frac{\partial T}{\partial X_3}\right)_P + RT\left(\frac{\partial \ln X_1^{\nu_1}X_2^{\nu_2}}{\partial X_3}\right) = 0. \quad (15)$$

Since $X_1 + X_2 + X_3 = 1$

$$\frac{\partial X_1}{\partial X_3} + \frac{\partial X_2}{\partial X_3} = -1 \quad (16)$$

If X_1/X_2 is kept constant, so that $X_1 = KX_2$

$$\frac{\partial X_2}{\partial X_3} = -\frac{1}{1+K} = -\frac{X_2}{X_1 + X_2}. \quad (17)$$

Combining Equations (16) and (17)

$$\frac{\partial X_2}{\partial X_3} = -\frac{1}{1+K} = -\frac{X_2}{X_1 + X_2}. \quad (18)$$

Thus,

$$\left[\frac{\partial(\ln X_1^{\nu_1}X_2^{\nu_2})}{\partial X_3}\right]_{P,X_1/X_2} = \left(\frac{\nu_1}{X_2} + \frac{\nu_2}{X_2}\right)\frac{\partial X_2}{\partial X_3} \quad (19)$$

Hence

$$\left(\frac{\partial T}{\partial X_3}\right)_{P,X_1/X_2} = -\frac{RT}{\Delta \bar{S}}\left(\frac{\nu_1 + \nu_2}{X_1 + X_2}\right) = -\frac{RT^2}{\Delta \bar{H}}\left(\frac{\nu_1 + \nu_2}{X_1 + X_2}\right) \tag{20}$$

We can map the entire T–X surface for a solid–gas equilibrium involving two reacting volatile species in an ideal three component gas mixture, such as CO_2–H_2O–CH_4, by first knowing the binary T–X relation, and then determining the effect of the third component at constant X_1/X_2 values, that is, at constant relative concentration of the first two components. For the particular case that $\nu_1 + \nu_2 = 0$, such as the reaction 2Zo + CO_2 = 3An + Cc + H_2O, the third component has no effect on the equilibrium temperature, $\partial T/\partial X_3$ ideal $= 0$. For the other cases, the qualitative effect of the third component can be readily evaluated by knowing the sign of $\nu_1 + \nu_2$ from the reaction stoichiometry, and of $\Delta \bar{H}$ or $\Delta \bar{S}$, which can be decided by noting the high and low temperature sides of the reaction ($\Delta \bar{S}$ and $\Delta \bar{H}$ are positive for equilibria driven to the right with increasing temperature). Since the net evolution of volatile is normally accompanied by evolution of heat, $\Delta \bar{H}$ and $(\nu_1 + \nu_2)$ are unlikely to have opposite signs. Indeed, no reaction with opposite signs of $\Delta \bar{H}$ and $(\nu_1 + \nu_2)$ is known to the author in metamorphism. Consequently, for ideal gas mixture, the third component would either lower the equilibrium temperature, or leave it unaffected.

It has been assumed in the above discussion that the fluid phase is at the same pressure as the rock. Under certain circumstances, however, rocks may have sufficient strength against nonhydrostatic stress to maintain open pores such that the fluid phase sustains a lower pressure than the load pressure, P_f(Pore) $< P_1$. Even though there is a pressure gradient on the solid increasing away from the pores, the pressure on a sufficiently small volume of solid in the *immediate surrounding* of the pores would essentially be the same as that of the pore fluid. Evidently, the *local* equilibrium temperature between the pore fluid and the above solid volume would be that corresponding to the P_f (Pore) on the equilibrium boundary determined under the usual conditions of $P_f = P_l$. Most solid–gas reaction boundaries have positive slopes in the $P_l = P_f$–T space. The equilibrium temperatures of these reactions would, thus, be depressed in the immediate neighborhood of such pores as P_f(Pore) is reduced below P_l(cf. Ellis and Fyfe, *in* Combs et al., 1959, p. 83). The reverse would be the case for solid–gas reactions having negative P–T slope. Studies on the strength of rocks, especially considering the weakening effect of an aqueous fluid (Brace et al., 1970), however, suggest that conditions of $P_f < P_l$ could be realized only under relatively shallow, near surface conditions, and thus seem to be excluded from most metamorphic environments (cf. Fyfe et al., 1958).

The composition of the fluid phase in metamorphism may be controlled locally by the buffering action of solid–gas reactions, or by an outside reservoir of fluid (cf. Gordon and Greenwood, 1971; Kerrick, 1974). In the first case, the mineralogical characteristics of the rock would evolve along the paths determined by the mineral buffers, such as discussed earlier, and by Greenwood (1975), whereas in the latter the mineralogical characteristics evolve along paths determined by the externally imposed fluid compositions. There can obviously be various possible gradations between the two extremes. These are illustrated in Figure 7. There is no a priori criteria by which to decide whether or not the activities of certain gaseous components during metamorphism are determined by communi-

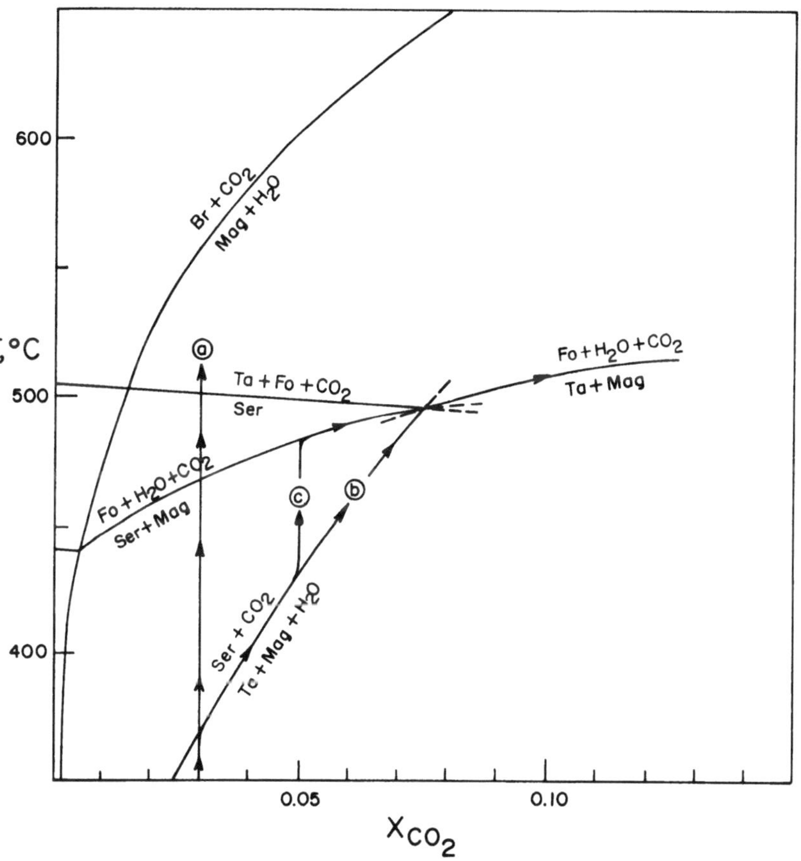

Fig. 7. Mineralogical evolution under two limiting conditions of the buffering of fluid composition; a: by an external reservoir; b: by the mineralogical assemblage of the rock; c: is an intermediate case. Progressive mineralogical assemblages beginning with a fluid composition of $X_{CO2} = 0.03$: a: Ta + Mag, Ser + Mag, Ser + Fo, Fo + Ta; b: Ta + Mag, Ta + Mag, Ta + Mag + Ser, Ta + Mag + Fo; c: Ta + Mag, Ta + Mag + Ser, Ser + Mag, Ser + Mag + Fo, Fo + Ta + Mag. The phase relations are adopted from Greenwood (1975). As shown by him, the amount of a phase that should be produced by any of the isobaric univariant reactions is extremely small

cations with external reservoirs, or are fixed by reaction with its immediate surroundings. The final conclusion must await detailed analysis of the mineralogical characteristics of the rock. Greenwood (1975) has concluded that unless the fluid phase is present in very large amounts, the mineral equilibria in many progressively metamorphosed terrains are capable of buffering the fluid composition, and thus driving the mineralogical evolution of rocks of same composition along certain systematic paths even when the compositions of the initial fluid were different.

The determination of the magnitudes and gradients of the activity of fluid components in a metamorphic terrain is a problem of considerable importance in the interpretation of the mineralogical characteristics of the rocks. Thus, Carmichael (1970) was able to interpret the intersection of mineral reactions or of the so called "isograds" in closely associated pelitic and calcareous rocks as due to variations of the H_2O and CO_2 activities. One method of determining the activity of the fluid components under isobaric-isothermal situation is through relating them to the composition of the mineral phases, which can be measured directly, by means of the equilibrium constants of the appropriate mineral equilibria, and a–X relation of the minerals (e.g., Mueller, 1966; Rumble, 1974). Thus, for an equilibrium assemblage of the minerals staurolite, chloritoid, kyanite, magnetite, and quartz, we can write the following independent solid-gas equilibria

$$\overset{\text{Chd}}{6FeAl_2SiO_5(OH)_2} + O_2 \rightleftarrows 6Al_2SiO_5 + 2Fe_3O_4 + 6H_2O \qquad \text{(a)}$$

$$4FeAl_2SiO_5(OH)_2 + Al_2SiO_5 \rightleftarrows \overset{\text{Staur}}{2Fe_2Al_9Si_4O_{23}(OH)} + SiO_2 + 3H_2O \qquad \text{(b)}$$

for which the equilibrium constants are

$$K_a(P,T) = \frac{a^6_{H_2O}}{a_{O_2}} \left(\frac{X^{Mt}_{Fe}}{X^{Chd}_{Fe}}\right)^6 (\gamma_a)^6 \qquad (21)$$

$$K_b(P,T) = a^3_{H_2O} \left(\frac{X^{St}_{Fe}}{X^{Chd}_{Fe}}\right)^4 (\gamma_b)^4. \qquad (22)$$

Here γ_a and γ_b take care of the departure of the solid solutions of Fe^{2+} from ideal ionic solution behavior, assuming for the sake of simplicity, that Fe^{2+} is the only ion to be substituted. Thus, by measuring the compositions of the coexisting chloritoid, staurolite and magnetite over restricted outcrop area which would seem to have formed under isothermal–isobaric condition, we can simultaneously determine the gradients of a_{H_2O} and a_{O_2} by simply differentiating the above expressions at constant P and T. If, in addition, K_a and K_b are known as functions of P and T, we can explicitly determine values of a_{H_2O} and a_{O_2} over extended regions over which the temperature and/or pressure may have varied. It may be noted, however, that at a given a_{H_2O}, P and T, the a_{O_2} varies as the inverse sixth power of $(X_{Fe}^{Mt}/X_{Fe}^{Chd})$, as well as of γ_b so that the determination of a_{O_2} is extremely sensitive to any error in the measured compositional property, and inferences relating to the a–X relations of the solid phases. This is typically the situation with respect to the relation of a_{O_2} to mineral composition. Caution should thus be exercised when silicate compositions are used to infer a_{O_2} and activities of any other vapor species (cf. Mueller, 1972). The expression for equilibrium constant should be examined carefully to determine the sensitivity of the activity of the vapor component to mineral compositions.

The differentiation of the expressions for the equilibrium constants, such as Equations (21) and (22), in the logarithmic form, permit us to graphically depict the mineral stabilities as functions of the activities or chemical potentials of certain components at constant P and T. Thus, from Equations (21) and (22), we have, at constant P, T, and a_{Fe} in the solid solutions

$$\left(\frac{\partial\mu_{H_2O}}{\partial\mu_{O_2}}\right)_d = \left.\frac{\partial \log a_{H_2O}}{\partial \log a_{O_2}}\right]_d = \frac{1}{6} = -\left(\frac{\nu_{O_2}}{\nu_{H_2O}}\right)_d \tag{23a}$$

$$\left(\frac{\partial\mu_{H_2O}}{\partial\mu_{O_2}}\right)_e = \left.\frac{\partial \log a_{H_2O}}{\partial \log a_{O_2}}\right]_e = 0 = -\left(\frac{\nu_{O_2}}{\nu_{H_2O}}\right)_e. \tag{24}$$

In general,

$$\frac{\partial\mu_i}{\partial\mu_j} = \frac{\partial \log a_i}{\partial \log a_j} = -\frac{\nu_j}{\nu_i} \quad (\nu > 0 \text{ for product, } < 0 \text{ for reactant}). \tag{23b}$$

By deriving analogous relations for other possible equilibria in the above assemblage ($54\text{Chd} + 5O_2 = 12\text{St} + \text{Mt} + 6\text{Q} + 48H_2O$ . . . (e); $6\text{St} + 3\text{Q} + 2O_2 = 27\text{Ky} + 4\text{Mt} + 3H_2O$ . . . (f)), we can represent the mineral stabilities as functions of μ_{H_2O} and μ_{O_2} (or $\log a_{H_2O}$ and $\log a_{O_2}$) for fixed a_{Fe}^{2+} in the solid solutions on the basis of the well-known Schreinemakers' principles. This is illustrated in Figure 8. Such diagrams were first introduced in geological literature by Korzhinskii (1959). The slopes of the reactions, relation (23b), were, however, derived by him from certain special thermodynamic formalisms which he developed, and were discussed, among others, by Thompson (1970).

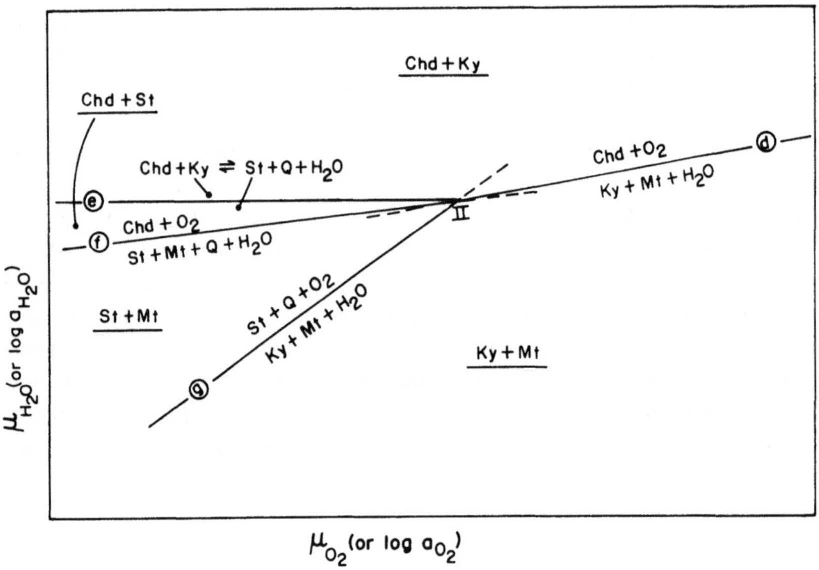

Fig. 8. Isothermal–isobaric stability relations of chloritoid, staurolite and kyanite as functions of chemical potentials (or activities) of H_2O and O_2. (Fig. 4 shows the isobaric log f_{O2}–T stability)

4 SOLID SOLUTIONS AND MINERAL EQUILIBRIA

Our present appreciation of the effects of solid solution on mineral equilibria owes greatly to the pioneering works of Ramberg (1944, 1952a,b), and of Ramberg and DeVore (1951). The effect of solid solution on mineral equilibria is manifested through its effect on the activity of the component, such as the conventional end-member formula unit, used to express the mineral reaction in the stoichiometrically balanced form. For the purpose of analytical treatment, it is evidently convenient to find expression of the activity of such component as function of composition of the solid solution phase (and, of course, of P and T) such that the activity coefficient(s) approaches unity, in other words, the solution obeys the "Raultion behavior," as closely as possible. It has been found most rewarding in this respect to treat silicate and oxide solid solutions as "ionic solutions," as was first suggested by Ramberg and DeVore (1951). In this model, the formula end members are considered to be dissociated in the solid solution into the smallest ionic units such that they can mix freely without any local imbalance of charge (Kretz, 1961; Bradley, 1962; Mueller, 1964; Saxena, 1973; Kerrick and Darken, 1975). Thus, for the garnet solid solution $(Ca,Mg)_3^{viii}(Al,Cr^{3+})_2^{vi}Si_3O_{12}$, we conceive the following scheme of ionic dissociation and mixing.

$$
\begin{array}{l}
Mg_3Al_2Si_3O_{12} \rightarrow 3 \\
Ca_3Al_2Si_3O_{12} \rightarrow 3 \\
Ca_3Cr_2Si_3O_{12} \rightarrow 3
\end{array}
\begin{bmatrix} Mg^{2+} \\ Ca^{2+} \\ Ca^{2+} \end{bmatrix}^{viii}
\begin{array}{c} +\ 2 \\ +\ 2 \\ +\ 2 \end{array}
\begin{bmatrix} Al^{3+} \\ Al^{3+} \\ Cr^{3+} \end{bmatrix}^{vi}
\begin{array}{c} +\ 3 \\ +\ 3 \\ +\ 3 \end{array}
\begin{bmatrix} SiO_4^{4-} \\ SiO_4^{4-} \\ SiO_4^{4-} \end{bmatrix}
$$

where the roman numerals indicate coordination site of the cations. Thus, the activity of the pyrope end member component is expressed as

$$a^{Gt}_{Mg_3Al_2Si_3O_{12}} = (a^{viii}_{Mg})^3 (a^{vi}_{Al})^2 = [(X^{viii}_{Mg})^3 (X^{vi}_{Al})^2] [(\gamma^{viii}_{Mg})^3 (\gamma^{vi}_{Al})^2] \tag{25a}$$

where,

$$X^{viii}_{Mg} = \left(\frac{Mg}{Mg + Ca}\right)^{viii}$$

and

$$X^{vi}_{Al} = \left(\frac{Al}{Al + Cr^{3+}}\right)^{vi}.$$

γ^{viii}_{Mg}, however, is not only a function of the composition of the eight coordinated site, but also of that of the octahedral site; similarly for γ^{vi}_{Al}. We call a^{viii}_{Mg} and a^{vi}_{Al} the partial ion activities. It may be noted that the statistics of dissociation demanded by the requirement of free mixing without any local imbalance of charge might, in some cases, lead to expression of the activity of a component in solution that is the same as for the undissociated component (e.g., Mueller, 1964; Ganguly, 1971). Thus, we have for the plagioclase solid solution (Mueller, 1964)

$$a^{Plag}_{NaAlSi_3O_8} = (X_{(NaSi)}\gamma_{(NaSi)})^{Plag} = (X_{Ab}\gamma_{Ab})^{Plag} \tag{25b}$$

where

$$X_{NaSi} = \frac{(NaSi)^{5+}}{(NaSi)^{5+} + (CaAl)^{5+} + (KSi)^{5+}} = X_{Ab}.$$

An integral part of the problem of the effect of solid solution on mineral equilibria is that of fractionation of elements between coexisting mineral phases, as determined by ion-exchange equilibrium (Ramberg and DeVore, 1951). Thus, in an assemblage consisting of the silicates garnet and cordierite, the fractionation of Fe^{2+} and Mg^{2+} between the two silicates is governed by the ion-exchange reaction

$$\tfrac{1}{3}\overset{Gt}{Fe_3Al_2Si_3O_{12}} + \tfrac{1}{2}\overset{Cord}{Mg_2Al_4Si_5O_{18}} \rightleftarrows \tfrac{1}{3}\overset{Gt}{Mg_3Al_2Si_3O_{12}} + \tfrac{1}{2}\overset{Cord}{Fe_2Al_4Si_5O_{18}}. \tag{c}$$

Expressing the activities of the end-member components according to the ionic solution model, as in Equation (25), we have at equilibrium

$$K_c = \exp\left(-\frac{\Delta G^*_c}{RT}\right) = \left[\frac{X^{Gt}_{Mg}/X^{Cord}_{Mg}}{X^{Gt}_{Fe}/X^{Cord}_{Mg}}\right]\left[\frac{\gamma^{Gt}_{Mg}/\gamma^{Cord}_{Fe}}{\gamma^{Gt}_{Fe}/\gamma^{Cord}_{Mg}}\right]. \tag{26}$$

We call the terms in the square brackets the distribution coefficient K_D and K_γ, respectively. It should be noted (cf. Mueller, 1972), that the expression such as Equation (26) holds whether or not the solid solutions consist of mixing in sites other than those involved directly in the exchange equilibrium under consideration. For example, the partial activity of ions other than Fe^{2+} and Mg^{2+}, such as $(a_{Al}^{vi})^{Gt}$ which would appear in the expressions for $a^{Gt}_{Mg_3Al_2Si_3O_{12}}$ and $a^{Gt}_{Fe_3Al_2Si_3O_{12}}$ if there were substitution in the octahedral site of garnet, would cancel out in the expression for the equilibrium constant for the exchange reaction (c). The expression [Eq. (26)] can be shown to reduce to

$$\bar{X}^{Cord}_{Fe} = \frac{\bar{X}^{Gt}_{Fe}}{K_D(1 - \bar{X}^{Gt}_{Fe}) + \bar{X}^{Gt}_{Fe}} \tag{27}$$

where $\bar{X}_{Fe} = Fe^{2+}/(Fe^{2+} + Mg)$, and $K_D = K(P,T)/K_\gamma(P,T,X)$. Similar relation can also be written for $\bar{X}_{Mg}$. If the divalent cation site of both cordierite and garnet consist of mixing of only Fe and Mg, then $\bar{X}_{Fe}$ becomes the total atomic fraction of Fe, X_{Fe}.

One of the petrologically important reaction relations between garnet and cordierite is as follows

$$\overset{\text{Cord}}{3Fe_2Al_4Si_5O_{18}} \rightleftarrows \overset{\text{Gt}}{2Fe_3Al_2Si_3O_{12}} + 4Al_2SiO_5 + 5SiO_2. \tag{d}$$

The reaction is driven to the right with both increasing P and T. It is completely a matter of convenience, dictated by the existence of the appropriate experimental and thermochemical data, whether or not the equilibrium relation among (Fe, Mg) cordierite, (Fe, Mg) garnet, sillimanite, and quartz is expressed in terms of ferrous or magnesium end members. Among the equilibria (c), (d), and the magnesium analog of (d), only two are independent. For the reaction (d), we have at equilibrium

$$\Delta G_d = 0 = \Delta G^0_d(1 \text{ atm}, T) + (P - 1)\,\Delta V^0_d + RT \ln Q_d, \tag{28}$$

assuming ΔV^0_d to be independent of pressure. If the solid solution is restricted only to the divalent cation sites in garnet and cordierite, the expression for Q_d reduces to

$$Q_d = \left(\frac{X^{Gt}_{Fe}}{X^{Cord}_{Fe}}\right)^6 \left(\frac{\gamma^{Gt}_{Fe}}{\gamma^{Cord}_{Fe}}\right)^6. \tag{29}$$

$\Delta G^0_d(T)$ has often to be determined, for the lack of adequate thermochemical data base, from experimentally determined equilibrium conditions for which Q is known:

$$\Delta G^0_d(1 \text{ atm}, T) = -(P_{ex} - 1)\,V^0_d - RT \ln Q_{d,ex} \tag{30}$$

where P_{ex} and $Q_{d,ex}$ are the experimentally determined values of P and Q at equilibrium at T. The effect of addition of MgO on the equilibrium condition among cordierite, garnet, sillimanite, and quartz, relative to that, say, in the pure iron system $FeO–Al_2O_3–SiO_2$, is determined by its effect on the relative activities of the Fe–end member components in the two minerals, Q_d ($Q_d = 1$ for the pure iron system). This effect is, however, governed by the fractionation of Fe^{2+} and Mg between coexisting garnet and cordierite, as determined by the Fe–Mg exchange equilibrium between the two minerals. The new equilibrium condition among cordierite, garnet, sillimanite and quartz is thus determined by the simultaneous solution of Equations (27) and (28) for the given bulk composition of the system, in this case the Fe/Fe + Mg ratio (cf. Olsen and Mueller, 1966). For the purpose of such calculation, we need to know the distribution coefficient K_D for the Fe–Mg exchange equilibrium as function of P, T, and X. (It is independent of X if both cordierite and garnet are ideal solid solutions.) It should be evident that greater the magnitude of the standard free energy change of the appropriate ion exchange equilibrium, the more unequal is the distribution of the ion between the coexisting phases, and hence greater is its effect on the mineral reactions such as (d). Thus Mg^{2+} has little effect on the equilibrium 4Chd + 5Ky = 2St + Q + $3H_2O$ (for which $Q = (X_{Fe}^{St}/X_{Fe}^{Chd})^4 f_{H_2O}^3$, assuming ideal solid solutions), because of its nearly equal distribution between chloritoid and staurolite at equilibrium (Ganguly, 1969), whereas the stability relations of staurolite are extremely sensitive to Zn^{2+} and Co^{2+} owing to their extremely biased substitution in staurolite (Ganguly, 1972; Ashworth, 1975). The above approach can be easily extended to treat mineral equilibrium involving multicomponent multiple solid solution phases. In actual practice, we often like to compare the equilibrium conditions among a particular mineral assemblage for two observed values of Q. This is a relatively straightforward problem which requires solution of the equation of equilibrium such as Equation (28) for the given values of Q. It is, however, convenient to find solutions for pressure as function of Q at fixed values of T (Ganguly, 1972).

The effect of solid solutions on mineral equilibria can be elucidated by constructing $(P–X)_T$ or $(T–X)_P$ sections analogous to those used in discussing the melting relations of solids (e.g., Ramberg, 1944; Hansen, 1971). As discussed originally by Ramberg (1944), the compositions of the coexisting mineral phases serve as continuous registers of the changing P, T conditions. The petrogenetic interpretation of metamorphic mineral assemblage requires simultaneous consideration of the effects of the composition of the vapor and condensed phases, along with those of P and T, on the mineral equilibria (e.g., Mueller, 1966; Thompson, 1975). Recognition of reactions relatively insensitive to compositional effects, on which we ought to base "isogradic mapping" as gross indicator of changing $P–T$ conditions, as well as of buffered paths along which the mineralogical evolution of rock might be constrained in progressive metamorphism, would be greatly facilitated by such analysis.

5 ORDER–DISORDER AND NONSTOICHIOMETRY

In many rock forming crystals, certain atoms are partitioned or ordered among nonequivalent crystallographic sites or sublattices. Such intracrystalline partitioning can be treated in terms of formalisms analogous to those developed for the heterogeneous ion exchange equilibrium (Mueller, 1962). The ordering of ions contributes toward the stability or lowering of the free energy of the crystal relative to that in the completely disordered state (or else, why should it happen?). For solid solutions, ideal mixing of the ions in the sublattices implies negative deviation from ideality for the mixing of the macroscopic components (cf. Mueller, 1962, Figs 2 and 3; Grover, 1974). In addition to such substitutional disorders, there are many other types of disorders in a crystal, such as those arising from lattice vacancies (Swalin, 1962), which affect the energetic properties of the crystal. Thus, for a crystal of fixed chemical composition $G = f(T,P,\sigma_1,\sigma_2, \ldots ,\sigma_n)$. The exact differential dG is then

$$dG = -SdT + VdP + \sum_{i=1}^{n} \left(\frac{\partial G}{\partial \sigma_i}\right)_{T,P,\sigma_j \neq \sigma_i} \partial\sigma \tag{31}$$

where σ_i, $\sigma_2, \ldots, \sigma_n$ are parameters related to the n types of independent disorders in the crystal. Since G is at minimum at equilibrium at constant P and T, it follows that G is minimized with respect to the variation of every σ_i under such condition:

$$\left(\frac{\partial G}{\partial \sigma_i}\right)_{T,P,\sigma_j \neq \sigma_i} = 0, \qquad \frac{\partial^2 G}{\partial \sigma_i^2} > 0. \tag{32}$$

The value of σ_i defined by Equation (32) is its equilibrium value at the given P and T. Figure 9 illustrates the variation of G as function of σ_i. If now the equilibrium ordering state of a crystal at one temperature, say T_2, persists metastably to some other temperature T_1, the Gibbs free energy of this mineral would be elevated by an amount δG over its equilibrium value at T_1. This might, under certain circumstances, cause the metastable breakdown of the mineral at conditions where it should have been stable if perfect equilibrium were attained. In nature, stable equilibrium or the state with the lowest free energy state would commonly be attained because of the large time available, the metastable reaction constituting only an intermediate step toward transformation to that state. Only under special circumstances, such as rapid quenching, would any imprint of the metastable reaction be preserved. However, such metastable reactions might in many cases be responsible for major discrepancies among experimentally determined reaction boundaries by different workers. Hlabse and Kleppa (1968) have discussed the effect of the tetrahedral Al–Si disordering in albite on the transformation of albite

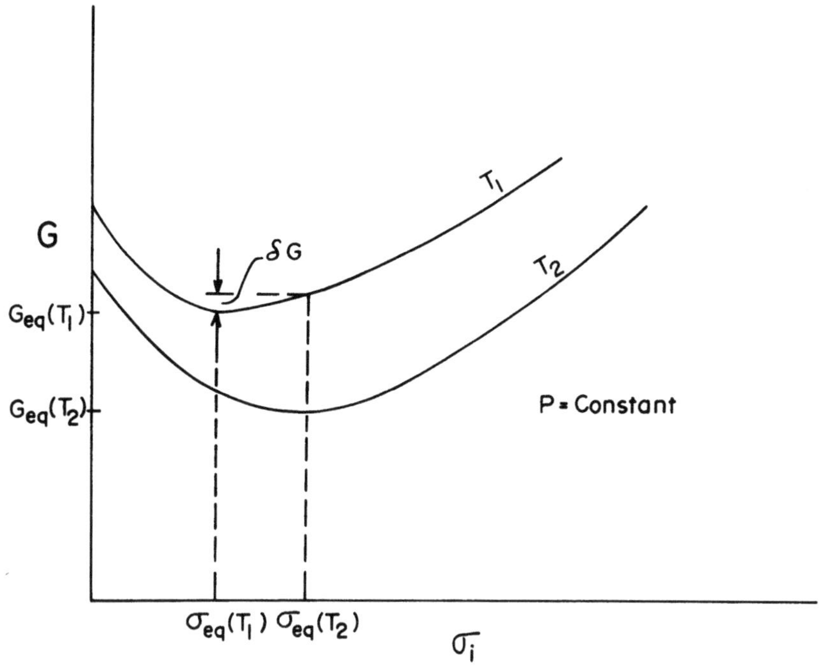

Fig. 9. Schematic illustration of the variation of Gibbs free energy G as function of an ordering parameter σ_i. σ_{eq} is the equilibrium value of σ at the specified temperature. Similarly for G_{eq}. $(G_{eq}(T_1) + \delta G)$ is the free energy at T_1 if the equilibrium ordering at T_2 persists metastably to T_1

to jadiete and quartz. The effect of extremely limited Al–Si disordering of sillimanite on the experimental determination of the aluminumsilicate stability relations have been discussed recently by several workers (Greenwood, 1972; Holdaway, 1971; Navrotsky et al., 1973). Greenwood has further suggested that the Al–Si disordering in sillimanite may provide a partial explanation of naturally coexisting aluminumsilicate polymorphs.

The experimental determination of an univariant reaction boundary within a given system is sometimes significantly affected by the disequilibrium nonstoichiometry of certain minerals within the system. An example is the pyrophyllite stability in the system Al_2O_3–SiO_2–H_2O discussed recently by Rosenberg (1974). Synthetic as well as natural pyrophyllites have been shown to exhibit substitution of the type $Al^{3+} + H^+ \rightleftarrows Si^{4+}$. Such substitution seem to be at least partly responsible for the widely descrepant experimentally determined univariant reaction boundary of Pyrophyllite = Andalusite + 3 Quartz + H_2O in the above system. A univariant reaction boundary exists for every given composition of pyrophyllite in this system. However, only one point of any such univariant curve is stable under equilibrium condition, and the thermodynamically stable univariant boundary is the locus of such points. The

$P{-}T$ condition of any of these stable points is that at which the free energy of pyrophyllite is minimized for the composition defined for the univariant curve.

6 CHEMICAL EQUILIBRIUM IN METAMORPHIC ROCKS

The application of equilibrium thermodynamics to the analysis of mineral parageneses in natural rocks requires an understanding of the attainment of chemical equilibrium in geologic processes. Whether an assemblage of minerals under consideration is in equilibrium or not can be determined by comparing the equilibrium constant K against Q [Eq. (3)] for the various possible reactions among the minerals under consideration. Whereas Q is readily measurable, the evaluation of K requires a knowledge of P, T condition of formation of the rock, and the appropriate thermodynamic data which are unfortunately often not available. Violation of the phase rule is a definite criteria of disequilibrium, but the nonviolation of phase rule is not a proof of equilibrium. Phase rule is hardly found to be violated even in rocks which may otherwise be shown to be out of internal equilibrium.

The presence of reaction textures, often considered as criteria of disequilibrium, does not necessarily imply disequilibrium. As discussed earlier, a field of coexistence between solid solution phases is thermodynamically permitted. If the rock is quenched from conditions within such field, then incomplete reaction, the extent of which is determined by Equation (5), is what is to be expected under equilibrium condition. By the requirement of chemical equilibrium, the chemical potential of all components which are free to move, and hence the chemical composition of a particular mineral, must be the same throughout a volume of rock subjected to the Same $P{-}T$ condition. The latter requirement is often not satisfied over a scale of more than a few centimeters (e.g., Phinney, 1963, Blackburn, 1968), and sometimes within the scale of single crystal dimension. This limits the volumes of rocks as a few cubic centimeters that may initially be presumed to be in internal equilibrium (also called "local equilibrium" by Thompson, 1959 and "mosaic equilibrium" by Korzhinskii, 1959), in the absence of any direct evidence to the contrary such as the presence of thermodynamically incompatible phase assemblage, and thus selected for further thermodynamic analysis. It should however, be stressed that physical contact between minerals is not necessary for their mutual equilibration, since diffusive transport of chemical species may be achieved over distances of several crystal dimensions, especially in the presence of solvent fluid such as H_2O, HCl etc. (Mueller, 1967). Equilibration between coexisting ideal or binary (whether or not ideal) mineral solid solutions is borne out by the orderly distribution of elements between them, as required by the equilibrium element distribution equation (27). For solid solution phases involving mixing of more than two components, the element distribution would be disorderly when the solid solutions are nonideal, since K_D would then depend on the concentration of these additional components.

However, the element distribution would become systematic ($\bar{X}_i^\alpha$ single-valued function of $\bar{X}_i^\beta$) for constant levels of these additional components if equilibrium is achieved. These relationships are illustrated in the Figure 10. Obviously, the domain over which an assemblage of minerals may be in equilibrium cannot exceed that defined by the compositional homogeneity of any of the minerals in the assemblage. In the case of zoned minerals, it should be the outermost zone that has more closely

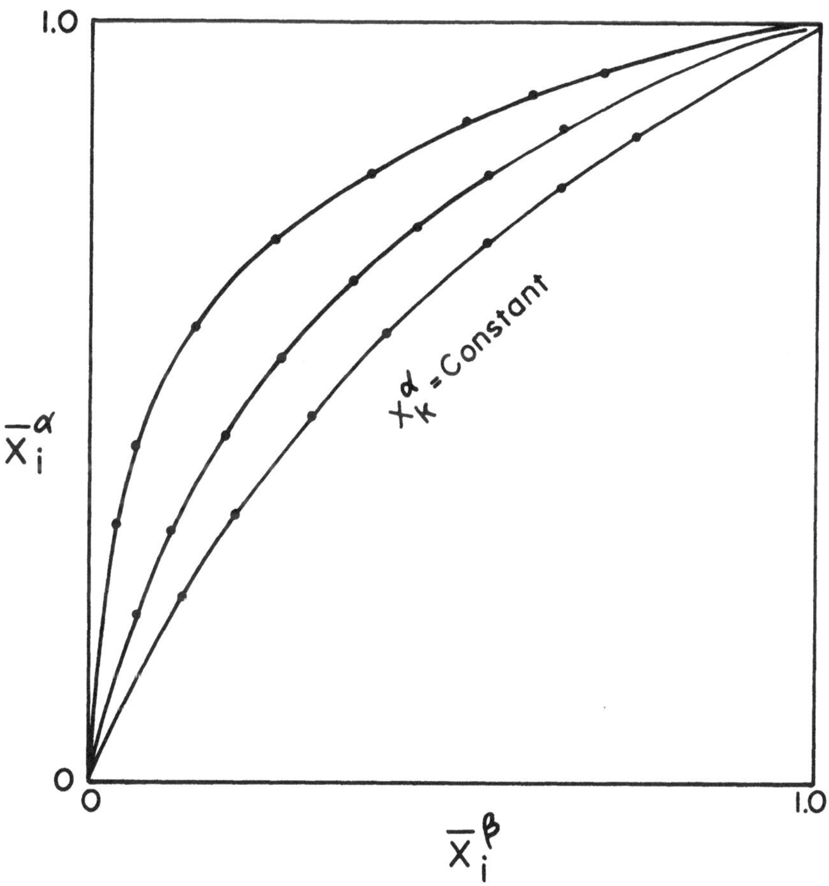

Fig. 10. Idealized equilibrium element distribution between two coexisting phases α and β. α is a nonideal ternary solution of *i, j,* and *k,* whereas β is a binary solution of *i* and *j*. $\bar{X}_i = i/(i + j)$. If both α and β were ideal ionic solid solution of *i* and *j,* then the distribution curves would have been symmetrical to 1.0–1.0 axis, as shown by Mueller (1961). If, in addition, the component *k* mixed ideally with both *i* and *j* in α, then all the distribution points would have fallen on one symmetrical curve under equilibrium conditions. (Example: distribution of Mg between coexisting garnet and orthopyroxene, Kretz, 1961)

approached equilibrium with the adjacent minerals. Between coexisting zoned minerals, however, there may be a few compatible zones of the two minerals which have equilibrated with respect to different sets of P–T conditions, thus enabling reconstruction of the P–T history of the rock (Råheim and Green, 1975). A volume of rock might be in equilibrium with respect to the diffusion of some components, but at the same time be out of equilibrium with respect to the diffusion of certain other components owing to their differential mobilities under the physiochemical environments to which they have been subjected. As discussed by Mueller (1967) on the basis of rate theory, lattice diffusion is facilitated by differential stress. This is corroborated by the works of Blackburn (1968), who found that in the volumes of rocks showing uniform compositions of garnet, the longest axis is almost always parallel to foliation and lineation, whereas the intermediate axis is parallel to foliation and perpendicular to lineation.

APPENDIX OUTLINE OF THE THERMODYNAMIC PRINCIPLES RELATED TO THE TEXT

At constant P and T, the chemical potential μ_i of the component i in a *simple system* is defined as

$$\mu_i = \bar{G}_i = \left(\frac{\partial G}{\partial n_i}\right)_{T,P,n_j \neq n_i} \tag{1}$$

where n is the number of moles of the specified component, G is the Gibbs free energy, and $\bar{G}$ is called the partial molal free energy. We define *simple systems* as "systems that are macroscopically homogeneous, isotropic, uncharged, and chemically inert, that are sufficiently large that surface effects can be neglected, and that are not acted on by electric, magnetic, or gravitational fields" (Callen, 1963). Our developments here are restricted to the simple systems.

The total differential of G can be expressed as

$$\begin{aligned} dG &= \left(\frac{\partial G}{\partial T}\right) dT_{P,n_1,\ldots,n_c} + \left(\frac{\partial G}{\partial P}\right) dP_{T,n_1,\ldots,n_c} + \sum_i \left(\frac{\partial G}{\partial n_i}\right) dn_{i\,P,T,n_j \neq n_i} \\ &= -SdT + VdP + \sum_i \mu_i \, dn_i \end{aligned} \tag{2}$$

where S is the entropy and V is the volume. From Equations (1) and (2), the partial molal entropy and partial molal volume of the component i can be expressed as

$$\bar{S}_i = \left(\frac{\partial S}{\partial n_i}\right)_{T,P,n_j \neq n_i} = -\left(\frac{\partial \mu_i}{\partial T}\right)_{P,n_1,\ldots,n_c} \tag{3a}$$

$$\bar{V}_i = \left(\frac{\partial V}{\partial n_i}\right)_{T,P,n_j \neq n_i} = \left(\frac{\partial \mu_i}{\partial P}\right)_{T,n_1,\ldots,n_c} \tag{3b}$$

Since

$$G = H - TS \tag{4a}$$

where H is the enthalpy of the system,

$$\mu_i = \bar{H}_i - T\bar{S}_i. \tag{4b}$$

G is a homogeneous function of the first degree in $n_1, \ldots, n_c$. Hence, according to Euler's theorem,

$$G = \sum_i n_i \mu_i. \tag{5}$$

Thus,

$$dG = \sum_i n_i \, d\mu_i + \sum_i \mu_i \, dn_i. \tag{6}$$

Combining (2) and (6), we get

$$SdT - VdP + \sum_i n_i \, d\mu_i = 0 \tag{7}$$

Equation (7) is known as Gibbs–Duhem relation. It provides a relation among the chemical potential of the various components at constant P and T that has many important applications in the thermodynamic study of solution.

The chemical potential of a component varies with its concentration according to

$$\mu_i(P,T,X) = \mu_i^* + RT \ln a_i \tag{8a}$$
$$= \mu_i^* + RT \ln \chi_i \gamma_i \tag{8b}$$

where a_i is the activity of the component i, γ_i is its activity coefficient, χ_i is a compositional parameter, not necessarily the mole fraction of i, that is, completely determinable from a knowledge of the concentration of i in the solution, and μ^*_i is the chemical potential of component i at some conveniently chosen state of fixed composition, which we call its standard state. Whatever be the choice of standard state, the activity of a component in its standard state, according to Equation (8a), is unity. It should be evident from Equation (8b) that the explicit expression of the coefficient γ_i, which is a function of P, T, and composition, depends on μ^*_i, that is, on the choice of the standard state, and also on the choice of

χ_i (χ_i = 1 for the pure component i). We find it mostly convenient to make $\gamma_i \rightarrow 1$, as $\chi_i \rightarrow 1$. This relation is satisfied, according to Equation (8.2), if we choose $\mu^*_i = \mu_i^0\ (P,T)$, where $\mu_i^0\ (P,T)$ is the chemical potential of the pure component i at the P and T concerned. With this choice of standard state, $\gamma_i \rightarrow$ constant, as $\chi_i \rightarrow 0$. (It may be mentioned incidentally that under certain circumstances it is convenient to make $\gamma_i \rightarrow 1$, as $\chi_i \rightarrow 0$, for which we have to choose a different standard state for the component i.) γ_{solute} is constant for the composition range for which γ_{solvent} is unity. A solution is defined as *ideal,* or said to obey the Rault's law, when the chemical potential of all its components follows Equation (8b) with $\gamma_i = 1$.

The free energy of a *solution* is given by

$$\begin{aligned} G &= \sum_i n_i\mu_i \\ &= \sum_i n_i\mu_i^* + RT\sum_i n_i \ln a_i, \end{aligned} \tag{9a}$$

or the molar free energy by

$$\begin{aligned} g &= \frac{G}{n} = \sum_i X_i\mu^*_i + RT\sum_i X_i \ln a_i \\ &= \sum X_i\mu^*_i + \Delta g^{\text{mix}} \end{aligned} \tag{9b}$$

where $n = \Sigma n_i$, and Δg^{mix} is the molar free energy of mixing, $\Delta G^{\text{mix}}/n$. Other thermodynamic properties of solution can be easily derived by performing the apprpriate thermodynamic operations on Equation (9). Thus,

$$s = -\frac{\partial g}{\partial T} = \sum_i X_i\left(s_i^* - R\ln a_i - RT\frac{\partial \ln \gamma_i}{\partial T}\right) = \sum_i X_i\bar{s}_i \tag{9c}$$

$$v = \frac{\partial g}{\partial P} = \Sigma X_i\left(v_i^* + RT\frac{\partial \ln \gamma_i}{\partial P}\right) = \Sigma X_i\bar{v}_i. \tag{9d}$$

It follows from Equation (9) that for an ideal solution $\Delta H^{\text{mix}} = \Delta V^{\text{mix}} = 0$. The difference between the real and the corresponding ideal expressions of the thermodynamic functions is known as the *excess* thermodynamic function.

The chemical potential of an ideal gaseous component can be expressed as a function of its partial pressure P_i as follows.

$$\mu_i(P,T,X) = \mu_i^0\,(1\ \text{atm},T) + RT\ln P_i \tag{10a}$$

where μ_i^0 (1 atm,T) is the chemical potential of pure i at 1 atm pressure and the given temperature. We fine P_i as $P_i = PX_i$, where P is the total pressure. Thus, $P = \sum_i P_i$. By analogy, we write for a real gas

$$\mu_i(P,T,X) = \mu_i^0\,(1\ \text{atm},T) + RT\ln f_i \tag{10b}$$

where f_i is called the *fugacity* of the component i at P, T, and X. It is related to P_i and a_i, according to

$$\frac{f_i}{P_i} = \frac{f_i^* a_i}{P_i} = \phi_i. \tag{11}$$

Also,

$$Lt_{P\to 0}\phi_i = 1. \tag{12}$$

The free energy change of a reaction can be expressed as

$$\begin{aligned}\Delta G &= \sum_i \nu_i \mu_i \\ &= \Delta G^* + RT\ln Q\end{aligned} \tag{13a}$$

where ν_i is the stoichiometric coefficient of the components or formula units in the reaction, conventionally taken as positive for the products and negative for the reactants,

$$\Delta G^* = \sum_i \nu_i \mu^*{}_i, \tag{14}$$

and

$$Q = \prod_i a_i^{\nu_i}. \tag{15}$$

Using fugacity instead of activity for the gaseous species, we can write

$$\Delta G = \Delta G^*{}_c + \Delta G_g^0\,(1\ \text{atm},T) + RT\ln Q' \tag{13b}$$

where c stands for the condensed and g for the gaseous species in the reaction, and

$$Q' = \prod_i (a_i^{\nu_i})_c\,(g_i^{\nu_i})_g. \tag{15b}$$

Choosing pure states at P and T as the standard states for the respective condensed phases

$$\Delta G_i^* = \Delta G_i^0(P,T) = \Delta G_i^0(1\ \text{atm},T) + \int_1^P \Delta V_c^0\,dP. \tag{16}$$

Equation (13b) then reduces to

$$\Delta G = \Delta G^0(1\ \text{atm},T) + \int_1^P \Delta V_c^0\,dP + RT\ln Q' \tag{13c}$$

ACKNOWLEDGMENT

This research was suppoted by a Cottrell Research Grant from Research Corporation. Thanks are due to Dr. Surendra Saxena for reading the manuscript.

REFERENCES

Ashworth, J. R.: Staurolite at anomalously high grade. Contr. Mineral. Petrol. *53,* 281–291 (1975)

Blackburn, W. H.: The spatial extent of chemical equilibrium in some high-grade metamorphic rocks from the Grenville of Southeastern Ontario. Contr. Mineral Petrol. *19,* 72–92 (1968)

Brace, W. F., Ernst, W. G., Kallberg, R. W.: An experimental study of tectonic overpressure in Franciscan rocks. Geol. Soc. Am. Bull. *81,* 1325–1338 (1970)

Bradley, R. S.: Thermodynamic calculations on Phase equilibria involving fused salts. Pt II. Solid solutions and application to olivines. Am. J. Sci. *260,* 550–554 (1962)

Brown, T. H., Skinner, B. J.: Theoretical prediction of equilibrium phase assemblages in multicomponent systems. Am. J. Sci. *274,* 961–986 (1975)

Callen, H. B.: Thermodynamics. New York: Wiley, 1963

Charmichael, D. M.: Intersecting isograds in the Whetstone lake area, Ontario. J. Petrol. *11,* 147–184 (1970)

Chinner, G. A.: Pelitic geneisses with varying ferrous-ferric ratios from Glen Clova, Angus, Scotland. J. Petrol. *1,* 178–217 (1960)

Condie, K. C.: Oxygen, carbon dioxide, and sulfur fugacities during diagenesis and low-grade metamorphism of late precambrian subgraywackes from Northern Utah. Am. Mineral. *52,* 1153–1160 (1967)

Coombs, D. S., Ellis, A. J., Fyfe, W. S., Taylor, A. M.: The zeolite facies with comments on the interpretation of hydrothermal synthesis, Geochim. Cosmochim. Acta *17,* 53–107 (1959)

Eugster, H. P., Wones, D. R.: Stability relations of ferruginous biotite, annite. J. Petrol. *3,* 82–125 (1962)

Eugster, H. P., Skippen, G. B.: Igneous and metamorphic reactions involving gas equilibria. *In:* Researches in geochemistry, vol. II, Abelson, P. H. (ed.). London-New York: Wiley (1967b), pp. 492–520

French, B. M.: Some geological applications of equilibrium between graphie and a C-H-O gas at high temperatures and pressures. Rev. Geophys. *4,* 223–253 (1966)

Fyfe, W. S., Turner, F. J., Verhoogen, J.: Metamorphic reactions and metamorphic facies. Mem. Geol. Soc. Am. *73,* (1958)

Ganguly, J.: Analysis of the stabilities of chloritoid and staurolite and some equilibria in the system $FeO–Al_2O_3–SiO_2–H_2O–O_2$. Am. J. Sci. *266,* 277–298 (1968)

Ganguly, J.: Chloritoid stability and related parageneses: theory, experiments, and applications. Am. J. Sci. *267,* 910–944 (1969)

Ganguly,J.: Staurolite stability and related parageneses. J. Petrol. *13,* 335–65 (1972)

Ganguly, J.: Activity-composition relation of jadeite in omphacite pyroxene. Earth. Planet. Sci. Letters *19,* 145–153 (1973)

Ganguly, J., Newton, R. C.: Thermal stability of chloritoid at high pressure and relatively high oxygen fugacity. J. Petrol. *9*, 444–466 (1968)

Gordon, T. M., Greenwood, H. J.: Stability of grossularite in H_2O-CO_2 mixtures. Am. Mineral., *56*, 1674–1688 (1971)

Greenwood, H. J.: The *N*-dimensional tie line problem. Geochim. Cosmochim. Acta *31*, 465–490 (1967a)

Greenwood, H. J.: Mineral equilibria in the system MgO–SiO_2–H_2O–CO_2. In: Researches in geochemistry, Vol. II. Abelson, P. H. (ed.) London-New York: Wiley (1967b), pp. 542–547

Greenwood, H. J.: Al^{IV}–Si^{IV} disorder in sillimanite and its effect on phase relations of the aluminum silicate minerals. Geol. Soc. Am. Mem. *132*, 533–571 (1972)

Greenwood, H. J.: Buffering of pore fluids by metamorphic reactions. Am. J. Sci. *275*, 573–593 (1975)

Grew, E. S.: Carbonaceous material in some metamorphic rocks of New England and other areas. J. Geol., *82*, 50–73 (1974)

Grover, J.: On calculating activity coefficients and other excess functions from the intracrystalline exchange properties of a double-site phase. Geochim. Cosmochim. Acta *38*, 1527–1548 (1974)

Hansen, B. J.: Theoretical phase relations involving cordierite and garnet in the system MgO–FeO–Al_2O_3–SiO_2. Contr. Mineral Petrol. *33*, 191–214 (1971)

Harte, B.: Displacement of the kyanite 'isograd' by the local variations of oxygen fugacity within pelitic schists in Glen Lethnot, Angus, Scotland. EOS Am. Geophys. Union *56*, 466 (1975)

Hlabse, T., Kleppa, O. J.: The thermochemistry of jadeite. Am. Mineral. *53*, 1281–1292 (1968)

Holdaway, M. J.: Stability of andalusite and the aluminum silicate phase diagram. Am. J. Sci. *271*, 97–131 (1971)

Holdaway, M. J.: Effect of rock and fluid composition on chloritoid-staurolite equilibria, Picuris, New Mexico. Geol. Soc. Amer., Abstracts with Programs, *7*, 1119 (1975)

Hounslow, A. M., Moore, J. M., Jr.: Chemical petrology of Grenville schists near Frenleigh, Ontario. J. Petrol. *8* 1–28 (1967)

Izawa, E.: Carbonaceous material in some metamorphic rocks in Japan. J. Geol. Soc. Japan, *74*. 427–432 (1968)

Kerrick, D. M.: Review of metamorphic mixed volatile (H_2O–CO_2) equilibria. Am. Mineral. *59*, 729–762 (1974)

Kerrick, D. M., Darken, L. S.: Statistical thermodynamic models for oxide and silicate solid solutions, with applications to plagioclase. Geochim. Cosmochim. Acta *39*, 1431–1442 (1975)

Korzhinskii, D. S.: *Physicochemical analysis of the paragenesis of minerals.* Translation in English, Consultants Bureau, Inc. New York (1959)

Kretz, R.: Some applications of thermodynamics to coexisting minerals of variable composition. Examples: orthopyroxene–clinopyroxene and orthopyroxene–garnet. J. Geol. *69*, 361–387 (1961)

Miyashiro, A.: Oxidation and reduction in the Earth's crust with special reference to the role of graphite. Geochim. Cosmochim. Acta *28*, 717–720 (1964)

Mueller, R. F.: Analysis of relations among Mg, Fe and Mn in certain metamorphic minerals. Geochim. Cosmochim. Acta *25*, 267–296 (1961)

Mueller, R. F.: Energetics of certain silicate solid solutions. Geochim. Cosmochim. Acta *26*, 581–598 (1962)

Mueller, R. F.: Theory of equilibria between complex silicate melts and crystalline solutions. Am. J. Sci. *262*, 643–652 (1964)

Mueller, R. F.: Stability relations of the pyroxones and olivine in certain high grade metamorphic rocks. J. Petrol. *1*, 363–374 (1966)

Mueller, R. F.: Mobility of the elements in Metamorphism. J. Geol. *75*, 565–582 (1967)

Mueller, R. F.: Stability f biolite: A discussion. Am. Mineral. *57*, 300–316 (1972)

Navrotsky, A., Newton, R. C., Kleppa, O. J.: Sillimanite-disordering enthalpy by calorimetry. Geochim. Cosmochim. Acta *37*, 2497–2508 (1973)

Newton, R. C.: An experimental determination of the high pressure stability limits of magnesian cordierite under wet and dry conditions. J. Geol. *80*, 398–420 (1974)

Nokleberg, W. J.: CO_2 as a source of oxygen in the metasomatism of carbonates. Am. J. Sci. *273*, 498–514 (1973)

Olsen, E. Mueller, R. F.: Stability of orthopyroxenes with respect to pressure, temperature, and composition. J. Geol. *74*, 620–625 (1966)

Phinney, W. C.: Phase equilibria in the metamorphic rocks of St. Paul Island and Cape North, Nova Scotia. J. Petrol. *4*, 90–130 (1963)

Prigogine, J., Defay, R.: Chemical Thermodynamics. London: Longmans, 1954

Råheim, A., Green, D. H.: P,T paths of natural eclogites during metamorphism—a record of subduction. Lithos *8*, 317–328 (1975)

Ramberg, H.: Patrologic significance of subsolidus phase transitions in mixed crystals. Norsk. Geol. Tidsskr. *24*, 43–74 (1944)

Ramberg, H.: The Origin of Metamorphic and Metasomatic Rocks. Chicago: Univ. Chicago, 1952a

Ramberg, H.: Chemical bonds and the distribution of cations in silicates. J. Geol. *60*, 331–355 (1952b)

Ramberg, H., DeVore, G. W.: The distribution of Fe^{++} and Mg^{++} in coexisting olivines and pyroxenes. J. Geol. *59*, 193–210 (1951)

Rosenberg, P. E.: Pyrophyllite solid solutions in the system Al_2O_3–SiO_2–H_2O. Am. Mineral. *59*, 254–260 (1974)

Rumble, D., III.: Fe-Ti oxide minerals from regionally metamorphosed quartizites of Western New Hampshire. Contr. Mineral. Petrol. *42*, 181–195 (1973)

Rumble, D., III.: Gibbs phase rule and its application in geochemistry. J. Washington Acad. Sci. *64*, 199–208 (1974)

Saxena, S. K.: Thermodynamics of Rock-Forming Crystalline Solutions. Berlin-Heidelberg-New York: Springer, 1973

Swalin, R. A.: Thermodynamics of Solids. New York-London: Wiley, 1962

Thompson, A. B.: Mineral reactions in calc-mica schist from Gassetts, Vermont, U.S.A. Contr. Mineral Petrol. *53*, 105–127 (1975)

Thompson, J. B. Jr.: Local equilibrium in metasomatic processes. In: Researches in Geochemistry. Abelson, P. H. (ed), New York-London: Wiley, 1959, pp. 427–457

Thompson, J. B. Jr.: Geochemical reactions and open systems. Geochim. Cosmochim. Acta *34*, 529–551 (1970)

Yui, S.: Results of equilibrium calculation on the reaction between graphite and a H_2O–CO_2 fluid at 300–500°C and 1000–2000 bars, and its implication to relict graphite in pyrometasomatic ore deposits. J. Mining College Akita Univ. Ser. A., *4*, 29–39 (1968)

14

L. L. PERCHUK

Abstract The distribution of cations in coexisting minerals of various metamorphic rocks has been studied, using the principles of Sobolev and Ramberg to explain the relationship between crystal energetic parameters such as polarization of oxygen ions in silicates of differing structures and the distribution of cations. Statistical studies of distribution coefficients in rocks of low to high metamorphic grade have resulted in the formulation of several geothermometers and geobarometers. This formulation should be regarded as an attempt to systematize the theory, data and application of the crystal-chemical and thermodynamic principles to understand metamorphic and igneous processes. The results from the different thermo- and baro-meters are internally consistent and have been used to evaluate partial pressures of O_2, H_2O, and CO_2 in the fluid phase in metamorphic processes. The metamorphic processes are dependent on the geotectonic processes and, therefore, the study of the *P–T* regimes of metamorphism cannot be separated from the study of the geodynamic evolution of the crust.

THERMODYNAMIC CONTROL OF METAMORPHIC PROCESSES

Symbols

I. Thermodynamic parameters:

a_i: activity of component i in the mixture
μ_i: chemical potential of component i in the mixture
X_i: mole content (fraction) of component i in a mineral
n_i: molar number of component i
C_{FeO}: FeO concentration (weight percent) in the given mineral
V: volume
S: entropy
K_D: distribution coefficient of two components between two phases
K_i: the partition coefficient of component i between two phases
T: temperature, °K
t: temperature, °C
P: overall pressure
P_s: pressure on solid phases and melts
G: Gibbs free energy
H: enthalpy
P_{fl}: fluid pressure
P_i^{fl}: the partial pressure of component i in fluid
P_i^0: the pressure of monocomponent fluid
R: a gas constant (1.987 cal/mol)
γ_i^0: fugacity coefficient of monocomponent fluid at the given P_i^0
γ_i^{fl} fugacity coefficient of component i in an ideal gas mixture at the given P_{fl}.

II. Mineral symbols:

Ab: albite
Am: monoclinic amphibole
An: anorthite
And: andalusite
Andr: andradite
Ar: aragonite
Bi: biotite
Cal: calcite
Car: carnegeite
Chd: chloritoid
Chl: chlorite
Cor: cordierite
Cpx: clinopyroxene
Cum: cummingtonite
Ep: epidote
Fsp: feldspar
Gr: garnet
Gros: grossularite
Hem: hematite

Hb: hornblande
Ilm: ilmenite
Ind: indialite
Ks: kalsilite
Ky: kyanite
Law: lawsonite
Mic: microcline
Mt: magnetite
Mu: muscovite
Ne: nepheline solid solution
Ol: olivine
Opx: orthopyroxene
Or: orthoclase
Px-Ca: clinopyroxene
Par: pargasite
$Pl_{0.25}$: plagioclase and the content of the molar albite component: Na:(Na + Ca)
Qz: quartz
San: sanidine
Sap: sapphirine
Ser: serpentine
Sil: sillimanite
Sp: ringwoodite (Rin)
St: staurolite
Ta: talc
Tur: tourmaline
Ves: vesuviane
Usp: ulvöspinel
Wol: wollastonite
Zo: zoisite

Subscripts in mineral symbols are for molar magnesium content in a mineral, e.g. $cor_{0.5}$ mean that $X_{Mg}^{cor} = 0.5$ in cordierite

1 INTRODUCTION

In this paper we are concerned with the estimates of temperature, overall pressure and partial pressures of volatile components during the formation of metamorphic and certain magmatic rocks at various depths. I believe that available experimental, theoretical and empirical data would enable me to outline the general tendencies in the change of these parameters.

The importance of such tendencies, if they exist, can hardly be overemphasized for a number of geophysical and geodynamical problems. However, these regularities can only be formulated on the basis of a strict physicochemical theory of mineral equilibria under the thermodynamic conditions of the earth crust and upper mantle. Ramberg (1949, 1952a) was among the first (Korzhinskii, 1936, 1937; Friedman, 1949) who attempted to establish the *P T* conditions of metamorphism and magmatism from equilibria of minerals of variable composition. Thermodynamics of olivine-pyroxene equilibrium (Ramberg, and De Vore, 1951) was the first step to a quantitative determination of paragenetic temperatures. Even now Ramberg's (1952b) theory on chemical bonds and distribution of cations in silicates continues to be important. Ramberg's rule enables us to compare qualitatively compositions of coexisting minerals (with respect to isomorphous components) on the basis of their interaction with the radicals of hypothetic polysilicic acids of differing strength. This rule was important for the theory of *phase correspondence* in natural systems (Perchuk, 1964; Perchuk and Ryabchikov, 1976), on the basis of which a number of mineral thermometers and barometers were calibrated (Perchuk, 1965, 1966, 1967a, b, 1968a, b, 1969a, b, c, 1970a, 1971c, 1973; Perchuk and Ryabchikov, 1968).

The present article is an attempt to discuss this theory and fluid thermodynamics and their application to the problem of thermodynamic regime of metamorphic processes.

2 THE FUNDAMENTALS OF THE THEORY OF PHASE CORRESPONDENCE

2.1 Sobolev and Ramberg's Rules

Sobolev (1948) postulated a rule, according to which salt pairs consisting of a strong acid and a strong base or a weak acid and a weak base are stable in a ternary reciprocal system at low polarization values. At high polarization values the energy effects accompanying the exchange reactions might vary and even change sign.

Ramberg's rule specifies the shift of the exchange equilibria towards the formation of silicic salts, in which strong (electropositive) cations are combined with the radicals of strong silicic acids, and weak (less electropositive) cations are combined with the radicals of weak silicic acids. The only exception are the minerals where Al has a pronounced effect on the oxygen electronegativity, and where this is the case, the exchange equilibria may become displaced in the reverse direction.

2.2 General Principles

The analysis of the distribution of isomorphous elements among coexisting minerals in igneous and metamorphic rocks showed that the relationships found by Ramberg change regularly with increasing *T* and *P* all the way to the shift of the exchange equilibria into opposite directions (e.g., $Mg_2Si_2O_6 + CaFeSi_2O_6 \rightleftharpoons CaMgSi_2O_6 + Fe_2Si_2O_6$). These relationships might change in the opposite direction when the solid solutions depart grossly from the ideality. Such is for example, the equilibrium $NaAlSiO_4$ + $KAlSi_3O_8$ = $NaAlSi_3O_8$ + $KAlSiO_4$ in its most sodium end (Perchuk and Ryabchikov, 1968). In the general case, with increasing temperature, the bonds between a strong cation and the hypothetical silicic acids become stronger (and those of a less electropositive isomorphous cation respectively weaker) in the following sequence: framework silicates → sheet dimetasilicates → band metasilicate → chain metasilicate → diorthosilicate → orthosilicate (separately for hydrous and anhydrous). It is believed that with temperature variations the strength of chemical bonds in the coexisting minerals will change in proportion to the difference between the silicates in this series. Consequently, the ΔS and ΔH values, and, correspondingly, the effects of redistribution will be greater in the exchange reactions.

Thus, the redestribution of the isomorphous elements among the coexisting minerals is governed by three principal factors: by relative strength of silicic radicals, the structure, and water content in the minerals. The greater the difference in the relative values of the factors, the more pronounced the redistribution effect would be. In the general case, *the degree of redistribution of two isomorphous components between the silicates with the temperature will be higher the greater the difference in the strengths of silicic radicals and in the symmetry categories of the minerals at equilibrium; the greatest effects have been noted for the equilibria between hydrous and anhydrous minerals.* This rule is illustrated in Tables 1 and 2. It is seen that the strength of the silicic radical and, accordingly, the oxygen electronegativity, diminish within each subgroup (anhydrous and hydrous silicates) down from the top of the tables (Ramberg, 1952b).

The tables show a rightward shift in the symmetry of mineral structures from triclinic to cubic. According to these characteristics, each rock-forming silicate occupies the corresponding cell of the table and is given a certain number; the polymorphs (or polytypes) of this silicate are given the same number and the respective number of marks. The sequence of two minerals from the same system and the same subclass is governed by the category of the symmetry of their polymorphs.

For example, Chd and St are monoclinic hydrous orthosilicates. However, Chd has a triclinic polymorph modification, while St shows the rhombic one (?). That is why Chd (No. 12 in Table 1) precedes St (No. 13). The same is true for spinel–ringwoodite $(Mg, Fe)_2SiO_4$ and Gr (No. 18, Table 1). If Al decreases the effective electronegativity of oxygen in a mineral, and the distribution of Fe and Mg (Ramberg, 1952b) becomes inverse, then the reverse relationships will be observed in the distribution of these elements upon temperature changes. Such, for example are the equilibria amphibole–chlorite, cordierite–biotite (distribution of Fe

Table 1. The Fe–Mg Rock-Forming Minerals (Silicates and alumino–silicates)

Type	Subclass	Radical	Singony, Space Group, Specific Gravity, Hardness									
			Triclinic	Monoclinic			Rombic	Trigonal	Tetragonal	Hexagonal	Cubic	
Hydrous silicates	Framework silicates	$[Al_nSi_{m-n}O_{2m}]^{-n}$					1 Cor (?) Cccm d = 2,52–2,78 h = 7			1′ Ind (?) C6/mcc		
	Sheet dimetasilicates	$[Al_nSi_{4-n}O_{10}]^{-(4+n)}$		2 Ta C2/c d = 2,58–2,83 h = 1	3 Ser Cm d = 2,55–2,6 h = 2,5–3,5							
			4′ Chl C1 d = 2,6–3,3 h = 2–3	4 Chl C2/m d = 2,6–3,3 h = 2–3	5 Bi_{1M} Cm(C2/m) d = 2,7–3,3 h = 2,5–3	5′ Bi_{2M_1} C2/c d = 2,7–3,3 h = 2,5–3	5″ Bi_{20} ?	5‴ Bi_{3T} $C3_112$		5⁗ Bi_{6H} ?		
	Band metasilicates	$[Al_nSi_{4-n}O_{11}]^{-(6+n)}$		6 Hb C2/m d = 2,97–3,45 h = 5–6	7 Na-Am C2/m d = 3,04–3,5 h = 5–6	8 Cum C2/m d = 3,1–3,6 h = 5–6	9 Am Pnma d = 2,98–3,6 h = 5,5–6,0					
	Ring metasilicates	$[SiO_3]^{-2}$						10 Tur R3m d = 3,03–3,25 h = 7				
	Diorthosilicates	$[Si_2O_7]^{-6}$										
	Orthosilicates	$[SiO_4]^{-4}$	12 Chd C1 (?)	12 Chd C2/c d = 3,5–3,8 h = 6,5		13 St C2/m d = 3,6–3,8 h = 7,5	13 St (?) Cmcm (?)		11 Ves P_4nnc d = 3,3–3,4 h = 6–7			
Anhydrous silicates	Chain metasilicates	$[SiO_3]^{-2}$			14 Ca-Px C2/c d = 3,21–3,56 h = 5,5–6,5	Cpx $P2_1/c$	15 Opx Pbca d = 3,21–3,96 h = 5–6					
	Orthosilicates	$[SiO_4]^{-4}$			16 Sap P2/c d = 3,4–3,6 h = 7,5		17 Ol Pbmn d = 3,2–4,39 h = 6,5–7,0				17′ Sp Ibmm d = 3,54–4,4	18 Gr Ia3d d = 3,58–4,32 h = 6–7,5

St—pseudorombic staurolite; the indexes ′,″,‴ and ′ are the polytypes and the polymorphs

Table 2. The K–Na Rock-Forming Alumino-Silicates

Type	Subclass	Radical	Singony, space group, specific gravity, hardness						
			Triclinic	Monoclinic		Rhombic	Hexagonal		Cubic
Hydrous	Sheet dimetasilicates	$[Al_nSi_{2-n}O_5]^{-(2+n)}$	[ ]	1 Mu C2/m d = 2,77 - 2,88 h = 2,5 - 3	2 $Bi_{(1M)}$ Cm (C2/m) d = 2,7 - 3,3 h = 2,5 - 3	2″ $Bi_{(20)}$ 0) ?	2‴ $Bi_{(6H)}$?		
	Band metasilicates	$[Al_nSi_{4-n}O_{11}]^{-(6+n)}$	[ ]	4 Hb C2/m d = 2,97 - 3,45 h = 5 - 6	5 Na-Am C2/m d = 3,04 - 3,5 h = 5 - 6				
Anhydrous	Framework silicates	$[Al_nSi_{m-n}O_{2m}]^{-n}$	6 Mic C1 d = 2,56 - 2,63 h = 6 - 6,5	6′ Or C2/m d = 2,55 - 2,63 h = 6 - 6,5	6″ San C2/m d = 2,56 - 2,63 h = 6 - 6,5	7′ Ne Ks	7 Ks_{Na} $P6_322$ d = 2,59 - 2,66 h = 6	8 Ne_K $P6_3$ d = 2,59 - 2,67 n = 5,5 - 6	8′ Car $P2_13$ d = 2,51

and Mg), plagioclase and Ca–Na pyroxene–Ca–Na amphibole. This is due to the heterogeneous isomorphism in the minerals, in which the overall electronegativity of two "weak" cations might be higher than the electronegativity of one "strong" cation. Besides, the exceptions from the general rule may be attributed to the fact that the solid solutions depart grossly from ideality. Cordierite is grouped with the framework silicates in Table 1. This follows directly from Ramberg's rule, because Cor is the most magnesian of all the Fe–Mg minerals associated with it (Marakushev, 1965). Moreover, the revision of the cordierite structure (Gibbs, 1966) classifies it with the framework rather than the ring metasilicates.

It follows from the general rule, that, as T rises, any mineral having a higher serial number would be enriched in Mg through the Fe exchange with a mineral of a lower serial number in Table 1, because Mg is less electronegative than Fe. In Table 2 similar relations show up for the K and Na redistribution between the minerals. It is seen, that the greater the redistribution effect, the larger the difference between the serial numbers. Empirical evaluation of this relationship has shown (Perchuk, 1971b) the difference in the serial numbers for two equilibrium minerals to be approximately equal to the ΔS value of the exchange reaction between the end members of the Fe and Mg or K and Na solid solutions (Tables 1 and 2). The experimental data available make it possible to verify the postulated rules. Thus, according to Akimoto and Fujisawa (1968), as the temperature rises, magnesium will transfer from Ol (No. 17 in Table 1) into Sp (No. 17′ in table) with a slight change in the entropy ($\Delta S \leqslant 1.5$ cal/K) of the exchange reaction $Ol_{Mg} + Sp_{Fe} = Sp_{Mg} + Ol_{Fe}$. The distribution of Na and K between Ne_{ss} and Fsp_{ss} of the varying degree of ordering is not ideal and depends on temperature (Zyrianov et al., 1976). In the 400–700°C range, the ΔS of exchange reactions vary within the limits 0.955 – 2.33 cal/K. The difference in the serial numbers between these minerals in Table 2 is from 1 to 2.

Table 1 reveals regular changes in specific gravity, hardness and other physical properties of the minerals. According to the rule postulated above, the redistribution of Fe and Mg is directly proportional to the difference in the mineral serial numbers (ΔN) and, hence, to the difference between their specific gravity values ($\Delta\rho$). It was found that the $\Delta\rho = 0.5[(\rho_{Mg} + \rho_{Fe})^A - (\rho_{Mg} + \rho_{Fe})^B]$ for the minerals A and B under consideration, depends linearly on $\Delta N \approx \Delta S$. Thus, for the exchange equilibria of garnet with hydrous silicates from Table 1 it was found that:

$$\Delta S \approx \Delta N = (1.364 + 12.121\ \Delta\rho) \pm 1.670\ (\text{cal/K}).$$

The high entropy values of exchange reactions are not (wholly) attributable to the redistribution of Fe and Mg upon temperature changes.

According to the general rule, *changes in T will bring about the redistribution of all the isomorphous elements according to their relative electronegativities, thus increasing the redistribution of* Fe *and* Mg. *As a result, we obtain the overall entropy of a great number of exchange reactions for any pairs of the isomorphous components between two natural silicates.*

2.3 The Distribution Coefficient

The analysis given above is the basis for computing a great number of diagrams of phase relationships—mineralogical thermometers. Detailed thermodynamic treatment of exchange equilibria has been given in many works (Ramberg and De Vore, 1951; Ramberg, 1952a; Kretz, 1961; Perchuk, 1969a, 1970a; Saxena, 1973, among others). I would like to reiterate that the relationship between the logarithm of the distribution coefficient and T, P and the equilibrium composition of a mineral is derived from the expression:

$$d \ln K_D = -\left[\frac{\Delta H^0 + \left(\frac{\partial H^e}{\partial X_2}\right)^A_{T,P} + \left(\frac{\partial H^e}{\partial X_1}\right)^B_{T,P}}{RT^2}\right] dT + \left[\frac{\partial V^0 + \left(\frac{\partial V^e}{\partial X_2}\right)^A_{T,P} + \left(\frac{\partial V^e}{\partial X_1}\right)^B_{T,P}}{RT}\right] dP, \quad (1)$$

where

$$K_D = \frac{X_1^B(1 - X_1^A)}{X_1^A(1 - X_1^B)} = \frac{X_1^B\, X_2^A}{X_1^A\, X_2^B} \quad (2)$$

and where ΔH^0 and ΔV^0 are the changes in the enthalpy and volume in the exchange reaction between the minerals

$$A_1 + B_2 = B_1 + A_2, \quad (3)$$

composed of the pure components; H^e and V^e are the excessive integral molar enthalpy and molar volume of the respective minerals.

2.4 The Partition Coefficient

The distribution coefficient [Eq. (2)] in the exchange reaction [Eq. (3)] includes two coefficients:

$$\bar{K}_1 = X_1^B/X_1^A; \quad (4)$$
$$\bar{K}_2 = X_2^B/X_2^A, \quad (5)$$

which hereafter will be referred to as "the partition coefficient" of components 1 and 2 between minerals A and B:

$$K_D = \bar{K}_1/\bar{K}_2. \tag{6}$$

These coefficients have somewhat different meaning than K_D. Specifically, if the solid solutions A and B are ideal, $\bar{K}_1$ and $\bar{K}_2$ will be influenced by the changes in the Gibbs free energy in the decomposition reactions of the type:

$$A_1 = B_1 + nC, \tag{7}$$
$$A_2 = B_2 + nC, \tag{8}$$
$$\bar{K}_1 = \frac{X_1^B}{X_1^A} = \frac{a_1^B}{a_1^A} = -\exp\left(\frac{G_1^B + nG^c - G_1^A}{RT}\right); \tag{9}$$
$$\bar{K}_2 = \frac{X_2^B}{X_2^A} = \frac{a_2^B}{a_2^A} = -\exp\left(\frac{G_2^B + nG^c - G_2^A}{RT}\right), \tag{10}$$

where the G values are the Gibbs total free energies of phases A, B, and C. Subtraction of Equation (8) from Equation (7) gives the equation of the exchange reaction [Eq. (3)]. However, the temperature and pressure derivatives $\ln \bar{K}_1$ and $\ln \bar{K}_2$ will be determined entirely by the volume and heat effect of Equations (7) and (8) rather than (3):

$$\left[\frac{\partial \ln \bar{K}_1}{\partial(1/T)}\right]_P = -\frac{H_1^B + nH^c - H_1^A}{R} = -\frac{\Delta H_{(7)}}{R}, \tag{11}$$
$$\left[\frac{\partial \ln \bar{K}_2}{\partial(1/T)}\right]_P = -\frac{H_2^B + nH^c - H_2^A}{R} = -\frac{\Delta H_{(8)}}{R}, \tag{12}$$
$$\left[\frac{\partial \ln \bar{K}_1}{\partial P}\right]_T = \frac{V_1^B + nV^c - V_1^A}{RT} = \frac{\Delta V_{(7)}}{RT}; \tag{13}$$
$$\left[\frac{\partial \ln \bar{K}_2}{\partial P}\right]_T = \frac{V_2^B + nV^c - V_2^A}{RT} = \frac{\Delta V_{(8)}}{RT}. \tag{14}$$

To illustrate the relationships between K_D and $\bar{K}_1$, we assume that $\Delta V_{(3)} \approx 0$ and $\Delta H_{(7)} \approx 0$, while $\Delta H_{(3)}$ and $\Delta V_{(7)}$ have the maximum possible value. Then, at T = constant and in the case of the ideal distribution of component 1 between the phases (minerals) A and B, $\bar{K}_1$ will vary with P at constant K_D. Let $K_D = 5$ and the compositions of the equilibrium phases change with increasing X_1 in such a manner that at three different pressures P_1, P_2, and P_3 we will find that $\bar{K}'_1 = 4$, $\bar{K}''_1 = 2.1$ and $\bar{K}'''_1 = 1.33$, respectively (Fig. 1). It means that mineral A will react with mineral B to form B' of differing composition. For example, when the pressure rises from 81 to 90 kbar, 1000°C, α olivine reacts with ringwoodite according to the following scheme:

$$\underset{\text{ringwoodite (1)}}{1{,}2\,Mg_{0.6}Fe_{1.4}SiO_4} + \underset{\alpha\text{ olivine}}{0.4Mg_{1.4}Fe_{0.6}SiO_4} \rightarrow \underset{\text{ringwoodite (2)}}{1.6Mg_{0.8}Fe_{1.2}SiO_4}$$

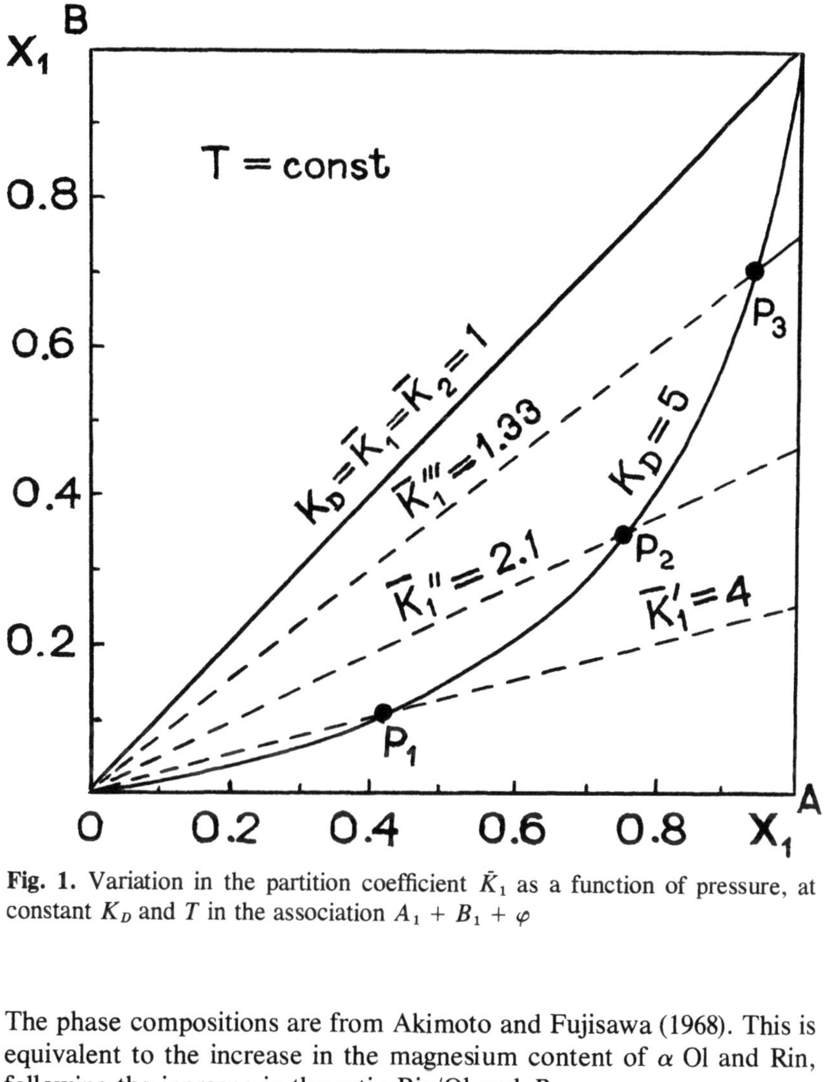

Fig. 1. Variation in the partition coefficient $\bar{K}_1$ as a function of pressure, at constant K_D and T in the association $A_1 + B_1 + \varphi$

The phase compositions are from Akimoto and Fujisawa (1968). This is equivalent to the increase in the magnesium content of α Ol and Rin, following the increase in the ratio Rin/Ol and P.

$$\underset{\alpha\ \text{olivine (1)}}{2.32Mg_{1.4}Fe_{0.6}SiO_4} + \underset{\text{ringwoodite (1)}}{0.88Mg_{0.6}Fe_{1.4}SiO_4} \rightarrow \underset{\alpha\ \text{olivine (2)}}{1.6Mg_{1.56}Fe_{0.44}SiO_4} + \underset{\text{ringwoodite (2)}}{1.6Mg_{0.8}Fe_{1.2}SiO_4} \qquad (15)$$

$\bar{K}_{Mg}$ changed from 2.33 to 1.95, i.e., the compositions of the minerals have come closer together, for which the appreciable volume effect of Equation (15) is responsible. According to Equation (13), this effect becomes:

$$\Delta V = \frac{RT}{P_2 - P_1} \ln \frac{\bar{K}_2}{\bar{K}_1} = -0.05\ \text{cal/bar}, \qquad (16)$$

or ~ 2.1 cm^3/mol. At the same time, the volume effect of the exchange reaction Fa + Rin_{Mg} = Rin_{Fe} + Fo becomes virtually zero, since K_D (≈ 5.35) has scarcely changed.

The more involved reactions [Eq. (7)–(8)] in the metamorphic rocks lead to the formation of the minerals of constant composition. For example, Cor decomposes with the formation of Gr, Al_2SiO_5, and Qz, and the association Opx + Sil + Qz is formed at the expense of Cor + Gr as P rises (Marakushev, 1965).

2.5 The Average Molar Contents

Reactions similar to Equation (15) will be hereinafter referred to as transfer reactions implying transfer of a solution component under the pressure and/or temperature effects. Their theoretical analysis might be performed using the partition coefficient method as well as the Gibbs–Korzhinskii method of the thermodynamic potential.

Let the sum of the mole fractions of the isomorphous components in the mineral crystallochemical formulae be identically equal to the number of atoms $X_1 + X_2 = 1$, and also let the system consist of K components and ϕ phases. It is sufficient to have $K + 2$ parameters $(T, P, \mu_1, \ldots, \mu_K)$, to describe thermodynamically this system. The change in these is expressed by the following group of ϕ equations:

$$\left.\begin{array}{l} -S^\alpha\, dT + V^\alpha\, dP - n_1^\alpha\, d\mu_1^\alpha - \ldots - n_k^\alpha\, d\mu_k^\alpha = 0 \\ -S^\beta\, dT + V^\beta\, dP - n_1^\beta\, d\mu_1^\beta - \ldots - n_k^\beta\, d\mu_k^\beta = 0 \\ \ldots \quad \ldots \quad \ldots \quad \ldots \quad \ldots \quad . \\ -S^\phi\, dT + V^\phi\, dP - n_1^\phi\, d\mu_1^\phi - \ldots - n_k^\phi\, d\mu_k^\phi = 0 \end{array}\right\}. \tag{17}$$

If the minerals (phases) $\alpha, \beta, \ldots, \phi$ interact, the correlation between intensive and extensive parameters of the system is convenient to be presented in the form of a matrix:

$$\begin{array}{c|cccccc} & -d\mu_1 & d\mu_2 & \ldots & d\mu_k & -dT & dP \\ \hline \alpha & X_1^\alpha & 1 - X_1^\alpha & \ldots & n_k^\alpha & \bar{S}^\alpha & \bar{V}^\alpha \\ \beta & X_1^\beta & 1 - X_2^\beta & \ldots & n_k^\beta & \bar{S}^\beta & \bar{V}^\beta \\ \ldots & \ldots & \ldots & \ldots & \ldots & \ldots & \ldots \\ \phi & X_1^\phi & 1 - X_3^\phi & \ldots & n_k^\phi & \bar{S}^\phi & \bar{V}^\phi \end{array} \tag{18}$$

where the above assumption on $n_1 + n_2 = X_1 + X_2 = 1$ is taken into account. The determinant of Equation (17) has the ϕth order, if components 1 and 2 enter only phases α and β, it can be expressed as:

$$\Omega = \begin{vmatrix} X_1^\alpha & 1 - X_1^\alpha & \ldots & n_k^\alpha \\ X_1^\beta & 1 - X_1^\beta & \ldots & n_k^\beta \\ \ldots & \ldots & \ldots & \ldots \\ 0 & 0 & \ldots & n_k^\phi \end{vmatrix}. \tag{19}$$

Solving this system for $d\mu_1$:

$$\begin{vmatrix} \bar{V}^\alpha & X_1^\alpha & \dots & n_k^\alpha \\ \bar{V}^\beta & X_1^\beta & \dots & n_k^\beta \\ \dots & \dots & \dots & \dots \\ \bar{V}^\Phi & 0 & \dots & n_k^\Phi \end{vmatrix} dP - \begin{vmatrix} \bar{S}^\alpha & 1 - X_1^\alpha & \dots & n_k^\alpha \\ \bar{S}^\beta & 1 - X_1^\beta & \dots & n_k^\beta \\ \dots & \dots & \dots & \dots \\ S^\Phi & 0 & \dots & n_k^\Phi \end{vmatrix} dT = \Omega\, d\mu_1 \quad (20)$$

Similarly, we find the solution of Equation (17) for $d\mu_2$:

$$\begin{vmatrix} \bar{V}^\alpha & X_1^\alpha & \dots & n_k^\alpha \\ \bar{V}^\beta & X_1^\beta & \dots & n_k^\beta \\ \dots & \dots & \dots & \dots \\ \bar{V}^\Phi & 0 & \dots & n_k^\Phi \end{vmatrix} dP - \begin{vmatrix} \bar{S}^\alpha & X_1^\alpha & \dots & n_k^\alpha \\ \bar{S}^\beta & X_1^\beta & \dots & n_k^\beta \\ \dots & \dots & \dots & \dots \\ \bar{S}^\Phi & 0 & \dots & n_k^\Phi \end{vmatrix} dT = \Omega\, d\mu_2, \quad (21)$$

Let us consider the simplest case, where the solid solutions of the components 1 and 2 in the minerals α and β are ideal. In this case the general condition is fulfilled (at constant T and P):

$$\begin{aligned} \mu_1 &= \mu_1^0 + RT \ln X_1 \\ \mu_2 &= \mu_2^0 + RT \ln (1 - X_1). \end{aligned} \quad (22)$$

Since the components 1 and 2 are contained in the minerals α and β alone, their mole fractions (X_1 and X_2) in the system might be taken as half sums of the mole fractions of these components in the given phases

$$\begin{aligned} X_1 &= \bar{X}_1 = 0.5(X_1^\alpha + X_1^\beta) \\ X_2 &= \bar{X}_2 = 0.5(X_2^\alpha + X_2^\beta). \end{aligned} \quad (23)$$

Substituting Equation (23) into (22) and making certain transformation we find

$$d\mu_1 = RTd \ln \bar{X}_1 \quad (24)$$
$$d\mu_2 = RTd \ln \bar{X}_2 = RTd(1 - \bar{X}_1). \quad (25)$$

Substituting Equation (24) into Equation (20), and Equation (25) into Equation (21) and solving them at constant T and P:

$$\left(\frac{d\ln\bar{X}_1}{dP}\right)_T = \frac{1}{\Omega RT}\begin{vmatrix} \bar{V}^\alpha & 1 - X_1^\alpha & \dots & n_k^\alpha \\ \bar{V}^\beta & 1 - X_1^\beta & \dots & n_k^\beta \\ \dots & \dots & \dots & \dots \\ \bar{V}^\phi & 0 & \dots & n_k^\phi \end{vmatrix} \tag{26}$$

$$\left(\frac{d\ln\bar{X}_1}{dT}\right)_P = -\frac{1}{\Omega RT}\begin{vmatrix} \bar{S}^\alpha & 1 - X_1^\alpha & \dots & n_k^\alpha \\ \bar{S}^\beta & 1 - X_1^\beta & \dots & n_k^\beta \\ \dots & \dots & \dots & \dots \\ \bar{S}^\phi & 0 & \dots & n_k^\phi \end{vmatrix} \tag{27}$$

$$\left(\frac{d\ln(1-\bar{X}_1)}{dP}\right)_T = \frac{1}{\Omega RT}\begin{vmatrix} \bar{V}^\alpha & 1 - X_1^\alpha & \dots & n_k^\alpha \\ \bar{V}^\beta & 1 - X_1^\beta & \dots & n_k^\beta \\ \dots & \dots & \dots & \dots \\ \bar{V}^\phi & 0 & \dots & n_k^\phi \end{vmatrix} \tag{28}$$

$$\left(\frac{d\ln(1-\bar{X}_1)}{dT}\right)_P = -\frac{1}{\Omega RT}\begin{vmatrix} \bar{S}^\alpha & 1 - X_1^\alpha & \dots & n_k^\alpha \\ \bar{S}^\beta & 1 - X_1^\beta & \dots & n_k^\beta \\ \dots & \dots & \dots & \dots \\ \bar{S}^\phi & 0 & \dots & n_k^\phi \end{vmatrix} \tag{29}$$

Entropies of the minerals of variables composition are not generally known. On the contrary, there is sufficient data on the molar volumes, and Equations (26) and (28) may be used in estimating the pressure effects on the content of certain isomorphous components in a given monovariant association.

2.6 Empirically Derived Regularities

Thus, we know that the T and P effects on the equilibrium of minerals of variable composition might be described via the $\ln K_D$, $\ln \bar{K}$, and $\ln \bar{X}_i$ variations. Usually, each of these provides information only on one of the two external parameters (T or P). For example, $\ln K_D^{Mg}$ in the paragenesis Cor + Gr depends strongly on the temperature (Table 1) but is virtually independent of the pressure. The $\ln \bar{K}_{Mg}$ value in the paragenesis Cor + Gr + Al_2SiO_5 + Qz depends greatly on P and to a less degree on T. The former equilibrium is called "exchange", the latter "transfer". These two types of equilibrium equally show up during metamorphism, which can be seen from simple examples.

For the estimates we selected from the handbook (Perchuk, 1971a) about 165 pairs of coexisting biotites and garnets from granulite, amphibolite, epidote–amphibolite, and greenschist facies of metamorphism (Perchuk, 1971a). This enabled us to establish to a first approximation the statistical averages for the compositions of the coexisting Bi and Gr in each of these facies and also to reveal the relationship between the degree of metamorphism and certain compositional parameters of these minerals. The results are given in Table 3, where $\ln K_D^{Mg} = \ln[X_{Mg}^{Gr}(1 - X_{Mg}^{Bi})/$

Table 3. Mean composition parameters of the coexisting Bi and Gr in the rocks from various metamorphic facies and in K_D^{Mg} which may be assumed to vary linearly with T^{-1}

		$\ln K_D^{Mg}$			
X_i*	Mineral	−0.883	−1.255	−1.949	−2.2
X_{Mg}	Bi	0.586	0.490	0.482	0.433
	Gr	0.369	0.215	0.117	0.078
$\ln K_{Mg}$		0.462	0.823	1.415	1.714
X_{Mn}	Bi	0.0013	0.0012	0.003	0.0057
	Gr	0.015	0.036	0.090	0.199
X_{Ti}	Bi	0.080	0.072	0.040	0.042
	Gr	0.004	0.005	0.007	0.008
X_{Al}	Bi	0.113	0.131	0.155	0.172
	Gr	0.415	0.415	0.433	0.453
X_K	Bi	0.047	0.042	0.044	0.037
X_{Ca}	Gr	0.044	0.073	0.146	0.187
Number of analyses		45	96	17	7

*X_{Mg} = Mg: (Mg + Fe + Mn); X_{Mn} = Mn: (Mg + Fe + Mn); X_{Ti} = Ti: (Ti + Fe +Mg+ Mn); X_{Al} = Al:(Al + Fe + Mn + Ti + Si);X_K = K:(K + Na);X_{Ca} = Ca: (Ca + Mg + Fe + Mn).

$X_{Mg}^{Bi}(1-X_{Mg}^{Gr})]$ and in Figure 2 (the upper diagrams). Similar work has been done with 35 samples of the cordierite–garnet paragenesis (with Qz and Al_2SiO_5) from the rocks of the amphibole facies. The results are in the lower diagrams in Figure 2.

The diagrams show, that the increase in the degree of metamorphism is accompanied by regular changes in certain parameters of the coexisting phases. In particular, $\ln K_D$ becomes larger, which means that the mineral compositions are getting closer together. For the Bi–Gr paragenesis this was first found by Frost (1962). The partition coefficients $\bar{K}$ and the average magnesium contents of the paragenesis $\bar{X}_{Mg}$ vary with the constants of exchange reactions. This is seen in the diagrams in Figure 3.

These two examples follow strictly the above rules: as the temperature rises, magnesium, which is a stronger base than iron and manganese, will transfer from hydrous silicates (Bi and Cor) into anhydrous (Gr) ones, from framework (Cor) and sheet (Bi) silicates into orthosilicate (Gr), from monoclinic (Bi) and rhombic (Cor) minerals into cubic (Gr)

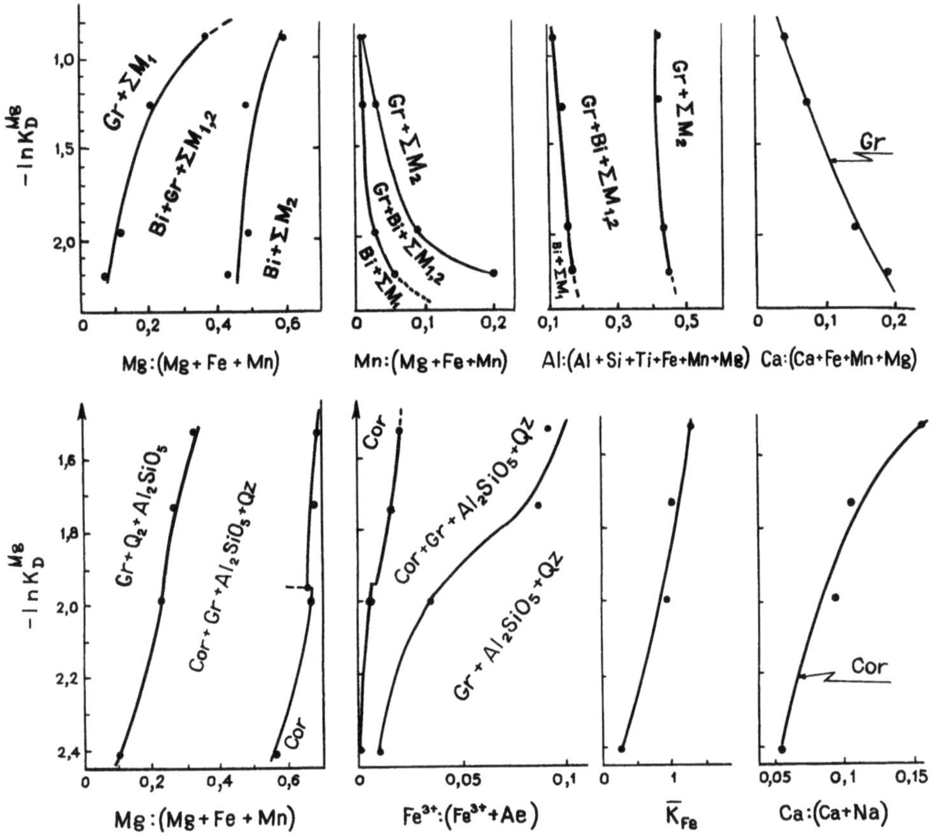

Fig. 2. Conjugate changes in the compositions of the coexisting Bi + Gr *(upper row)* and Cor + Gr *(lower row)* as a function of Mg and Fe distribution coefficient.

ones. Besides, the degree of Mg and Fe redistribution between Gr and Cor is higher than for Bi and Gr. This becomes evident when comparing the ln K_D^{Mg} values for these minerals within the amphibolite and granulite facies of metamorphism (Fig. 3).

Having the general rules of the effect of T and P on the equilibrium of rockforming minerals established, we can try to calibrate these effects.

3 MINERALOGICAL THERMOMETERS AND BAROMETERS

3.1 General Outline

The diagrams of Barth (1962), Buddington and Lindsley (1964), Perchuk (1965) and Iiyama (1966) were used to estimate the effect of temperature and pressure on the equilibrium of natural minerals of variable composition.

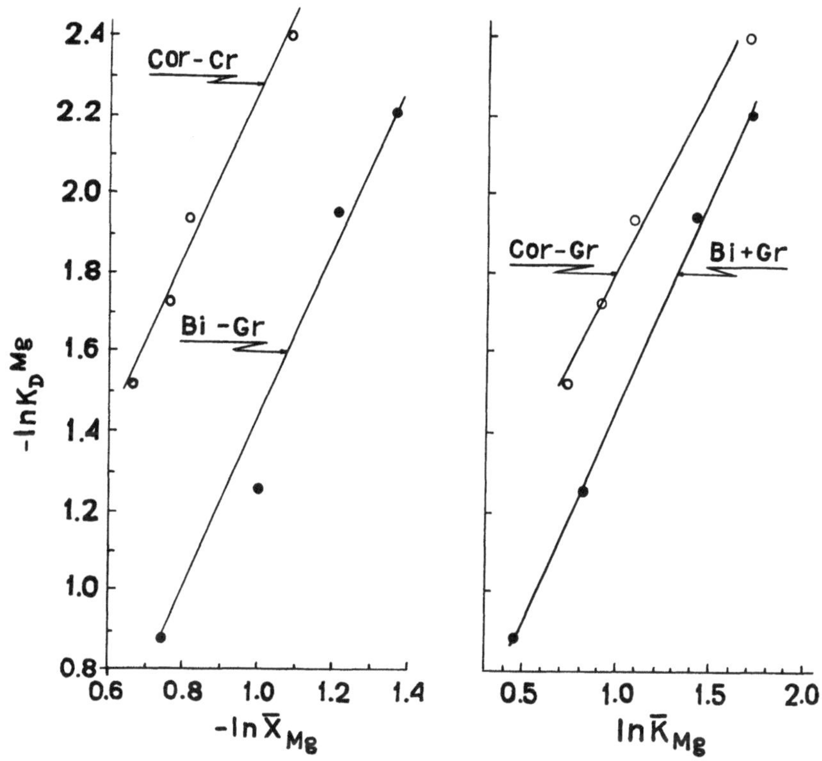

Fig. 3. Relationship between K_D^{Mg} and $\bar{X}_{Mg}$ and $\bar{K}_{Mg}$ for the natural parageneses Bi + Gr and Cor + Gr

To do this, we selected specimens with the following parageneses: Ne + Fsp + Pl + Mt + Il + Am, Ne + Fsp + Pl + Bi + Cpx, Ne + Fsp + Pl + Am + Cpx and others (Perchuk and Pavlenko, 1967; Perchuk, 1970a). Statistical and thermodynamic treatment of several hundred chemical analyses (Perchuk, 1971a) made it possible to estimate the equilibrium temperature for Am_{Ca} + Ab $\leftrightarrows$ An + Am_{Na}, Cpx_{Mg} + Am_{Fe} $\leftrightarrows$ Am_{Mg} + Cpx_{Fe}, Bi_{Mg} + Cpx_{Fe} $\leftrightarrows$ Cpx_{Mg} + Bi_{Fe}, and so on, in the wide range of compositions and *PT*-conditions. This was done according to the following procedure. First we considered the relationship between the composition of solid solutions and their molecular volumes. This enabled us, using the empirical data to establish the degree to which every given solution (isomorphous mixture) will depart from ideality, to calculate the molecular volumes of the solid solution end members as well as to estimate the volume effects of the numerous exchange equilibria from Tables 1 and 2. The overwhelming majority of these equilibria was found to depend little on the pressure. With the exception of garnet-pyroxene equilibria, most of the volume effects are within the range 0.1–3 cm^3. At the same time, the minerals from the various depth facies differ in their compositions. Thus, biotites from granulites contain more Ti than the biotites from amphibolites; Cpx and Am from eclogites and glaucophane schists differ sharply in their Al, Na and Ti contents from pyroxenes and amphiboles from the rocks of other metamorphic and magmatic facies. We accounted for these features of mineral compositions in constructing the phase diagrams—mineralogical thermometers. The statistical treatment of the data on the relationship between the compositions of the coexisting minerals and their equilibrium temperatures enabled us to establish the excess partial molar values for Opx, Am, Bi, Pl, Cpx, and other minerals, using the regression analysis. The estimates of these values were made by the power series expansion (Margules, method) in accordance with the mole fractions of the isomorphous components in the T range established empirically. After that, from the formulas $\alpha_0^G = \alpha_0^H - \alpha_0^S T$, $\alpha_1^G = \alpha_1^H - \alpha_1^S T$, $\alpha_2^G = \alpha_2^H - \alpha_2^S T$ etc. where $\alpha_0, \alpha_1, \alpha_2, \ldots$ are the coefficients of the power series expansion or the respective excess functions (G^e, H^e, S^e) (Wagner, 1952), we found G_i^0, G_i^e for a wider temperature range. Thus, the dependence of the mixing energy and ln K_D on the phase compositions over a wide temperature range could be established. Most of the diagrams calculated by this method have been published (Perchuk, 1966, 1967b, 1968a, 1969a, 1970, 1971c, 1973; Perchuk and Andrianova, 1968; Perchuk and Ryabchikov, 1968) and found wide petrologic application both in this country and abroad (Waldbaum and Thompson, 1968, 1969; Thompson and Waldbaum, 1968, 1969; Mottana, 1970; Delbove, 1971; Hubregtse, 1973; Saxena, 1973; Bakun-Czubarow, 1974 among others). Therefore, we shall only confine ourselves to the contribution that the phase correspondence diagrams can make in establishing the temperature regime of petrogeneses. Since each of the diagrams has been derived empirically, the ln K_D values depend on the compositions of the coexisting minerals, and these dependences follow certain patterns (Fig. 2). It means that every temperature range in the diagram corresponds to definite statistical compositions of the coexisting phases at the variable Mg:(Fe + Mg + Mn) ratio. It is evident that if the compositions of the minerals differ from these statistical averages, the information given here on the equilibrium temperature will prove insufficient for the description of the actual phase relations.

Consider now, as an example, the derivation of the phase diagrams for the pairs Bi–Gr, and Cor–Gr. According to Table 1, the temperature must have an appreciable effect on the redistribution of Fe and Mg between these minerals, i.e., the changes in ln K_D in the diagrams of Figure 2 result from the effect of the temperature on the exchange equilibria:

$$Bi_{Mg} + Gr_{Fe} \leftrightarrows Gr_{Mg} + Bi_{Fe} \qquad (30)$$
$$Cor_{Mg} + Gr_{Fe} \leftrightarrows Gr_{Mg} + Bi_{Fe}. \qquad (31)$$

These effects were used to derive the biotite–garnet and cordierite–garnet geothermometers (Perchuk, 1967b, 1969a). A great deal of new data have been accumulated in recent years, which offer the opportunity to ascertain these thermometers, especially in their high-temperature range.

3.2 The Biotite-Garnet Thermometer

The effect of the temperature on Equation (30) can be estimated using the amphibole-garnet geothermometer (Perchuk, 1967b, 1969a). Table 4 gives the mole fractions of magnesium:

$$X_{Mg} = \frac{Mg}{Mg + Fe^{+2} + Fe^{+3} + Mn} \tag{32}$$

in the three coexisting minerals from metamorphic rocks at various equilibrium temperatures (470–700°C). These data were used in constructing the diagram in Figure 4, revealing the linear relationship between ln K_D^{Mg} and inverse temperature; the scatter of points is negligible. These two facts seem to suggest that Mg redistribution between Bi and Gr is close to the ideal. The dependence of ln K_D on T in such distributions is generally given by empirical formulas:

$$\ln K_D = \frac{A}{t + 273} + B = \frac{A}{T} + B \tag{33}$$

$$t(°C) = \frac{A}{\ln K_D - B} - 273 \tag{34}$$

$$X_{Mg}^{\alpha} = N/(1 + N), \tag{35}$$

$$N = \frac{X_{Mg}^{\beta}}{X_{Mg}^{\beta}} \exp\left[\frac{A}{T} + B\right]. \tag{36}$$

For the biotite–garnet equilibrium

$$A = 3650; \qquad B = -2.57. \tag{37}$$

Table 4. The data for calibration of a biotite–garnet thermometer using the temperatures t found from the amphibole–garnet equilibrium (Perchuk, 1976b, 1968a)

No.	Mg: (Mg + Fe + Mn) Am	Bi	Gr	t_1	T_1	$10^3/T_1$	ln K_D	T_T-T_1
1	0.410	0.427	0.148	635	908	1.101	1.456	−2
2	0.448	0.482	0.168	635	908	1.101	1.528	−17
3	0.457	0.477	0.182	635	908	1.101	1.411	+9
4	0.263	0.266	0.060	600	873	1.145	1.737	−25
5	0.416	0.436	0.148	630	903	1.074	1.493	−5
6	0.447	0.437	0.153	620	893	1.120	1.456	+14
7	0.478	0.442	0.176	625	898	1.114	1.323	+39
8	0.486	0.529	0.187	625	898	1.114	1.586	−19
9	0.601	0.681	0.397	700	973	1.028	1.176	+1
10	0.498	0.563	0.112	470	743	1.346	2.324	+3
11	0.500	0.558	0.122	480	753	1.328	2.207	+11
12	0.397	0.372	0.096	550	823	1.215	1.839	+5
13	0.424	0.467	0.144	620	893	1.120	1.476	+9

1–8: Coexisting minerals from the gneisses of Quebec (Kretz, 1959).
9: Granulite, Yakutia (Perchuk, 1971a).
10–12: Epidote-amphibolites, Scotland (Matthews and Cheenney, 1968).
13: Amphibolite, North Karelia (Perchuk, 1971a).

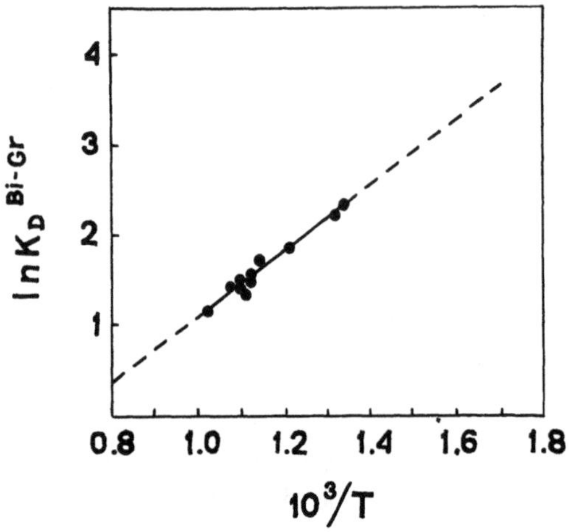

Fig. 4. The ln K_D^{Mg}—inverse temperature relationship in the biotite–garnet equilibrium (Table 4)

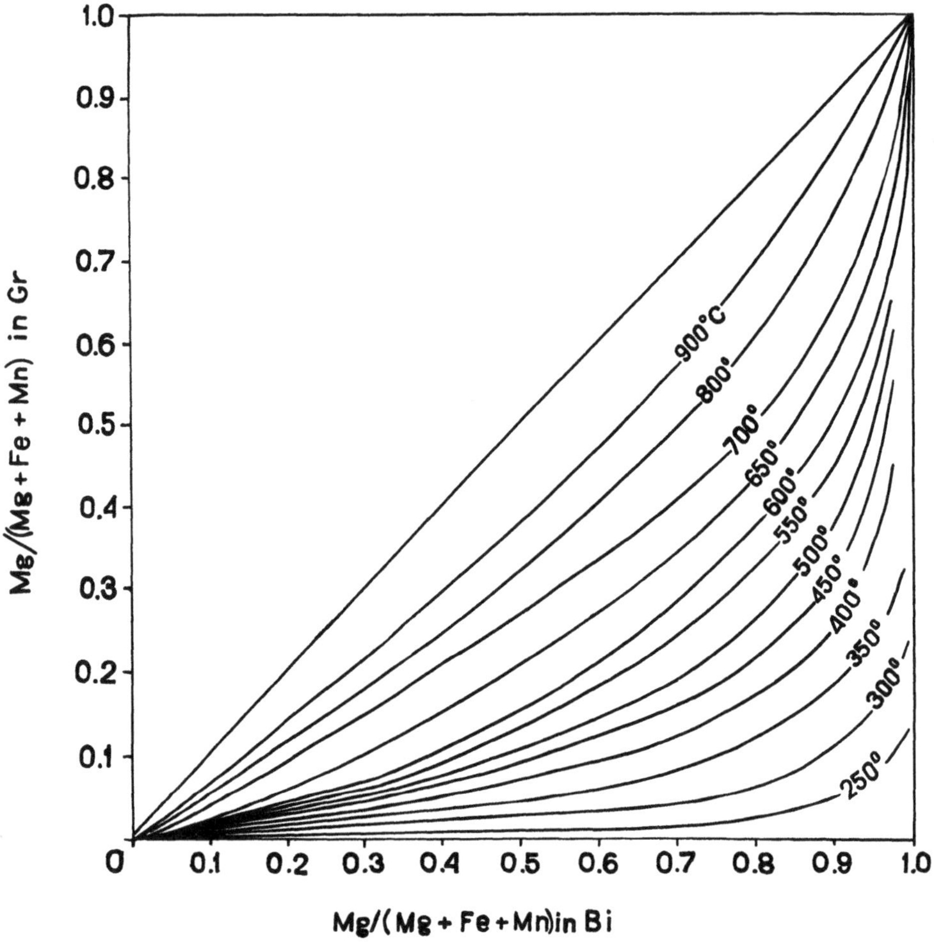

Fig. 5. Isotherms for Mg distribution between biotite and garnet from metamorphic rocks (nonideal distribution)

However, the analysis of Fe and Mg distribution between Bi and Gr in more than 200 parageneses (Perchuk, 1968b) and experimental data of Wones and Eugster (1965) show that biotite is not an ideal solid solution. The negative departure from ideality was found in investigating the concentrational dependence of the excess integral molar volume of biotite. This enabled us to calculate the isotherms for the Mg distribution between Bi and Gr with the allowance made for nonideality of the biotite solid solution (Fig. 5).

3.3 Cordierite–Garnet Thermometer

To calculate the temperature dependence of $\ln K_D$ in the cordierite–garnet equilibrium [Eq. (31)], we used the data from Table 5. Analysis of the diagram in Figure 6 suggests, that below ~750°C this relationship is linear and Equations (34)–(36) have the parameters:

$$A = 3421; \qquad B = -1.747. \tag{38}$$

Above ~750°C, as the entropy effect of Equation (31) becomes greater, deviation from the straight line relation becomes apparent. We believe, that the marked point scatter over the range 700 to 800°C and the change

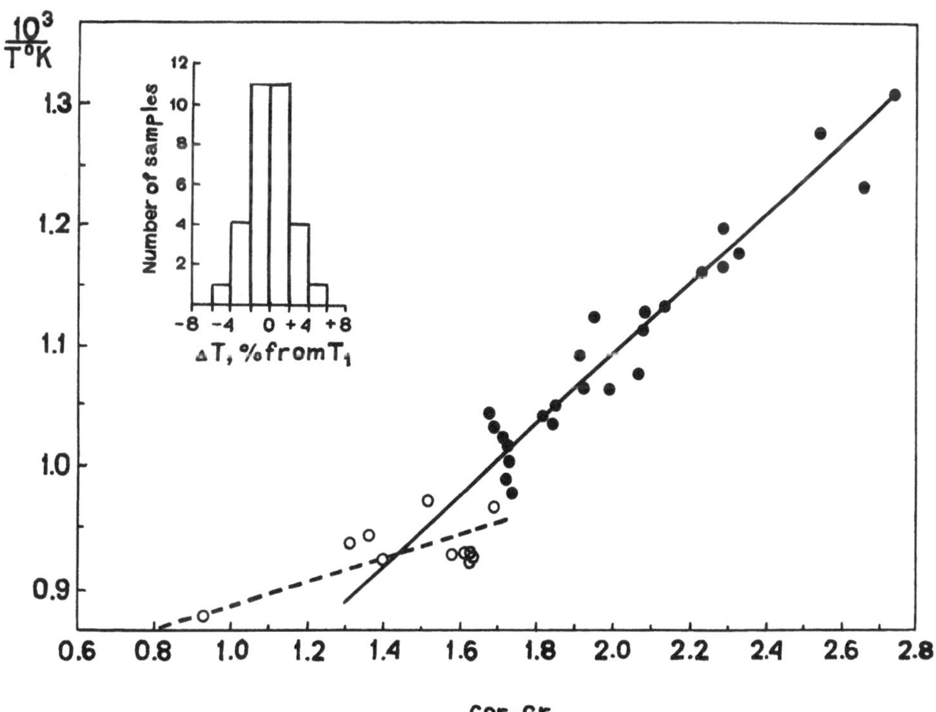

Fig. 6. The $\ln K_D^{(31)}$—inverse temperature relationship for the cordierite–garnet equilibrium. The break in the curve may well depend on the Cor phase transition *(solid circles)* into Ind *(open circles)*

Table 5. The data* for calibrating a cordierite-garnet thermometer using the temperatures (t_1) found from the biotite-garnet equilibrium (Figure 4)

	Mg: (Mg + Fe + Mn)							
No.	Bi	Gr	Cor	t_1	T_1	$10^3/T_1$	ln K_D	T_Σ-T_1
1	0.513	0.114	0.620	508	781	1.280	2.538	+17
2	0.443	0.113	0.645	536	809	1.236	2.659	−32
3	0.341	0.078	0.452	560	833	1.200	2.283	+16
4	0.508	0.158	0.656	580	853	1.172	2.323	−12
5	0.414	0.117	0.565	587	860	1.163	2.283	−11
6	0.357	0.095	0.494	589	862	1.160	2.235	−3
7	0.545	0.197	0.672	605	878	1.139	2.129	+4
8	0.510	0.177	0.645	607	880	1.136	2.137	0
9	0.568	0.219	0.691	614	887	1.127	2.079	+7
10	0.421	0.139	0.563	623	896	1.114	2.079	−2
11	0.659	0.317	0.758	640	913	1.095	1.910	+22
12	0.429	0.153	0.559	630	913	1.074	1.952	−10
13	0.636	0.321	0.763	668	941	1.063	1.917	−7
14	0.312	0.110	0.475	670	943	1.060	1.995	−29
15	0.095	0.027	0.150	675	948	1.054	1.849	+3
16	0.562	0.265	0.694	678	951	1.051	1.839	+3
17	0.496	0.221	0.632	684	957	1.045	1.820	+2
18	0.347	0.133	0.447	684	957	1.045	1.666	+45
19	0.406	0.165	0.609	685	923	1.084	2.064	−25
20	0.651	0.354	0.774	689	962	1.039	1.839	+8
21	0.638	0.360	0.752	695	968	1.033	1.682	+30
22	0.583	0.300	0.697	700	973	1.028	1.682	+26
23	0.775	0.517	0.855	703	976	1.025	1.709	+14
24	0.529	0.272	0.677	721	994	1.006	1.726	−9
25	0.405	0.196	0.579	742	1014	0.986	1.732	−30
26	0.777	0.565	0.855	753	1026	0.975	1.510	+24
27	0.288	0.135	0.458	763	1036	0.965	1.687	−48
28	0.591	0.346	0.662	789	1062	0.941	1.309	+33
29	0.498	0.303	0.689	802	1075	0.930	1.627	+2
30	0.514	0.317	0.691	803	1076	0.929	1.575	+4
31	0.401	0.229	0.606	806	1079	0.927	1.645	−3
32	0.637	0.441	0.800	810	1083	0.923	1.625	−6
33	0.484	0.293	0.625	840	1077	0.928	1.394	+0
34	0.404	0.262	0.474	862	1135	0.881	0.931	0

*The parageneses and chemical analyses of the coexisting minerals are from Table 5 of the Handbook (Perchuk, 1971a).

in slope of the straight line are due to the cordierite–indialite phase transition. Schreyer and Yoder (1959) fixed this transition in the same temperature range at $P_{H_2O} = 1$–1.5 kbar.

We have calculated the coefficients of the possible Ind–Gr equilibrium from empirical Equations (32)–(34). They are $A = 9.420$ and $B = 7.357$. The entropy of Equation (31) involving Ind is $\Delta S \approx 15$ cal/K which is approximately equal to the difference in Cor and Gr number in Table 1.

The cordierite–garnet thermometer (Fig. 6) possesses relatively high accuracy. This is seen from the histogram in Figure 6, which gives the statistical estimates of the differences of T_2 from T_1 obtained with the biotite–garnet thermometer for the same paragenesis.

3.4 Cordierite–Garnet–Quartz–Sillimanite Barometer

The volume effects on the exchange reactions [Eqs. (30) and (31)] are slight (Perchuk, 1970a) and, therefore, the compositions of the coexisting Bi, Gr, and Cor only provide information on the temperature. However, the magnesium contents of these minerals in certain parageneses can depend on the pressure. For example, with increasing P, $Cor_{0.5}$ becomes unstable in the association with Gr, Qz, and Sil:

$$Cor_{0.50} \rightarrow Cor_{>0.50} + Gr_{0.33} + Qz + Sil$$

or

$$(3n + 1)Mg_1Fe_1Al_4Si_5O_{18} \rightarrow Mg_{1+n}Fe_{1-n}Al_4Si_5O_{18} + 5nSiO_2 + 2nMg_1Fe_2Al_2Si_3O_{12} + 4nAl_2SiO_5 \quad (39)$$

$\Delta V \approx -3.848$ cal/bar.

Another reaction takes place alongside this one:

$$Cor_{0.50} + Gr_{0.33} \rightarrow Gr_{>0.33} + Sil + Qz$$

or

$$3nMg_1Fe_1Al_4Si_5O_{18} + (1 - 2n)Fe_2Mg_1Al_2Si_3O_{12} \rightarrow 5nSiO_2 + Fe_{2-n}Mg_{1+n}Al_2Si_3O_{12} + 4nAl_2SiO_5. \quad (40)$$

The volume change in this reaction has the same value $\Delta V = -3.848$ cal/bar. These reactions describe the shift in equilibrium with the pressure at T = constant:

$$Cor \rightarrow Gr + Sil + Qz \quad (41)$$

or

$$(Fe,Mg)Al_2Si_{2.5}O_9 = (Fe,Mg)Al_{2/3}SiO_4 + 0.666Al_2SiO_5 + 0.834SiO_2$$

$$\bar{K}_{(41)} = \frac{a_{Mg}^{Gr}}{a_{Mg}^{Cor}} \approx \frac{X_{Mg}^{Gr}}{x_{Mg}^{cor}} = \exp\left[\frac{G^0_{Mg-cor} - G^0_{Pyr} - 0.666G^0_{Sil} - 0.834G^0_{Qz}}{RT}\right] \quad (42)$$

The volume change, as calculated per 3.5 molecules involved in this reaction, is:

$$\left(\frac{RTd \ln \bar{K}_{(41)}}{dP}\right)_T = \Delta V_{(41)} \approx -0.637 \text{ cal/bar} \tag{43}$$

and when calculated per one "averaged" molecule, it is $\Delta V_{(41)} \approx -0.182$ cal/bar. Equation (43) can be integrated, provided that the limits on integration are known. As before, we shall use the data of the mineralogical thermo- and baro-metry. These are given in Table 6. Equilibrium temperatures in this table were found from the cordierite–garnet and biotite–garnet thermometers, and the pressure from the phase transitions in Al_2SiO_5 and the orthopyroxene–garnet barometer (Perchuk, 1967a, 1968a, 1969c, 1973). It must be noted, however, that the P_s values found from this barometer are not very accurate.

The PX_{Mg} diagram in Figure 7 was constructed from the data in Table 6. This diagram gives the correlation of the compositions for the coexisting Cor and Gr in association with Qz and Al_2SiO_5 as the function of temperature and pressure. It is seen that while the pressure shifts the equilibrium into the region of high magnesium content, the changes in the temperature bring the compositions closer together. This is the case which was considered in the previous section (Fig. 1). The diagram shows that the Qz-free but Sap-bearing parageneses do not follow the general tendency in variation of the Cor and Gr compositions with the pressure, though they approach this tendency.

Thermodynamic treatment of the data in Table 6 and Figure 7 (with certain extrapolation to the low-temperature region) made it possible to calculate the approximate relationships of $\ln \bar{K} = \ln(X_{Mg}^{Cor}/X_{Mg}^{Gr})$ and $\ln \bar{X}_{Mg}$ on P and $10^3/T$. The results are shown in Figure 8 together with the temperature dependence $\ln K_D^{(31)}$ (at $\Delta V_{(31)} \approx$ min). These diagrams can be used as thermometers and barometers for estimating the PT conditions of mineral equilibria in the rocks containing Cor + Gr + Al_2SiO_5 + Qz ± Bi ± Opx. The discrepancies in the P values, as determined from the diagrams A and B are not conspicuous.

To illustrate the use of diagrams in Figure 8 we have selected from the Handbook (Perchuk, 1971a) about 30 parageneses from the Precambrian metapelites and found P and T for them from the diagram A in Figure 8. We attempted the evaluation of these parameters for two sapphirine parageneses (without Qz), designated 11 and 12 in Table 6. The results are presented in the PT diagram of Figure 9. The dashed lines for the phase transitions in Al_2SiO_5 are plotted from the geothermobarometry data (Perchuk, 1967a, 1973). The overwhelming majority of the points fall within And and Sil field, which corresponds to these parageneses.

Table 6. *TP* conditions of mineral equilibria in certain cordierite–garnet metamorphic rocks (The compositions are from Perchuk, 1971a)

No.	Mg: (Mg + Fe + Mn)					Mu	Qz	And	Sil	Ky	$\ln X^{Gr}_{Cor}$	$\ln K_{(41)}$	t°C	P, kbar
	Cor	Gr	Bi	Opx	Sap									
1	(0.15)*	0.027	0.095	−	−	−	+	+	−	−	−2.424	1.849	675	2
2	(0.57)	0.186	0.424	−	−	+	+	+	+	−	−0.973	1.120	665	4
3	(0.67)	0.254	0.516	−	−	+	+	−	+	−	−0.772	0.970	670	4
4	(0.57)	0.175	(0.41)	−	−	+	+	+	+	−	−0.989	1.181	665	4
5	(0.57)	0.185	(0.42)	−	−	+	+	+	+	−	−0.973	1.125	665	4
6	0.691	0.219	0.568	−	−	−	+	(+)†	(+)	−	−0.787	1.149	625	4
7	(0.72)	0.376	0.562	0.547	−	−	+	−	+	−	−0.601	0.649	800	10.5
8	0.800	0.441	0.637	0.612	−	−	+	−	−	−	−0.478	0.596	800	10.5
9	0.800	0.421	−	0.625	−	−	+	−	−	−	−0.494	0.642	720	9
10	(0.82)	0.540	0.775	0.675	+	−	−	−	+	+	−0.386	0.418	770	9.1
11	0.855	0.517	0.775	0.775	0.823	−	−	−	−	−	−0.377	0.503	730	10.7
12	0.855	0.565	0.777	0.721	0.735	−	−	−	+	−	−0.342	0.414	765	10.8
13	0.774	0.754	0.651	−	−	−	+	−	(+)	(+)	−0.572	0.782	685	7.9
14	0.726	0.321	−	−	−	−	+	−	+	−	−0.652	0.809	720	7.5

*The compositions determined by optical properties of a mineral (in parentheses).
†A mineral occurs in a given metamorphic series (in parentheses).

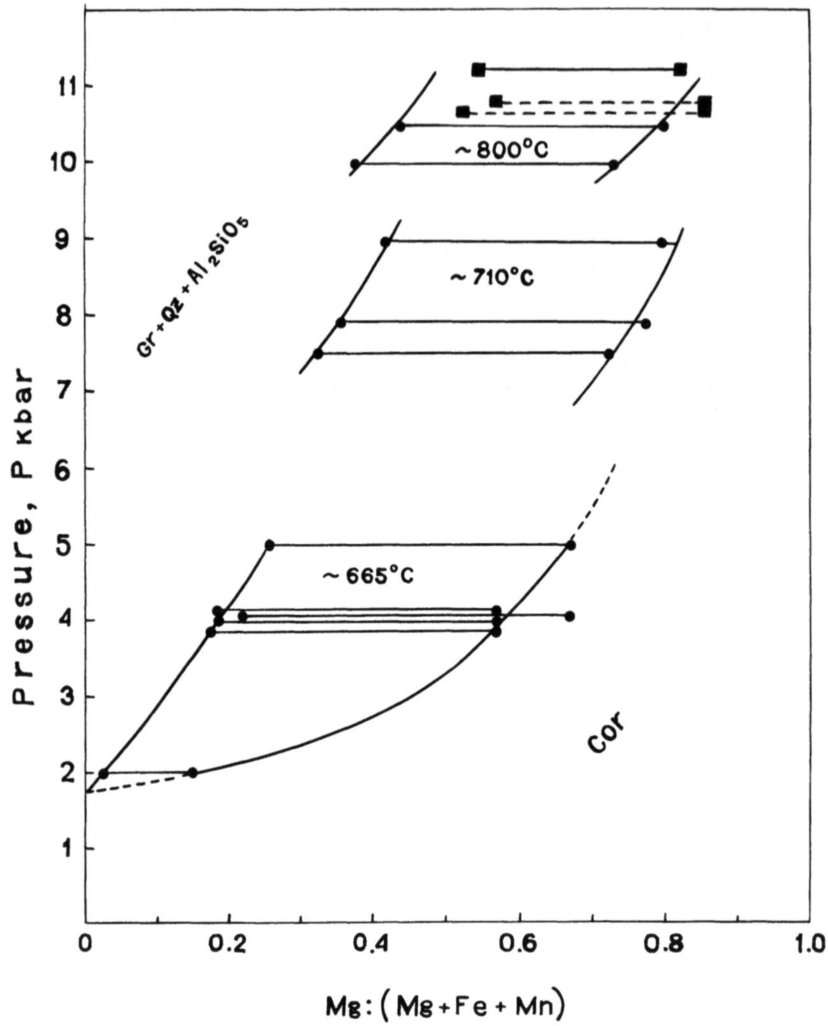

Fig. 7. The PX_{Mg} diagram for the equilibrium Cor = Gr + Qz + Al_2SiO_5 derived from the mineralogical thermobarometry data (Table 6). *Solid circles:* parageneses with Qz without sapphirine; *solid blocks:* parageneses with sapphirine without Qz

3.5 Some of the Biotite–Garnet Equilibria as Geothermobarometers

The diagrams in Figures 5 and 8 may form the basis for calculating new mineralogical geothermobarometers. The first to be mentioned are the parageneses involving Bi and Gr. It is clear that the *PT* conditions of $\bar{K}_{Mg} = X_{Mg}^{Bi}/X_{Mg}^{Gr}$ and $\bar{X}_{Mg} = 0.5(X_{Mg}^{Bi} + X_{Mg}^{Gr})$ can be calculated from the data in Tables 5 and 6. For example, the association Bi + Gr + Cor ± Fsp + Al_2SiO_5 + Qz ± Mu occurs frequently in metapelites. Although it is described in many metamorphic complexes of the world, however, it received most thorough study in Nagaitsev's works (1974) on the

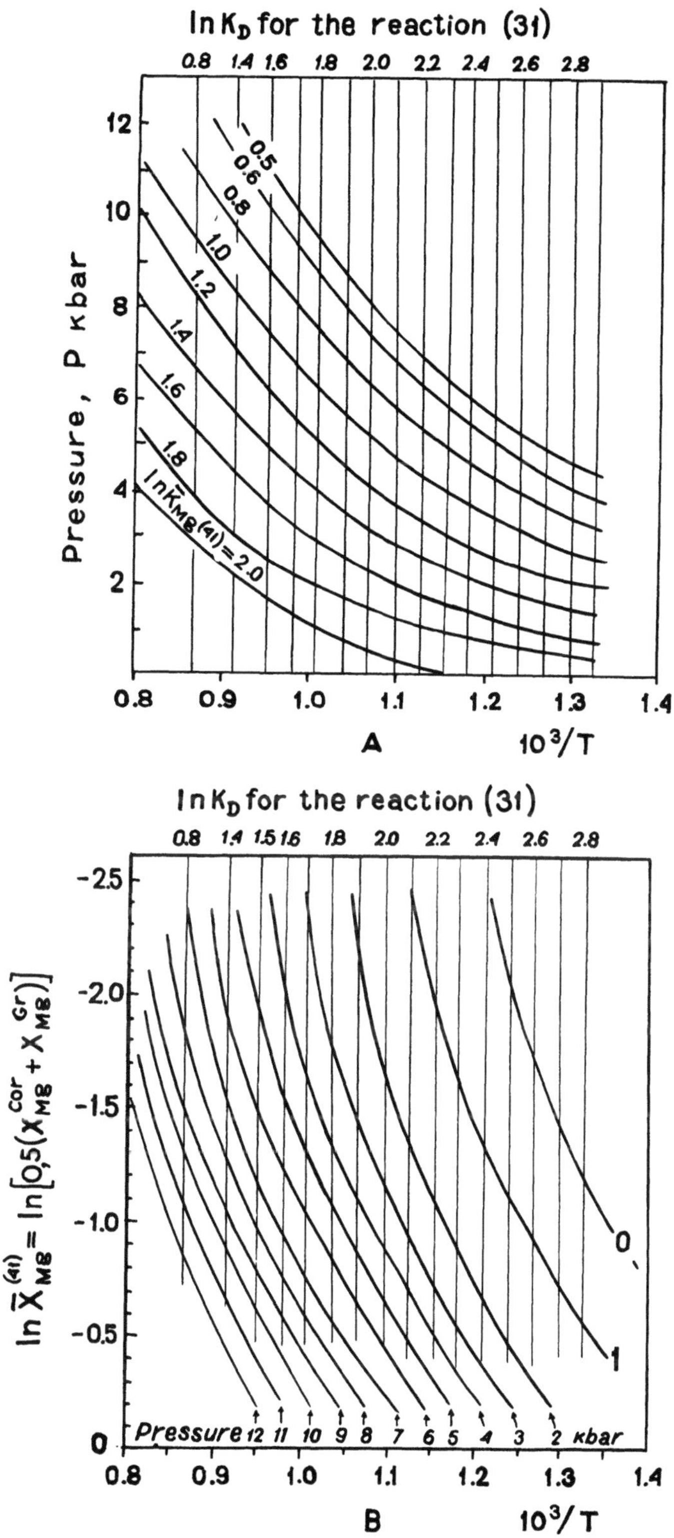

Fig. 8A, B. The *PT* dependences of ln $\bar{K}_{Mg}^{(41)}$ (A) and ln $\bar{X}_{Mg}^{(41)}$ (B) and their relations with ln $K_D^{(31)}$ in the paragenesis Cor + Gr + Al_2SiO_5 + Qz ± Bi ± Opx

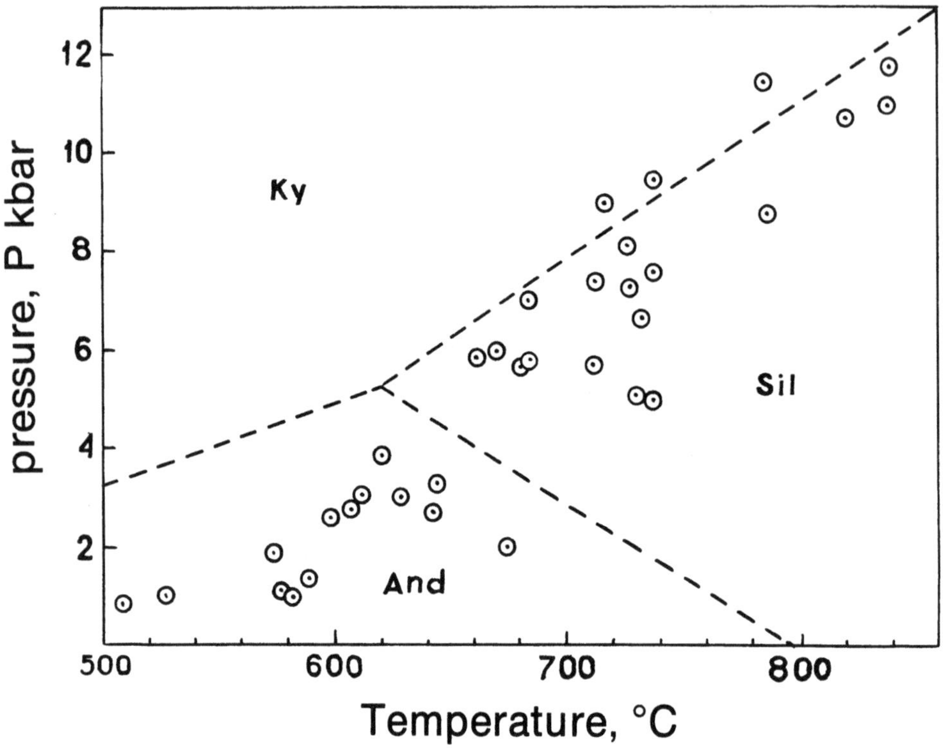

Fig. 9. Thermodynamic equilibrium conditions during formation of the cordierite–garnet metapelites in the earth's crust. Thermobarometry as established from the diagram in Figure 8

Ladoga formation showing distinct metamorphic zoning from greenshist facies to granulite facies. The following reaction is possible in the muscovite-free formation

$$\begin{aligned}(r-n)H_{2/3}K_{1/3}(Fe_{1-m}Mg_m)Al_{1/3}SiO_4 &+ (0.833m + 0.667r \\ &- 1.5m)SiO_2 + (0.333r + 0.667m - n)Al_2SiO_5 \\ = (r-m)(Fe_{1-n}Mg_n)Al_{2/3}SiO_4 &+ (m-n)(Fe_{1-r}Mg_r)Al_2Si_{2.5}O_9 \\ &+ \tfrac{1}{3}(r-n)KAlSi_3O_8 + \tfrac{1}{3}(r-n)H_2O,\end{aligned}$$

or

$$Bi + Qz + Al_2SiO_5 = Gr + Cor + Or + H_2O. \tag{44}$$

It is clear that similar reaction can be written for muscovite-bearing parageneses as well. In this case it will only be governed by the *TP* conditions and will not depend on the chemical potential of water. For example, with Fe and Mg as isomorphous components, the number of components is diminished by 1 and, we can write the reaction involving Mu (without Cor) as follows:

$$K_{0.4}(Fe,Mg)Al_{0.8}SiO_4(OH)_{0.8} + 0.533Al_2SiO_5 + 0.667SiO_2 = (Mg,Fe)Al_{0.6}SiO_4 + 0.4KAl_3Si_3O_{10}(OH)_2, \quad (45)$$

or

$$Bi + Al_2SiO_5 + Qz = Gr + Mu.$$

This is a "shifted" equilibrium. It means that upon changes in the *PT* conditions the biotite of certain composition decomposes with the formation of Bi and Gr of differing composition (in the presence of Qz, Al_2SiO_5, and Mu). The temperature of every equilibrium can be obtained with the biotite–garnet thermometer. Let us assume that K_D^{Bi-Gr}, as in the case of the cordierite–garnet exchange equilibrium, does not depend on pressure within 1–12 kbar. Then $\bar{K}_{Mg}^{(45)}$ or $\bar{X}_{Mg}^{(45)}$ (Perchuk, 1973) are indicators of *P* at each given temperature. Thus, there is linear relationship between $\ln \bar{X}_{Mg}^{(41)}$ and $\ln \bar{X}_{Mg}^{(45)}$in the cordierite–bearing parageneses:

$$\ln \bar{X}_{Mg}^{Bi-Gr} = 1.2272 \ln \bar{X}_{Mg}^{Cor-Gr} + 0.0182. \quad (46)$$

Therefore, if a rock contains Cor in equilibrium with Gr, Bi, Fsp, Al_2SiO_5 and Qz, it is only sufficient to know X_{Mg} for Bi and Gr to estimate the *PT* parameters of this equilibrium. The $\ln \bar{X}_{Mg}^{Cor-Gr}$ value is estimated from Equation (46); the temperature is found from the diagram in Figure 5 and the pressure from the diagram in Figure 8. For example, Kulish (1964) found the paragenesis $Bi_{0.562} + Gr_{0.376} + Opx_{0.547}$ + Sil + Cor + Qz + Fsp in the "quartzite" from the Archean granulites from the Aldan Schield. From Equation (46) $\ln \bar{X}_{Mg}^{Cor-Gr} = -0.602$, or $X_{Mg}^{Cor} = 0.719$ and the temperature is 800°C, according to Figures 5 and 6. From diagram *B* in Figure 8 it was found that $P \approx 10.5$ kbar. Nearly the same value ($P \approx 10.8$ kbar) was established using the orthopyroxene–garnet geobarometer (Perchuk, 1973).

Equation (44) is governed by *T, P,* phase composition, and μ_{H_2O}. The latter did not greatly affect $\bar{X}_{Mg}^{(45)} = 0.5\,(X_{Mg}^{Bi} + X_{Mg}^{Gr})$; its influence is virtually comparable to the errors in determining *P* and *T*. However, thermodynamic treatment of the data on the $\ln \bar{K}_{(45)} - \ln \bar{K}_{(41)}$ relationship revealed at least two such relationships (Fig. 10), other things being equal and neither the variations in the phase composition nor their parageneses and *PT* parameters can account for these relations. They might reflect different μ_{H_2O} levels in the formation of metamorphic rocks: line 1 in Figure 10 corresponds in this case to μ_{H_2O} = min, and the position of line 2 is determined by the maximum value of μ_{H_2O}.

The P_{H_2O} value does not influence only the equilibrium shift [Eq. (44)] in either direction during metamorphism. It is known that cordierites con-

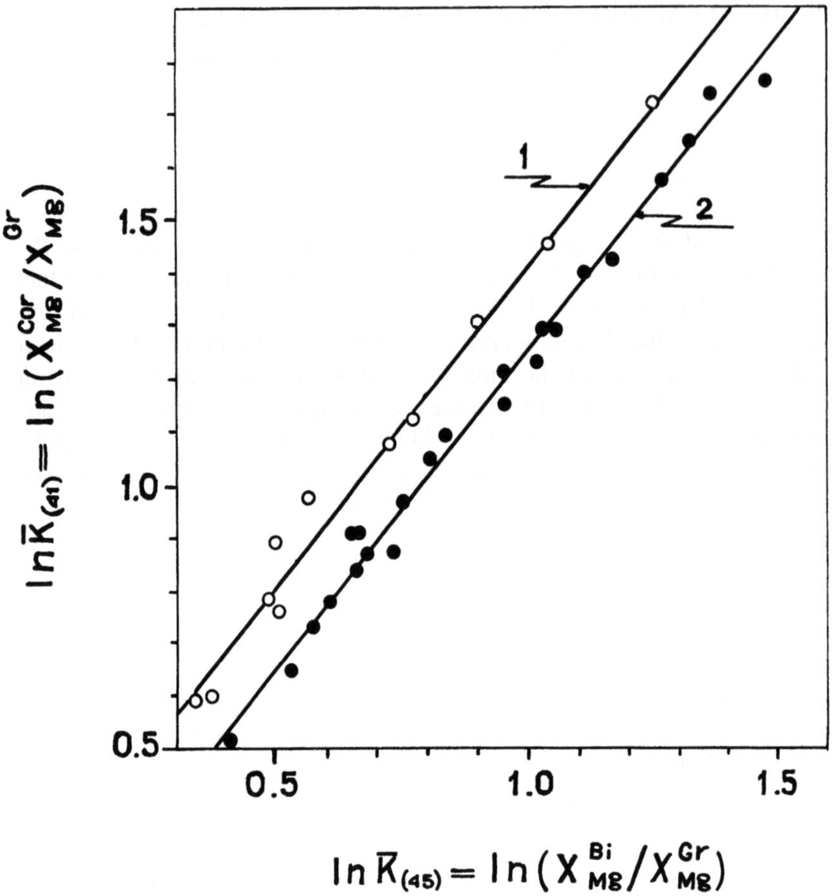

Fig. 10. Correlation between the logarithms of partition coefficients in Equations (41) and (45) in the parageneses Cor + Bi + Gr + Qz + Al_2SiO_5 ± Fsp ± Mu (Table 5). The symbols are in the text

tain variable quantities of nonstoichiometric water which also determine the Cor equilibria (Perchuk, 1969a; Newton, 1972; Wood, 1973; Newton et al., 1974).

Let us consider the cordierite-free parageneses, which occur most often in metapelites:

$$\text{Bi} + \text{Gr} + Al_2SiO_5 + \text{Fsp} + \text{Qz} \quad (47)$$
$$\text{Bi} + \text{Gr} + Al_2SiO_5 + \text{Fsp} + \text{Qz} + \text{Mu} \quad (48)$$
$$\text{Bi} + \text{Gr} + Al_2SiO_5 + \text{Mu} + \text{Qz} \quad (49)$$
$$\text{Bi} + \text{Gr} + \text{Mu} + \text{Qz} + \text{Fsp}. \quad (50)$$

The paragenesis [Eq. (47)] is described by:

$$K_{0.4}(Fe,Mg)Al_{0.8}SiO_4(OH)_8 + 0.133Al_2SiO_5 + 1.067SiO_2$$
$$= (Fe,Mg)Al_{0.6}SiO_4 + 0.4KAlSi_3O_8 + 0.4KAlSi_3O_8 + 0.4(H_2O). \quad (51)$$

The magnesium contents of the coexisting Bi and Gr in this case are governed by temperature, pressure P_s and the H_2O partial pressure in the fluid. If all these characteristics of the metamorphic processes are interrelated, the number of degrees of freedom will diminish and Equation (51) will become monovariant. We have shown the interrelation of P and T parameters of metamorphism using the diagram in Figure 10. Now it can also be seen from the Equations (47) through (50).

Table 7 presents 21 mineral associations with the known chemical compositions of Bi and Gr, as well as the thermodynamic conditions of their equilibria. The temperature was established from the Bi–Gr thermometer (Fig. 5), and the pressure was found from the Opx–Gr barometer and the stability of the polymorph varieties of Al_2SiO_5 (Perchuk, 1973). These data are plotted in the diagrams PX_{Mg} in Figure 11.

At P = 7 kbar and 665°C the equilibrium ratios of Bi and Gr compositions are governed by Equation (51), and above 7 kbar by the $P^{fl}_{H_2O}$ -

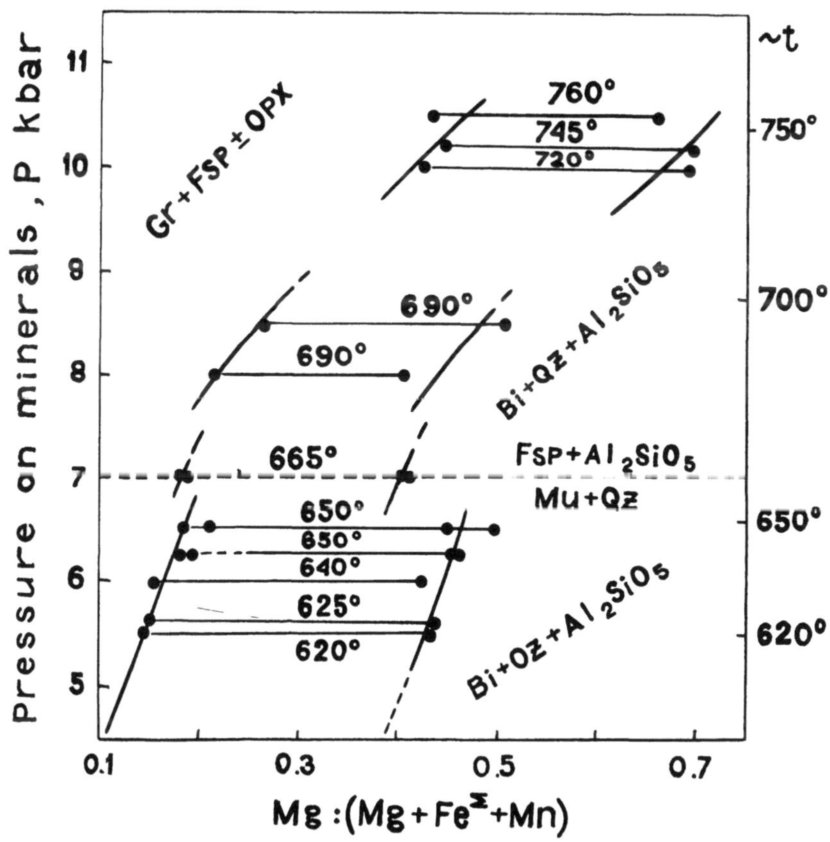

Fig. 11. The composition ratios for equilibrium Bi and Gr in association with Fsp, Mu, Qz and Al_2SiO_5, depending on P and t. Established from the mineralogical thermobarometry data (Table 7) at $P^{fl}_{H_2O}$ ≈ 0.3–3 kbar

Table 7. Composition parameters and PT conditions of the Bi and Gr equilibria in association with Al_2SiO_5 + Qz + Fsp + Pl and Al_2SiO_5 + Qz + Mu + Pl*

No.	X_{Mg}^{Bi}	X_{Mg}^{Gr}	$\ln \bar{K}_{Mg}$	t (°C)	$10^3/T$	P (kbar)	Sil	Ky	Cor	Opx	Fsp	Mu	Qz	Pl
1	0.562	0.376	0.402	800	0.931	10.8	+	−	+	+	+	−	+	+
2	0.690	0.426	0.482	720	1.007	10.0	+	−	+	−	+	−	+	+
3	0.661	0.437	0.414	760	0.968	10.5	+	−	+	+	+	−	+	+
4	0.700	0.448	0.466	745	0.982	10.2	+	−	+	−	+	−	+	+
5	0.508	0.264	0.655	690	1.038	8.5	+	+	−	−	+	−	+	+
6	0.405	0.214	0.638	690	1.038	8.0	+	+	−	−	+	−	+	+
7	0.451	0.186	0.885	650	1.083	6.5	+	+	−	−	+	−	+	+
8	0.409	0.189	0.832	665	1.066	7	+	+	−	−	+	+	+	+
9	0.411	0.178	0.837	665	1.066	7	+	+	−	−	+	+	+	+
10	0.461	0.194	0.866	650	1.083	6.7	+	+	−	−	−	+	+	+
11	0.435	0.145	1.098	620	1.119	5.5	+	+	−	−	−	+	+	+
12	0.459	0.143	1.166	620	1.119	5.7	+	+	−	−	−	+	+	+
13	0.451	0.160	1.036	625	1.114	5.7	+	+	−	−	−	+	+	+
14	0.435	0.143	1.113	625	1.114	6.2	+	+	−	−	−	+	+	+
15	0.441	0.159	1.020	640	1.095	5.5	+	+	−	−	−	+	+	+
16	0.454	0.151	1.100	620	1.119	5.5	+	+	−	−	−	+	+	+
17	0.451	0.152	1.087	620	1.119	5.5	+	+	−	−	−	+	+	+
18	0.428	0.142	1.103	620	1.119	5.5	+	+	−	−	−	+	+	+
19	0.444	0.144	1.126	620	1.119	5.5	+	+	−	−	−	+	+	+
20	0.445	0.153	1.068	625	1.114	5.6	+	+	−	−	−	+	+	+
21	0.398	0.145	1.01	640	1.095	6.2	+	+	−	−	−	+	+	+

*Mineral chemical analyses, parageneses and references are to be found in Perchuk, 1971a, 1973, p. 125.

dependent decomposition reaction Bi + Qz + Al_2SiO_5 giving Cor + Fsp ± Opx (Fig. 11). The $P^{fl}_{H_2O}$ values are not known so far but later we are going to describe the methods for their evaluation. Below $P_s = 7$ kbar the ratio of Bi and Gr compositions is governed by Equation (45) which does not depend on P_{H_2O} when the mineral association is in the Mu and Qz stability field. There is distinct shift towards the more iron Bi and Gr compositions in Figure 11. This shift can possibly be explained by the μ_{H_2O} effect on Equation (51).

The data in Table 7 were used to derive the diagram in Figure 12, which reveals the distinct P_s – ln $\bar{K}_{Mg}$ relationship in Equations (47) to (50). The point with the average ln $\bar{K}_{Mg}$ value for the Bi–Gr parageneses (with Cor) from the Ladoga metamorphic formation (Table 6, No. 2–6) is plotted in this diagram for comparison.

Unlike the cordierite–garnet shifted equilibria, the reactions like Equations (45) and (51) have large entropy effects. It means that ln $\bar{K}_{Mg}$ must be strongly temperature-dependent, which is seen from the diagram in Figure 13 derived on the basis of the Bi–Gr thermometer (Fig. 5) using the data on the compositions of the coexisting Bi and Gr (Perchuk, 1971a) from Equations (47) to (50). This diagram is self-explaining. Together with the diagrams in Figures 5 and 12 it provides information on the PT conditions of mineral equilibria in metapelites.

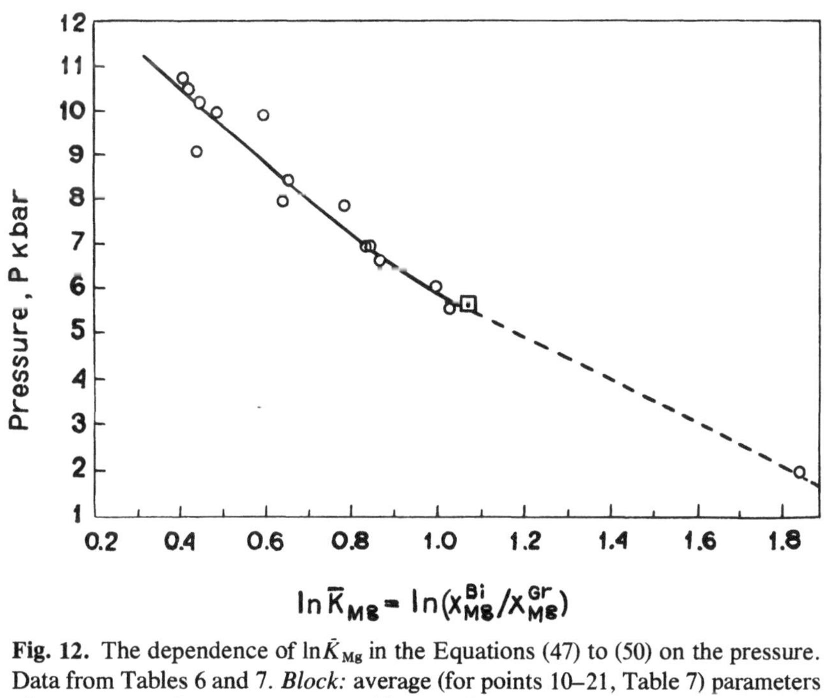

Fig. 12. The dependence of ln$\bar{K}_{Mg}$ in the Equations (47) to (50) on the pressure. Data from Tables 6 and 7. *Block:* average (for points 10–21, Table 7) parameters of parageneses from one metamorphic complex

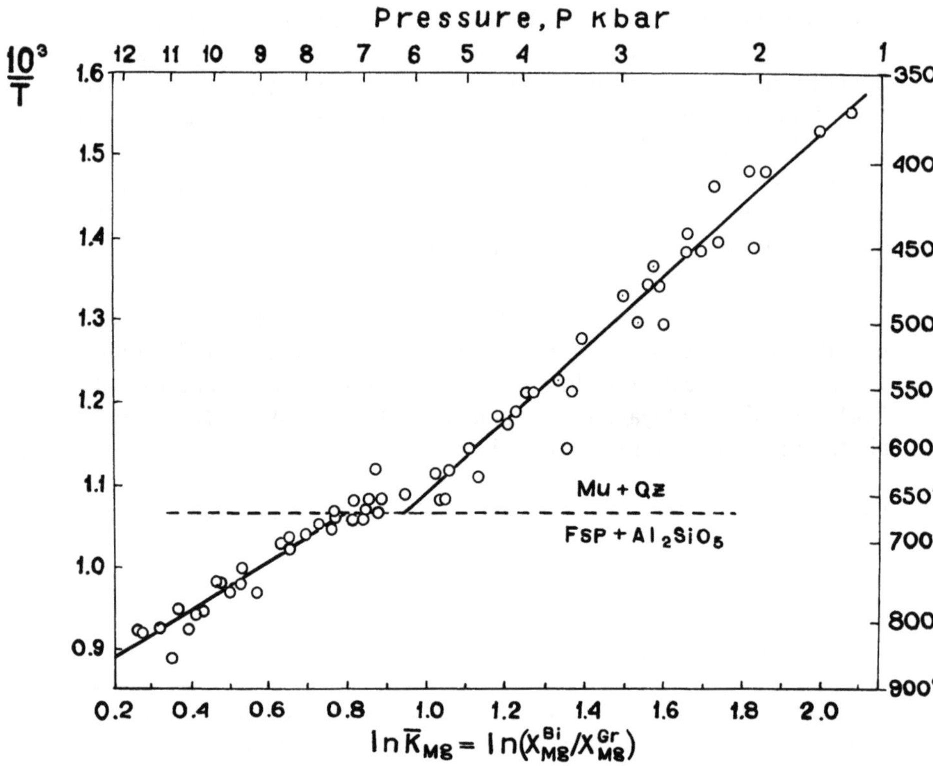

Fig. 13. The dependence of ln $\bar{K}_{Mg}$ on inverse temperature in Equations (47) to (50)

3.6 Staurolite–Garnet Equilibria as Geobarometers

Two staurolite–garnet equilibria were found very interesting and important:

$$Gr_{Fe} + St_{Mg} = St_{Fe} + Gr_{Mg} \tag{52}$$

and

$$St + Qz + Mu = Bi + Gr + Al_2SiO_5 + H_2O. \tag{53}$$

The first of these equilibria only depends on temperature and mineral composition, the second on T, P_s, and μ_{H_2O}. The following relationship of ln $K_{D(52)}^{Mg}$ to reverse temperature was obtained by 36 points for the parageneses Bi + Gr + St with the known mineral compositions (Perchuk, 1969b, 1970a; Perchuk and Fed'kin, 1974; Fed'kin, 1975), using the diagram in Figure 5

$$\ln K_{D(52)}^{Mg} = \ln \frac{X_{Mg}^{St}(1 - X_{Mg}^{Gr})}{X_{Mg}^{Gr}(1 - X_{Mg}^{St})} = \exp \frac{\Delta G_{(52)}}{RT} = \frac{3.425}{T(K)} - 3.974, \tag{54}$$

where $\Delta G_{(52)}$ is the change in the Gibbs free energy in Equation (52). From the compositions of the associations Bi + Gr + Al_2SiO_5 + Mu + Qz coexisting with staurolite, we can try to establish the relationship of $\ln \bar{K}_{Mg}^{(53)} = \ln(X_{Mg}^{St}/X_{Mg}^{Gr})$ to the pressure and temperature on the assumption that P_{H_2O} depends on these parameters. From the data on 19 such parageneses (Table 8), it was found in a very rough approximation that P_s (kbar) = 6.2 − 4 ln $\bar{K}_{Mg}^{(53)}$. More accurate theoretical analysis of Equation (53), which accounted for the thermodynamic properties of the polymorph varieties of Al_2SiO_5 and revealed a break in the isolines ln $\bar{K}_{Mg}^{(53)}$. in the *PT* diagram, near the phase transition lines (Fig. 14). For example, in Equation (55) as presented below, the derivative $(\partial P_s/\partial \mu_{H_2O})_T$ changes sharply both its meaning and sign, depending on the Al_2SiO_5 structure. This is determined by the large number of moles of alumina silicate in Equation (55) and by the sign-variable quantity $\Delta V_{(55)}$.

$$1.4KAl_3Si_3O_{10}(OH)_2 + 14.625SiO_2 + 9.25Mg_{0.4}Fe_{1.6}Al_9Si_4O_{23}(OH) \\ = 1.4Mg_{1.25}Fe_{1.25}Al_2Si_{2.5}O_{10}(OH)_2 + 5Mg_{0.39}Fe_{2.61}Al_2Si_3O_{12} \\ + 37.325Al_2SiO_5 + 4.625H_2O. \qquad (55)$$

Table 8. *TP* conditions of the equilibrium St + QZ + Mu = Bi + Gr + Al_2SiO_5 + H_2O

No.	Mg:(Mg + Fe + Mn)* Bi	Gr	St	$\ln \frac{X_{Mg}^{St}}{X_{Mg}^{Gr}}$	$\ln \frac{X_{Mg}^{Bi}}{X_{Mg}^{Gr}}$	*t* (°C)	*P* (kbar)	Al_2SiO_5	$\ln K_D^{Mg}$
1	0.667	0.169	0.294	0.552	1.373	490	4.2	Ky	0.717
2	0.686	0.163	0.306	0.624	1.437	470	3.9	Ky	0.814
3	0.645	0.157	0.275	0.561	1.413	490	4.0	?	0.734
4	0.400	0.111	0.155	0.339	1.282	600	4.6	Sil	0.385
5	0.485	0.088	0.189	0.764	1.707	450	2.6	Ky	0.875
6	0.761	0.306	0.370	0.190	0.910	560	6.5	Ky	0.288
7	0.499	0.137	0.212	0.437	1.292	555	4.55	Sil	0.527
8	0.392	0.095	0.143	0.409	1.417	550	4.2	Ky	0.462
9	0.533	0.132	0.216	0.492	1.395	520	4.1	Ky	0.594
10	0.507	0.146	0.190	0.263	1.245	560	4.8	Ky	0.316
11	0.460	0.153	0.168	0.094	1.100	620	5.5	Ky	0.111
12	0.446	0.154	0.162	0.050	1.064	625	5.6	Ky	0.060
13	0.452	0.160	0.168	0.050	1.038	625	5.7	Ky + Sil	0.058
14	0.471	0.144	0.164	0.130	1.145	600	5.25	Ky + sil	0.154
15	0.452	0.091	0.164	0.589	1.603	500	3.1	And	0.672
16	0.506	0.113	0.198	0.561	1.499	500	3.6	Ky	0.662
17	0.529	0.133	0.217	0.490	1.544	520	3.4	Ky	0.591
18	0.455	0.150	0.158	0.050	1.110	615	5.35	Ky	0.061
19	0.518	0.174	0.201	0.144	1.091	600	5.5	Ky	0.190

*Data from Perchuk, 1971a; Fed'kin, 1975.

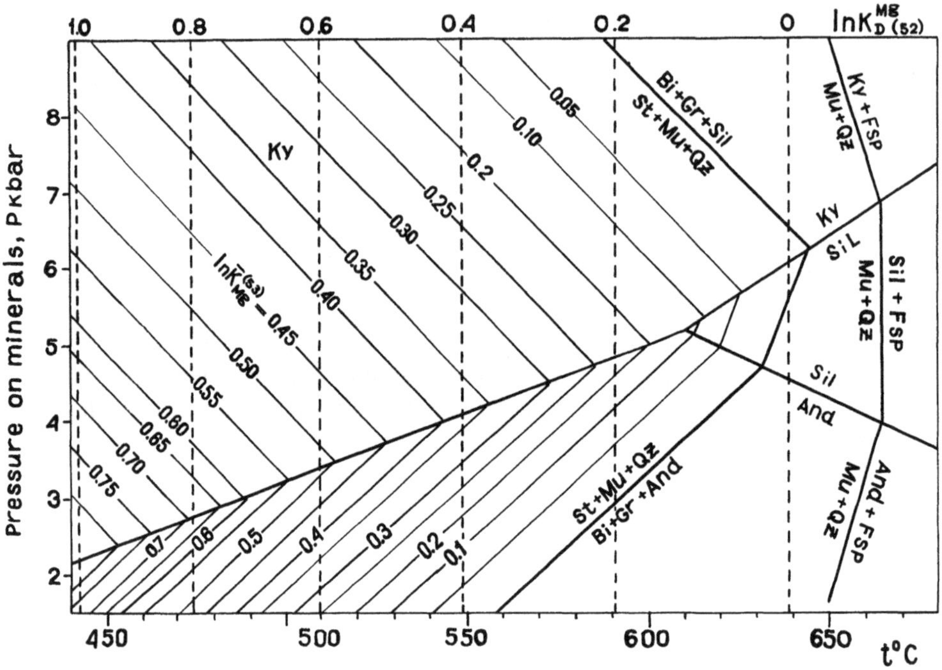

Fig. 14. The $\ln K_{D(52)}^{Mg} - \ln \bar{K}_{Mg}^{(53)}$ relationship in the PT diagram, which may serve as a mineralogical thermobarometer

The diagram in Figure 14 corresponds approximately to the actual ratio $(\partial P_s/\partial T)$ at some arbitrary but P_s - dependent value of

$$\mu_{H_2O} = \int_{P_1}^{P_2} V_{H_2O}\, dP_{H_2O}.$$

3.7 Orthopyroxene–Quartz–Olivine Geobarometer

The Opx + Ol + Qz association occurs frequently in alumina-free but FeO rich high-temperature silicate rocks (eulisites). It was felt that the decomposition of Opx to Ol and Qz in iron-rich compositions must depend strongly on the pressure, because, all other things being equal, the magnesium olivine is not stable with quartz. That is why this equilibrium has attracted investigators' attention. Smith (1971) studied this equilibrium as a function of pressure at 900°C, and Wood and Strens (1971) revealed the dependence of this reaction on temperature under the atmospheric pressure. The experiments showed that with increasing P and T the Fe/Mg increases in equilibrated Ol and Opx. This implies, that the reaction

$$(0.2 - n)Fe_{0.7}Mg_{0.3}SiO_3 + 0.5n(Fe_{1.8}Mg_{0.2})SiO_4 + 0.5nSiO_2 \rightarrow 0.2Fe_{0.7+n}Mg_{0.3-n}SiO_3 \quad (56)$$

shifts to the right as P and T increase.

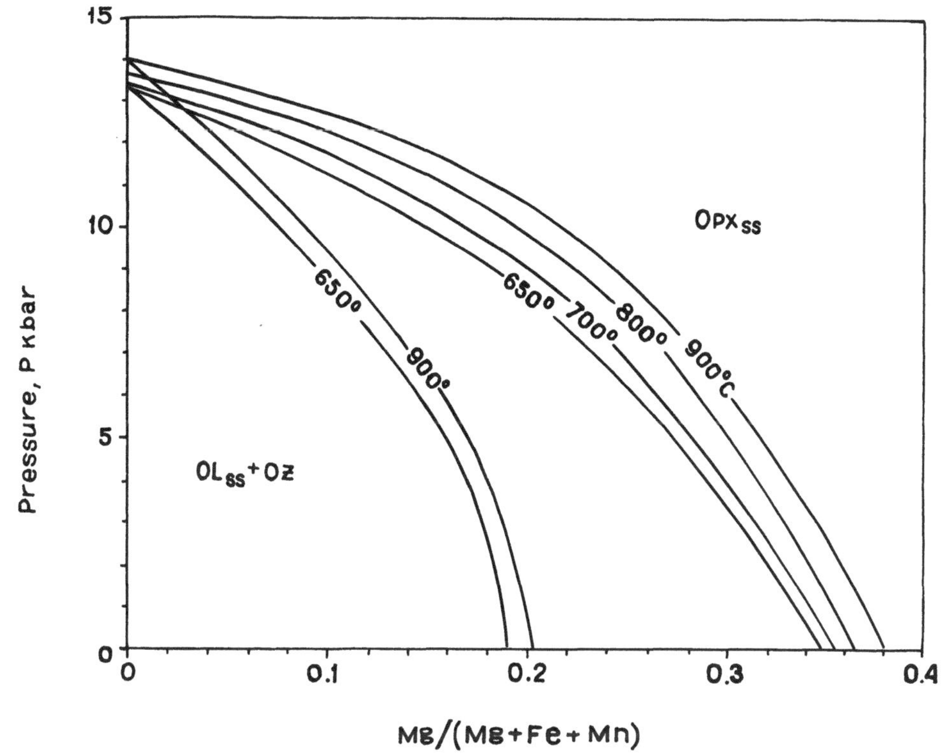

Fig. 15. The PX_{Mg} diagram for the equilibrium $Opx_{ss} = Ol_{ss} + Qz$ for 650–900°C

We treated these experimental data thermodynamically and used the P_s and t values for the Opx + Gr and Bi + Gr + Qz parageneses in iron silicate rocks. The results (Perchuk, 1973) are presented in Figure 15 which can also be used for the purpose of geothermobarometry. We discussed here several diagrams of phase relationship–mineralogical thermometers and barometers for metamorphic rocks. Earlier works translated into English in the USA (Perchuk, 1968b, 1969a), present a sufficiently large number of similar diagrams for various types of crystalline rocks.

3.8 Geothermometers for Evaluating the Liquidus and Solidus Temperatures of Basaltoids

Our recent investigations (Perchuk and Vaganov, 1977) enabled us to attempt the evaluation of liquidus and solidus temperatures of mafic and ultramafic rocks. This information is very important for distinguishing the magmatic formation of rocks from metamorphic formation. This applies primarily to slightly changed metavulcanites.

If the liquidus of natural basalt or gabbro or of some other rocks contains Ol of know composition, the initial crystallization temperature may be found from:

$$t(°C) = 1039.62 + 43.7368X_{Mg}^{Ol} + 353.86y + 0.001182(100X_{Mg}^{Ol})^2 + 0.125186y(100X_{Mg}^{Ol})^2 - 0.000031y(100X_{Mg}^{Ol})^3, \quad (57)$$

where $y = (Mg + Fe + Mn)/0.50$ under natural conditions, and the Fo molar content in Ol can be established from an approximate empirical equation:

$$X_{Mg}^{Ol} = 3.15X_{Mg}^{rock}/(1 + 2.15X_{Mg}^{rock}), \quad (58)$$

obtained on the basis of Roeder's magnificent experimental data (1974). Using Equation (57), we obtained the liquidus temperatures for several hundred specimens of magmatic rocks from various differentiation series (Kuno, 1968) and we also demonstrated the shift in the cotectics with the pressure (depth) during initiation of crystallization in the diagram for alkali–alumina. If the compositions of the coexisting Opx and Cpx are known, their equilibrium temperatures can be found from the diagrams (Perchuk, 1969c, 1973) or from equation

$$t(°C) = 10^3/b - 273, \quad (59)$$

where $b = 2.525658 - 3.615266\bar{X}_{Mg} + 0.089303 \log \bar{K}_{Mg} + 1.595062(\bar{X}_{Mg})^2 - 1.43(\bar{X}_{Mg})^2 \log \bar{K}_{Mg} + 0.633388(\bar{X}_{Mg})^3 \log \bar{K}_{Mg}$ and where $\bar{X}_{Mg} = 0.5\,(X_{Mg}^{Opx} + X_{Mg}^{Cpx})$ and $\bar{K}_{Mg} = C_{FeO}^{Cpx}/C_{FeO}^{Opx}$, C_{FeO} is the concentration of FeO (in wt%) in this mineral.

This brings to an end the discussion of mineralogical thermometers and barometers, derived on the basis of phase correspondence. I do not evaluate their absolute accuracy; this is the purpose of the experimental studies under way now in a special laboratory of the Institute of Experimental Mineralogy of the USSR Academy of Science. However, future results must account for their good correlation (relative accuracy is ~ ±10%). The validity of these P and t estimates obtained from the phase diagrams will be verified in future studies.

4 THE METHODS OF EVALUATING FUGACITY AND PARTIAL PRESSURE OF VOLATILE COMPONENTS IN FLUIDS

4.1 Thermodynamic Models

Fugacities of volatile components in natural fluids can be estimated by two methods. *The first* involves compositional determinations of the primary gas–liquid inclusions in the rocks, as well as collection of data on the composition of volcanic gases during modern eruptions. From the bulk composition of the gaseous mixture and *PT* conditions of mineral equilibria the fugacities and partial pressures of the components of the gaseous mixture can be evaluated with confidence. *The second method* is by calculating gas-mineral equilibria. To me, this is the only possible method, because there are no sufficient data on the composition of gas-liquid inclusions in metamorphic and magmatic rocks at various depths. Therefore, we shall discuss here the methods of calculating gas–mineral equilibria.

Two models of calculation are possible. According to the first model (Marakushev, 1968; Perchuk, 1970a, 1973) at given T and P_s the fluid is monocomponent, with $P_s \geqslant P_i^0$. In this case the thermodynamic equilibrium is specified by the equation, obtained by integrating the characteristic function. Gibbs thermodynamic potential. For the reaction proceeding with the intake of gas component i, this equation takes up the form:

$$\mu_i = RT \ln f_i^0 = \int_1^{P_i=P_s} V_i \, dP_i^0 = \Delta G_T^0 + \Delta V_s P_s, \tag{60}$$

where μ_i is the relative thermodynamic potential of component i at the given T and P_i^0; f_i^0 is the fugacity of this component and P_i^0 is its pressure; ΔG_T^0 is the change in Gibbs free energy at given T and $P = 1$ bar; ΔV_s is the change in volume in the course of this reaction on condition that the volume does not depend on T and P_s. It is evident that the coefficients of isobaric expansion and isothermal compression have opposite signs and their increments will not change ΔV_s greatly with T and P_s. Therefore, if we know these parameters and free energies of formation of minerals and gases involved in this reaction, the calculation of f_i^0 is not difficult:

$$f_i^0 = \exp\left(\frac{\Delta G_T^0 + \Delta V_s P_s}{RT}\right). \tag{61}$$

This model is very well known: it is generally used in estimating the fugacities of O_2 in oxidation reactions. For the simplest reactions of this type, say, $Fe + \frac{1}{2}O_2 = FeO$ or $3Fe + 2O_2 = Fe_3O_4$, etc., the $f_{O_2}^0$ values are identically equal to the constants of these reactions at given T and P_s.

To calculate the hydration reactions one has either to know the mineral constants, or have experimental data for $P_s = P_{H_2O}^0$ (Marakushev, 1968). Thermodynamics of water steam has been thoroughly studied (Burnham et al., 1969) up to 1000°C and 10 000 bar, therefore the estimation of $f_{H_2O}^0$ and $P_{H_2O}^0$ for many reactions does not present any difficulty.

To calculate the carbonation reactions one needs the data on the *PVT* properties of CO_2 over a wide range of temperature and pressure. We obtained the necessary data (Perchuk, 1973; Perchuk and Karpov, 1975) by linear extrapolation of the molar CO_2 volumes to the high temperature region. Later findings showed that the extrapolation was within the experimental accuracy (Shmonov and Shmulovich, 1974). Table 9 presents the $\gamma^0_{CO_2}$ fugacity coefficients for $t = 400–1000°C$ and $P^0_{CO_2} = 500–10\,000$ bars. Using these values we can find $f^0_{CO_2} = \gamma^0_{CO_2} P^0_{CO_2}$ and $\mu^*_{CO_2}$. We calculated all thermodynamic properties (S_i, V_i, μ_i, etc.) of some other gases, too, H_2, N_2, CH_4, and CO (Perchuk, 1973). At the same time and independently of our studies Melnik (1972) calculated the fugacity coefficients for the gases on the basis of the law of corresponding states. He obtained γ^0_i values which were very similar to ours.

The natural fluids are not generally monocomponent, and the above model may appear inaccurate. Therefore, we must choose another model which would make allowances for multicomponent composition of a fluid, but not accounting for the rockforming components (SiO_2, K_2O, Na_2O, Al_2O_3, etc.) dissolved in this fluid.

We can assume for this model the ideality of the gaseous mixture and that $P_s = P_{fl} \geqslant \sum_i P^{fl}_i$ at constant T. Then the calculation of gas–mineral equilibria will be done by:

$$\sum_i n_i \mu^*_i \approx \Delta G^0_T + \Delta V_s P_{fl},$$

where n_i is the number of moles of the component i participating in the reaction. For example, if $P_{fl} = P_{CO_2} + P_{H_2O}$, this equation, as calculated per one mole CO_2 becomes:

$$\mu^*_{CO_2} = \Delta G^0_T + \Delta V_s P_{fl} - n_{H_2O}\mu_{H_2O} \tag{62}$$

It is generally used in calculating hydration and carbonation reactions. If only hydration takes place in the presence of the water–carbon dioxide fluid, equation (60) can be used, but $P^0_{H_2O}$ will be obtained from $\gamma^0_{H_2O}$ for a given P_{fl}. This is particularly important for the components of a gas mixture with large fugacity coefficients. Thus, the CO_2 partial value in the gaseous mixture can be obtained from:

$$P^{fl}_{CO_2} = \exp\left[\frac{\Delta G^0_T + \Delta V_s P_{fl}}{RT} - \ln \gamma^{fl}_{CO_2}\right], \tag{63}$$

where $\gamma^{fl}_{CO_2}$ is the CO_2 fugacity coefficient at given T and P_{fl}. If water becomes involved in the reaction together with CO_2, $P^{fl}_{H_2O}$ and $P^{fl}_{CO_2}$ are found by solving two equations with two unknowns (if $n_{H_2O} = n_{CO_2} = 1$):

$$\begin{aligned} P^{fl}_{CO_2} + P^{fl}_{H_2O} &= P_{fl} \\ RT(\ln \gamma^{fl}_{CO_2}\gamma^{fl}_{H_2O} + \ln P^{fl}_{CO_2} + \ln P^{fl}_{H_2O}) &= \Delta G^0_T + \Delta V_s P_{fl}. \end{aligned} \tag{64}$$

Table 9. Carbon dioxide fugacity coefficient γ_{CO_2} at elevated temperatures and pressures

P	Temperature (°C)									
	100	200	300	400	500	600	700	800	900	1000
100	0.729	0.885	0.980	1.031	1.057	1.069	1.075	1.077	1.079	1.082
200	0.649	0.850	0.971	1.037	1.072	1.090	1.099	1.103	1.106	1.109
300	0.599	0.831	0.970	1.049	1.091	1.112	1.123	1.129	1.132	1.135
400	0.564	0.820	0.977	1.065	1.113	1.138	1.150	1.157	1.160	1.163
500	0.533	0.817	0.989	1.086	1.139	1.167	1.180	1.187	1.190	1.193
600	0.513	0.818	1.004	1.111	1.168	1.198	1.213	1.220	1.223	1.225
700	0.502	0.822	1.024	1.139	1.201	1.233	1.248	1.255	1.258	1.259
800	0.490	0.831	1.047	1.171	1.237	1.271	1.287	1.293	1.295	1.295
900	0.481	0.842	1.073	1.206	1.276	1.312	1.328	1.333	1.334	1.333
1000	0.473	0.856	1.103	1.244	1.319	1.356	1.371	1.376	1.376	1.374
1100	0.468	0.873	1.136	1.286	1.365	1.403	1.418	1.422	1.420	1.417
1200	0.465	0.892	1.172	1.331	1.414	1.453	1.468	1.470	1.467	1.462
1300	0.463	0.914	1.211	1.380	1.467	1.506	1.521	1.522	1.517	1.510
1400	0.462	0.938	1.254	1.432	1.523	1.564	1.577	1.576	1.569	1.560
1500	0.494	0.991	1.315	1.493	1.582	1.619	1.629	1.625	1.614	1.603
2000	0.708	1.329	1.688	1.859	1.924	1.934	1.918	1.891	1.860	1.831
2500	1.046	1.822	2.205	2.346	2.366	2.330	2.273	2.211	2.150	2.095
3000	1.572	2.540	2.925	3.044	2.949	2.844	2.727	2.614	2.511	2.421
3500	2.370	3.548	3.889	3.857	3.690	3.488	3.292	3.115	2.961	2.830
4000	3.608	4.994	5.199	4.974	4.633	4.291	3.984	3.720	3.497	3.311
4500	5.527	7.053	6.965	6.420	5.818	5.275	4.816	4.436	4.123	3.866
5000	8.492	9.973	9.329	8.278	7.293	6.470	5.805	5.272	4.844	4.497
5500	13.067	14.095	12.476	10.649	9.115	7.909	6.972	6.242	5.667	5.209
6000	20.109	19,898	16.649	13.660	11.355	9.633	8.341	7.359	6.601	6.005
6500	30.927	28.042	22.166	17.474	14.101	11.692	9.942	8.643	7.658	6.895
7000	47.515	39.449	29.444	22.293	17.460	14.147	11.812	10.116	8.853	7.889
7500	72.906	55.400	39.033	28.380	21.569	17.077	13.998	11.811	10.208	9.002
8000	111.709	77.685	51.665	36.072	26.603	20.579	16.561	13.765	11.751	10.254
8500	170.929	108.811	68.318	45.808	32.785	24.782	19.580	16.033	13.518	11.674
9000	261.219	152.305	90.311	58.169	40.410	29.851	23.158	18.684	15.560	13.299
9500	398.785	213.151	119.435	73.926	49.863	36.007	27.434	21.811	17.944	15.179
10000	608.314	298.430	158.141	94.118	61.663	43.540	32.588	25.536	20.757	17.382

This system can easily be solved both graphically (on the P_{fl} isobar in the isothermal section of the diagram $\mu^*_{H_2O}$ $\mu^*_{CO_2}$) and analytically (using a computer). However, Equation (64) does not account for the composition of the fluid involved in natural mineral-forming processes: composition of the fluid is generally more complicated. Therefore, hydration-dehydration and carbonation-decarbonation reactions which go separately have to be chosen. In this case, assuming ideal mixing for the gaseous components, we can compute H_2O and CO_2 partial pressures. The difference between $P_{fl} \approx P_s$ and $(P^{fl}_{H_2O} + P^{fl}_{CO_2})$ makes up the sum of partial pressures of the more reduced components in the fluid (H_2, CO, CH_4, N_2, etc.).

4.2 Examples of the P_{CO_2}, P_{H_2O} and f_{O_2} Calculation

Thus we have discussed two different models for evaluating the pressure of gaseous components. The first of these models describes the behaviour of monocomponent fluid ($P_s \geqslant P_i$), while the second ($P_s \geqslant P_{fl} = \sum_i P^{fl}_i$) refers to the multicomponent fluid. It is clear that it is easier to determine the P^0_i by using the first model. It was shown earlier (Perchuk, 1976) that simple equation makes the transition to the second model quite easily:

$$P^{fl}_i = \frac{f^0_i}{\gamma^{fl}_i} = \frac{P^0_i \gamma^0_i}{\gamma^{fl}_i}, \tag{65}$$

where γ^0_i is the fugacity of the monocomponent gas i at given T and $P^0_i \leqslant P_s$, and γ^{fl}_i is the fugacity coefficient for the same gas with the partial pressure P^{fl}_i in the mixture where $P_{fl} \approx P_s$ at the same temperature. For example, using the first model [Eq. (60)] we found that at $t = 800°C$ and $P = 10$ kbar the CO_2 pressure is $P^0_{CO_2} = 7000$ bar, $\gamma^0_{CO_2} P^0_{CO_2} = f^0_{CO_2} = 10.116 \times 7000 = 70\,812$ bar; see Table 9). We have to find $\gamma^{fl}_{CO_2}$ at $P_{fl} = P_s = 10\,000$ bars before we can go over to the second model (Table 9) for 800°C and this pressure gives $\gamma^{fl}_{CO_2} = 25.536$. Then from Equation (65) we find:

$$P^{fl}_{CO_2} = \frac{70\,812}{25.536} = 2773 \text{ bars.}$$

All calculations were carried out by this method, because it makes it possible to evaluate the fluid regime from the two specified models. This is particularly important for the cases where the $P_s \approx P_f$ condition is actually met, and the fluid approaches ideal gas mixture.

Next some examples of calculating hydration-dehydration and carbonation-decarbonation reactions are presented and then the P_{H_2O} and P_{CO_2}. will be estimated. We selected several dozens of mineral parageneses from silicate-carbonate rocks and granites over a wide range of T and P.

Metamorphic rocks of the Aldan shield (Eastern Siberia) show three depth facies: Sutam facies, Iengrsk facies and Aldan facies, and the temperatures of mineral equilibria change accordingly (Marakushev, 1965; Perchuk, 1969c, 1970a). In the first two facies grossular occurs instead of An + Cal + Qz. The garnet from a carbonate–silicate rock, has the following unit cell parameters: $a = 11.849 \pm 0.001$Å, $V = 1663.585$ Å^3; the refraction index is 1.740. These parameters correspond to the composition $Gross_{90}Andr_{10}$. Wollastonite and garnet with variable grossular content occur widely in the Aldan formation. Wollastonite occurs sometimes in the paragenesis with Qz and Cal. We have numerous chemical analyses for coexisting minerals from all these rocks. Their compositions and P and t estimates are available in other papers (Perchuk, 1971a, 1973, 1975), and the following are refined values (average values):

Facies	t(°C)	P(kbar)
Sutam	825 ± 20	11 ± 1
Iengrsk	750 ± 30	10 ± 0.5
Aldan	700 ± 20	7 ± 0.5

Using the reference data on the thermodynamic constants of minerals (Karpov et al., 1971) and the CO_2 fugacity coefficients from Table 9, we calculated the following reactions

$$\text{Gros} + CO_2 = \text{An} + \text{Cal} + \text{Qz} \tag{66}$$

$$\text{Wol} + CO_2 = \text{Cal} + \text{Qz}. \tag{67}$$

The results of calculations by the first model [Eq. (60)] are plotted on the diagram in Figure 16. The CO_2 partial pressure in a fluid can be readily obtained for Equation (65). They are ~2250 bar for the Sutam facies, ~1260 bar for the Iengrsk facies and ~690 bar for the Aldan facies. It is easy to see that the P_{CO_2} is two to three times higher than $P^{fl}_{CO_2}$. Besides, if $P_{fl} \approx P_s$, then $P_{fl} - P^{fl}_{CO_2} \approx 8.5$ kbar in case of the Sutam facies, for example. It means that other components (H_2O, CO, CH_4, N_2, H_2, F_2, Cl_2, etc.) account for more than 75% of the fluid composition. Now metamorphic condition of the terrigene rock masses, corresponding to the Aldan depth facies have been studied thoroughly.

We used empirical data on rock parageneses from Connecticut USA (Lundgren, 1966), Darjeeling (Sen and Chakraborty, 1968) and Nanga-Parbat (Misch, 1964), regions in Himalii, and southwest Pamir (Zotov, 1967) in our calculations. The characteristic feature of aluminosilicate rocks from most of these regions is the occurrence of the Mu + Qz + Fsp + Ky + Sil + Bi + Gr paragenesis. This makes it possible to evaluate the PT conditions of metamorphism and to find $P^{fl}_{H_2O}$ from the equilibria Mu + Qz = Fsp + Al_2SiO_5 + H_2O. Moreover, according to Misch's data (1964) a very interesting metamorphic zoning shows up in

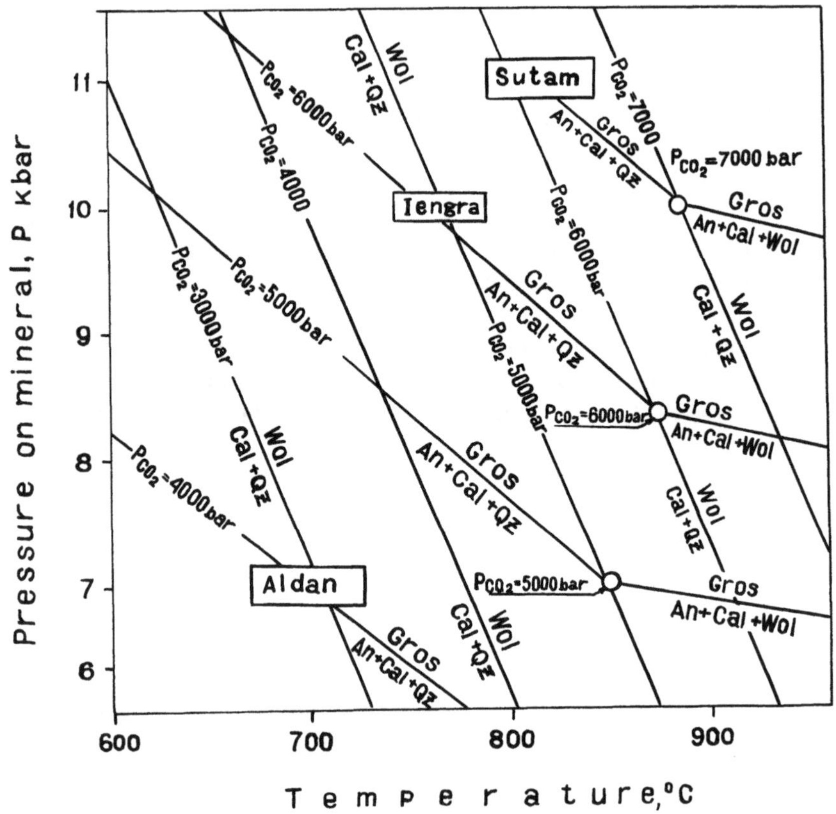

Fig. 16. Metamorphic conditions of the rocks from the Aldan shield and its southern provinces. The calculation was done by Equation (60)

carbonate–silicate rocks of Nanga-Parbat, with the parageneses appearing in the following sequence: (1) An + Qz + Wol; (2) Cal + Wol + Gros or An + Qz + Gros; (3) Cal + An + Qz or Cal + Qz + Gros and (4) Cal + Zo + Qz.

The reactions responsible for the zoning are shown in Figure 17, and their importance can hardly be overestimated, because they are indispensible for separate and simultaneous evaluation of the H_2O and CO_2 partial pressure in the fluid. However, we have no data on the compositions of the coexisting minerals in these rocks, though their conditions of metamorphism appear to be similar to the conditions, established for the formations of Darjeeling, south-west Pamir and Connecticut from the great body of data: $P \approx 6.5 - 7$ kbar and $t = 650$–700°C. Having accepted this hypothesis, we calculated by Equation (60) a number of equilibria involving H_2O and CO_2 characteristic of the sequence metamorphic zone in Figure 17. The temperatures of metamorphism in Figure 17 were obtained from the analysis of Figure 18. This diagram gave us P_{CO_2} and P_{H_2O}. Now we can find the CO_2 and H_2O partial pressures, using equation (65): $P^{fl}_{CO_2} = 350$–600 bar, $P^{fl}_{H_2O} = 2400$–2470 bar. The

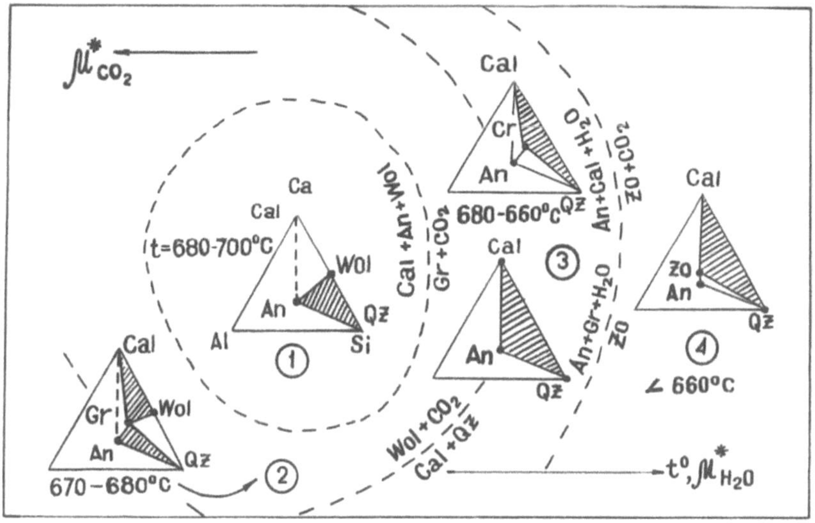

Fig. 17. Paragenetic scheme of the metamorphic zoning sequence in the carbonate–silicate rocks from Nanga-Parbat, Himalaii. According to Misch (1964) and Gordon and Greenwood's data (1971)

first value is very close to $P^{fl}_{CO_2}$ = 690 bar, of the Aldan facies ($P_s \approx 7$ kbar). And again, $P^{fl}_{CO_2} + P^{fl}_{H_2O}$ is two to three times lower than $P_{fl} \approx P_s$, which means that the composition of the fluid is more complex in these zones of metamorphism as well.

The position of the Ladoga metamorphic formation in which several parageneses with Mu + Qz + Sil + And + Fsp + Bi + Gr occur is plotted in Figure 18. We could not detect wollastonite, but according to Glebovitskii (personal communication), a paragenesis Woll + Cal + Qz was found in the same formation in Finland. On the western shore of Lake Ladoga there is only one stable association Cal + Qz, and therefore the $P^{fl}_{CO_2}$ must be a little higher here. The diagram in Figure 18 and Equation (65) give 590 bar at $P^{fl}_{H_2O}$ = 3097 bar. In this the sum of partial pressures approaches $P_{fl} \approx P_s$. We shall take into account this very important finding in discussing the fluid regime of metamorphism in the next section of the paper. The metapelites in the south of central Connecticut, USA, contain thin interlayers and lenticles of mica-calcareous rocks. The mineralogy of this, so-called Wepawaug formation has been studied very thoroughly by Hewitt (1973a, 1973b). He made a detailed study of the parageneses and mineral compositions on the contacts of these rocks with metapelites, and also of the reaction Mu + Cal + 2Qz = An + Fsp + H_2O + CO_2 over a wide range of thermodynamic parameters and H_2O/CO_2 ratios in the fluid. Experiments and paragenetic analyses of the mineral associations enabled him to estimate thermodynamic conditions of metamorphism in this region:

$P_{fl} = 5.5 - 6$ kbar, $t = 510 - 590°C$, $X^{fl}_{CO_2} \approx 0.1$.

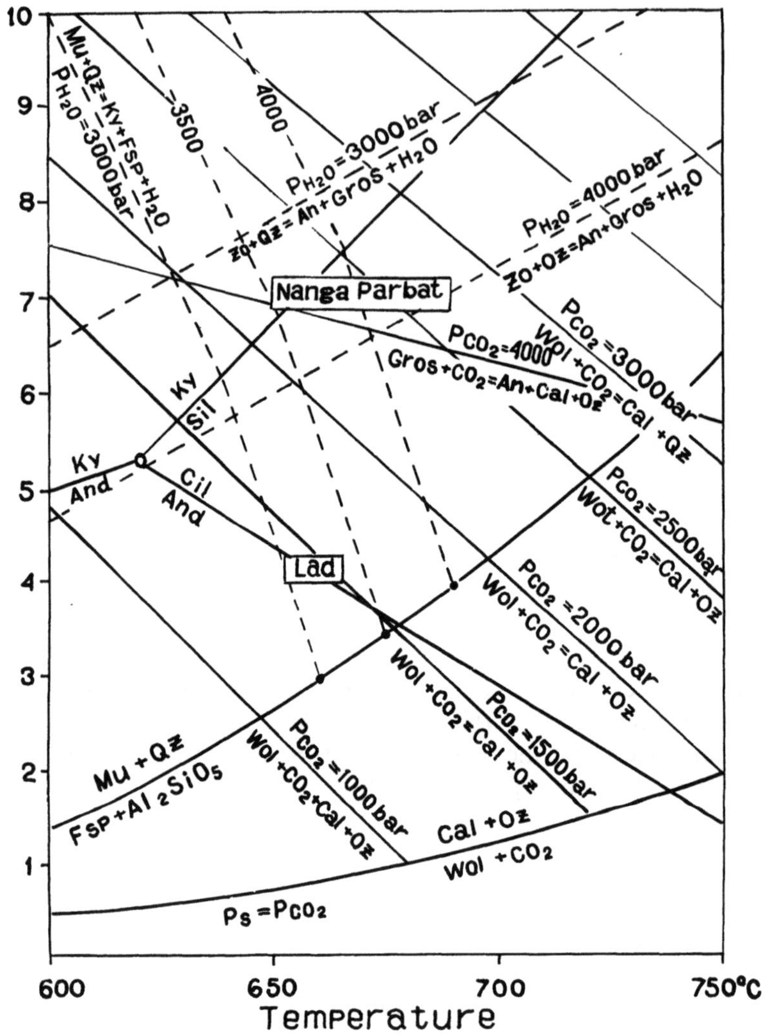

Fig. 18. Thermodynamic conditions of mineral equilibria in the Darjeeling, Nanga-Parbat, Connecticut and Ladoga formation. Calculation by Equation (60)

Using Hewitt's data (1973b) on the compositions of the coexisting phases, we can estimate the *TP* parameters of mineral equilibria and calculate $P^{fl}_{H_2O}$ and $P^{fl}_{CO_2}$. We obtained t = 525°C and P = 4.8 kbar for the sample Wep–16c using the biotite–garnet equilibrium (Figs. 5 and 12), and the average value t = 650°C for the sample Wep–8g using three mineralogical thermometers (Bi–Am, Am–Cpx, Am–Pl, and Bi–Cpx; Perchuk, 1969a). We correlated our measurements with the result of the gas–mineral equilibrium estimates in Figure 19. The intersection of monovariant lines at different temperatures yields the point with 5.1 kbar. The H_2O molar contents in the fluid were found graphically. For the first sample $X^{fl}_{H_2O}$ = 0.873 at 525°C, and for the second sample it was 0.863, that is, $P^{fl}_{H_2O}$ = 4.4 to 4.45 kbar. Although our estimates

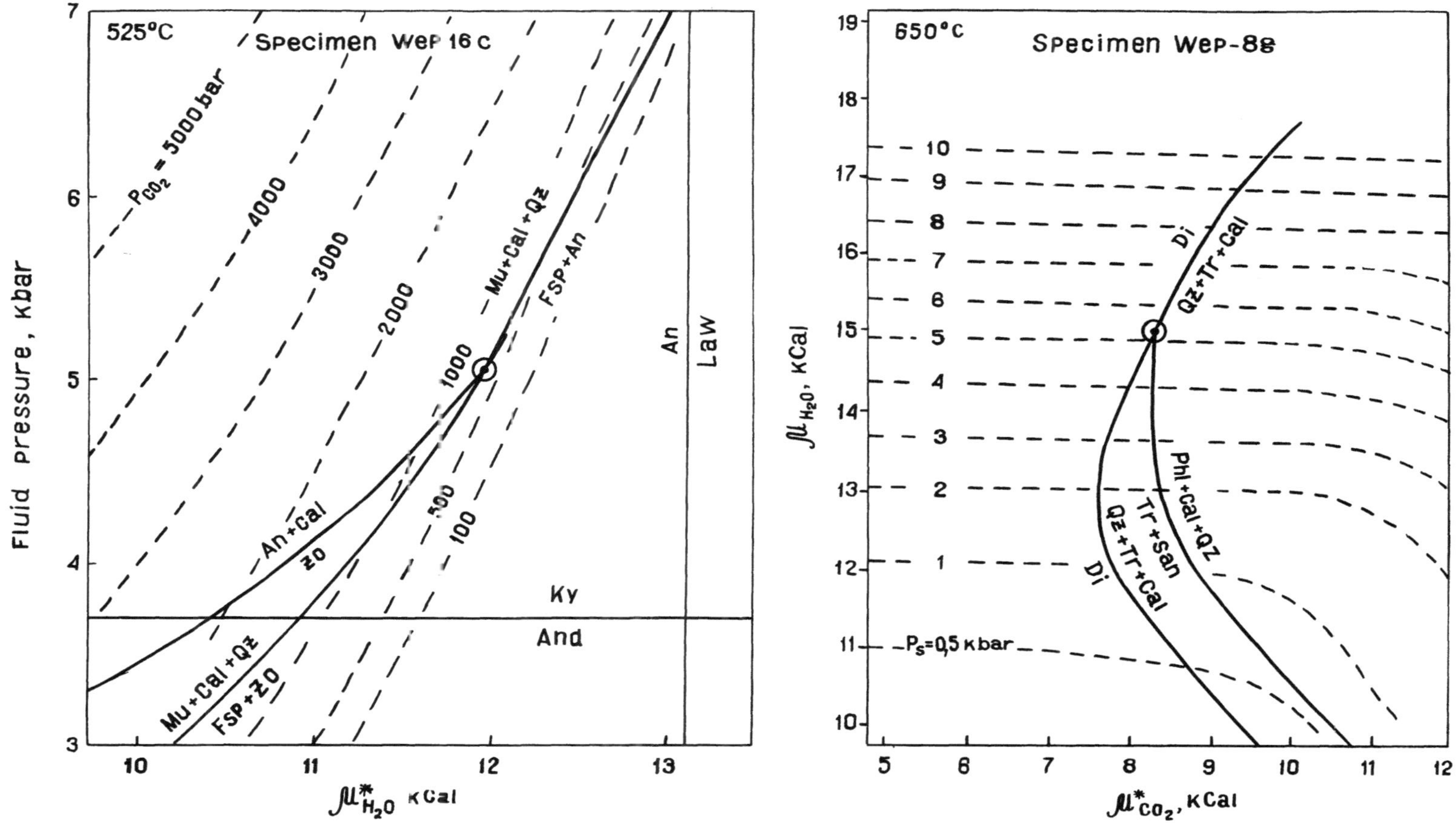

Fig. 19. Thermodynamic metamorphic conditions of the rocks from the Wepawaug formation, as evaluated from mineral compositions in the samples Wep-16c and Wep-18 (Hewitt, 1973a and by reactions using thermochemical data (Karpov et al., 1971)

approach Hewitt's data (1973b), they are not quite accurate because they were based on bicomponent fluid composition. Meanwhile, these calculations revealed the possibilities in the methods used to calculate partial pressures of volatile components and *PT* conditions in the gas-mineral equilibria in metamorphic rocks.

Metamorphism is intimately related to the granitization processes in the broad sense of the term (ultrametamorphism, migmatization, granitization, etc.). The evaluation of P_{H_2O} in these processes is, to our opinion, one of the most important petrological objectives because the increase in the H_2O pressure in the fluid accounts for the occurrence of granite melt, as the temperature of the regional metamorphism goes down

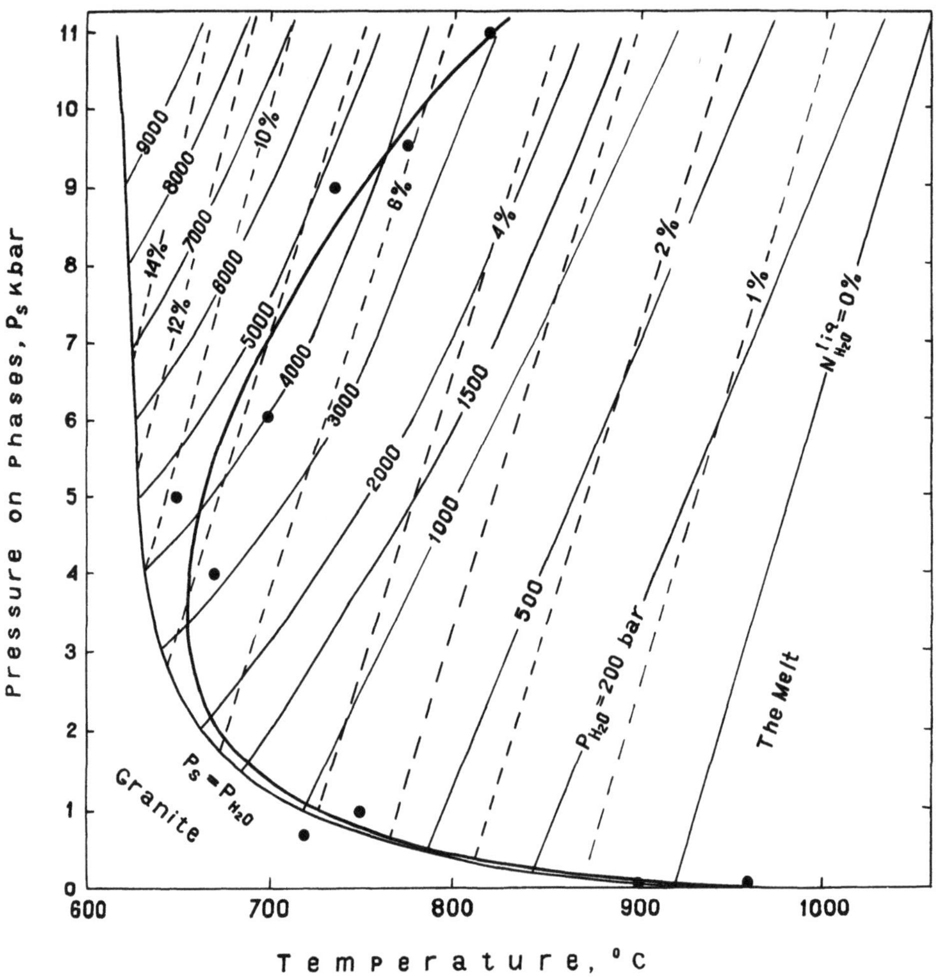

Fig. 20. A generalized melting curve for granite of eutectic composition under natural conditions with $P_{H_2O} \leqslant P$ (Perchuk, 1973, p. 258) in the P_sT diagram, with the P_{H_2O} and $N^{Liq}_{H_2O}$ isolines on the liquidus. Calculation by Equation (61)

(Perchuk, 1970b). Therefore, we tried to calculate water isobars in the P_sT diagram for the eutectic natural granite. The method of calculating these isobars (cotectic minima, or liquidus projections on the P_sT plane) was discussed in detail in a separate paper (Perchuk, 1973). Figure 20 shows these isobars, lines of the constant H_2O content in the melt and the points of granite eutectics, for which the PT-crystalization parameters were established by the methods of thermobarometry (Perchuk, 1973). Rough estimates of the water pressure during granitization and the water content near the eutectic can be obtained using the diagram in Figure 20. Heavy line in this diagram represents variations in temperature and water pressure (monocomponent fluid) with the change in the overall pressure (depth) during crystallization of the granitic melt. Equation (65) gives readily the H_2O partial pressure in the ideal gas mixture.

As stated above, calculating the redox reactions allows us to determine f_{O_2} during metamorphism. It is clear that if the paragenesis Il + Mt occurs in a rock, f_{O_2} can be found from the diagram of Buddington and Lindsley (1964). Besides, the association Mt + Qz + Fa is not unusual in eulisites and $\ln f_{O_2}$ can easily be calculated from Equation (61) at the given T and P.

Graphite-bearing rocks deserve particular notice. Their importance in the f_{O_2} evaluations has been discussed more than once in the literature (Sobolev, 1947; Miyashiro, 1964; Perchuk and Suvorova, 1973; Perchuk, 1973; Marakushev and Perchuk, 1974, etc.). The graphite oxidation reactions calculated by the same Equation (61) at the given T and P_s, can serve as a reliable criteria of the equilibum oxygen fugacity in metamorphic processes. The discussion of these will be undertaken in the next section of this paper.

5 RELATIONSHIPS BETWEEN TEMPERATURE AND FLUID REGIMES OF METAMORPHISM

Let us assume the overall pressure to be the depth parameter in metamorphic and magmatic processes and try to establish empirical relationships between P_s and the temperature and partial pressures of the volatile components in a fluid. These relationships, if there are any, will determine thermodynamic regime of rock formation in the earth crust.

5.1 The Temperature Regime

The positive correlation between P_s and t during metapelite formation (Fig. 9) was first noticed in derivation Bi–Gr, Cor–Gr, St–Gr barometers. But we have used more than ten thermometers and 6 barometers in evaluating the *PT-stabilization* parameters of rocks of the many different composition. I italicize the word "stabilization," because these parameters only reflect the thermodynamic conditions of equilibria in these rocks in any period of geological history of the given region. Irreversibility of natural processes results in that the composition ratios of the coexisting minerals "remember" these conditions. It does not imply that the given rocks have been formed at the estimated P and t values. For example, in intrusive rocks the equilibrium mineral relations are always detected at the PT conditions of solidus, or subsolidus and not of liquidus. Therefore, one necessary and sufficient condition of equilibrium fixation in rocks must be met: a well-known principle of the chemical thermodynamics according to which the rate of component diffusion between the phases must be infinitely small relative to the rate of changes in the external parameters of the system (P, T, P_i, etc.). Ramberg showed (1952a) that the rates of component diffusion in minerals in concentration gradient become negligible as the temperature decreases. Introduction of the fluid in the system increases this rate by several orders of magnitude, which makes the experimental modelling of mineral equilibria possible.

Consequently, equilibrium fixation in rocks must be concurrent with the withdrawal of pore or intragrain ("dislocation") fluid from them. This principle must be taken as the basis for devising the mineral facies system of magmatic and metamorphic rocks.

We have found the PT parameters of mineral equilibria for several hundred samples from the earth crust for which analyses of coexisting minerals are available (Perchuk, 1971a, 1976). When these data were plotted in the P–T diagram it was covered most heavily with dots, particularly in the middle and in the area of low t and high P, and, conversely, in the region of high t and low P. It would be difficult to understand anything from this diagram unless certain provisions for the specific nature of geotectonic evolution of the earth crust are made, and so we divided the samples into three "geotectonic" groups:

(1) Metamorphic belts formed in the early stages in the evolution of geocynlines, or in the zones where continental plates are underthrust by oceanic plates; here belong many prehnite–pumpellyite and lawsonite rocks, some green and all blueschists, certain metamorphic eclogites and garnet amphibolites;
(2) Metamorphic and ultrametamorphic formations, developed in the terrigene geosyncline zones are usually composed of metapelites, amphibolites, various crystalline schists, marbles and carbonate–silicate rocks;
(3) Contact-metamorphic rocks and volcanic provinces of the stabilized plates of the earth's core, volcanic islands, plutonometamorphic formations of the subplate periods in the evolution of the earth's crust.

This division made it possible to establish the specific features of temperature regimes in every stage in the evolution of the earth's crust:

in the early stage, in the period of intense folding, and during the platform stabilization.

Diagrams in Figure 21 give the data on temperature regime during metamorphism of the rocks in the early stages in the evolution of the earth. Chemical analyses of coexisting minerals from 40 samples of green and blue schists, eclogites and garnet amphibolites were used to establish *t* and *P* (Perchuk, , 1968a, 1971a, 1976; Perchuk and Ushakov, 1973). The curves for the temperature regime of metamorphism in these

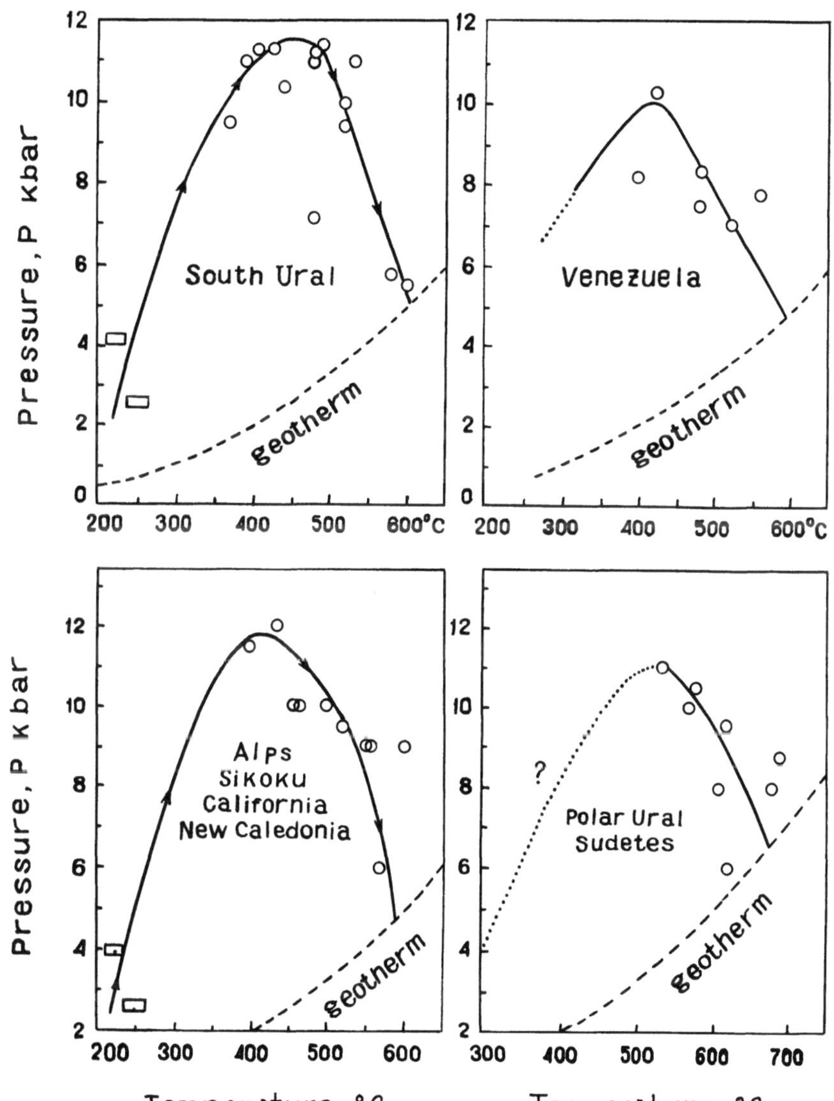

Fig. 21. The possible temperature and pressure regimes during metamorphism of the early geosyncline complexes of rocks from various regions of the world (mineralogical thermo- and barometry data)

diagrams show maxima at 400–600°C. Usually these maxima appear in the regions of glaucophane schists and eclogites. They are shown schematically in the diagram (Fig. 22) as generalized curvees with two maxima relative to the curve for the ancient geothermal gradient.

The curves in Figure 21 show metamorphic zoning usual for the glaucophane schist provinces described by Ernst in his excellent work (1971). We evaluated the thermodynamic conditions of mineral equilibria in the first two zones by calculating the corresponding reactions in the P–T diagram (Perchuk, 1976); they were verified experimentally by Ivanov and Gurevich (1975).

The diagrams in Figures 21 and 22 definitely indicate that metamorphism in the early stages in the geosyncline evolution (in the protogeosyn-

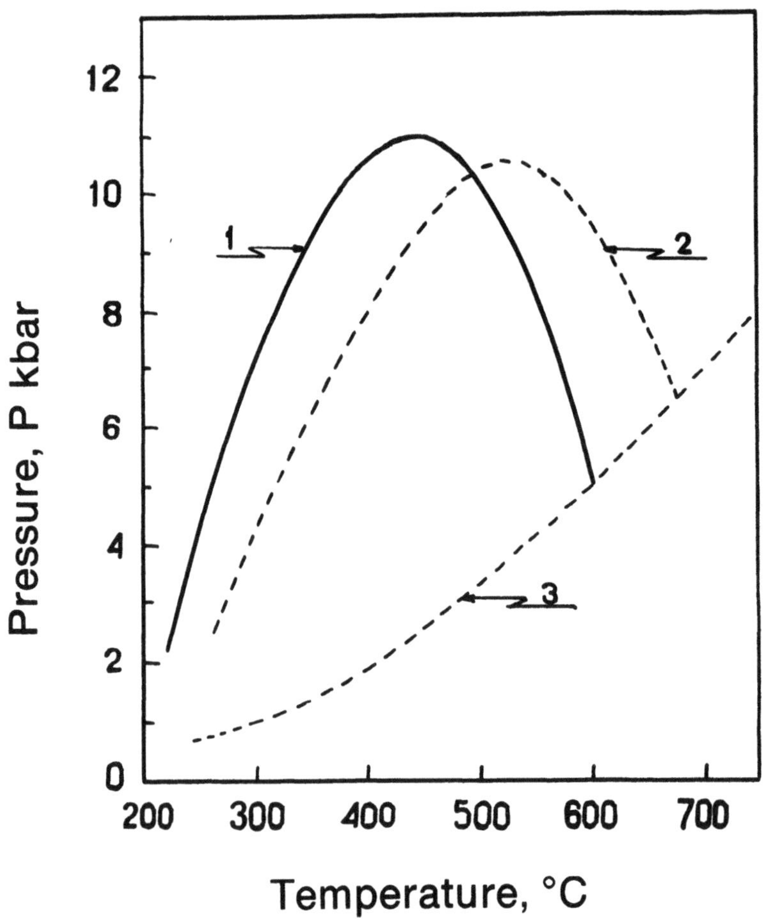

Fig. 22. Generalized curves for the metamorphic temperature regime in early stages in the geosyncline evolution: (1) glaucophane schist provinces; (2) eclogite complexes; (3) the ancient geothermal gradient in the developed metamorphic areas (the systems with the developed folding-shields and miogeosyncline metamorphosed structures)

clines) begins at very low temperatures, but sufficiently high pressures ($P_s \geqslant 3$ kbar). This means that these structures are being developed in the powerful compression zones with P_s having an infinitely small gradient on every given point, i.e., it approaches the hydrostatic pressure. The P_s gradient for the zone as a whole can have sufficiently great values. Such conditions appear to develop in the underthrust zones where the pressure gains its maximum (Ushakov, 1974) and the temperature increases continuously.

Protostructures can then serve as the basis for the development of eugeosyncline structures with the marked content of terrigene sediments. They generally form the outer layers (on the ocean side) of the island arcs (Miyashiro, 1961; Marakushev, 1965). The quantity of the terrigene material increases sharply upon transition from the eugeosynclinic to miogeosynclinal structures. The metamorphic temperature regime of these rocks is related to the geothermal gradient in every given region of the earth's crust. It is seen in Figure 23, which, using the data of mineralogical thermobarometry for more than 100 rock (Pre-Cambrian) samples (schists, gneisses, marbles, migmatites, etc.), shows the

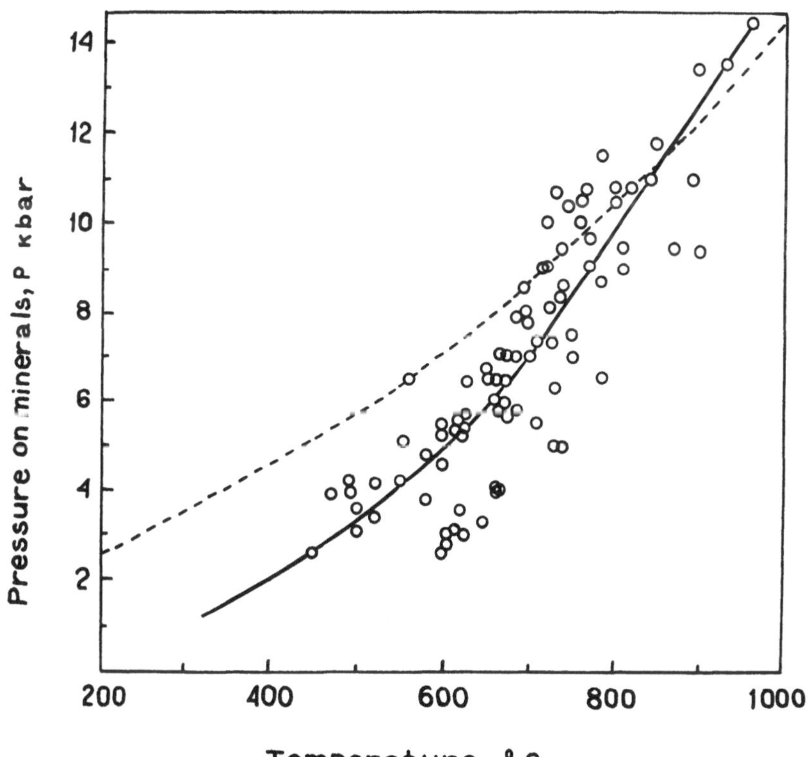

Fig. 23. Metamorphic temperature regime of the terrigene sediments in the developed fold structures (miogeosynclines). *Dots* are from the mineralogical thermobarometry; *solid line:* from statistical data; *dotted line:* geotherm under the modern fold structures

change in t with the overall pressure (depth). Notably, we did not use the contact-metamorphic rocks in deriving this diagram, because they generally belong to a later period in the development of mobile belts of the earth. The thermodynamic regime of the ancient, primarily Pre-Cambrian rock is compared in Figure 23 with the PT-gradient under Zakarpatie, where metamorphic processes must occur at the present time. The P–T-gradient has been measured by a heat current (Gordienco and Kutas, 1970). The choice of this region was not casual: it has the highest $dP/dt \approx 12.5$ bar/°C. However this ratio is markedly higher for the Pre-Cambrian geotherm: $dP/dt \approx 20$ bar/°C. It means that in the pre-Cambrian higher-temperature metamorphism has been occurring at shallower depths. There is no appreciable difference between ancient and modern gradients at great depths, as clearly seen in the diagram in Figure 23. It is not impossible that the temperature regime of the actual metamorphism on Venus should be similar to the Pre-Cambrian metamorphism in earth's crust.

Intrusive magmatism occurs commonly during the stabilization of the core and in its platform parts, while "plutonometamorphism" at great depths (to 15 km) and contact metamorphism at shallow depths (less than 10 km) occur concurrently. Unfortunately, we have only a score of analyses of the PT-parameters for such samples, and they are given in Figure 24. It appears that the temperature regime in these portions of the crust is connected with either plutonometamorphism ($P_s \approx 1 - 4$ kbar and $t \leqslant 700$°C) or with contact metamorphism mainly of trap intrusions ($P < 1$ kbar and $t > 700$°C).

5.2 Oxygen Regime

To evaluate the oxygen regime during metamorphism and granitization we have selected about 20 mineral parageneses from the rocks of various depth facies. These rocks generally contained the mineral associations, useful in the analyses of the equilibrium PT-conditions. These data are summarized in special tables (Marakushev and Perchuk, 1974); the points plotted in the diagram (Figure 25) reflect the $\log f_{O_2}$ variations with depth, or more exactly, with pressure. In general the inverse relationship exists between these values. But the diagram shows three groups of paragenesis; from metapelites (open circles), from ferruginous quartzites (triangles) and from magmatic rocks (black circles). It is seen that their P and $\log f_{O_2}$ parameters differ greatly. The more reduced fluids were involved in the regional metamorphism of depth facies. However, the stable paragenesis Mt + Hem occurs in the ferrugineous quartzites of the amphibolite facies, which suggests a higher O_2 chemical potential in their metamorphism.

The highest $\log f_{O_2}$ values are found for the magmatic rocks, which is not surprising because they are the highest temperature rocks and contain Mt + Il. Similar oxygen regime must occur in the contact–metamorphic

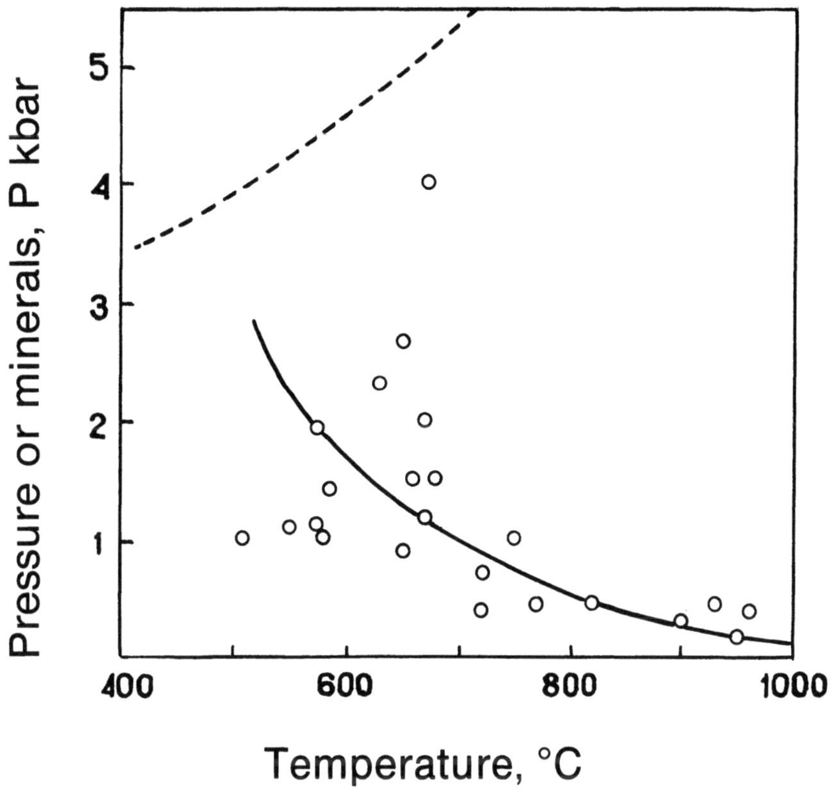

Fig. 24. Temperature regime of contact metamorphism and plutonometamorphism in the stabilized areas of the earth's crust. *Dotted line:* ancient geotherm (from Figure 23)

rocks. In effect, the Black-Hill hornfels (cordierite–garnet–biotite hornfels) in Glen Muin, Scotland ($Bi_{0.288}$ + $Gr_{0.135}$ + $Cor_{0.458}$ + Qz + Pl + Or) contain hematite and occasionally, magnetite (Chinner, 1962). Consequently, as the pressure (depth) of regional and contact metamorphism increases, the oxygen fugacity diminishes by several orders of magnitude. This decrease is regular, as the distribution of the points in Figure 25 indicates. Such an oxygen regime allows us to consider the μ_{O_2} as a factor of metamorphism, i.e., oxygen is a perfectly mobile component (Korzhinskii, 1959).

The analyses of the coexisting minerals in glaucophane schists, serpentinites and other rocks in the embryo stages in the geosyncline evolution gave very low $\log f_{O_2}$ values: from −33 to −37 at 415–490°C and 9.2–11.8 kbar (Perchuk, 1973). However, these data are insufficient to establish the oxygen regime in these stages.

5.3 The CO_2 and H_2O Regimes

In a previous section of this paper we discussed the methods of evaluating the CO_2 and H_2O partial pressures and gave examples of their

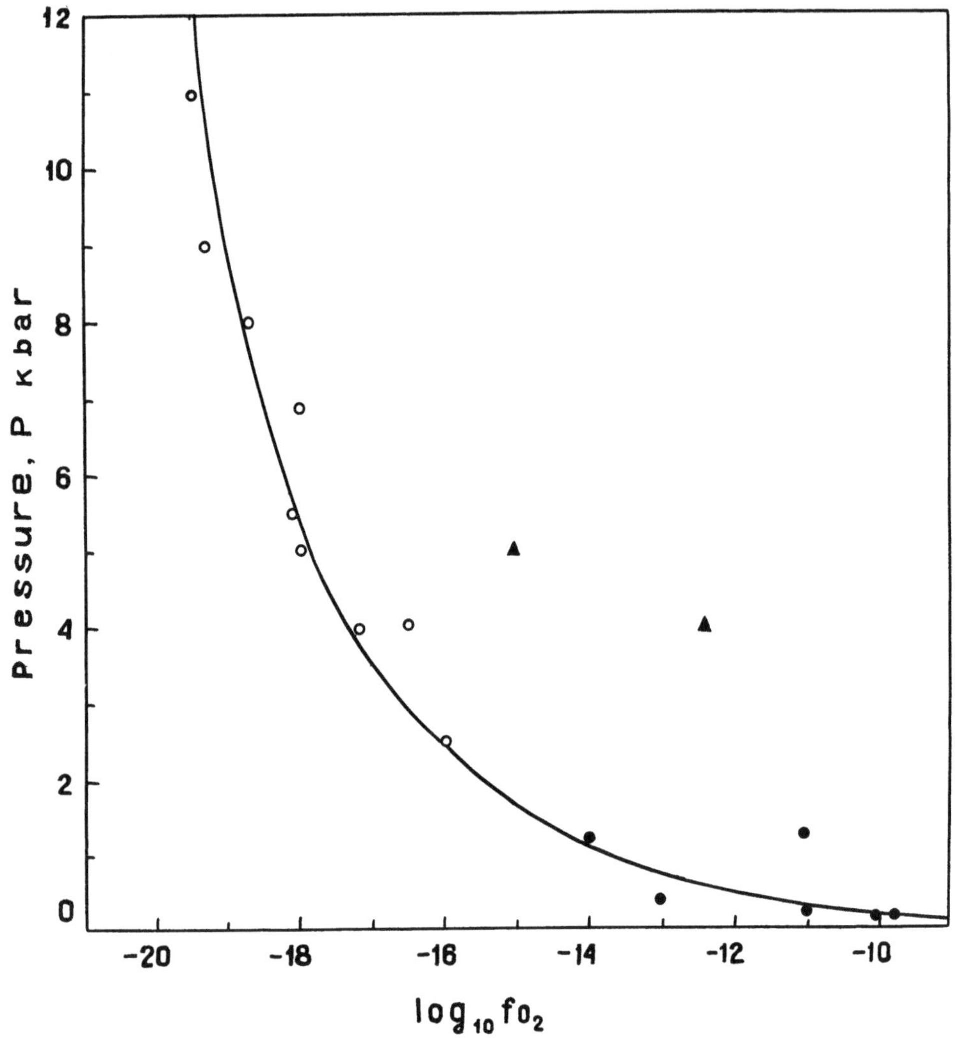

Fig. 25. Oxygen regime during regional and contact metamorphism. See symbols in the text

calculations. Many $P^{fl}_{H_2O}$ and $P^{fl}_{CO_2}$ values have been obtained for the rocks of various depth facies. The data are summarized in diagrams in Figure 26. They reflect the CO_2 and H_2O regimes only in the regional and contact metamorphism, and not in the early stages. We do not have sufficient empirical data for these stages, especially for the rocks from the zeolite and prehnite metamorphic facies. The main difficulties in estimating the *PT*-conditions of mineral equilibria are due to lack of mineralogical thermometers and barometers for this range of temperature. We believe that further developments in the isotopic thermometry (O^{18}/O^{16}) will bring more satisfactory results. However, even now we can conclude that $P^{fl}_{CO_2}$ in these stages is one order of magnitude lower than at P_s during regional metamorphism of the terrigene geosynclines.

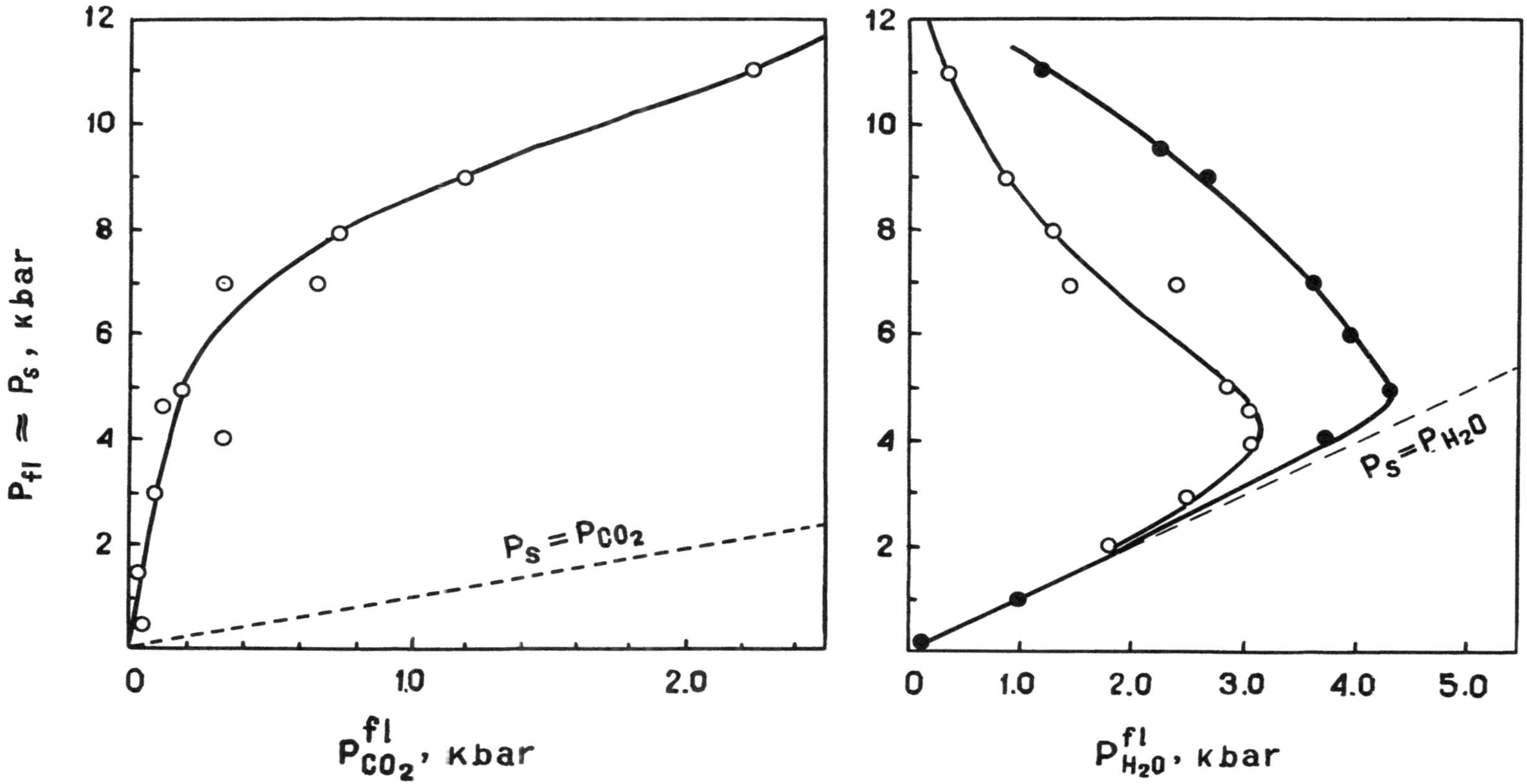

Fig. 26. The CO_2 and H_2O regimes during (A) regional and (B) contact metamorphism *(open circles)* and during granitization *(solid circles)*

For example, in the quartz–pumpellyite zone of the glaucophane-schist formation, the Penjinskii Range, Kamchatka (Dobretsov, 1974), the CO_2 partial pressure does not exceed 20 bar, and the H_2O pressure is about 3.5 kbar at $P_s \geqslant 4$ kbar and $t \approx 200$°C. Similar metamorphic conditions are noted for the rocks from the Fransiscan Province in California and the Maksutov formation in the southern Urals.

Let us go back to the diagrams in Figure 26. The first of these shows that the $P^{fl}_{CO_2}$ increases exponentially with the $P_s \approx P_{fl}$ increase, i.e., with the metamorphic depth. The CO_2 regime during metamorphism was first qualitatively established by Korzhinskii (1940) who followed the sequence of the carbonation in the Ca-bearing minerals depending on metamorphic depth. Korzhinskii concluded that $P^{fl}_{CO_2}$ increases with the depth of metamorphic processes. The possible reason for that will be considered at the end of this paper. Note, that it is not unusual for the abyssal rocks (alkali basalts, kimberlites; etc.) to contain liquid carbon dioxide inclusions (Roedder, 1965). The increase in P_{CO_2} with the depth suggests the mantle nature of metamorphic fluid.

The second diagram in Figure 26 presents the H_2O regime during metamorphism and granitization. The $P^{fl}_{H_2O}$ values were calculated for the natural associations to which the discussed methods of mineralogical thermobarometry can be applied; the polymineral equilibria are described by the hydration–dehydration reactions. With P_s, T, and ln $\bar{K}_{Mg}$ known, one can find $f^0_{H_2O}$:

$$f^0_{H_2O} = \exp\left(\ln \bar{K}_{Mg} + \frac{\Delta G^0_T + \Delta V_s P_s}{RT}\right). \qquad (68)$$

The calculation by this method does not present any difficulty if mineral constants are known or the position of the line for monovariant equilibrium involving solid solution end members is established. Then, using Equation (65) we move on to $P^{fl}_{H_2O}$ estimations.

We used this method only for the high temperature granulites (Bi + Gr + Sil + Cor + Fsp + Qz); the $P^{fl}_{H_2O}$ value was established from the stability of the Bi + Sil + Qz association (Marakushev, 1965, 1968). In all other cases the equilibria Mu + Qz = Fsp + Al_2SiO_5 + H_2O, Zo + Qz = An + Gros + H_2O and others were used.

To establish the $P^{fl}_{H_2O}$ during granitization we have recalculated the data from Figure 20, using Equation (65). The corresponding dots are plotted in the $P_s - P^{fl}_{H_2O}$ diagram as solid circles.

This diagram leads us to the following conclusions:

(1) The H_2O regime during metamorphism and granitization exhibits extreme behavior: maximum $P^{fl}_{H_2O}$ are at $P \approx 4$–5 kbar;

(2) Granitization occurs at higher $P^{fl}_{H_2O}$ than regional metamorphism of terrigene rock masses.

The first of these conclusions will be explained on the basis of the detailed analysis of fluid regime during metamorphism and granitization. The second conclusion is self-explanatory: *the granitization process, i.e., initiation and growth of the magmatic melt results from the increase in* P_{H_2O} *in metamorphic fluid* (Perchuk, 1970a, b).

5.4 The Regime of Fluid Oxidation

If $P \approx P_{fl}$, then the changes in the partial pressures of fugacities of O_2, CO_2, and H_2O suggest the oxidation of abyssal fluids within the earth's crust. In spite of the marked P_{CO_2} increase with the fluid pressure, the fugacity ratios CO_2/CO and H_2O/H_2 diminish with the depth (Perchuk, 1972). These data were obtained for a number metamorphic rocks with the P_s and t parameters determined by mineralogical thermobarometry. Most of the samples contained graphite, and therefore the reactions $CO_2 = C + O_2$ and $CO = C + \frac{1}{2}O_2$ have been calculated. However, two other gas reactions are important for solving the problem (Marakushev and Perchuk, 1974):

$$H_2O = H_2 + \tfrac{1}{2}O_2; \qquad (69)$$
$$CO_2 = CO + \tfrac{1}{2}O_2, \qquad (70)$$

where

$$\ln (f_{H_2O}/f_{H_2}) = \frac{\Delta G^0_T(69)}{RT} + 0.5 \ln f_{O_2} \qquad (71)$$
$$\ln (f_{CO_2}/f_{CO}) = \frac{\Delta G^0_T(70)}{RT} + 0.5 \ln f_{O_2}. \qquad (72)$$

The data of Robie and Waldbaum (1968) were used in these calculations. These reactions have been reduced to a standard pressure. We have established not only P and t, but $\log f_{O_2}$ as well for all the samples analyzed (Perchuk, 1973; Marakushev and Perchuk, 1974). The results are in the diagrams in Figure 27 and they show a decrease in the logarithms of the oxidized–reduced gas ratios. In other words, *the degree of fluid oxidation decreases with the depth of metamorphism and/or the overall pressure.*

This conclusion is based on the calculation. Unfortunately, we have no data on the actual changes in the composition of metamorphic fluid with depth. Let's assume the metamorphic fluid to comprise both oxidized and reduced gases of the system C–O–H–N–F–Cl. Then

$$P_{fl} = P_{CO_2} + P_{CO} + P_{CH_4} + P_{H_2O} + P_{H_2} + P_{N_2} + P_{F_2} + P_{Cl_2} + \ldots \qquad (73)$$

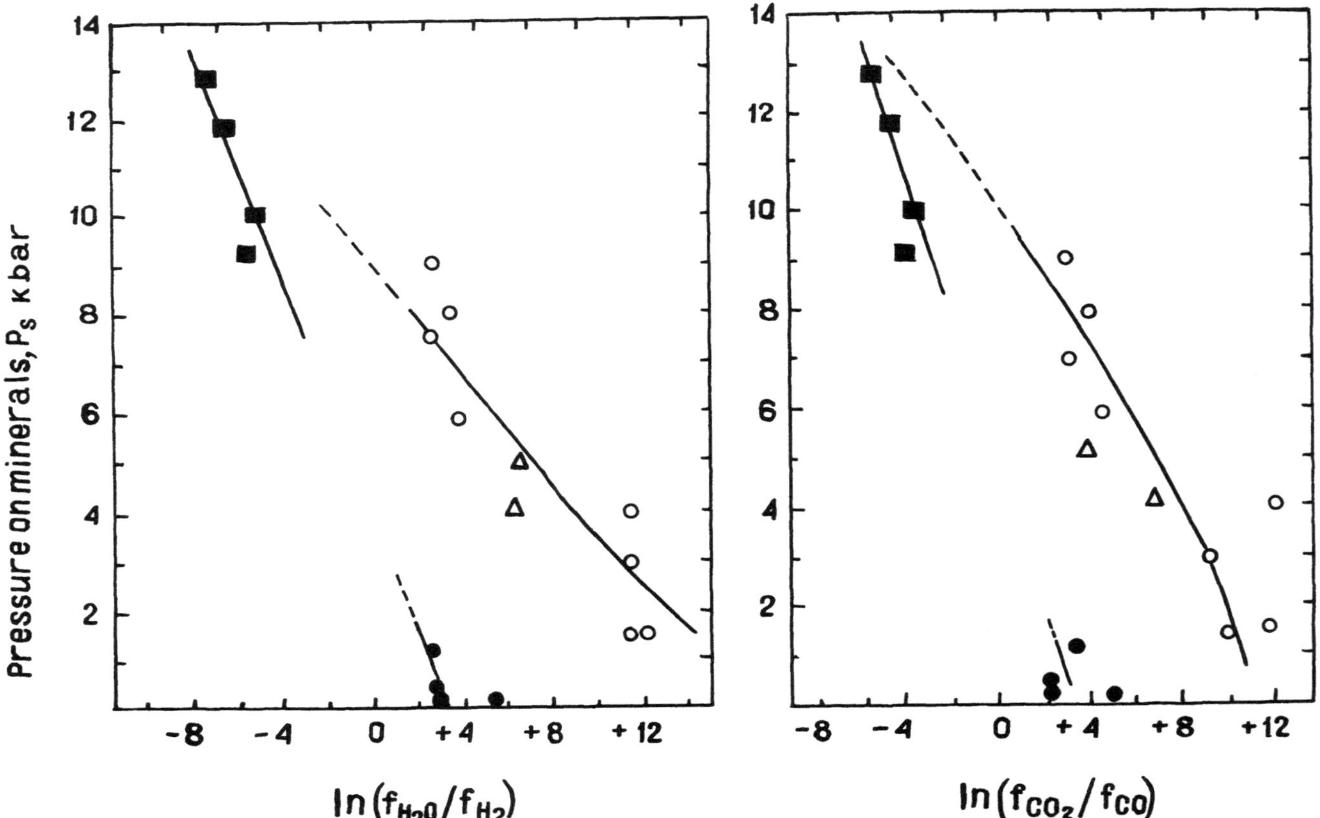

Fig. 27. Changes in ln (f_{H_2O}/f_{H_2}) and ln (f_{CO_2}/f_{CO}) with the pressure during metamorphism. *Filled blocks:* samples from metamorphic rocks in the early stages in the evolution of geosynclines (almandine eclogites with jadeite from glaucophane schist formations). Other symbols as in Figure 25

We know the relationships between P_{fl} and P_{CO_2} and P_{H_2O} that is, the most oxidized forms of carbon and hydrogen. Consequently, all other gases represent the "reduced" portion of the fluid equilibrium composition at the given T and P_{fl}. From Equation (73) we find:

$$P_{red} = P_{CO_2} + P_{CH_4} + P_{H_2} + P_{N_2} + P_{F_2} + P_{Cl_2} = P_{fl} - P_{H_2O} - P_{CO_2}. \quad (74)$$

Using the diagrams in Figures 23, 25, 26, and Equation (74) we can calculate the P_{red} regime, i.e., the change in this parameter with the depth of metamorphism of terrigene sediments. The results are in Table 10 and in Figure 28. The composition of metamorphic fluid is seen to change sharply within $8 > P_s > 3$ kbar; the fluid enriches in H_2O and CO_2 at the expense of the reduced forms of the gas components in the system under consideration. Such changes can only be the result of fluid oxidation. Indeed, up to 25–20 km the fluid composition remained relatively constant ($X^{fl}_{H_2O+CO_2} \approx 0.24$). At shallower depths the increase in f_{O_2} from 10^{-19} to 10^{-17} resulted in the sudden fluid oxidation and at shallow depths the fluid became virtually aqueous. It is interesting to note that the Konrad boundary corresponds to a depth of 20–25 km. It appears that above this boundary, where the terrigene sediments predominate, different oxygen regime of metamorphism exists.

The established regularity can be compared with the changes in the gas mixture composition in the system H_2–O_2–H_2O at differing f_{O_2} but with constant T and P_{fl}. The results of calculation are plotted in Figure 28B. The comparison of these two diagrams reveals their similarity. Thus, metamorphism proceeds in the presence and under the influence of the abyssal fluids of the mantle origin. These fluids are mainly composed of hydrogen, methane, nitrogen, fluorine, chlorine, and other reduced

Table 10. Variations in fluid composition with temperature, pressure and oxygen fugacity during metamorphism of terrigene rocks (statistical data)

P_{fl} (kbar)	t (°C)	$-\log f_{O_2}$	$X^{fl}_{CO_2}$	$X^{fl}_{H_2O}$	$\sum X^{fl}_{red}$
12	875	19.5	0.221	0.015	0.764
11	840	19.35	0.204	0.030	0.764
10	800	19.2	0.172	0.060	0.760
9	770	19.0	0.122	0.118	0.760
8	740	18.8	0.092	0.156	0.751
7	700	18.5	0.071	0.250	0.679
6	650	18.2	0.050	0.383	0.567
5	600	17.8	0.040	0.570	0.390
4	550	17.3	0.035	0.787	0.178
3	480	16.6	0.030	0.867	0.103
2	400	16.0	0.020	0.890	0.090

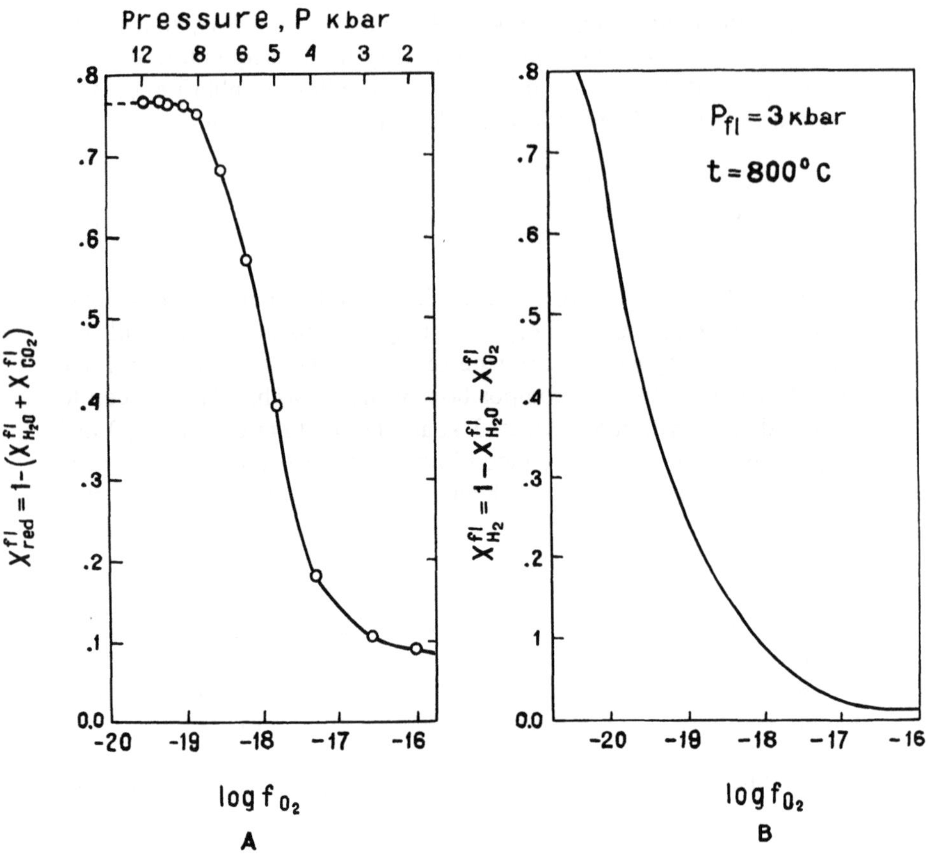

Fig. 28A, B. (A) The oxidation regime of the natural metamorphic fluid and (B) changes in the gas mixture composition as a function of f_{O_2} in the system H_2–O_2–H_2O

gases. Above Konrad's boundary in the earth's crust these fluids undergo abrupt oxidation to form aqueous–carbon dioxide mixtures.

These abyssal fluids are generally of alkaline character. In this connection a hypothesis has been put forward that the nature of these fluids must be sought in the decomposition of oxides, hydrides, carbides, sulfides, nitrides, and halogens of alkali metals at great depths. The reactions of these gas mixtures as they ascend to the surface, result in many important petrological and geophysical processes. This problem is discussed elsewhere (Marakushev and Perchuk, 1974; Perchuk, 1973, 1974, 1975).

6 CONCLUSION

The results from this work are based entirely on the natural laws revealed by applying the theory and methods of chemical thermodynamics to enormous factual material.

We have extended Ramberg's analysis of component distribution and developed a system of mineralogical thermometers and barometers using Tables 1 and 2 and minimum experimental data.

In spite of the good correlation between different mineralogical thermometers and barometers, it should be remembered that they are based on statistics. It means that individual parageneses can give substantial scatter of t or P values in different diagrams (more than 10%). This can be caused by a large variety of factors: disequilibrium of association, the specific nature of the composition, inaccuracy of chemical or X-ray analysis, impurities in monomineral fraction sampling for composition analysis etc. The gas–mineral equilibria calculations were based on the PT-parameter statistics, after which the O_2, H_2O, CO_2 partial pressures were evaluated. Analyzing changes in these parameters and the temperature with depth and pressure enabled us to establish the fluid and temperature regimes of metamorphic processes. These regimes depend on the degree of geotectonic development of the given area of the crust. Consequently, the geotectonic and thermodynamic regimes are unseparable, and more work in the theory and field studies is needed.

The oxygen regime during metamorphism remains one of the most important theoretical problems. The reasons of the f_{O_2} increase as fluid ascends to the surface must be determined unambiguously.

ACKNOWLEDGMENT

The author would like to express his special gratitude to G. B. Lakoza for the literary arrangement of the English text.

REFERENCES:

Akimoto, S., Fujisawa H.: Olivine-spinel solid solution equilibria. J. Geophys. Res. *73,* 1467–1479 (1968)

Bakun-Gzubarov, N.: On equilibration conditions of the sudetic eclogites, Poland. I.M.A. Meeting: Collected Abstracts, W. Berlin: 1974, p. 190

Barth, T. F. W.: The measurements of paleotemperatures of granitic rocks. Ac. Sci. USSR, Moscow: Nauka, 1962

Buddington, A. F., Lindsley, D. H.: Iron–titanium oxide minerals and synthetic equivalents. J. Petrol. *5,* 310–357, (1964)

Burnham, C. W., Holloway, J. R., Davis, N. F.: Thermodynamic properties of water to 1000°C and 10000 bars. Geol. Soc. Am. Spec. Paper: 1969, p. 96

Chinner, G. A.: Almandine in thermal aureoles. J. Petrol., *3,* 316–340, (1962)

Delbove, F.: Équilibre d'échange d'ions entre feldspaths alcalins et halogénures sodi-potassiques fondus. Application an calcul des propriétés thermodynamiques de la série des feldspaths alcalins. Bull. Soc. Franc. Minér. Crist. *94,* 456–466 (1971)

Dobretsov, N. L.: Glaucophane Schist and Eclogite Glaucophane Schist Complexes of the USSR. Novosibirsk: Nauka, 1974, p. 429

Ernst, W. G.: Metamorphic zonations on presumably subducted lithospheric plates from Japan, California and the Alps. Contr. Mineral. Petrol., *34*, 43–59 (1971)

Fed'kin, V. V.: Staurolite. Moscow: Nauka, 1975, p. 270

Friedman, I. I.: A proposed method for the measurement of geologic temperatures. J. Geol. *57*, N 6 (1949)

Frost, M. Y.: Metamorphic grade and iron-magnesium distribution between coexisting garnet-biotite and garnet-hornblende. Geol. Mag., *99*, 427–438 (1962)

Gibbs, G. V.: The polymorphism of cordierite. I. The crystal structure of low cordierite. Am. Mineral. *51*, 1068–1087 (1966)

Gordienko, V. V., Kutas, R. I.: Depth temperatures in some regions of the Ukraine. Acad. Nauk USSR. Doklady, Ser. B, *18*, 702–705 (1970)

Gordon, T. M., Greenwood, H. J.: The stability of grossularite in H_2O-CO_2 mixtures. Am. Mineral. *56*, 1674–1688 (1971)

Hewitt, D. A.: Stability of the assemblage muscovite-calcite-quartz. Am. Mineral. *58*, 785–791 (1973a)

Hewitt, D. A.: Metamorphism of micaceous limestones from south-central Connecticut. Am. J. Sci., *273A* (Cooper volume), 444–468 (1973b)

Hubregtse, J. J. M. W.: Distribution of Elements in Some Basic Granulite-Facies Rocks. Amsterdam-London: North-Holland, 1973

Iiyama, J. T.: Contribution a. l'etude des equilibres subsolidus du systeme ternaire orthose-albite-anorthite a l'aide reactions d'echanges d'ions Na-K au contact d'une solution hydrothermale. Bull. Soc. France Miner. Crist. *89*, 442–454 (1966)

Ivanov, I. P., Gurevich, L. P.: Experimental study of T–X_{CO_2} boundaries of metamorphic zeolite facies. Contrib. Mineral. Petrol. *53*, 55–60 (1975)

Karpov, I. K., Kiselev, A. I., Letnikov, F. A.: Chemical Thermodynamics in Petrology and Geochemistry. Irkutsk: 1971, p. 384

Korzhinskii, D. S.: Mobility and inertness of components in metasomatism. Akad. Nauk USSR. Izv., Ser. Geol. *1*, 3–57 (1936)

Korzhinskii, D. S.: Relation between mineral formation and depth. Vses. Mineral. Obsch. Zap. *66*, 369–384 (1937)

Korzhinskii, D. S.: The Factors in Mineral Equilibria and Abyssal Mineralogical Facies. Moscow: Inst. Geol. Sc., 1940, p. 136

Korzhinskii, D. S.: Physico-Chemical Basis of the Analysis of Parageneses of Minerals. New York: Consultants Bureau, 1959

Kretz, R.: Chemical study of garnet, biotite and hornblende from gneisses of south-western Quebec with emphasis on distribution of elements in coexisting minerals. J. Geol. *67*, 371–402 (1959)

Kretz, R.: Some applications of thermodynamics to coexisting minerals of variable compositions. Examples: orthopyroxene–clinopyroxene and orthopyroxene–garnet. J. Geol. *69*, 361–387 (1961)

Kulish. E. A.: Quartzites in Archean Rocks of Southern Aldan Shield. Magadan: 1964, p. 112

Kuno, H.: Differentiation of basalt magmas. In: Basalts. Hess, H. H., Poldevaart, A. (eds.). New York: Wiley, 1968, pp. 623–689

Lundgren, L. W.: Muscovite reactions and partial melting in south-eastern Connecticut. J. Petrol. *7*(3), 421–453 (1966)

Matthews, D. W., Cheeney, R. F.: The metamorphic evolution of the Moin Nappe in Skye. Scottish J. Geol. *4*(1), 20–30 (1968)

Marakushev, A. A.: The Problems of Mineral Facies in Metamorphic and Metasomatic Rocks. Moscow: Nauka, 1965, p. 312

Marakushev, A. A.: The Thermodynamics of Metamorphic Mineral Hydration. Moscow: Nauka, 1968, p. 199

Marakushev, A. A., Perchuk, L. L.: The thermodynamic model of the fluid regime of the Earth. In: Contributions to Physico-Chemical Petrology. Zharikov, V. A. (ed.), Moscow: Nauka, 1974, pp. 102–129

Melnik, Ju. P.: Thermodynamic properties of compressed gases and some peculiarities of metamorphic reactions involving water and carbon dioxide. Geokhimia *6*, 654–662 (1972)

Misch, P.: Stable association wollastonite-anortite, and other calc–silicate assemblages in amphibolite–facies crystalline schists of Nanga Parbat, northwest Himalayas. Beitr. Mineral. Petrogr. *10*, 315–356 (1964)

Miyashiro, A.: Evolution of metamorphic belts. J. Petrol. *2*, 277–311 (1961)

Miyashiro, A.: Oxidation and reduction in the Earth's crust with special reference to the role of graphite. Geochim. Cosmochim Acta *28*, 717–732 (1964)

Mottana, A.: Distribution of elements among coexisting phases in amphibole-bearing eclogites. Neues. Jahrb. Mineral Abhandl. *112*, 161–187 (1970)

Nagaitsev, Ju. V.: Petrology of Metamorphic from the Lake Ladoga and the White Sea Complexes. Leningrad: Univ. Press, 1974, p. 160

Newton, R. C.: An experimental determination of the high-pressure stability limits of magnesian cordierite under wet and dry condition. J. Geol. *80*, 398–420 (1972)

Newton, R. C., Charlu, T. V., Kleppa, O. J.: A calorimetric investigation on the stability of anhydrous magnesium cordierite with application to granulite facies metamorphism. Contr. Miner. Petrol. *44*, 295–313 (1974)

Perchuk, L. L.: The Physico-Chemical Petrology of Granitic and Alkaline Intrusions of Central Turkestan-Alay. Moscow: Nauka, 1964, p. 236

Perchuk, L. L.: Paragenesis of nepheline with alkali feldspar as a possible indicator of the thermodynamic conditions of mineral equilibrium. Akad. Nauk USSR, Dokl, *161*, 932–935 (1965)

Perchuk, L. L.: Temperature dependence of coefficient of calcium distribution between coexisting amphiboles and plagioclases. Akad. Nauk USSR, Dokl. *169*, 1436–1438 (1966)

Perchuk, L. L.: Biotite-garnet geothermometer. Akad. Nauk USSR, Dokl. *177*, 411–414 (1967a)

Perchuk, L. L.: The analysis of thermodynamic conditions of mineral equilibria in amphibole–garnet rocks. Akad. Nauk USSR, Izv., Ser. Geol. *3*, 52–83 (1967b)

Perchuk, L. L.: Biotite–garnet equilibrium in metamorphic rocks. In: Theoretical and Experimental Study of Mineral Equilibria. Nekrasov, I. Ya. (ed.), Moscow: Nauka, 1968a, pp. 3–36

Perchuk, L. L.: Pyroxene–garnet equilibrium and the depth facies of eclogites. Intern. Geol. Rev. *10*, 280–318 (1968b)

Perchuk, L. L.: The effect of temperature and pressure on the equilibrium of natural iron-magnesium minerals. Intern. Geol. Rev. *11*, 875–901 (1969a)

Perchuk, L. L.: The staurolite-garnet thermometer. Akad. Nauk USSR, Dokl. *186*, 1405–1407 (1969b)

Perchuk, L. L.: Paragenesis of rhombic pyroxene with garnet in metamorphic rocks. In: Contribution to Physico-Chemical Petrology. Zharikov, V. A. (ed.). Moscow: Nauka, 1969c, pp. 261–285

Perchuk, L. L.: Equilibria of Rock Forming Minerals. Moscow: Nauka, 1970a, p. 301

Perchuk, L. L.: Thermodynamic conditions of granitization of metapelitic rock masses. In: Contributions to Physico-Chemical Petrology. Zharikov, V. A. (ed.), Moscow: Nauka, 1970b, Vol. II, pp. 188–213

Perchuk, L. L.: Coexisting Minerals (Collection of Chemical Compositions and Paragenesis). Moscow: Nedra, 1971a, p. 424

Perchuk, L. L.: Crystallochemical problems in theory of phase relations. Geokhimiya *1*, 23–38 (1971b)

Perchuk, L. L.: Mineral equilibria and problems of thermo- and barometry. Vses. Min. Obsch. Zap. *100*, 3–19 (1971c)

Perchuk, L. L.: Thermodynamic regime of metamorphism, Akad. Nauk USSR, Isv., Ser. Geol., *12*, 46–60 (1972)

Perchuk, L. L.: Thermodynamic Regime of Depth Petrogenesis. Moscow: Nauka, 1973, p. 312

Perchuk, L. L.: Thermodynamic regime of metamorphism and granitization of terrigenic rock masses. Collected abstracts. I.M.A. Meeting. Ninth general meeting. Berlin: 1974, p. 106

Perchuk, L. L.: Gas–mineral equilibria, geophysical measurements and a proposed chemical model of the inner structure of the Earth. XYI General Assembly. Abstracts of papers presented at the interdisciplinary symposia. Grenoble: 1975

Perchuk, L. L.: Parageneses and Compositions of Coexisting Minerals. Moscow: Nauka, 1976, p. 205

Perchuk, L. L., Andrianova, Z. S.: Thermodynamics of equilibrium of alkali feldspar (K,Na)$AlSi_3O_8$ with aqueous solution (K,Na)Cl at 500–800°C and 2000–1000 bar pressure. In: Theoretical and Experimental Studies of Mineral Equilibria. Nekrasov, I. Ya. (ed.). Moscow: Nauka, 1968, p. 37–72

Perchuk, L. L., Fed'kin, V. V.: The derivation of the PT-diagrams for equilibria of minerals of variable composition. In: Contributions to Physico-Chemical Petrology, Zharikov, V. A. (ed.). Moscow: Nauka, 1974, Vol. IV, pp. 162–175

Perchuk, L. L., Karpov, I. K.: Thermodynamic properties of carbon dioxide (CO_2) at $100 < P < 10000$ bar and $100 < t < 1000$°C. In: Contributions to Physico-Chemical Petrology, Zharikov, V. A. (ed.). Moscow: Nauka, 1975, Vol. V, pp. 221–235

Perchuk, L. L., Pavlenko, A. S.: The temperature influence on distribution some isomorphic components between coexisting minerals from alkali rocks. Geochemistry *9*, 1063–1082 (1967)

Perchuk, L. L., Ryabchikov, I. D.: Mineral equilibria in the system nepheline-alkali feldspar–plagioclase and their petrological significance. J. Petrol. *9*, 123–167 (1968)

Perchuk, L. L., Ryabchikov, I. D.: Phase correspondence in mineral systems (textbook for students). Moscow: Nedra, 1976, p. 287

Perchuk, L. L., Suvorova, V. A.: Thermodynamic calculation of CO and CO_2 fugacities in the area of graphite–diamond phase transition. In: Phase Equilibria and the Processes of Mineral Formation, Ivanov, I. P. (ed.). Moscow: Nauka, 1973, Pt. 3, pp. 15–19

Perchuk, L. L., Ushakov, S. A.: Thermodynamic regime of metamorphism in the subducted lithospheric platforms. Vestnic, Ser. Geol. *6*, 19–29 (1973)

Perchuk, L. L., Vaganov, V. I.: Temperature regime of differentiation and crystallisation of basic and ultrabasic magmas. J. Petrol. (1976, in press)
Ramberg, H.: The facies classification of rocks: a clue to the origin of quartzo-feldspathic massifs and veins. J. Geol. *57,* 17–31 (1949)
Ramberg, H.: The Origin of Metamorphic and Metasomatic Rocks. Chicago: Univ. Chicago, 1952a, p. 317
Ramberg, H.: Chemical bonds and the distribution of cations in silicates. J. Geology *60,* 331–355 (1952b)
Ramberg, H., DeVore, G.: The distribution of Fe^{++} and Mg^{++} in coexisting olivines and pyroxenes. J. Geol. *59,* 193–210 (1951)
Robie, R. A., Waldbaum, D. R.: Thermodynamic Properties of Minerals and Related Substances at 298.15°K (25.0°C) and One Atmosphere (1.013 bars) Pressure and at Higher Temperatures. Washington: U.S. Govt. Print. Office, 1968, p. 256
Roedder, E.: Liquid CO_2 inclusions in olivine-bearing nodules and phenocrysts from basalts. Am. Miner. *50,* 1746–1782 (1965)
Roeder, P. L.: Activity of iron and olivine solubility in basaltic liquids. Earth Planet. Sci. Lett. *23,* 397–410 (1974)
Saxena, S. K.: Thermodynamics of Rockforming Crystalline Solutions. Berlin-Heidelberg-New York: Springer, 1973
Schreyer, W., Yoder, H. S.: Cordierite–water system. Ann. Rep. Direct. Geop. Labor., Carnegie Inst. Washington, *58,* 83–85 (1959)
Sen, S. R., Chakraborty, K. R.: Magnesium-iron exchange equilibrium in garnet–biotite and metamorphic grade. N. Jb. Miner. Abh., *108,* 181–207 (1968)
Smith, D.: Stability of the assemblage iron-rich orthopyroxene–olivine–quartz. Am. J. Sci. *271,* 370–382 (1971)
Shmonov, V. M., Shmulovich, K. I.: Molar volumes and equations of state for CO_2 at 100–1 000°C and 2 000–10 000 bars. Akad. Nauk USSR, Dokl. *212,* 935–938 (1974)
Sobolev, V. S.: The petrography of Botogolsky alkaline massif. In: Botogolsky Graphite Deposit and Its Prospective Use. Irkutsk: 1947, p. 310
Sobolev, V. S.: The energy of crystalline lattice and iron distribution laws in minerals. Miner. Sbornik (Lvov), *2,* 25–42 (1948)
Thompson, J. B., Waldbaum, D. R.: Mixing properties of sanidine crystalline solutions: I. Calculations based on ionexchange data. Am. Miner. *53,* 1965–1999, (1968)
Thompson, J. B., Waldbaum, D. R.: Mixing properties of sanidine crystalline solutions: III. Calculations based on two phase data. Am. Miner. *54,* 811–838, (1969)
Ushakov, S. A.: Physics of the Earth. Structure and Development of the Earth. Moscow: Viniti, 1974, Vol. I, p. 296
Wagner, C.: Thermodynamics of Alloys. Cambridge, Mass: Addison-Welsley, 1952, p. 136
Waldbaum, D. R., Thompson, J. B.: Mixing properties of crystalline solutions: II. Calculations based on volume data. Am. Miner. *53,* 2000–2017, (1968)
Waldbaum, D. R., Thompson, J. B.: Mixing properties of sanidine crystalline solutions: IV. Phase diagrams from equations of state. Am. Miner. *54,* 1274–1298, (1969)
Wones, D. R., Eugster, H. P.: Stability of biotite: experiment, theory and application. Am. Miner. *9,* 1228–1272 (1965)

Wood, B. J.: Fe^{2+}–Mg^{2+} partition between coexisting cordierite and garnet. A discussion of the experimental data. Contr. Mineral. Petrol. *40,* 253–258 (1973)

Wood, B. J., Strens, R. G. J.: The orthopyroxene geobarometer. Earth Planet. Sci. Lett. *11,* 1–9 (1971)

Ziryanov, V. N., Perchuk, L. L., Podlesskii, K. V.: Nepheline-feldspar equilibrium. J. Petrol. (1976, in press)

Zotov, I. A.: Temperature and pressure regime during metamorphism of the rocks from the crystalline complexes from south-westest Pamirs. Akad. Nauk USSR, Dokl. *177,* 1170–1173 (1967)

15

A. C. LASAGA, S. M. RICHARDSON
AND H. D. HOLLAND

Abstract Neighboring crystals of garnet and cordierite in metamorphic rocks at Phillipston, Massachusetts, are chemically homogeneous except near their common boundaries. Mg^{2+}–Fe^{2+} exchange occurred across these boundaries during the waning stages of the Acadian orogeny. Fe^{2+} was transferred from cordierite to garnet and an equivalent number of moles of Mg^{2+} were transferred from garnet to cordierite. The compositional profiles of these elements in adjacent crystals of the two phases are potential sources of information regarding cooling rates during retrograde metamorphism. The equations governing the interdiffusion of Mg^{2+} and Fe^{2+} between the two phases during a linear decrease in temperature were solved in closed form. Although there are adequate data for the distribution coefficient of Mg^{2+} and Fe^{2+} between garnet and cordierite, too little is known about the diffusion constants of these ions in the critical temperature range between 675°C and 475°C. No absolute values for the cooling rates at Phillipston can, therefore, be extracted from the measured composition profiles. We have used our model and

THE MATHEMATICS OF CATION DIFFUSION AND EXCHANGE BETWEEN SILICATE MINERALS DURING RETROGRADE METAMORPHISM

the measured composition profiles, however, to refine our meager knowledge of the temperature dependence of the diffusion coefficients. Furthermore, the observed chemical uniformity of all but the rims of the garnet crystals permits a calculation of the minimum value of the interdiffusion coefficient of Mg^{2+} and Fe^{2+} in garnet and hence a minimum value of 100°C per million years for the cooling rate of the Phillipston rocks between 675°C and 475°C. This implies a very abrupt ending to the Acadian orogeny accompanied by an exceptionally high rate of uplift in the Phillipston area.

1 INTRODUCTION

There are very few techniques for determining the cooling history of regionally metamorphosed rock units. The most useful technique to date has been developed by Jäger and her colleagues (Jäger et al., 1967; Clark and Jäger, 1969; Hunziker and Bearth, 1969; Hunziker, 1970). They have demonstrated that muscovite and biotite have "closing" temperatures of 500°C and 300°C respectively for cation diffusion, and that the difference between the Rb/Sr age of individual grains of muscovite and biotite from a hand specimen is close to the time interval during which the containing rock unit cooled from 500°C to 300°C.

The method which we have developed is also based on diffusion during cooling. Chemically uniform crystals of garnet and cordierite are frequently found in high grade metamorphosed pelitic rocks. During cooling, cation exchange between adjacent grains of almandine and cordierite takes place in response to falling temperatures. This exchange is usually incomplete. Concentration gradients are therefore preserved on both sides of the boundary between adjacent grains of almandine and cordierite. Zoning in garnets which can be attributed to such a process, rather than to growth, is common in high-grade rocks, and has been described in a number of localities (Grant and Weiblen, 1968, 1971; Hess, 1969, 1971; Gable and Sims, 1969). The shape of such composition profiles is determined by the time–temperature history of the grains, and can be used to reconstruct the thermal evolution of metamorphic rock units. In principle, our approach is similar to that first used by Wood (1964), who set limits on the cooling history of iron meteorites by analyzing nickel and iron profiles in adjacent crystals of taenite and kamacite. The method is, therefore, quite general and can be applied to any pair of crystals of known initial composition, provided they have not equilibrated completely during cooling.

The extraction of time–temperature information from composition profiles produced by such incomplete equilibration requires the solution of a rather difficult form of the diffusion equation, as well as data for the diffusion coefficients of cations within the adjacent crystals and for the distribution coefficients of the cations between the adjacent crystals. This paper deals largely with the mathematics of cation exchange between neighboring crystals during a fall in temperature. The results are applied to Mg^{2+}–Fe^{2+} exchange between crystals of almandine and cordierite in a high-grade metamorphic unit near Phillipston, in central Massachusetts. The physical parameters for this cation exchange process are not well known. Nevertheless, the estimated cooling rates are geologically reasonable, there are grounds for optimism that the proposed method for obtaining cooling rates will be rather widely applicable, and that the analysis can be used to set limits on the diffusion constants and their temperature dependence.

2 ALMANDINE–CORDIERITE ROCKS OF THE PHILLIPSTON AREA, CENTRAL MASSACHUSETTS

The rock units of the Phillipston area were subjected to high-grade metamorphism during the Acadian orogeny between 380 and 410 million years B.P. (Faul et al., 1963; Lyons and Faul, 1968; Naylor, 1971) and to a weaker, late Paleozoic metamorphism. During the later event temperatures probably did not exceed 400°C. Although reheating probably did not affect the chemical composition of garnets and cordierites formed during the Acadian orogeny, it did affect the composition of the biotites. This is demonstrated by the 265 million year K/Ar age of two biotite samples from the Phillipston area.

Our samples were taken from a fresh roadcut on Barre Road, approximately 100 yards north of its intersection with state route 101 (Queen Lake Road), about one-third of a mile west of the Templeton town line.

Almandine, plagioclase, biotite, cordierite, sillimanite, ilmenite, zircon, quartz and graphite are the dominant minerals. Neither muscovite nor alkali feldspar were observed in thin sections, hand specimens, or in mineral separates. However, the geographic setting of the outcrop places it unequivocally on the high-temperature side of the regional metamorphic isograd defined by the reaction

$$\text{albite}_{ss} + \text{muscovite}_{ss} + \text{quartz} \leftrightarrows \text{K–feldspar}_{ss} + \text{sillimanite} + H_2O$$

and on the high-pressure side of the aluminosilicate triple point (Fig. 1). These relationships and the presence of migmatites indicate that peak metamorphic conditions were in the vicinity of 675°C and 5.0 kbar (Thompson, 1974).

The rocks are coarse-grained and faintly gneissic. The largest dimension of the garnet and cordierite grains is generally about 1 mm. Fracturing,

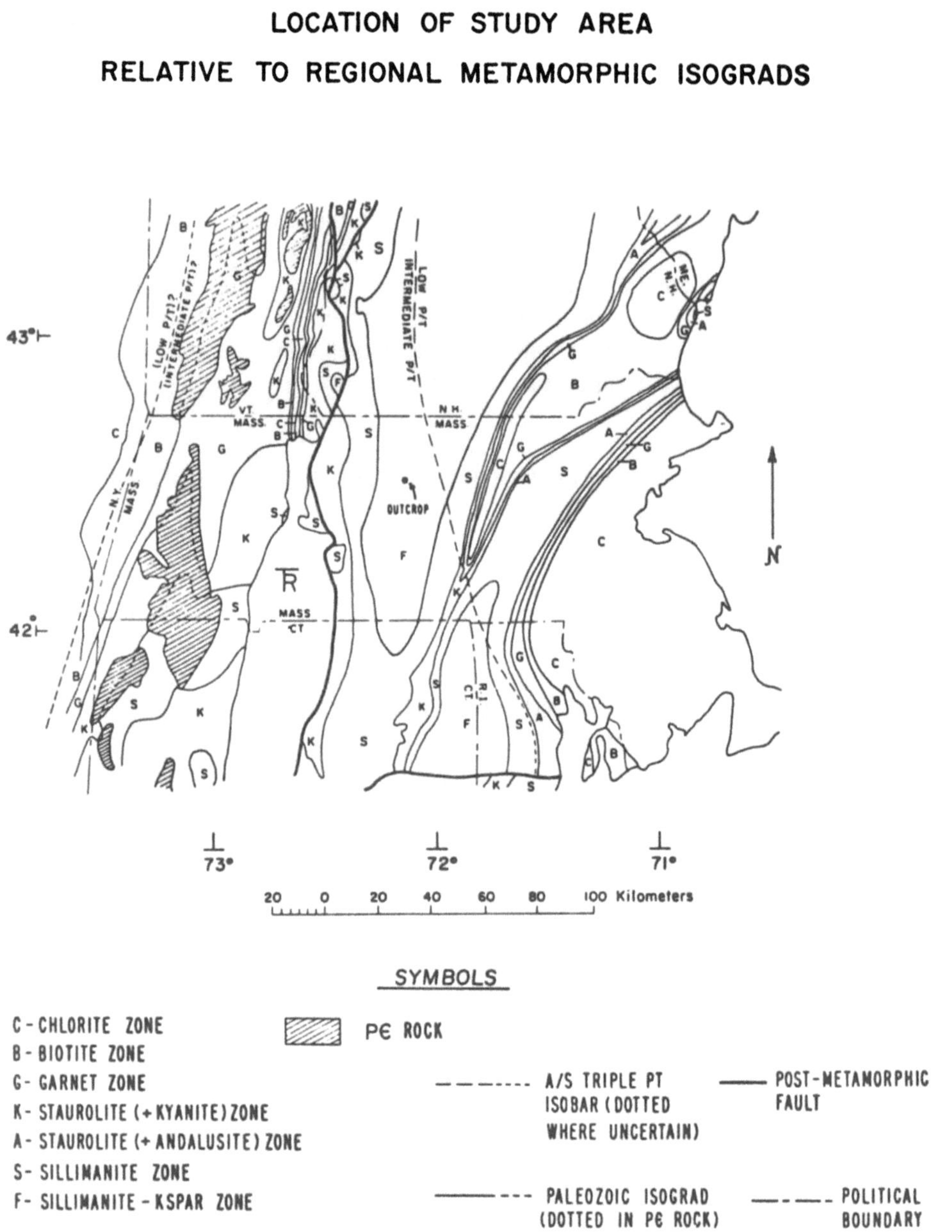

Fig. 1. Regional metamorphic isograds in central New England (after Thompson and Norton, 1968)

particularly microfracturing along grain boundaries, is minor, and no products of retrograde hydration reactions or other chemical alteration were observed. Some representative electron microprobe analyses of garnet and cordierite grains are shown in Table 1. Concentrations are given as mole percent of total oxides present. This convention is used throughout the paper.

Grains of both minerals were found to be chemically homogeneous except near mutual grain boundaries. To a first approximation, the garnets are binary solid solutions of almandine ($Fe_3Al_2Si_3O_{12}$) and pyrope ($Mg_3Al_2Si_3O_{12}$), with an average composition near Alm_{82}. CaO and MnO, the major impurities, are present in amounts less than 2.0 mol % and do not participate in the zoning. The average cordierite composition is approximately $(Fe_{0.3}Mg_{0.7})_2Al_4Si_5O_{18}$. No elements other than Fe, Mg, Al, and Si were observed in cordierites above the lower detection limit of the microprobe. Although garnet and cordierite crystals from the same thin section have similar compositions, they are not identical. The FeO/(FeO + MgO) ratio varies slightly from one area to another, suggesting the existence of small "equilibrium volumes" of the type discussed by Blackburn (1968). These regions are, however, much larger ($\geq$1 cm) than the grain boundary zones adjacent to garnet-cordierite contacts (<100 μm).

Zoning near garnet-cordierite grain boundaries was investigated by fixing a variable step scan motor to the sample stage drive of the electron microprobe and performing straightline scans perpendicular to the grain boundaries. Most scans were produced with a step interval of 1.5 or 2.0

Table 1. Representative mineral compositions (mol %)

	Garnets				Cordierites		
	#37	#15-1	#20	#8-6	#29	#9-1	#24
SiO_2	42.09	41.96	42.98	43.70	56.11	56.30	56.24
TiO_2	0.05	—	—	0.0	—	—	—
Al_2O_3	14.64	14.58	14.30	14.39	21.63	21.83	21.50
FeO	32.90	32.74	32.96	32.85	7.17	5.74	6.66
MgO	7.48	6.60	7.54	7.36	14.71	15.84	15.38
MnO	1.30	1.30	1.30	1.26	0.00	0.00	0.00
CaO	1.67	1.84	1.83	1.68	0.00	0.72	0.00
K_2O	0.0	—		0.0	—	—	—
Na_2O	0.0	—	—	0.0	—	—	—
H_2O	—	—	—	—	—	—	—
Total	100.12	99.02	100.91	101.24	99.62	100.44	99.77
Atoms per:		12 oxygens				18 oxygens	
Si	2.9443	2.9600	2.9902	3.0187	5.0754	5.0574	5.0863
Ti	0.0033	—	—	0.0000	—	—	—
Al	2.0481	2.0562	1.9897	1.9878	3.9129	3.9214	3.8889
Fe	2.3013	2.3091	2.2930	2.2696	0.6488	0.5152	0.6027
Mg	0.5234	0.4652	0.5245	0.5085	1.3310	1.4229	1.3913
Mn	0.0909	0.0915	0.0906	0.0867	0.0000	0.0000	0.0000
Ca	0.1169	0.1299	0.1271	0.1161	0.0000	0.0650	0.0000
K	0.0001	—	—	0.0000	—	—	—
Na	0.0000	—	—	0.0000	—	—	—

μm; the uncertainty in the location of analysis points with respect to the grain boundary was approximately 0.5 μm.

The precision of our analyses was estimated by performing several analyses of randomly selected points during single analytical sessions and processing the data separately. Comparison of the replicate analyses after computer reduction indicated a consistent uncertainty of ±0.1 mol % for magnesium and ±0.25 mol % for iron. The precision of analyses of other elements is not crucial to this study, and was not calculated.

The results of a typical step scan are shown in Figure 2. Figure 3 is an expanded representation of the two best Mg profiles in garnet. Zoning is only detectable in the distribution of iron and magnesium; it is restricted to the outermost 30 μm of the garnet grains and to the outermost 80–100 μm of the adjacent cordierite grains. The concentration of all other

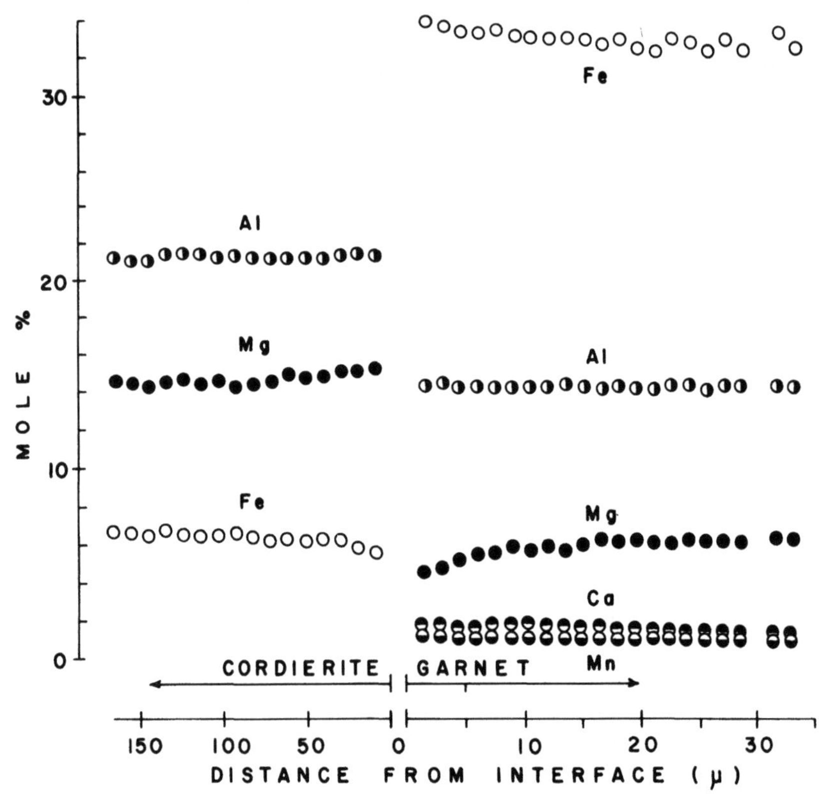

Fig. 2. Step scan chemical analyses across a garnet–cordierite grain boundary. Ca and Mn are below the lower detection limit in cordierite

elements which are present above the lower detection limit of the microprobe are constant within the analytical uncertainty. Within each grain the iron and magnesium profiles are mirror images of each other. Careful material balance calculations (to be presented in a forthcoming paper) indicate that the magnesium or iron excess on one side of a grain boundary is balanced by an equivalent magnesium or iron deficit on the other side of the same grain boundary. Simple Mg^{2+}–Fe^{2+} exchange, therefore, accounts for all of the observed chemical zoning near boundaries between garnet and cordierite grains.

The concentration of magnesium increases and the concentration of iron decreases toward the center of the garnet grains in all of the composition profiles. The reverse is true of the concentration of these elements in cordierites. Iron has, therefore, been transferred from cordierites to garnets, and an equivalent quantity of magnesium has been transferred from garnets to cordierites. The transfer took place almost certainly in response to a decrease in temperature. Its direction is in accord with predictions based on the temperature variation of the distribution coefficient of Mg^{2+} and Fe^{2+} between garnet and cordierite observed in the laboratory (Hensen and Green, 1971, 1972) and in the field (Thompson, 1976). The magnitude of the exchange and the shape of the concentration

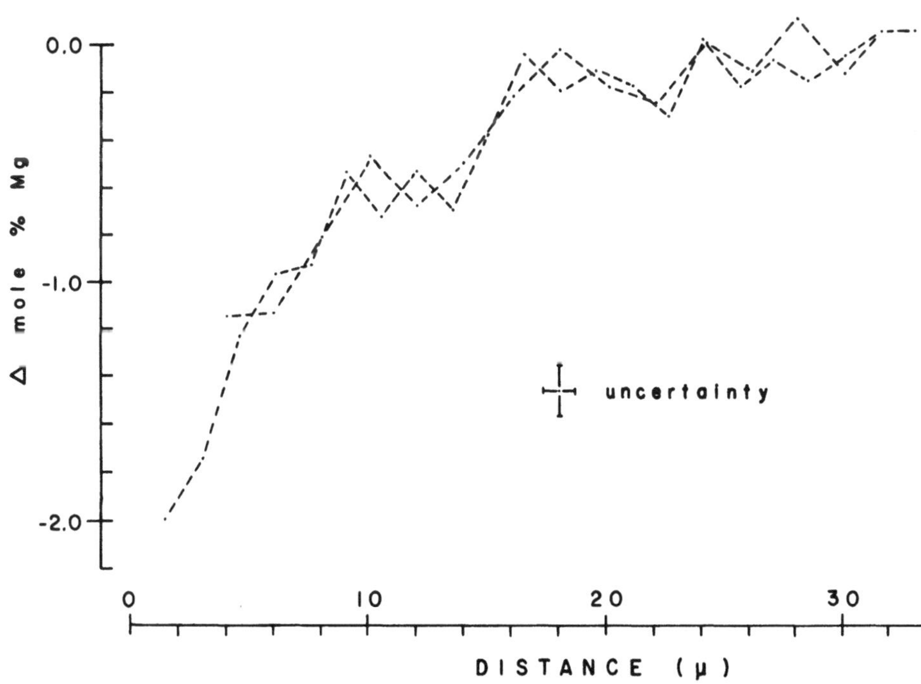

Fig. 3. Magnesium content as a function of distance from the grain boundary in two garnets adjacent to cordierites. For ease of comparison, concentrations in the two grains have been normalized by setting Δ mole percent Mg = 0.0 at their cores

profiles supply the observational data for our estimates of time–temperature relations during cooling of the Phillipston area at the end of the Acadian orogeny.

3 THE PHYSICS OF BINARY CATION EXCHANGE DURING RETROGRADE METAMORPHISM

The compositional data for garnets and cordierites in rocks from the Phillipston area suggest that the following propositions are reasonable starting points for the interpretation of the composition profiles near the grain boundaries between the two minerals.

(1) At the peak temperature T_0 of the Acadian orogeny each grain of garnet and cordierite attained a constant composition, although not all grains of garnet and cordierite attained precisely the same composition.
(2) At the peak of metamorphism adjacent crystals of garnet and cordierite were in chemical equilibrium.
(3) During cooling chemical equilibrium was maintained at the garnet-cordierite grain boundary.
(4) Only Mg^{2+} and Fe^{2+} were exchanged across the garnet-cordierite boundary during cooling. The concentrations of all of the other ions in each phase remained constant. If $c_{i\phi}(x,t)$ is defined as the mole percent of component i in phase ϕ at a distance x μm from the grain boundary at a time t million years after the onset of cooling, then proposition (1) implies that $c_{i\phi}(x,0) = c_{i\phi}^{I}$. The concentration of Mg^{2+} and Fe^{2+} in garnet and cordierite crystals at mutual grain boundaries were related by the equation

$$\frac{c_{\mathrm{Mg,gar}}(0,t)\, c_{\mathrm{Fe,cord}}(0,t)}{c_{\mathrm{Mg,cord}}(0,t)\, c_{\mathrm{Fe,gar}}(0,t)} = K_D(T) \tag{1}$$

where K_D is the distribution coefficient, as first defined by Ramberg and DeVore (1951), of Mg^{2+} and Fe^{2+} between the two phases. K_D is independent of the concentration of garnet and cordierite only when the two phases behave ideally. However, in the Phillipston rocks the variation of the Mg^{2+} and Fe^{2+} concentration in the two phases is only a few mole percent. The effects of deviations from nonideality in the Mg^{2+}–Fe^{2+} substitution in the two phases on $K_D(T)$ were therefore small, and errors introduced into the calculations by the assumption of a constant K_D at any given temperature are almost certainly negligible.
The effect of pressure on $K_D(T)$ is also probably small, since the differences in molar volume of the Fe^{2+} and Mg^{2+} end members of the garnet and cordierite series are less than 1%.
Unfortunately Equation (1) is a nonlinear function of the cation concentrations; this is the major cause for the mathematical complexity of time dependent cation exchange problems.

Variations of the distribution coefficient K_D with temperature are given by the standard thermodynamic formula

$$K_D(T) = K_D^{\infty}\, e^{-\Delta H^0/RT} \tag{2}$$

where K_D^{∞} is the extrapolated coefficient at $T = \infty$ and $\Delta \bar{H}_r^0$ is the standard molar enthalpy for the ion exchange reaction. $\Delta \bar{H}_r^0$ is almost certainly a function of temperature and pressure, and of the composition of the coexisting phases. However, the composition range of the Phillipston materials is small, and the temperature range of interest turns out to be $475° \leqslant T \leqslant 675°$C; we believe, therefore, that setting $\Delta \bar{H}_r^0$ constant in our calculations has not introduced any sizeable errors.

(5) Mass and charge were conserved at the garnet–cordierite boundary during retrograde metamorphism. The flux $J_{i\phi}(x)$ of an ion i in phase ϕ at a distance x from the boundary is given by Fick's first law

$$J_{i\phi}(x) = -D_{i\phi}(T)\frac{\partial c_{i\phi}}{\partial x} \tag{3}$$

where $D_{i\phi}(T)$ is the diffusion coefficient of ion i in phase ϕ at temperature T. Thus

$$\begin{aligned} D_{\mathrm{Mg,gar}}(T)\left.\frac{\partial c_{\mathrm{Mg,gar}}}{\partial x}\right|_{x=0} &= -D_{\mathrm{Mg,cord}}(T)\left.\frac{\partial c_{\mathrm{Mg,cord}}}{\partial x}\right|_{x=0} \\ &= D_{\mathrm{Fe,gar}}(T)\left.\frac{\partial c_{\mathrm{Fe,cord}}}{\partial x}\right|_{x=0} \\ &= -D_{\mathrm{Fe,cord}}\left.\frac{\partial c_{\mathrm{Fe,gar}}}{\partial x}\right|_{x=0} \end{aligned} \tag{4}$$

where x is always taken to be a positive distance from the boundary, and where the diffusion coefficients have been normalized to take into account the different molar volume and cation content of garnet and cordierite.

(6) Diffusion between garnet and cordierite can be described by a one-dimensional model. This assumes that the radius of curvature of the grain boundary is large compared to the diffusion distance, and that the concentration gradients to which the ions are responding are essentially perpendicular to the interface. This assumption is justified for the Phillipston material and eliminates the need for a diffusion tensor.

The rate of change of the concentration of ion i at x and t is, therefore, given by the expression

$$\frac{\partial c_{i\phi}}{\partial t} = \frac{\partial}{\partial x}\left(D_{i\phi}(T)\frac{\partial c_{i\phi}}{\partial x}\right). \tag{5}$$

Since both Fe^{2+} and Mg^{2+} are codiffusing in a given grain, one cannot use the independent self-diffusion coefficients $D^0_{\mathrm{Mg},\phi}$ and $D^0_{\mathrm{Fe},\phi}$ in Equations (4) and (5), but rather a single interdiffusion coefficient D_ϕ for the pair of ions in each mineral ϕ. D_ϕ is given by the expression

$$D_\phi = \frac{D^0_{\mathrm{Fe},\phi}\, D^0_{\mathrm{Mg},\phi}}{N_{\mathrm{Fe},\phi} D^0_{\mathrm{Fe},\phi} + N_{\mathrm{Mg},\phi}\, D^0_{\mathrm{Mg},\phi}} \tag{6}$$

where $N_{i\phi}$ is the mole fraction $c_{i\phi}/(c_{i\phi} + c_{j\phi})$ in phase ϕ, and $D^0_{i\phi}$ is the self-diffusion coefficient of ion i at infinite dilution in the binary solid solution (Smigelskas and Kirkendall, 1947). D_ϕ is, therefore, a function of the composition of phase ϕ. However, in the Phillipston material the differences between the composition of garnet and cordierite at the grain boundaries and at the center of the grains are so small, that the compositional dependence of the diffusion constants D_ϕ can probably be neglected. A rough calculation indicates that variations of 2 to 3 mol % of Fo or Fa in the composition of olivine change the interdiffusion coefficient of Mg^{2+} and Fe^{2+} in this phase by less than 50% (Buening and Buseck, 1973).
The temperature dependence of D_ϕ is of the form

$$D_\phi = D^\infty_\phi\, e^{-(\Delta \bar{H}_D)_\phi/RT} \tag{7}$$

where D^∞_ϕ probably depends slightly on temperature. In our calculations we have taken both D^∞_ϕ and $(\Delta \bar{H}_D)_\phi$ to be constant within the range of temperature and composition that are of interest at Phillipston.
(7) Electrical neutrality was maintained at all distances x from the grain boundaries. This implies that

$$c_{\mathrm{Fe},\phi}(x,t) + c_{\mathrm{Mg},\phi}(x,t) = \text{constant}$$

for all ϕ at all x and t. The concentration profiles of Fe^{2+} and Mg^{2+} are, therefore, mirror-images within each phase, and allow us to deal with the concentration of a single element in the two phases. In our analysis we have chosen Mg^{2+} rather than Fe^{2+} because the analytical data for Mg^{2+} are somewhat more precise.

4 THE MATHEMATICS OF BINARY CATION EXCHANGE DURING RETROGRADE METAMORPHISM

It was shown above that the interdiffusion constant D_ϕ is sufficiently independent of composition that it can be taken to be independent of x. Equation (5), therefore, becomes

$$\frac{\partial c_\phi}{\partial t} = D_\phi(t)\,\frac{\partial^2 c_\phi}{\partial x^2}. \tag{8}$$

The subscript i has been dropped, since it was shown that the concentration of only one ion need be considered in formulating the mathematics of Mg^{2+}–Fe^{2+} exchange between almandine and cordierite. c_ϕ in Equation (8) and henceforth refers to the mole percent of Mg^{2+} in phase ϕ.

$c_{\text{gar}}(x,t)$ and $c_{\text{cord}}(x,t)$ must satisfy the conditions

$$D_{\text{gar}}(t)\left.\frac{\partial c_{\text{gar}}}{\partial x}\right|_{x=0} = -D_{\text{cord}}(t)\left.\frac{\partial c_{\text{cord}}}{\partial x}\right|_{x=0} \tag{9}$$

and the nonlinear equation

$$\frac{c_{\text{gar}}(0,t)(c^{\text{I}}_{\text{Fe,cord}} - (c_{\text{cord}}(0,t) - c^{\text{I}}_{\text{Mg,cord}}))}{c_{\text{cord}}(0,t)(c^{\text{I}}_{\text{Fe,gar}} - (c_{\text{gar}}(0,t) - c^{\text{I}}_{\text{Mg,gar}}))} = K_D(T(t)). \tag{10}$$

We must now find a function $c_\phi(x,t)$ which satisfies condition (1) in Section II and Equations (8), (9), and (10). To accomplish this, we first obtain a general solution of Equation (8), using the diffusion Green's function (Sneddon, 1975), and then demand that the solution satisfy conditions (9) and (10). We can deal with the variation of the diffusion coefficient with time by defining a new time variable (which really has units of area):

$$t'_\phi = \int_0^t D_\phi(\tau)\, d\tau. \tag{11}$$

Hence

$$\frac{\partial c_\phi}{\partial t'_\phi} = \frac{1}{D_\phi(t)}\frac{\partial c_\phi}{\partial t}$$

and thus

$$\frac{\partial c_\phi}{\partial t'_\phi} = \frac{\partial^2 c_\phi}{\partial x^2}. \tag{12}$$

The initial conditions with the new variables are still given by $c_\phi(x,0) = c^I_\phi$, since $t'_\phi = 0$ when $t = 0$. However, to use Equation (12), we must now give the boundary conditions in terms of t'_ϕ, not in terms of t. If $g_\phi(t'_\phi)$ is the boundary concentration of Mg^{2+} (in garnet or cordierite) at the time t corresponding to t'_ϕ in Equation (11), we can write the general solution of Equation (12) satisfying $c_\phi(x,0) = c^{\text{I}}_\phi$ as

$$\begin{aligned} c_\phi(x,t) = c_\phi(x,t'_\phi) &= \frac{x}{2\sqrt{\pi}}\int_0^{t'} g_\phi(t'')\,\frac{e^{-x^2/4(t'_\phi-t'')}}{(t'_\phi - t'')^{3/2}}\,dt'' \\ &\quad + c^{\text{I}}_\phi \operatorname{erf}\left(\frac{x}{2\sqrt{t'_\phi}}\right). \end{aligned} \tag{13}$$

However, in our case, we are not given $g_\phi(t'_\phi)$ explicitly as a boundary condition, but we are given more complex and nonlinear boundary conditions. To satisfy these complex conditions, it is necessary to derive the general form of the flux term $\partial c_\phi/\partial x|_{x=0}$, from the general Equation (13). Unfortunately, the evaluation of $\partial c_\phi/\partial x$ at $x = 0$ turns out to be plagued by mathematical diseases (see appendix for details). We give here the final simplified solution:

$$\left.\frac{\partial c_\phi}{\partial x}\right|_{x=0} = \frac{1}{\sqrt{\pi t'_\phi}}[c^I_\phi - g_\phi(t'_\phi)] + \frac{2}{\sqrt{\pi}}\int_0^{t'_\phi} \frac{g_\phi(z) - g_\phi(t'_\phi)}{4(t'_\phi - z)^{3/2}}\, dz \tag{14}$$

which demands that $g(z)$ be analytic at $z = t'_\phi$.

To proceed further with Equations (13) and (14), we must know the explicit dependence of D_ϕ and K_D on time. From Equations (2) and (7) it follows that

$$D_\phi(T) = D^\infty_\phi\, e^{-(\Delta H_D)_\phi/RT} \tag{15a}$$

$$K_D(T) = K^\infty_D\, e^{-\Delta H^0_r/RT} \tag{15b}$$

If we define the more useful D^0_ϕ and K^0_D as the values of the diffusion coefficient and the distribution coefficient at the peak metamorphic temperature T_0 we can rewrite Equations (15a) and (15b) in the form

$$D_\phi(T) = D^0_\phi\, e^{-(\Delta H_D)_\phi[1/T - 1/T_0]/R}$$

or

$$D_\phi(T) = D^0_\phi\, e^{-\Delta(H_D)_\phi[\Delta T/TT_0]/R} \tag{16a}$$

Similarly,

$$K_D(T) = K^0_D\, e^{-\Delta H^0_r[\Delta T/TT_0]/R} \tag{16b}$$

where $\Delta T \equiv T_0 - T$. If ΔT is small compared to T_0, we can replace T in Equations (16a) and (16b) by T_0. Finally, if the temperature decreases linearly with time, so that

$$T = T_0 - st \tag{17}$$

then,

$$D_\phi(t) = D^0_\phi\, e^{-\gamma_\phi t} \tag{18a}$$

and

$$K_D(t) = K^0_D\, e^{-\beta t} \tag{18b}$$

where

$$\gamma_\Phi = (\Delta\bar{H}_D)_\Phi s/(RT_0^2) \tag{18c}$$
$$\beta = \Delta\bar{H}_r^0\, s/(RT_0^2). \tag{18d}$$

We could have made the approximate Equations (18a) and (18b) exact, if we had chosen the time–temperature relationship

$$T = \frac{T_0}{1 + \dfrac{st}{T_0}}\ . \tag{19}$$

If $st << T_0$, Equation (19) can be expanded to give $T = T_0 - st$; that is, Equation (19) agrees with Equation (17) for all times for which $st << T_0$. Equation (19) is actually more realistic than Equation (17), since $T \rightarrow 0$ as $t \rightarrow \infty$, whereas in Equation (17) $T \rightarrow -\infty$ as $t \rightarrow \infty$.

The ultimate parameter of geologic interest, naturally, is the cooling rate s. Given $D_\Phi(t)$ as in Equation (18) we obtain that

$$t'_\Phi = \frac{D_\Phi^0}{\gamma_\Phi}[1 - e^{-\gamma_\Phi t}]. \tag{20}$$

We can now rewrite the flux Equation (14) in terms of t, rather than t'_Φ. To do this we must write $g_\Phi(t'_\Phi)$ in terms of $f_\Phi(t) \equiv c_\Phi(0,t)$.

By definition, $g_\Phi(t'_\Phi) = f_\Phi(t)$ as long as t'_Φ and t are related by Equation (20). If we change variables in Equation (14) to

$$u = \frac{-1}{\gamma_\Phi}\ln\left(1 - \frac{\gamma_\Phi}{D_\Phi^0}z\right)$$

or

$$z = \frac{D_\Phi^0}{\gamma_\Phi}(1 - e^{-\gamma_\Phi u}),$$

we can rewrite Equation (14) in the form

$$\left.\frac{\partial c_\Phi}{\partial x}\right|_{x=0} = \frac{1}{\sqrt{\pi t'_\Phi}}(c_{\mathrm{Mg},\Phi}^{\mathrm{I}} - f_\Phi(t)) + \frac{\gamma_\Phi^{3/2}}{2\sqrt{\pi D_\Phi^0}}\int_0^{t_\Phi -} \frac{(f_\Phi(u) - f_\Phi(t))e^{-\gamma_\Phi u}}{(e^{-\gamma_\Phi u} - e^{-\gamma_\Phi t})^{3/2}}\,du \tag{21}$$

We still have the unknown $f_\Phi(t)$ in the integrand. This can be remedied by breaking up the cooling time into small intervals of magnitude h.

If we know $f_\phi(u)$ for u up to $t-h$ and we seek the next value of f_ϕ (i.e., $f_\phi(t)$), we can break down Equation (21) into two parts:

$$\left.\frac{\partial c_\phi}{\partial x}\right|_{x=0} = \frac{1}{\sqrt{\pi\dfrac{D_\phi^0}{\gamma_\phi}}(1-e^{-\gamma_\phi t})^{1/2}}[c^{\mathrm{I}}_{\mathrm{Mg},\phi} - f_\phi(t)]$$
$$+ \frac{\gamma_\phi^{3/2}}{2\sqrt{\pi D_\phi^0}}\int_0^{t-h}\frac{(f_\phi(u)-f_\phi(t))e^{-\gamma_\phi u}}{(e^{-\gamma_\phi u}-e^{-\gamma_\phi t})^{3/2}}\,du \tag{22}$$
$$+ \frac{\gamma_\phi^{3/2}}{2\sqrt{\pi D_\phi^0}}\int_{t-h}^{t-}\frac{(f_\phi(u)-f_\phi(t))e^{-\gamma_\phi u}}{(e^{-\gamma_\phi u}-e^{-\gamma_\phi t})^{3/2}}\,du.$$

Breaking the first integral into two further parts we can define $\tilde{A}_\phi$ and $\tilde{B}_\phi$ as

$$\frac{\gamma_\phi^{3/2}}{2\sqrt{\pi D_\phi^0}}\int_0^{t-h}\frac{f_\phi(u)e^{-\gamma_\phi u}}{(e^{-\gamma_\phi u}-e^{-\gamma_\phi t})^{3/2}}\,du \equiv \tilde{A}_\phi \tag{23a}$$

and

$$-\frac{\gamma_\phi^{3/2}}{2\sqrt{\pi D_\phi^0}}\left[\int_0^{t-h}\frac{e^{-\gamma_\phi u}}{(e^{\gamma_\phi u}-e^{-\gamma_\phi t})^{3/2}}\,du\right]f_\phi(t) \equiv -\tilde{B}_\phi f_\phi(t), \tag{23b}$$

where $\tilde{A}_\phi$ and $\tilde{B}_\phi$ can be calculated from the known values of $f_\phi(u)$. Finally, in the second integral of Equation (22) we can write

$$f_\phi(u) - f_\phi(t) = \frac{f_\phi(t-h)-f_\phi(t)}{h}(t-u)$$

in the small region $t-h \leqslant u \leqslant t$; thus,

$$\frac{\gamma^{3/2}}{2\sqrt{\pi D_\phi^0}}\int_{t-h}^{t-}\frac{f_\phi(t-h)-f_\phi(t)}{h}\,\frac{(t-u)e^{-\gamma_\phi u}}{(e^{-\gamma_\phi u}-e^{-\gamma_\phi t})^{3/2}}\,du. \tag{24}$$

Using Equation (24) we define $\tilde{C}_\phi$ and $\tilde{D}_\phi$ as

$$\frac{f_\phi(t-h)}{h}\,\frac{\gamma_\phi^{3/2}}{2\sqrt{\pi D_\phi^0}}\int_{t-h}^{t-}\frac{(t-u)e^{-\gamma_\phi u}}{(e^{-\gamma_\phi u}-e^{-\gamma_\phi t})^{3/2}}\,du = \tilde{C}_\phi \tag{25a}$$

and

$$-\frac{1}{h}\,\frac{\gamma_\phi^{3/2}}{2\sqrt{\pi D_\phi^0}}\int_{t-h}^{t-}\frac{(t-u)e^{-\gamma_\phi u}}{(e^{-\gamma_\phi u}-e^{-\gamma_\phi t})^{3/2}}\,du\,f_\phi(t) = -\tilde{D}_\phi f_\phi(t), \tag{25b}$$

where $\tilde{C}_\phi$ and $\tilde{D}_\phi$ can, again, be calculated from the known values of $f_\phi(u)$.

It turns out that we can do the integrals in Equations (23a), (25a), and (25b) analytically and thus altogether we have

$$\left.\frac{\partial c_\phi}{\partial x}\right|_{x=0} = \sqrt{\frac{\gamma_\phi}{\pi D_\phi^0}} \frac{1}{(1-e^{-\gamma_\phi t})^{1/2}} [c_{\mathrm{Mg},\phi}^{\mathrm{I}} - f_\phi(t)] + \tilde{A}_\phi - \tilde{B}_\phi f_\phi(t) + \tilde{C}_\phi - \tilde{D}_\phi f_\phi(t), \qquad (26)$$

where

$$\tilde{A}_\phi = \frac{\gamma_\phi^{3/2}}{2\sqrt{\pi D_\phi^0}} \int_0^{t-h} \frac{f_\phi(u) e^{-\gamma_\phi u}}{(e^{-\gamma_\phi u} - e^{-\gamma_\phi t})^{3/2}} du$$

$$\tilde{B}_\phi = \frac{\gamma_\phi^{3/2}}{2\sqrt{\pi D_\phi^0}} \left(\frac{2}{\gamma_\phi}\right) [e^{\gamma_\phi t/2}(e^{\gamma_\phi h} - 1)^{-1/2} - (1 - e^{-\gamma_\phi t})^{-1/2}]$$

$$\tilde{C}_\phi = \frac{\gamma_\phi^{3/2} f_\phi(t-h) e^{\gamma_\phi t/2}}{2\sqrt{\pi D_\phi^0}\, h} \left[\frac{4}{\gamma_\phi^2} \tan^{-1}(e^{\gamma_\phi h} - 1)^{1/2} - \frac{2h}{\gamma_\phi (e^{\gamma_\phi h} - 1)^{1/2}}\right]$$

$$\tilde{D}_\phi = \frac{\gamma_\phi^{3/2} e^{\gamma_\phi t/2}}{2\sqrt{\pi D_\phi^0}\, h} \left[\frac{4}{\gamma_\phi^2} \tan^{-1}(e^{\gamma_\phi h} - 1)^{1/2} - \frac{2h}{\gamma_\phi (e^{\gamma_\phi h} - 1)^{1/2}}\right].$$

We can finally write

$$D_\phi(t) \left.\frac{\partial c_\phi}{\partial x}\right|_{x=0} = D_\phi^0 e^{-\gamma_\phi t} \left.\frac{\partial c_\phi}{\partial x}\right|_{x=0} = A_\phi - B_\phi f_\phi(t), \qquad (27)$$

where

$$A_\phi \equiv \left(\sqrt{\frac{\gamma_\phi}{\pi D_\phi^0}} \frac{c_{\mathrm{Mg},\phi}^{\mathrm{I}}}{(1 - e^{-\gamma_\phi t})^{1/2}} + \tilde{A}_\phi + \tilde{C}_\phi\right) D_\phi^0 e^{-\gamma_\phi t}$$

$$B_\phi \equiv \left(\sqrt{\frac{\gamma_\phi}{\pi D_\phi^0}} \frac{1}{(1 - e^{-\gamma_\phi t})^{1/2}} + \tilde{B}_\phi + \tilde{D}_\phi\right) D_\phi^0 e^{-\gamma_\phi t}.$$

Now we do this for the Mg profile in each phase ϕ and obtain as our flux condition

$$A_{\mathrm{gar}} - B_{\mathrm{gar}} f_{\mathrm{gar}}(t) = -(A_{\mathrm{cord}} - B_{\mathrm{cord}} f_{\mathrm{cord}}(t)). \qquad (28)$$

The equilibrium boundary condition then becomes

$$\frac{f_{\mathrm{gar}}(t)\,[c_{\mathrm{Fe,cord}}^{\mathrm{I}} - f_{\mathrm{cord}}(t) + c_{\mathrm{Mg,cord}}^{\mathrm{I}}]}{f_{\mathrm{cord}}(t)\,[c_{\mathrm{Fe,gar}}^{\mathrm{I}} + c_{\mathrm{Mg,gar}}^{\mathrm{I}} - f_{\mathrm{gar}}(t)]} = K_D(t) \qquad (29)$$

where again we have used the mirror image assumption for Fe^{2+} profiles.

We can solve the equations above for $f_{\text{gar}}(t)$ and $f_{\text{cord}}(t)$ by using

$$f_{\text{gar}}(t) = \frac{-b \pm \sqrt{b^2 - 4ac}}{2a} \tag{30}$$

where

$$\begin{aligned}
a &\equiv \frac{-B_{\text{gar}}}{B_{\text{cord}}}(1 - K_D(t)) \\
b &\equiv \left(\frac{A_{\text{cord}}}{B_{\text{cord}}} + \frac{A_{\text{gar}}}{B_{\text{cord}}} - c^{\text{I}}_{\text{Fe,cord}} - c^{\text{I}}_{\text{Mg,cord}}\right) \\
&\quad - \left[(c^{\text{I}}_{\text{Fe,gar}} + c^{\text{I}}_{\text{Mg,gar}})\frac{B_{\text{gar}}}{B_{\text{cord}}} + \frac{A_{\text{gar}}}{B_{\text{cord}}} + \frac{A_{\text{cord}}}{B_{\text{cord}}}\right] K_D(t) \\
c &\equiv \frac{A_{\text{cord}} + A_{\text{gar}}}{B_{\text{cord}}}(c^{\text{I}}_{\text{Fe,gar}} + c^{\text{I}}_{\text{Mg,gar}}) K_D(t),
\end{aligned}$$

and

$$f_{\text{cord}}(t) = \frac{A_{\text{cord}} + A_{\text{gar}}}{B_{\text{cord}}} - \frac{B_{\text{gar}}}{B_{\text{cord}}} f_{\text{gar}}(t). \tag{31}$$

In this manner, we can obtain $f_{\text{gar}}(t)$ and $f_{\text{cord}}(t)$ quickly and exactly, and hence a solution of the original problem.

It is important to know whether there are any simple analytic forms for the function $c_\phi(0,t)$, i.e., the concentration of Mg^{2+} in garnet and cordierite at the garnet–cordierite boundary, during retrograde metamorphism. If the temperature–time relationship is basically linear [Eq. (19)], then it can be shown that the concentrations at the grain boundary change linearly with time for a nontrivial fraction of the cooling interval, i.e.,

$$c_\phi(0,t) \equiv f_\phi(t) = c^{\text{I}}_{\text{Mg},\phi} - \alpha_\phi t \tag{32}$$

where α_ϕ is an as yet unspecified constant. While this is true, we can insert Equation (32) for $f_\phi(t)$ into Equation (20), and obtain a boundary flux given by

$$D_\phi(t) \left.\frac{\partial c_\phi}{\partial x}\right|_{x=0} = \frac{2\alpha_\phi}{\sqrt{\pi \gamma_\phi}} \sqrt{D^0_\phi}\, e^{-\gamma_\phi t/2} \tan^{-1}(e^{\gamma_\phi t} - 1)^{1/2}. \tag{33}$$

This flux reaches a sharp maximum early during the cooling interval (Fig. 4), with the maximum at $t_{\max} = 0.85/\gamma_\phi$. If we let $\gamma_\phi t << 1$, we can reduce Equation (33) to

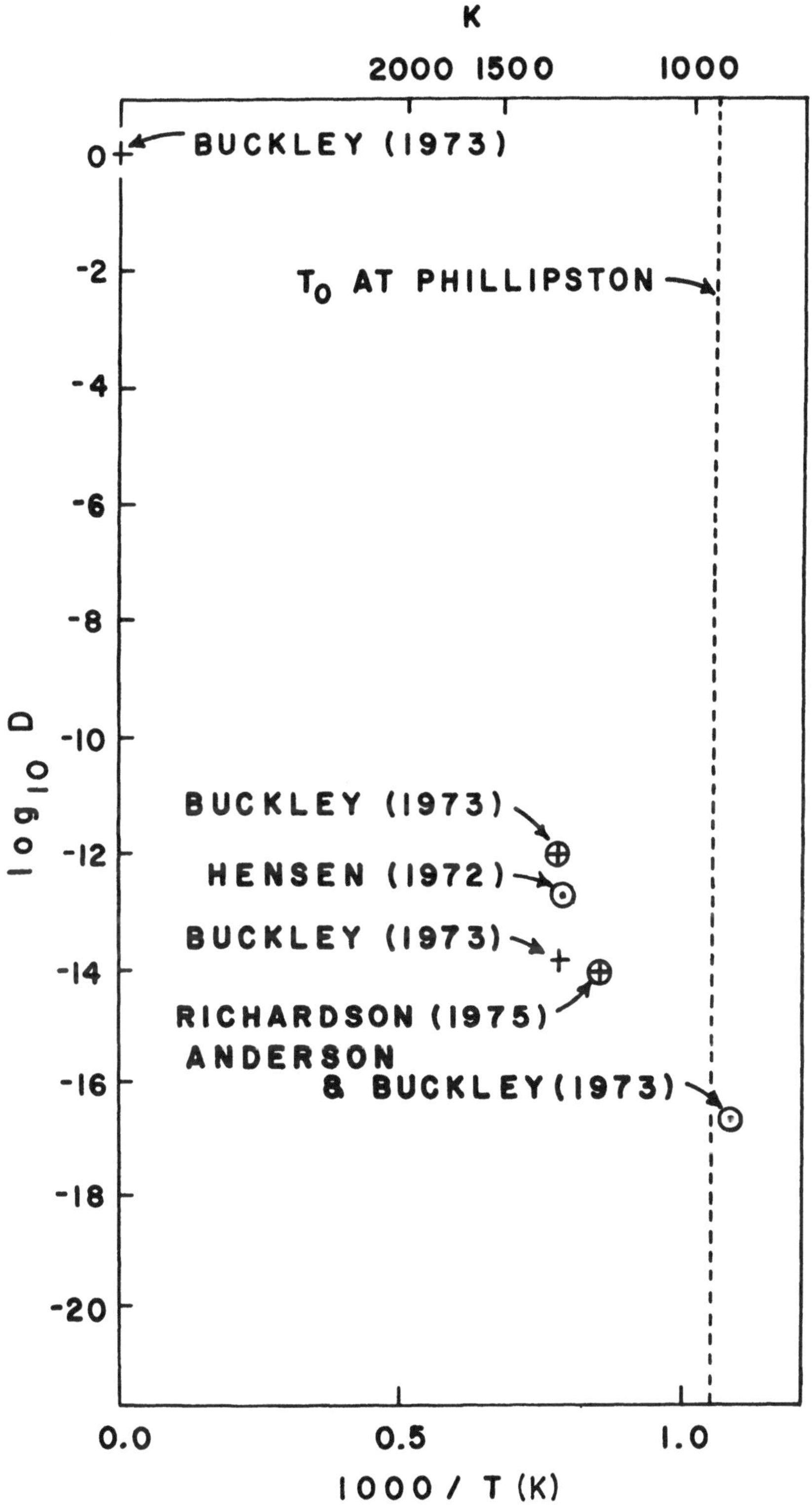

Fig. 4. Diffusive flux of material at a garnet–cordierite grain boundary as a function of temperature during cooling

$$D_\phi(t) \frac{\partial c_\phi}{\partial x}\bigg|_{x=0} \cong 2\alpha_\phi \sqrt{\frac{D^0_\phi t}{\pi}}. \tag{34}$$

We can use Equation (34) to solve for α_ϕ in this small time regime ($\gamma_\phi t << 1$). To do this we rewrite K_D as

$$K_D = K^0_D\, e^{-\beta t} \cong K_D(1 - \beta t). \tag{35}$$

If we insert

$$\begin{aligned} c_{\text{gar}}(0,t) &= c^{\text{I}}_{\text{Mg,gar}} - \alpha_{\text{gar}}\, t \\ c_{\text{cord}}(0,t) &= c^{\text{I}}_{\text{Mg,cord}} + \alpha_{\text{cord}}\, t \end{aligned} \tag{36}$$

and Equation (35) into Equation (10) and if we linearize the left-hand side with respect to t, we find that

$$\alpha_{\text{gar}} \left(\frac{1}{c^{\text{I}}_{\text{Mg,gar}}} + \frac{1}{c^{\text{I}}_{\text{Fe,gar}}}\right) + \alpha_{\text{cord}} \left(\frac{1}{c^{\text{I}}_{\text{Mg,cord}}} + \frac{1}{c^{\text{I}}_{\text{Fe,cord}}}\right) = \beta. \tag{37}$$

Using Equation (34), we can also rewrite the flux condition [Eq. (9)] as

$$2\alpha_{\text{gar}} \sqrt{\frac{D^0_{\text{gar}}\, t}{\pi}} = 2\alpha_{\text{cord}} \sqrt{\frac{D^0_{\text{cord}}\, t}{\pi}}. \tag{38}$$

Equations (37) and (38) together yield

$$\begin{aligned} \alpha_{\text{gar}} &= \frac{\beta\sqrt{D^0_{\text{cord}}/D^0_{\text{gar}}}}{\sqrt{D^0_{\text{cord}}/D^0_{\text{gar}}}\left(\frac{1}{c^{\text{I}}_{\text{Mg,gar}}} + \frac{1}{c^{\text{I}}_{\text{Fe,gar}}}\right) + \left(\frac{1}{c^{\text{I}}_{\text{Mg,cord}}} + \frac{1}{c^{\text{I}}_{\text{Fe,cord}}}\right)} \\ \alpha_{\text{cord}} &= \frac{\beta}{\sqrt{D^0_{\text{cord}}/D^0_{\text{gar}}}\left(\frac{1}{c^{\text{I}}_{\text{Mg,gar}}} + \frac{1}{c^{\text{I}}_{\text{Fe,gar}}}\right) + \left(\frac{1}{c^{\text{I}}_{\text{Mg,cord}}} + \frac{1}{c^{\text{I}}_{\text{Fe,cord}}}\right)}. \end{aligned} \tag{39}$$

Equations (39), though only exact for small times, are very illuminating. Clearly, the size of β (hence $\Delta\bar{H}^0_r$) directly correlates with the rate of variation of the boundary concentration with time. Thus we expect that variations in $\Delta\bar{H}^0_r$ will have sizable effects on the concentration profiles near the grain boundary. The T_0 values of the diffusion coefficients, D^0_{gar}, D^0_{cord}, only influence $c(0,t)$, in this approximation, by the square root of their ratio, $(D^0_{\text{cord}}/D^0_{\text{gar}})^{1/2}$. However, this means that if, for example, one of the pair of minerals has a much higher diffusion coefficient, the actual value of either of the diffusion coefficients will have a negligible effect on the behavior of $c_\phi(0,t)$ in the other mineral;

e.g., if $D^0_{\text{cord}} >> D^0_{\text{gar}}$, α_{gar} is independent of both D^0_{cord} and D^0_{gar}. This information is very useful in understanding the effects of uncertainties in D^0_ϕ on the calculated profiles.

In general we have found that the exact functions $f_\phi(t)$ are reasonably linear functions of time for widely different values of the thermodynamic parameters. This has led us to believe that a refinement of the simple formula in Equations (36) and (39) can lead to a very satisfactory treatment of the whole cooling process. If we use Equation (10) and set $c_{\text{gar}}(0,t) = f_{\text{gar}}(t)$ and $c_{\text{cord}}(0,t) = f_{\text{cord}}(t)$, we can solve for $f_{\text{gar}}(t)$ in terms of $f_{\text{cord}}(t)$:

$$f_{\text{gar}}(t) = \frac{f_{\text{cord}}(t)\,[c^{\text{I}}_{\text{Fe,gar}} + c^{\text{I}}_{\text{Mg,gar}}]K_D(t)}{c^{\text{I}}_{\text{Mg,cord}} + c^{\text{I}}_{\text{Fe,cord}} - f_{\text{cord}}(t) + f_{\text{cord}}(t)K_D(t)}. \tag{40}$$

Equation (40) is exact. The refinement relies, first of all, on the slow variation of the boundary concentration with time in the mineral with the higher diffusion coefficient, as predicted by Equation (39). In our case, this mineral would be cordierite. Thus, we can treat $f_{\text{cord}}(t)$ approximately and derive $f_{\text{gar}}(t)$ from Equation (40). If we now approximate $f_{\text{cord}}(t)$ in Equation (40) by $c^{\text{I}}_{\text{Mg,cord}} + \alpha_{\text{cord}}t$ with α_{cord} given by Equation (39), we obtain an improvement for $f_{\text{gar}}(t)$ over that given by $f_{\text{gar}}(t) = c^{\text{I}}_{\text{Mg,gar}} - \alpha_{\text{gar}}t$. In fact, we have extended our refinement to include the possible difference between $(\Delta\bar{H}_D)_{\text{gar}}$ and $(\Delta\bar{H}_D)_{\text{cord}}$ by replacing the ratio $D^0_{\text{cord}}/D^0_{\text{gar}}$ in Equation (39) with the exact time-dependent ratio $D_{\text{cord}}(t)/D_{\text{gar}}(t)$. Thus,

$$\alpha_{\text{cord}}(t) = \frac{\beta}{\sqrt{\dfrac{D^0_{\text{cord}}}{D^0_{\text{gar}}}}\, e^{1/2(\gamma_{\text{gar}} - \gamma_{\text{cord}})t}\left(\dfrac{1}{c^{\text{I}}_{\text{Mg,gar}}} + \dfrac{1}{c^{\text{I}}_{\text{Fe,gar}}}\right) + \left(\dfrac{1}{c^{\text{I}}_{\text{Mg,cord}}} + \dfrac{1}{c^{\text{I}}_{\text{Fe,cord}}}\right)}. \tag{41}$$

Equation (39) implies that $df_{\text{cord}}/dt = \alpha_{\text{cord}}(0)$. We now set $df_{\text{cord}}/dt = \alpha_{\text{cord}}(t)$ and use Equation (41) to solve for f_{cord};

$$f_{\text{cord}}(t) = c^{\text{I}}_{\text{cord}} + \frac{2\beta}{(\gamma_{\text{cord}} - \gamma_{\text{gar}})K_2} \ln \frac{K_1 + K_2 e^{1/2(\gamma_{\text{cord}} - \gamma_{\text{gar}})t}}{K_1 + K_2} \tag{42}$$

where

$$K_1 \equiv \sqrt{\frac{D^0_{\text{cord}}}{D^0_{\text{gar}}}}\left(\frac{1}{c^{\text{I}}_{\text{Mg,gar}}} + \frac{1}{c^{\text{I}}_{\text{Fe,gar}}}\right);$$

$$K_2 = \frac{1}{c^{\text{I}}_{\text{Mg,cord}}} + \frac{1}{c^{\text{I}}_{\text{Fe,cord}}}.$$

Note that if $(\gamma_{cord} - \gamma_{gar})t << 1$, Equation (42) reduces exactly to $f_{cord}(t) = c^{I}_{cord} + \alpha_{cord}t$; with α_{cord} given by Equation (39).

5 COOLING RATES AT PHILLIPSTON, MASSACHUSSETTS, U.S.A.

The results of the mathematical analysis in the previous section can be applied to extract cooling rate information from the composition profiles in garnets and cordierites from the Phillipston area. Unfortunately this information is somewhat incomplete and rather imprecise, because data for the distribution coefficient K_D and particularly for diffusion rates in garnet and cordierite are inadequate. In spite of these handicaps, reasonably useful bounds can be set on the cooling rate in the Phillipston area at the end of the Acadian orogeny.

We have been able to find or to calculate from published data only a few values for $D_{gar}(T)$ and not all of these are for garnets on the almandine–pyrope binary. Plotted as $\log_{10} D_{gar}(T)$ versus $1000/T$ (K) in Figure 5, these values yield an estimate of $10^{-18\pm1.5}$ cm²/second for $D_{gar}(T_0)$ at Phillipston's estimated peak metamorphic temperature of 950 K, and an estimate of 78 ± 10 kcal/mol for $(\Delta\bar{H}_D)_{gar}$. There are no diffusion data for Mg^{2+} and Fe^{2+} in cordierite. We observed, however, that the average penetration distance $\bar{x}$ at which $c_{i\phi}(\bar{x},t) = 1/2[c_{i\phi}(\bar{x},0) + c_{i\phi}(0,t)]$ on measured concentration profiles in cordierite is approximately five times the average penetration distance in adjacent garnets. We can extract a crude estimate of D_{cord} from this information by applying the "square root" relationship [Crank, 1956, Eq (3.40)] to the two sets of profiles and observing that if $\bar{x} = \sqrt{4D_\phi t}$ in each mineral, then the ratio of $\bar{x}_{cord}/\bar{x}_{gar}$ should be approximately equal to the ratio of the square root of the diffusion coefficients in the two phases. We estimate, therefore, that $D_{cord} \cong 25\ D_{gar}$. Since the diffusion profiles in the Phillipston material developed over a fair temperature range, the estimated D_{cord}/D_{gar} ratio of 25 refers to some mean value during the cooling process, and should be taken to imply little more than that the rate of diffusion of Mg^{2+} and Fe^{2+} in cordierites was rather more rapid than in garnet during the early evolution of the diffusion profiles.

Fortunately, the kinetic models are relatively insensitive to uncertainties in the diffusion parameters for cordierite, and hence even our extremely limited knowledge of D_{cord} is only a minor handicap. It is evident, however, that more precise values of all of the diffusion parameters are much needed.

The temperature coefficient of K_D depends on the molar enthalpy $\Delta\bar{H}^0_r$ for the cation exchange reaction. This is now reasonably well known. We have used Thompson's value of 6.5 kcal/mol as our best initial value for $\Delta\bar{H}^0_r$. The uncertainty in this enthalpy is ca. ± 0.5 kcal/mol (Thompson, pers. comm.). All of the values used in our analysis of the diffusion profiles fall between 5.5 and 7.0 kcal/mol.

Our garnet-cordierite composition profiles were measured at room temperature. The first question to be answered is whether these profiles reflect the entire cooling history of the mineral pairs down to room temperature. This would be unfortunate, since many of the simplifying assumptions made in the previous sections would surely be invalid if applied throughout the temperature range 675°C > T > 25°C. Fortunately the very high value of $(\Delta\bar{H}_D)_\phi$ (~ 70 kcal/mole) implies that the diffusion constants $D_\phi(T)$ decrease so markedly with decreasing temperature that after a drop of a few hundred degrees there is no further change in the concentration profiles, except within a distance from the interface which is smaller than the distance to the first measured point on the profile. At lower temperatures the concentration profiles are "frozen," except at distances less than 1 micron from the interface. Hence, the profiles that we see today are the same as those which existed when the system passed below some critical temperature T^*. Clearly, T^*

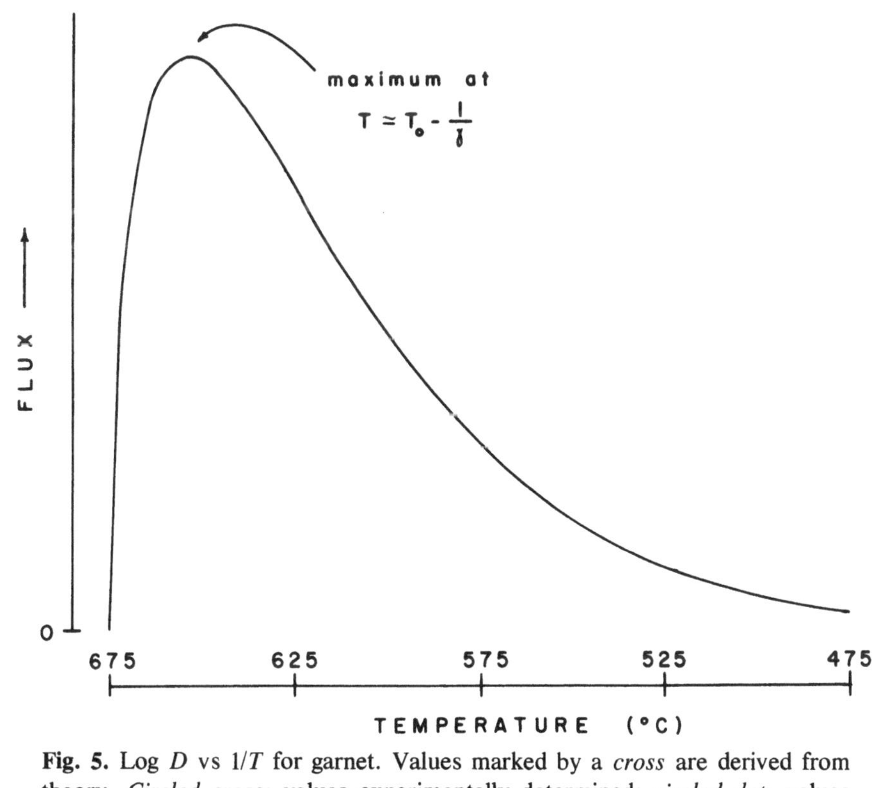

Fig. 5. Log D vs $1/T$ for garnet. Values marked by a *cross* are derived from theory. *Circled cross:* values experimentally determined; *circled dot:* values derived from data in the petrologic literature

depends on the closest distance from the interface to which element concentrations can be measured accurately. For our garnet-cordierite profiles, $T^* \approx 450°C$.

Figure 6 shows the evolution of a typical diffusion profile in garnet, and demonstrates the approach to an invariant profile as $T \rightarrow T^*$. Table 2 confirms that the profiles at 450°C and 400°C are virtually identical. To define the cooling rate at lower temperatures, we would have to measure the composition of garnets accurately at distances less than 1 μm from the interface. If we could do this, perhaps with an ion microprobe, the profiles would begin to reveal the low-temperature history of the mineral pair. The best estimate of the maximum temperature attained by the Phillipston rocks during their second, i.e., Permian, metamorphism is approximately 400°C. If the parameters used in computing the profiles in Figure 6 are correct, then the effect of the Permian metamorphism on the observed composition profiles was probably very small.

Table 3 shows the variation of the concentration of Mg^{2+} in garnet at the interface with cordierite during cooling from 675°C to 475°C at a rate of

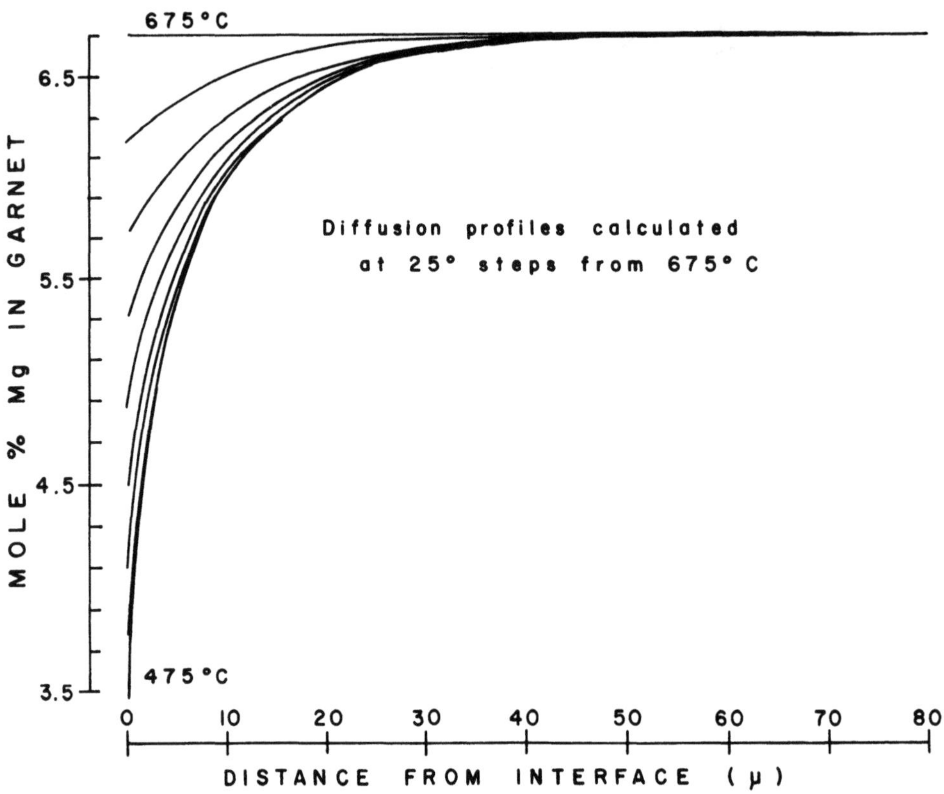

Fig. 6. Evolution of a diffusion profile in garnet

Table 2. Behavior of profiles below T*

Distance from interface	T = 450°C Concentration	T = 400°C Concentration
1.5 μ	4.629	4.623
3.0 μ	5.115	5.113
4.5 μ	5.405	5.404
6.0 μ	5.609	5.609
7.5 μ	5.765	5.765
9.0 μ	5.889	5.889
10.5 μ	5.991	5.991
12.0 μ	6.076	6.076
13.5 μ	6.149	6.149
15.0 μ	6.211	6.211

*Note: Parameters used:
$D^0_{gar} = 3.35 \times 10\ ^{19}$cm²/second
$D^0_{cord} = 7.75 \times 10^{-18}$ cm²/second
$(\Delta \bar{H}_D)_{gar} = 61.6$ kcal/mol
$(\Delta \bar{H}_D)_{cord} = 60.5$ kcal/mol
$\Delta \bar{H}^0_R = 6.5$ kcal/mol
$s = 50$ K/million year
units of concentration in mole percent of Mg^{2+} in Garnet.

Table 3. The concentration of Mg^{2+} in mole percent in garnet at the garnet–cordierite boundary during cooling from 675° to 475°C

Time (million years)	$f_1(t)$ (linear)	$f_1(t)$ (refined)	$f_1(t)$ (extact)
0	6.700 mol %	6.700 mol %	6.700 mol %
0.5	6.304	6.322	6.326
1.0	5.908	5.980	5.995
1.5	5.512	5.672	5.705
2.0	5.117	5.400	5.452
2.5	4.721	5.161	5.235
3.0	4.325	4.956	5.051
3.5	3.929	4.784	4.898
4.0	3.533	4.643	4.773

Note: Parameters used:
$D^0_{gar} = 2.779 \times 10^{-19}$ cm²/second
$D^0_{cord} = 7.608 \times 10^{-18}$ cm²/second
$(\Delta \bar{H}_D)_{gar} = 55$ kcal/mol
$(\Delta \bar{H}_D)_{cord} = 85$ kcal/mol
$\Delta \bar{H}^0_R = 6.0$ kcal/mol
$s = 50$ K/million year
$c^i_{Mg,cord} = 14.6$ mol % $c^i_{Fe,cord} = 6.8$ mol %
$c^i_{Fe,gar} = 32$ mol %
$c^i_{Mg,gar} = 6.7$ mol %

50°C per million years. The parameters used in the calculations are included in the Table. The variation of the Mg^{2+} concentration in garnet during cooling turns out to be rather insensitive to the values chosen for D^0_{cord}, as long as $D_{\text{cord}}/D_{\text{gar}} >> 1$. The calculated value of α_{gar} is 0.79 per million years using the diffusion coefficients listed in Table 3. A doubling of D^0_{cord} only increases α_{gar} to 0.84 per million years. The interpretation of the measured composition profiles is therefore not particularly sensitive to the choice of D^0_{cord}.

The column $f_1(t)$ (linear) gives the variation of the Mg^{2+} content of garnet at the garnet–cordierite boundary calculated with the linear model approximation of Equation (39). It is evident that this model gives a reasonably accurate description of $f_{\text{gar}}(t)$ during the first 1–1½ million years of the cooling. During the later stages of cooling the nonlinearity of $f_{\text{gar}}(t)$ becomes quite significant and the disagreement between the linear and the exact solutions becomes large. The nonlinearity of $f_{\text{gar}}(t)$ would be less pronounced if a value of $(\Delta\bar{H}_D)_{\text{cord}}$ closer to $(\Delta\bar{H}_D)_{\text{gar}}$ had been chosen. The linear approximation would then have given much more acceptable results. The refined analytic formula for $f_{\text{gar}}(t)$ [Eqs. (40) and (42)] was used to calculate $f_1(t)$ (refined) in Table 3. The results describe the actual boundary concentration exceedingly well during the entire cooling period from 675°C to 475°C. Hence Equations (40) and (42) should be very useful in describing systems similar to ours; their simplicity compared to the exact equation allows a clearer insight into the physical effects of the several thermodynamic quantities on the cooling profiles observed in our samples.

The application of the exact or of the refined model to extract cooling rates from composition profiles in garnets and cordierites is rather severely limited by the absence of reliable data for the diffusion constants of Mg^{2+} and Fe^{2+} in these minerals. Happily, however, the inversion of the problem can provide meaningful constraints on the cooling rate, s, and also on the permissible range of the physical parameters governing diffusion between garnet and cordierite. As a preliminary it is useful to discuss a pertinent "scaling" law. In the solution of the diffusion model, the parameters that determine the evolution of the profiles are γ_{gar}, γ_{cord}, β [Eq. (18b)], D^0_{gar}, and D^0_{cord}, where convenient units of γ_ϕ and β are per million years, and the units of the diffusion coefficients D^0_ϕ are $\mu_m{}^2$/per million years. Notice that all of the parameters vary as inverse time. From the earlier discussion, we know that below the critical temperature T^* our profiles do not change. Suppose that we now choose two temperatures, T_1 and T_2, such that T_1, $T_2 < T^*$. Then, if we rescale time so that one unit of time is not one million years, but the time interval elapsed between $T = T_0$ and $T = T_1$ [i.e., a unit equal to $T_0(T_0 - T_1)/T_1 s$ million years, from Equation (19)], the profile at $t=1$ will now correspond to that for T_1, if γ_ϕ, β, and D^0_ϕ are all scaled accordingly. If we perform another rescaling, this time defining a unit of time as the time needed to reach T_2 from T_0, the profile at $t = 1$ will correspond to that at T_2, with γ_ϕ, β, and D^0_ϕ rescaled to values appropriate to the new time unit. But, since the profiles at T_1 and T_2 are the same, the profiles obtained from the two different sets of parameters at $t = 1$ are the same also. Hence, the "freezing" of our profiles below T^* leads to the conclusion that our model cannot distinguish between two sets of values for our parameters $\{\gamma^1_\phi, \beta^1, D^{01}_\phi\}$ and $\{\gamma^2_\phi, \beta^2, D^{02}_\phi\}$ such that $\gamma^1_\phi = \eta\gamma^2_\phi, \beta^1 = \eta\beta^2$, and $D^{01}_\phi = \eta D^{02}_\phi$, where η is a scaling factor. The scaling, of course, cannot proceed to units of time less than $t^* = T_0(T_0 - T^*)/T^*s$, i.e., the time needed to reach T^* [from Eq. (19)]. This scaling law implies that in fitting our profiles we have to keep one of our parameters fixed and may then allow the others to vary. Notice that s is clearly involved in the definitions of γ_ϕ, β, and D^0_ϕ, and hence that fixing the best values of these parameters fixes s.

In order to follow the dependence of the calculated profiles on all the parameters and optimize the fits to $c_\phi(x,t)$, we have found it useful to employ the iterative method of steepest descents. We define a mean square deviation for our model by the quantity

$$L_\phi = \sum_j [c_\phi(x_j,t) - y_j]^2\omega_j \tag{43}$$

where x_j and y_j represent the measured distances from the interface and the measured concentrations of Mg^{2+} ions in phase ϕ at that distance. $c_\phi(x_j,t)$ is the calculated "frozen" concentration of Mg^{2+} in phase ϕ at the measured distance x_j (and where t can be any time $t > t^*$). The ω_j represent the statistical weights attached to each of the measured points. We clearly want to vary our parameters so as to make L_ϕ a minimum. In the method of steepest descents, we choose some initial values for the parameters and calculate the derivatives $\partial L_\phi/\partial D^0_\phi$, $\partial L_\phi/\partial\gamma_\phi$, and $\partial L_\phi/\partial\beta$ at these values. For example, from Equation (43):

$$\frac{\partial L_\phi}{\partial D^0_{\text{gar}}} = \sum_j 2\omega_j [c_\phi(x_j,t) - y_j] \frac{\partial c_\phi(x_j,t)}{\partial D^0_{\text{gar}}} \tag{44}$$

we evaluate the derivatives $\partial c_\phi(x_j,t)/\partial D^0_{\text{gar}}$ numerically. (Note: we can derive these derivatives analytically if we use the accurate analytical expressions for $f_\phi(t)$ developed earlier.) Having calculated the first derivatives, we can write

$$L_\phi = L^0_\phi + \frac{\partial L_\phi}{\partial D^0_{\text{gar}}}\Delta D^0_{\text{gar}} + \frac{\partial L_\phi}{\partial D^0_{\text{cord}}}\Delta D^0_{\text{cord}} + \frac{\partial L_\phi}{\partial\gamma_{\text{gar}}}\Delta\gamma_{\text{gar}} + \frac{\partial L_\phi}{\partial\gamma_{\text{cord}}}\Delta\gamma_{\text{cord}} + \frac{\partial L_\phi}{\partial\beta}\Delta\beta \tag{45}$$

where L_Φ^0 is the value of L_Φ at the initial values of the parameters and ΔD_Φ^0, $\Delta\gamma_\Phi$, and $\Delta\beta$ refer to changes in these values. Using Equation (45), we demand that the Δ values be chosen in the direction which maximizes the rate of decrease of L_Φ. However, the scaling law does not allow all five values to vary independently. Hence we really want the dependence of L_Φ on the ratios, i.e.

$$\frac{\partial L_\Phi}{\partial(p_k/D_{\text{gar}}^0)} = \frac{D_{\text{gar}}^0}{p_k}\left[p_k\frac{\partial L_\Phi}{\partial p_k} - D_{\text{gar}}^0\frac{\partial L_\Phi}{\partial D_{\text{gar}}^0}\right] \tag{46}$$

where we have arbitrarily chosen to divide by D_{gar}^0 and where p_k stands for any of the other four parameters. Instead of Equation (45) we have that

$$L_\Phi = L_\Phi^0 + \sum_{k=1}^{4}\frac{\partial L_\Phi}{\partial(p_k/D_{\text{gar}}^0)}\Delta\left(\frac{p_k}{D_{\text{gar}}^0}\right). \tag{47}$$

We then maximize $|L_\Phi - L_\Phi^0|$, allowing any variations in the ratios $\Delta\,(p_k/D_{\text{gar}}^0)$ within the constraints imposed by the relations

$$\left(\frac{\Delta\left(\dfrac{D_{\text{cord}}^0}{D_{\text{gar}}^0}\right)}{\dfrac{D_{\text{cord}}^0}{D_{\text{gar}}^0}}\right)^2 + \ldots + \left(\frac{\Delta\left(\dfrac{\beta}{D_{\text{gar}}^0}\right)}{\dfrac{\beta}{D_{\text{gar}}^0}}\right)^2 = q^2 \tag{48}$$

where q is the step size for the current iteration. The result is that we change the old ratios by

$$\Delta\left(\frac{p_k}{D_{\text{gar}}^0}\right) = \left(\frac{p_k}{D_{\text{gar}}^0}\right)^2 q\,\frac{\partial L_\Phi/\partial(p_k/D_{\text{gar}}^0)}{\left[\sum_{k=1}^{4}\left(\dfrac{p_k}{D_{\text{gar}}^0}\right)^2\left(\dfrac{\partial L_\Phi}{\partial(p_k/D_{\text{gar}}^0)}\right)^2\right]^{1/2}}. \tag{49}$$

We then use the new values p_k to repeat the process. We have found that this method works quite well for our model.

We have analyzed the dependence of our measured Mg^{2+}-garnet profiles on each of the parameters by means of the method of steepest descent. Since the uncertainty in $\Delta\bar{H}_r^0$ was smaller than that of the other parameters, we chose to keep $\Delta\bar{H}_r^0$ fixed at the observed value of 6.5 kcal/mol. This fixed the value of β at

$$\beta = \frac{\Delta\bar{H}_r^0\,\Delta T}{RT_0(T_0 - \Delta T)} \tag{50}$$

where $T_0 - \Delta T$ is a temperature at which the profile is "frozen." As discussed earlier, we used Equations (18b) and (19), and defined a unit of time as that needed to go from T_0 to $T_0 - \Delta T$, so that the profile which we calculate is that at $t = 1$. Since β is constant, we can determine the other four parameters from the values of the ratios reached by the steepest descent method. The values of γ_{gar} and γ_{cord} will depend only on $(\Delta \bar{H}_D)_{\text{gar}}$ and $(\Delta \bar{H}_D)_{\text{cord}}$ as in Equation (50). The D_ϕ^0 in these time units are given by

$$D_\phi^0 = \frac{\mathbf{D}_\phi^0 \times 3.11 \times 10^{21} \times T_0\, \Delta T}{s(T_0 - \Delta T)} \tag{51}$$

where $\mathbf{D}_\phi^0$ is in units of centimeter squared per second. Thus, the parameters D_ϕ^0 depend on the ratio $\mathbf{D}_\phi^0/s$. We are interested in the range of values of our parameters for which calculated profiles match the measured profiles within experimental error. We have found that these ranges are coupled in such a way that it is impossible to vary D_{gar}^0 or $(\Delta \bar{H}_D)_{\text{gar}}$ very much from the calculated optimum values without rapidly raising L_ϕ to unacceptable values. A typical good fit is given in Table 4. For the given fixed value of β we can safely say that

$$\frac{s}{50}\, 3 \times 10^{-19}\ \text{cm}^2/\text{second} \leqslant \mathbf{D}_{\text{gar}}^0 \leqslant \frac{s}{50}\, 6 \times 10^{-19}\ \text{cm}^2/\text{second}$$

and that

$$60\ \text{kcal/mol} \leqslant (\Delta \bar{H}_D)_{\text{gar}} \leqslant 75\ \text{kcal/mol}. \tag{52}$$

Table 4*.

Distance (μ)	Concentration (calculated)	Concentration (measured)
1.5	4.629 mole %	4.690 mole %
3.0	5.115	4.910
4.5	5.405	5.390
6.0	5.609	5.630
7.5	5.765	5.670
9.0	5.889	6.050
10.5	5.991	5.860
12.0	6.076	6.050
13.5	6.149	5.890
15.0	6.211	6.210
18.0	6.313	6.360
19.5	6.354	6.440
21.0	6.391	6.380
22.5	6.424	6.260
24.0	6.453	6.570
25.5	6.479	6.380
27.0	6.502	6.470
28.5	6.523	6.410
31.5	6.558	6.580
33.0	6.573	6.600

*Fit using parameters from Table 2.

Furthermore, high values of $\mathbf{D}^0_{gar}$ are coupled to high values of $(\Delta\bar{H}_D)_{gar}$, as expected. Thus, we cannot have $\mathbf{D}^0_{gar} = 5 \times 10^{-19}$ cm²/second, $s =$ 50°C per million years *and* $(\Delta\bar{H}_D)_{gar} = 60$ kcal/mol.

The permitted ranges for $\mathbf{D}^0_{cord}$ and $(\Delta\bar{H}_D)_{cord}$ are much wider because, as explained earlier, $D^0_{cord} >> D^0_{gar}$ and thus the Mg^{2+} profile in garnet is relatively insensitive to these two parameters. Thus, we can obtain good fits with the values

$$\begin{aligned} \mathbf{D}^0_{gar} &= 3.35 \times 10^{-19} \text{ cm}^2/\text{second} \\ \mathbf{D}^0_{cord} &= 2.25 \times 10^{-17} \text{ cm}^2/\text{second} \\ (\Delta\bar{H}_D)_{gar} &= 61.6 \text{ kcal/mol} \\ (\Delta\bar{H}_D)_{cord} &= 60.5 \text{ kcal/mol} \\ s &= 50^\circ\text{C per million year} \\ (\sigma &= 0.0265 \text{ mol } \%) \end{aligned} \tag{53a}$$

and

$$\begin{aligned} \mathbf{D}^0_{gar} &= 4.63 \times 10^{-19} \text{ cm}^2/\text{second} \\ \mathbf{D}^0_{cord} &= 2.38 \times 10^{-17} \text{ cm}^2/\text{second} \\ (\Delta\bar{H}_D)_{gar} &= 68.1 \text{ kcal/mol} \\ (\Delta\bar{H}_D)_{cord} &= 46.4 \text{ kcal/mol} \\ s &= 50^\circ\text{C per million year} \\ (\sigma &= 0.0258 \text{ mol } \%). \end{aligned} \tag{53b}$$

Hence, $(\Delta\bar{H}_D)_{cord}$ can vary a good deal without affecting the composition profiles. If we were to use profiles of Mg^{2+} in cordierite instead of those in garnet, $(\Delta\bar{H}_D)_{cord}$ and $\mathbf{D}^0_{cord}$ would be more narrowly constrained, and if we had available profiles for garnet–cordierite pairs with different initial compositions, we could obtain even better control on the values of all of the parameters.

The remarkable sensitivity of our model for $(\Delta\bar{H}_D)_{gar}$ and D^0_{gar} can be used to improve our knowledge of these parameters. Within the range of $\Delta\bar{H}^0_r$, 5.5 kcal/mol $\leqslant \Delta\bar{H}^0_r \leqslant$ 7.0 kcal/mol, the highest value allowed for $(\Delta\bar{H}_D)_{gar}$ by our results is 80 kcal/mol. On the other hand, the probable experimental lower bound for $(\Delta\bar{H}_D)_{gar}$, 68 kcal/mol, and the scaling law, imply that $\Delta\bar{H}^0_r \geqslant$ 6.0 kcal/mol. Our model thus couples these otherwise unrelated thermodynamic quantities in such a way that we can constrain their range and thus obtain rather useful thermodynamic information. We have thus reduced the ranges of $\Delta\bar{H}^0_r$ and $(\Delta\bar{H}_D)_{gar}$ to

$$\begin{aligned} 6 \leqslant \Delta\bar{H}^0_r \leqslant 7 \text{ kcal/mol} \\ 68 \leqslant (\Delta\bar{H}_D)_{gar} \leqslant 80 \text{ kcal/mol.} \end{aligned}$$

The fit of calculated composition profiles to observed composition profiles depends quite critically on the ratio of the cooling rate s to the diffusion constant D^0_{gar}. This is demonstrated in Table 5. $\mathbf{D}^0_{gar}$ has been set equal to 3.3×10^{-19} cm²/second; the data in Table 5 show that the best fit between the calculated and observed profiles is very close to s = 50°C per million year. Thus, from Equation (52) it follows that

$$\frac{s}{\mathbf{D}^0_{gar}} \approx \frac{50}{3.3 \times 10^{-19}} = 1.5 \times 10^{20} \frac{°\text{C/million year}}{\text{cm}^2/\text{second}}. \quad (54)$$

Unfortunately, the uncertainty in $\mathbf{D}^0_{gar}$ is very large. The data in Figure 4 permit a range of ca. 3 orders of magnitude for $\mathbf{D}^0_{gar}$ at 675°C, and it is not entirely certain that the value of $\mathbf{D}^0_{gar}$ for the almandines at Phillipston is contained within this range.

The calculation of meaningful values of the cooling rate s for the Phillipston rocks therefore requires either better laboratory measurements of $\mathbf{D}^0_{gar}$ or relatively accurate estimates of $\mathbf{D}^0_{gar}$ based on field observations. Fortunately the homogeneity of all but the rims of the garnets at Phillipston can be used to demonstrate that the Phillipston rocks cooled very rapidly.

Table 5*. Comparison of profiles for different cooling rates

Distance from interface (μ)	Concentration in mole percent (Mg^{2+})					
	s = 30 (°C/million years)	s = 40 (°C/million years)	s = 50 (°C/million years)	s = 60 (°C/million years)	s = 70 (°C/million years)	Measured
1.5	4.458	4.554	4.629	4.692	4.745	4.690
3.0	4.933	5.035	5.115	5.180	5.235	4.910
4.5	5.222	5.325	5.405	5.470	5.525	5.390
6.0	5.428	5.530	5.609	5.673	5.727	5.630
7.5	5.586	5.688	5.765	5.827	5.880	5.670
9.0	5.714	5.814	5.889	5.950	6.000	6.050
10.5	5.821	5.917	5.991	6.049	6.098	5.860
12.0	5.911	6.005	6.076	6.133	6.179	6.050
13.5	5.988	6.080	6.149	6.203	6.247	5.890
15.0	6.056	6.145	6.211	6.263	6.305	6.210
18.0	6.168	6.252	6.313	6.360	6.398	6.360
19.5	6.215	6.296	6.354	6.400	6.436	6.440
21.0	6.257	6.335	6.391	6.434	6.468	6.380
22.5	6.295	6.370	6.424	6.465	6.496	6.260
24.0	6.330	6.402	6.453	6.491	6.521	6.570
25.5	6.361	6.430	6.479	6.515	6.543	6.380
27.0	6.389	6.456	6.502	6.536	6.562	6.470
28.5	6.415	6.479	6.523	6.555	6.579	6.410
31.5	6.460	6.518	6.558	6.586	6.607	6.580
33.0	6.480	6.535	6.573	6.599	6.619	6.600
	$\sigma = 0.044$	$\sigma = 0.028$	$\sigma = 0.026$	$\sigma = 0.033$	$\sigma = 0.043$	

*Parameters used are those in Table 2.

All of the Phillipston garnets which we have studied are chemically homogeneous except near their contacts with other ferromagnesian minerals. It is very likely that these garnets were originally zoned and that they became homogeneous close to the peak of metamorphism. This inference implies a minimum value for D^0_{gar} near or at the peak of metamorphism. The largest garnets studied were 1.5 mm in diameter. Thus

$$\sqrt{4\,\mathbf{D}^0_{gar}\,\Delta t} \geqslant 0.075 \text{ cm},$$

where Δt is the time interval close to maximum metamorphic grade. Δt is not known, but must be less than the duration of the entire Acadian orogeny. Naylor (1971) has shown that the orogeny was contained within the time interval 410 million years b.p. and 380 million years b.p. Thus

$$\Delta t < 30 \text{ million years}$$

and the minimum value of $\mathbf{D}^0_{gar}$ at or near the peak of the Acadian orogeny must have been

$$\mathbf{D}^0_{gar} > \frac{(0.075 \text{ cm})^2}{4 \times 30 \times 3.1 \times 10^{13}\text{second}}$$

$$> 1.5 \times 10^{-18}\text{cm}^2/\text{second}.$$

The actual value of $\mathbf{D}^0_{gar}$ could have been considerably greater.

If the calculated minimum value of $\mathbf{D}^0_{gar}$ is inserted into Equation (54) we obtain

$$s > 1.5 \times 10^{20} \times 1.5 \times 10^{-18} = 225°\text{C per million year}.$$

An absolute lower limit of 100°C per million year can be obtained for the cooling rate by choosing the minimum value for the ratio $s/\mathbf{D}^0_{gar}$ consistent with the geometry of the diffusion profiles at the garnet–cordierite contacts.

The pressure during the peak of metamorphism at Phillipston was probably between 5 and 6 kbar (Richardson 1975). The depth of burial was, therefore, on the order of 15–18 km. If cooling kept pace with uplift, a cooling rate of 100°C per million year corresponded to an uplift rate of about

$$\frac{100^0}{675^0}\,\frac{(15\text{–}18 \text{ km})}{1 \text{ million year}} \approx 2.5 \text{ km/million year} = 2.5 \text{ mm/yr}$$

during the waning stage of metamorphism between 675°C and 475°C. The actual uplift rate was probably at least twice as fast.

These calculated rates of uplift are extraordinarily rapid. Robinson (1966) has estimated that uplift rates in the Himalayas today are approximately 1 mm/yr. Jäger et al. (1967) found differences of 6.3–8.9 million years between the Rb/Sr age of a number of coexisting muscovites and biotites in the Alps. If the closing temperatures of these minerals were 500°C and 300°C respectively, cooling rates of ca. 30°C per million year are obtained for this 200°C temperature interval during the waning of Alpine metamorphism. The Acadian orogeny, therefore, seems to have ended more rapidly than Alpine metamorphism and with an uplift rate in excess of the present Himalayan rate. These results agree well with Naylor's (1971) view of the Acadian orogeny as a short, abrupt event.

APPENDIX

To obtain $\partial c_\Phi/\partial x|_{x=0}$ we must be careful in taking the limit where $x \rightarrow 0$. From Equation (13):

$$\begin{aligned}\frac{\partial c_\Phi}{\partial x} &= \frac{1}{2\sqrt{\pi}}\int_0^{t'_\Phi} g_\Phi(t'')\,\frac{e^{-x^2/4(t'_\Phi - t'')}}{(t'_\Phi - t'')^{3/2}}\,dt'' \\ &- \frac{x^2}{4\sqrt{\pi}}\int_0^{t'_\Phi} g_\Phi(t'')\,\frac{e^{-x^2/4(t'_\Phi - t'')}}{(t'_\Phi - t'')^{5/2}}\,dt'' \\ &+ \frac{2}{\sqrt{\pi}}\,c_\Phi^I\,\frac{1}{2\sqrt{t'_\Phi}}\,e^{-x^2/4\,t'_\Phi}.\end{aligned} \tag{A1}$$

Clearly we cannot just let $x = 0$ in Equation (A1) since we would get $\infty - \infty$ for the first two terms. It is advantageous to change variables of integration to

$$u = \frac{1}{2\sqrt{t'_\Phi - t''}} \qquad du = \frac{1}{4\,(t'_\Phi - t'')^{3/2}}\,dt''$$

so then

$$\begin{aligned}\frac{\partial c_\Phi}{\partial x} &= \frac{2}{\sqrt{\pi}}\int_{1/2(t'_\Phi)^{1/2}}^{\infty} g\,(t'_\Phi - \frac{1}{4\,u^2})e^{-u^2x^2}(1 - 2\,u^2x^2)\,du \\ &+ \frac{1}{\sqrt{\pi\,t'_\Phi}}\,c_\Phi^I\end{aligned} \tag{A2}$$

where we have combined the first two integrals in Equation (A1) into one integral and let $x = 0$ in the third term, since that is well-behaved at $x = 0$.

Now we pick a value, call it $\bar{v}$, and a value of x (which, since, we are taking the limit $\rightarrow 0$, can be made as small as we please) such that for $u < \bar{v} : u^2 x^2 <<< 1$ and for $u > \bar{v} : g_\phi \times [t'_\phi - 1/(4u^2)] = g_\phi(t'_\phi) \pm \epsilon$.

$$\frac{\epsilon}{g_\phi(t'_\phi)} <<< 1.$$

[Clearly we can do this; for example, let $\bar{v}$ be defined so that

$$\frac{\left| g_\phi\left(t'_\phi - \frac{1}{4\bar{v}^2}\right) - g_\phi(t'_\phi)\right|}{g_\phi(t'_\phi)} < 10^{-8}.$$

$\bar{v}$ is a large number. But no matter how large $\bar{v}$ is, pick an even smaller x so that

$$\bar{v}^2 x^2 < 10^{-8}$$

then we satisfy the above two requirements.] Given such a $\bar{v}$ and x we can rewrite Equation (A2) as

$$\frac{\partial c_\phi}{\partial x} = \frac{2}{\sqrt{\pi}} \int_{1/2(t'_\phi)^{1/2}}^{v} g_\phi\left(t'_\phi - \frac{1}{4u^2}\right) du + \frac{2g_\phi(t'_\phi)}{\sqrt{\pi}} \int_v^\infty e^{-u^2x^2}(1 - 2u^2x^2)\, du + \frac{1}{\sqrt{\pi t'}} c^{\mathrm{I}}_\phi. \tag{A3}$$

Look at the second integral and change variables to $\omega = ux$ so that

$$\int_v^\infty e^{-u^2x^2}(1 - 2u^2x^2)\, dx = \int_{vx}^\infty e^{-\omega^2}(1 - 2\omega^2)\frac{1}{x}\, d\omega.$$

Now we want the limit as $x \rightarrow 0$, but note that it can be checked that

$$\int_0^\infty e^{-\omega^2}(1 - 2\omega^2)\, d\omega = 0$$

so

$$\lim_{x\rightarrow 0} \frac{\int_{\bar{v}x}^\infty e^{-\omega^2}(1 - 2\omega^2)}{x} d\omega$$

has both numerator and denominator going to zero. Thus we can use l'Hospital's rule to obtain

$$\lim_{x\to 0} \frac{\int_{vx}^{\infty} e^{-\omega^2}(1-2\omega^2)}{x}\, d\omega = \lim_{x\to 0} \frac{-e^{-v^2x^2}(1-2(\bar{v}x)^2)\,\bar{v}}{1} = -\bar{v}.$$

We can, therefore, write $\frac{\partial c_\phi}{\partial x}$ as

$$\left.\frac{\partial c_\phi}{\partial x}\right|_{x=0} = \frac{2}{\sqrt{\pi}} \int_{1/2(t'_\phi)^{1/2}}^{v} g_\phi\left(t'_\phi - \frac{1}{4u^2}\right) du - \frac{2g_\phi(t'_\phi)}{\sqrt{\pi}}\,\bar{v} + \frac{1}{\sqrt{\pi\, t'_\phi}}\, c^{\mathrm{I}}_\phi. \tag{A4}$$

Now look at the first integral in Equation (A4) and change variables to

$$z = t'_\phi - \frac{1}{4u^2}$$

$$dz = \frac{1}{2u^3}\, du$$

or

$$dz = 4\,(t'_\phi - z)^{3/2}\, du,$$

so that

$$\left.\frac{\partial c_\phi}{\partial x}\right|_{x=0} = \frac{2}{\sqrt{\pi}} \int_{0}^{t'_\phi - 1/4v^2} g_\phi(z)\, \frac{1}{4\,(t'_\phi - z)^{3/2}}\, dz - \frac{2g_\phi\,(t'_\phi)}{\sqrt{\pi}}\,\bar{v} + \frac{1}{\sqrt{\pi\, t'_\psi}}\, c^{\mathrm{I}}_\phi. \tag{A5}$$

Now we note that

$$\int_0^{t'_\phi - 1/4\bar{v}^2} \frac{1}{4\,(t'_\phi - z)^{3/2}}\, dz = \bar{v} - \frac{1}{2\sqrt{t'_\phi}}$$

so we may write the second term of Equation (A5) as

$$-\frac{2g_\phi(t'_\phi)}{\sqrt{\pi}}\,\bar{v} = -\frac{2g_\phi(t'_\phi)}{\sqrt{\pi}}\left[\int_0^{t'_\phi - 1/4\bar{v}^2} \frac{1}{4\,(t'_\phi - z)^{3/2}}\, dz + \frac{1}{2\sqrt{t'_\phi}}\right]$$

or, altogether

$$\left.\frac{\partial c}{\partial x}\right|_{x=0} = \frac{1}{\sqrt{\pi\, t'_\phi}}\left[c^{\mathrm{I}}_\phi - g_\phi(t'_\phi)\right] + \frac{2}{\sqrt{\pi}}\int_0^{t'_\phi} \frac{g_\phi(z) - g_\phi(t'_\phi)}{4\,(t'_\phi - z)^{3/2}}\,dz. \tag{A6}$$

This is the general answer that we sought.

ACKNOWLEDGMENTS

The authors would like to thank Professors J. B. Thompson, A. B. Thompson, and J. F. Hays for numerous helpful discussions. The senior author is grateful for the support of a Parker Fellowship, the third author for the support of a Guggenheim Fellowship. The support of the Committee on Experimental Geology and Geophysics at Harvard University is also gratefully acknowledged.

REFERENCES

Anderson, D. C., Buckley, G. R.: Zoning in garnets—diffusion models. Contrib. Mineral. Petrol. *40*, 87–104 (1973)

Blackburn, W. H.: The spatial extent of chemical equilibrium in some high-grade metamorphic rocks from the Grenville of southeastern Ontario. Contrib. Mineral. Petrol. *19*, 72–92 (1968)

Buckley, G. R.: The effect of diffusion on garnet zoning. Univ. of Illinois Ph.D. Thesis (1973)

Buening, D. K., Buseck. P. R.: Fe–Mg lattice diffusion in olivine. J. Geophys. Res. *78*, 6852–6862 (1973)

Clark, S. P., Jäger, E.: Denudation rate in the Alps from geochronologic and heat flow data. Am. J. Sci. *267*, 1143–1160 (1969)

Crank, J: Mathematics of Diffusion. London: Oxford Univ., 1956

Faul, H., Stern, T., Thomas, H., Elmore, P.: Ages of intrusion and metamorphism in the northeastern Appalachians. Am. J. Sci. *261*, 1–19 (1963)

Gable, D. J., Sims, P. K.: Geology and regional metamorphism of some high grade cordierite gneisses, Front Range, Colorado. Geol. Soc. Am. Spec. Paper *128*, 1–87 (1969)

Grant, J. A., Weiblen, P. W.: Distribution of iron and magnesium in garnet–biotite pairs in high-grade metamorphic rocks. (abstr.) Geol. Soc. Am. Spec. Paper *115*, 80–81 (1968)

Grant, J. A., Weiblen, P. W.: Retrograde zoning in garnet near the second sillimanite isograd. Am. J. Sci. *270*, 281–296 (1971)

Hensen, B. J.: Cordierite-garnet equilibrium as a function of pressure, temperature, and iron–magnesium ratio. Carnegie Inst. Wash. Yearbook *71*, 418–421 (1972)

Hensen, B. J., Green, D. H.: Experimental study of the stability of cordierite and garnet in pelitic compositions at high temperatures and pressures, I. compositions with excess aluminosilicate. Contrib. Mineral. Petrol. *33*, 309–330 (1971)

Hensen, B. J., Green, D. H.: Experimental study of the stability of cordierite and garnet in pelitic compositions at high temperatures and pressures, II. compositions without excess aluminosilicate. Contrib. Mineral. Petrol. *35*, 331–354 (1972)

Hess, P.: The paragenesis of cordierite, garnet, and biotite in the Brimfield area. Harvard Univ. Ph.D. Thesis (1969)

Hess, P.: Prograde and retrograde equilibria in garnet–cordierite gneisses in south-central Massachusetts. Contrib. Mineral. Petrol. *30*, 177–195 (1971)

Hunziker, J. C.: Polymetamorphism in the Monte Rosa, western Alps, Eclogae Geol. Helv. *63*, 151–161 (1970)

Hunziker, J. C., Bearth, P.: Rb–Sr–Altersbestimmungen aus den Walliser Alpen Biotitalterswerte und ihre Bedeutung für die Abkühlungsgeschichte der Alpinen Metamorphose. Eclogae Geol. Helv. 62, 205–222 (1969)

Hurley, P. M., Fairbairn, H. W., Pinson, W. H., Faure, G.: K–Ar and Rb–Sr minimum ages for the Pennsylvania section in Narragansett Basin. Geochim. Cosmochim. Acta *18*, 247–248 (1960)

Jäger, E., Niggli, E., Wenk, E.: Rb–Sr Altersbestimmungen an Glimmern der Zentralalpen. Beitr. Geol. Karte Schweitz. N. F. *134*, 67 (1967)

Lin, T. H., Yund, R. A.: Potassium and sodium self-diffusion in alkali feldspar. Contrib. Mineral. Petrol. *34*, 177–184 (1972)

Lyons, J. B., Faul, H.: Isotope geochronology of the northern Appalachians. In: Studies of Appalachian Geology, Northern and Maritime. Zen, E., White, W. S., Hadley, J. B., Thompson, J. B. (eds.). New York: Wiley Interscience, 1968, pp. 305–318.

Misener, D. J.: Interdiffusion studies in the system Fe_2SiO_4–Mg_2SiO_4. Carnegie Inst. Wash. Yearbook *71*, 516–520 (1972)

Naylor, R. S.: Acadian orogeny: an abrupt and brief event. Science *172*, 558–560 (1971)

Ramberg, H., DeVore, D. G. W.: The distribution of Fe^{2+} and Mg^{2+} in coexisting olivines and pyroxenes. J. Geol. *59*, 193–210 (1951)

Richardson, S. M.: Fe–Mg exchange among garnet, cordierite and biotite during retrograde metamorphism. Harvard Univ. Ph.D. Thesis (1975)

Robinson, P.: Aluminosilicate polymorphs and Paleozoic erosion rates in central Massachusetts. (abstr.) EOS, Trans. Am. Geophys. Union *47*, 424 (1966)

Smigelskas, A., Kirkendall, E.: Zinc diffusion in alpha brass. Trans. A.I.M.E. *171*, 130–134 (1947)

Sneddon, I. N.: Elements of Partial Differential Equations. New York: McGraw-Hill, 1975

Thompson, A. B.: Calculation of muscovite–paragonite–alkali feldspar phase relations. Contrib. Mineral. Petrol. *44*, 173–194 (1974)

Thompson, A. B.: Mineral reactions in pelitic rocks: II. calculation of some *P–T–X* (Fe–Mg phase relations. Am. J. Sci. *276*, 425–454 (1976)

Thompson, J. B., Norton, S. A.: Paleozoic regional metamorphism in New England and adjacent areas. In: Studies of Appalachian Geology, Northern and Maritime. Zen, E., White, W. S., Hadley, J. B., Thompson, J. B. (eds.). New York: Wiley Interscience, 1968, pp. 319–328.

Wood, J. A.: Cooling rates and parent planets of several iron meteorites. Icarus *3*, 429–459 (1964)

Zartman, R., Hurley, P., Krueger, H., Giletti, B. J.: A Permian disturbance of K–Ar radiometric ages in New England: its occurrence and cause. Geol. Soc. Am. Bull. *81*, 3359–3374 (1970)

Zartman R., Snyder, G., Stern, T., Marvin, R., Bucknam, R.: Implications of new radiometric ages in eastern Connecticut and Massachusetts. U.S.G.S. Professional Paper *525D*, D1–D10 (1965)

16

P. J. WYLLIE

Abstract There are many lines of evidence to be evaluated in considering the identity and origin of primary magmas; conclusions reached cannot be valid unless they satisfy the constraints imposed by phase equilibrium experiments. Interpretation of laboratory phase equilibrium experiments is not always unambiguous. Experiments in small crucibles in the presence of water under pressure satisfied many petrologists that batholithic granites involved partial fusion of crustal rocks during the culmination of regional metamorphism. Then, the concept of subducted oceanic lithosphere provided a relatively high-silica source material for magma generation at mantle depths, and dehydration of the sinking slab provided water to lower the melting temperature of overlying mantle peridotite. Batholiths may include material derived from magmas whose genesis was initiated in all three environments. Phase equilibrium data are available to explore the melting products of subducted micaceous sediments and calcareous oozes trapped within basalt. Limestones or siliceous limestones could escape complete dissociation or melting to considerable depths, possibly for long-term storage in the mantle. Phase relationships in the system basalt–andesite–rhyolite–H_2O through the pressure interval from depths where metamorphosed basaltic ocean crust melts, to the near-surface levels where batholiths are emplaced and andesites are erupted, is fundamental for understanding this magmatic system. On balance, the phase equilibrium data do not favor the concept of primary granite or tonalite from mantle or subducted crust. Primary water-undersaturated granite magma is a normal product of partial fusion of the crust, but temperatures of normal regional metamorphism

FROM CRUCIBLES THROUGH SUBDUCTION TO BATHOLITHS

are too low to generate tonalite liquids. The water content of large batholithic bodies is probably less than 1.5%. Uprise and crystallization produces water-saturated liquids in the upper regions and margins of magma chambers, for satellite intrusions or eruption. Gravity drives magma and energy upward from subducted oceanic crust, and the final, uppermost expression of the process is represented by the batholiths and attendant volcanoes.

1 INTRODUCTION

The concept of primary magmas has dominated petrogenesis. A primary magma is a liquid derived by partial fusion of a crystalline rock at depth, followed by separation from its source and independent uprise. Petrologists have argued about which magmas were primary, and which igneous rocks were derived from primary magmas by various differentiation processes. The problem is to unravel the history of a rock from its petrography, chemistry and mineralogy. This is not an easy task, especially when one considers that magmas could experience continuous fractionation from source to surface eruption—the other extreme from primary magmas.

I am very fond of a statement by C. E. Wegmann (1963, p. 5): "commonly *the notions, concepts and hypotheses control the selection of facts recorded by the observers*. They are nets retaining some features as useful, letting pass others as of no immediate interest. The history of geology shows that a conceptual development in one sector is generally followed by a harvest of observations, since many geologists can only see what they are asked to record by their conceptual outfit." It is a fact that interpretation of field data is rarely unambiguous. Most readers have probably shared the rather disconcerting experience of being in a field party where three geologists examing the same outcrop come up with three different interpretations. Their interpretations are guided not only by what they see, but also by their conceptual spectacles, or blinkers. Interpretations of petrography and mineralogy may be guided in the same way.

There are many factors to be evaluated in considering the identity and origin of primary magmas, and there are many lines of evidence including the study of isotopes, rare earth elements, and other trace elements. Conclusions reached from these and other lines of evidence cannot be valid unless they also satisfy the requirements and constraints imposed by phase equilibrium experiments. Laboratory experiments conducted under controlled conditions could be expected to provide unambiguous data relevant to the origin of rocks. For example, a primary magma must be in equilibrium with the residual minerals in the source rock at its place of origin. Therefore, the near-liquidus minerals for a primary magma must correspond to major minerals in the source material at the pressure and temperature of origin, with allowances being made for reaction relationships. This requirement can be tested directly by experiments on magma compositions and possible source rocks. Such experiments, and related criteria and arguments for primary magmas, have been reviewed in detail by Mysen et al. (1974) and Nicholls (1974). These papers also show that interpretation of the controlled laboratory experiments is not unambiguous.

It is known that many formerly deep-seated rocks did not attain equilibrium, or that equilibrium was confined to small, microscopic units of the rock, and this information is of course useful in unraveling the history of a rock. Nevertheless, the point of departure for study of the physical chemistry of petrological processes is to determine the equilibrium phase relationships first, and then to evaluate the factors leading to failure of equilibrium under natural conditions. Walker et al. (1976) have recently demonstrated that the temperature and order of precipitation of crystalline phases from a cooling silicate melt can be altered by changing the cooling rate.

One of the problems in the laboratory experiments is the attainment and confirmation of a state of stable equilibrium. Despite this, and the problems cited above, experimental determination of the equilibrium phase relationships for igneous rocks provides an essential framework for petrogenesis. In this chapter, I will review some of the implications of laboratory experiments made in small crucibles, as they relate to the generation and emplacement of large batholiths.

2 HISTORICAL DEVELOPMENTS

When I was an undergraduate I was given many reading lists dealing with the granite problem and the origin of batholiths. Through the 1950s the arguments between magmatists and transformists remained bitter. Many magmatists maintained that the diorites, granodiorites, and granites comprising the batholiths, as well as the eruptive andesites, were derived by fractional crystallization of primary basalts. Others concluded that the enormous volumes of granitic magma involved in batholiths had been augmented by contamination of the parent basalts by assimilation of crustal material. Transformists were convinced that the batholiths were produced by metasomatism and mobilization of preexisting rocks in the solid state. The necessary components were introduced into and subtracted from the original rock either by diffusion of ions through crystal lattices or by the flow of interstitial pore fluids. The systematic studies of J. J. Sederholm at the turn of the centry on the anatexis of metamorphic rocks fathered the view that batholiths were produced by partial fusion of the crust.

Magmatists found comfort in crystal-liquid phase equilibria from synthetic silicate systems studied at atmospheric pressure, and they had no difficulty interpreting most of their field observations in these terms. Transformists found comfort in the field relationships that demonstrated to their satisfaction that the rocks had been transformed in situ, without the intervention of a molten magmatic phase.

In his 1952 book, Ramberg noted the frequent occurrence of quartzofeldspathic gneisses, pegmatite veins and granitic bodies in rocks that had been regionally metamorphosed at temperatures of 300°C and above. He wrote (p. 244): "The hypothesis of partial anatexis and palingenesis is simple and attractive, but closer studies reveal that it is difficult to reconcile with the observations in field and laboratory . . . At a maxium content of water and other fugitive substances, the eutectic granitic melt may originate somewhere between 550° and 600°C . . . There seems to be no way to make a hydrous silicate melt responsible for this low-temperature mobility," and on page 245 he concluded: "Again we reach the conclusion that granitization, migmatization, and pegmatitization to a great extent take place without the cooperation of a silicate melt." In this book, Ramberg also discussed the influence of gravity in causing the upward intrusion of solid granitic bodies to produce domed gneisses, diapire granites, and many batholiths. In his later book (Ramberg, 1967), he demonstrated the reality of gravity-induced solid-state diapiric intrusion with centrifuged, scale models.

From among my reading lists I remember two statements by famous geologists, although I am not sure which famous geologists—was one of them H. H. Read? They do have the ring of many fine Read comments. The statements, as I recall them, were that no one could gaze into a small platinum crucible and expect to see a granite intrusion, and that petrology could not be represented or illustrated by points and lines on triangular diagrams. It is a fact, however, that the bitterness of the international granite controversy subsided into intermittent local arguments as a result of the small crucibles and triangular diagrams used and published by N. L. Bowen and O. F. Tuttle.

For decades, petrologists had expressed the need for experiments involving mineral and melting reactions in the presence of water vapor under pressure. Tuttle designed two pressure vessels which led to the routine study of such reactions. The first successful melting results on the alkali feldspar system (Bowen and Tuttle, 1950) were followed by Tuttle and Bowen's (1958) classic memoir on the *Origin of Granite in the Light of Experimental Studies in the System* $NaAlSi_3O_8$–$KAlSi_3O_8$–

SiO_2–H_2O. They developed an anatectic model, using the experimental results to show how the partial melting of crustal rocks in the presence of small amounts of aqueous pore fluid offers a mechanism for producing large batholithic masses of granite. Similar conclusions were reached by H. G. F. Winkler and associates in a series of melting studies on sedimentary rocks and metamorphic rocks (see Winkler, 1967, for review; see also Wyllie and Tuttle, 1961). The feasibility of Sederholm's early interpretation of field occurrences in terms of partial fusion processes had been confirmed by experiments in small platinum or gold crucibles, and the results were displayed unambiguously on triangular diagrams for those who knew how to read them. But proof of feasibility is not proof of reality for all occurrences.

In many Precambrian terranes it is a few short steps from a migmatite into granitic gneiss and thence into gneissic granite, and a short conceptual step to the conclusion that the reasonably homogeneous masses of granitic rocks were derived from the fused portion of the migmatites. Haller's (1955, 1971) magnificent photographs and three-dimensional reconstructions of granitic gneiss domes and mushrooms in northeast Greenland demonstrate the intimate involvement of migmatites with intrusive material. Hutchison (1970) described diapiric and recumbent granodiorite and quartz diorite plutons that are rooted in the migmatites of the Central Gneiss Complex in the Price Rupert region of the Coast Plutonic Complex in British Columbia. The field relationships between individual plutons and migmatites are less clear in the southern part, probably because here erosion has dissected the batholith less deeply (Roddick and Hutchison, 1974).

Do migmatites necessarily prove the presence of a molten component? The experimental data make this conclusion inescapable for H_20-bearing rocks of high metamorphic grade. But as Ramberg (1952) pointed out, migmatites are found in lower grade metamorphic rocks where temperatures could not have been high enough to cause partial melting. If pore fluids were present in the metamorphic rocks at these lower temperatures, they must certainly have been potent agents for metasomatism. The fluid components were surely mobile, as a result of the tectonic kneading that accompanies metamorphism, and the field examples outlined by Ramberg and many others indicate that one of the metasomatic products consists of rocks with the appearance and mineralogy of migmatites and pegmatites, rocks which have a perfectly satisfactory explanation in terms of magmatic components, *if* the metamorphic temperatures had been high enough.

Thirteen years after Bowen and Tuttle's (1950) melting experiments in the alkali feldspar–water system. Orville (1963) published the results of alkali ion-exchange reactions between alkali feldspars and alkali chloride solution at high pressures, and reported a surprising similarity with the equilibria between alkali feldspar and its water-saturated melt. The experimental data show that in the presence of a thermal gradient, a potential is developed for reciprocal alkali transfer within the vapor phase. There is a tendency for potassic feldspar to be replaced by sodic feldspar in the higher temperature environment, and for sodic feldspar to be replaced by alkali feldspar in the cooler environment. This provides an explanation for the potash metasomatism that occurs near the margins of many granitic intrusions. In the presence of a pore fluid during metamorphism, it also provides an explanation for the solution of alkali feldspar from one volume of rock and reprecipitation at shallower depths in cooler rock volumes; this could lead to granitization of rocks with suitable compositions.

Experimental phase equilibrium data can therefore be used to argue the case for metasomatic granites as well as for anatectic granites. However, relatively low temperature metasomatic granitization *must* be followed by the generation of magmatic migmatites as a normal consequence of regional metamorphism. For the small quantities of aqueous pore fluids anticipated in metamorphic rocks, the initiation of partial melting is accompanied by the solution of all of the water. The interstitial silicate liquid is more viscous than the pore fluid, and some petrologists argue that if metamorphic temperatures rise high enough for partial melting to occur, then most of the interesting reactions cease because the mobility of the interstitial magmatic fluid is insignificant compared with that of the lower temperature pore fluid (personal communication from J. A. Roddick, September 1975, during field excursion through the Coast Plutonic Complex of British Columbia).

Following the melting experiments, many petrologists found at least partial satisfaction during the 1950s and 1960s in the model that produced magmas by partial fusion of crustal rocks during the culmination of regional metamorphism, and the fierce arguments between magmatists and transformists were heard less frequently at scientific meetings. But conclusions continue to be based on conceptual outfits, as always, and the concept of granitization on the scale of batholithic dimensions had not disappeared.

Satisfaction with the anatectic model was disturbed during the 1960s by the accumulation of strontium isotope data indicating that a large fraction of batholithic granites could contain only minor amounts of older sialic crustal material (Faure and Powell, 1972). Values for the lead isotopes of many granitic rocks are consistent with this conclusion (Carmichael et al., 1974, p. 592–595). The whole concept of the origin of batholiths underwent reevaluation with the formulation of plate tectonic theory in 1968, and subducted oceanic crust was introduced as an additional source of magmas for batholiths and andesites. Armstrong (1968) examined the evolution of strontium isotopes and lead isotopes in magmatic processes ranging from subducted crust to near-surface emplacement or eruption.

There are many factors to be evaluated in considering the origin and evolution of batholiths. Detailed knowledge of the field relationships is essential, and to this must be added the chemistry, petrography, and mineralogy of the rocks comprising the batholith. It is essential to know the phase relationships of the major rock series and the possible source rocks at crustal pressures, and through the range of pressures where partial fusion could occur in the magma source rocks.

3 EXPERIMENTAL STUDIES

Exploration in experimental petrology has established the main variables that influence magma temperatures and crystallization sequences. These are pressure, water content (or water fugacity), and oxygen fugacity. Carbon dioxide becomes important with increasing pressure in more basic rock compositions. Experimental studies have by now provided us with fairly complete frameworks for the effects of P, T, f_{O_2}, and H_2O content on the degree of crystallization of major magma types, and the influence of CO_2 has been the subject of much recent attention (see Boettcher, 1975, for review).

Experimental approaches to petrogenetic problems include the following:

(1) Determination of the phase diagrams for individual minerals,
(2) Determination of the phase relationships of mineral assemblages representing simplified rocks,
(3) Determination of the phase relationships in ternary or quaternary systems that include the rock-forming minerals,
(4) Determination of the phase relationships in whole rocks, including possible source rocks as well as possible derivative magma compositions, and
(5) Determination of the phase relationships in genetically related rock series.

Phase equilibrium relationships in silicate systems form an important basis for interpretive petrology. The simple silicate systems that can be treated theoretically in terms of the phase rule establish patterns of behavior that must be considered in the development of petrogenetic theories. However, it is a long extrapolation from these systems to the complex rock systems, and the study of natural rock samples in the presence of one or more volatile components provides useful stepping stones between the synthetic and natural systems. The study of a single rock sample is analogous to the study of a single bulk composition within a phase diagram, whereas the study of a series of genetically associated igneous rocks approximates more closely to the study of a complete phase diagram.

The design of experimental programs may vary depending upon the philosophy of the investigators. There are at least three distinct approaches leading to petrogenetic insights:

(1) Detailed studies of significant material. For example, 63% of the crust is composed of feldspar plus quartz, so let us determine the phase relationships in feldspar–quartz–water systems. Similarly, basalt is the dominant magma derived from the mantle, so we must determine the position of the gabbro-eclogite phase transition.
(2) Detailed studies aimed at understanding the origin of particular rocks. For example, in order to learn about basalt or granite, let us determine the phase relationships of basalt, or of granite, under all conceivable conditions, and see what emerges from the study.
(3) Studies aimed at testing theories. For example, I have a theory about the origin of andesites, so what experiments can I conduct to test the validity of this theory? Similarly, I do not believe that granites can be primary magmas from subduction zones, so what experiments can I devise to confirm this? Two other approaches have less chance of providing petrological insights, but they have sometimes paid off.
(4) Studies aimed at gaps. For example, no one has tried high pressure experiments in carbonate-borate-phosphate systems, so let us try and see what happens.
(5) Studies aimed at research budgets. For example, the Explo Funding Agency has expressed interest in tin, so let us apply for a grant to work on high pressure experiments in tin-bearing systems.

Many petrologists have asked me why it is that experimental petrologists disagree about something as fundamental as the compositions of liquids produced by the partial melting of peridotite. They are accustomed to arguing about interpretations of field observations, but they expect experimental petrologists to be able to determine phase relationships and liquid compositions unambiguously. After all, the experiments are conducted under controlled conditions, using various black boxes. In fact, there are controversies among experimental petrologists that are quite as bitter as the arguments that revolved around the granite controversy. They do not engulf the profession, as the granite debate did, but they do involve a large proportion of experimental petrologists at one time or another.

There are many practical factors leading to the publication of different results and interpretations for similar studies conducted in different laboratories. So many, in fact, that it is impossible to review them all here. These may relate to starting materials, reaction rates, design of experiments, loss of iron from sample to capsule wall during the experiments, and reactions occurring durin the quench (see, for example, Yoder and Tilley, 1962; Nehru and Wyllie, 1975).

There are in addition controversies that arise from early publication of preliminary data, published in attempts to establish priority for an idea or discovery. Fertile minds are faster than the slow burners in pressure vessels. Ideas and the generation of new petrogenetic hypotheses tend to leap-frog ahead of confirmation or adequate testing by reliable, substantiated phase equilibrium experiments. This produces a situation where one or more experimental groups tend to publish a handful of runs at a time, and to dispute the interim conclusions published by other groups on the basis of their own handful of runs. Sometimes, it seems that protection of an original idea becomes temporarily more important than the satisfactory completion of an experimental program.

Unsuspecting petrologists and geophysicists seeking answers to their problems are often unaware that published experimental data do not necessarily merit the high degree of confidence implied by enthusiastic investigators. The only certain test for equilibrium is the demonstration of reversibility (bearing in mind the caveat that metastable as well as stable equilibrium can be reversed). For this reason, a claim of reversibility in an experimental paper tends to convey a certain magic—the results must be correct! Experimentalists are well aware of the fact that not all claims of reversibility demonstrate the attainment of equilibrium, and other petrologists should be aware of this. The literature is crowded with contradictory phase diagrams for which the authors claimed reversibility (see Zen, 1969, Fig. 1), so it behooves the users of experimental data to check precisely what reversibility tests were made. For complex systems with volatile components that must be sealed within noble-metal capsules, it may be impossible to establish reversibility within narrow temperature limits (Biggar and O'Hara, 1972; Stern and Wyllie, 1975).

As Fyfe (1960) pointed out long ago, an important test for a phase diagram is to evaluate it with whatever thermodynamic data are available. In this volume honoring H. Ramberg, it is appropriate to recall the pioneering paper by Ramberg and DeVore (1951) on the distribution of elements between coexisting minerals. This was followed in the 1960s by significant theoretical advances in the thermodynamics of crystalline solutions, which are reviewed in the monograph by Saxena (1973). Many studies now routinely combine experiment and theory in the determination of distribution coefficients. Experimental studies on mineral reactions in mixed volatile systems are incomplete without consideration of the theoretical treatment developed by Greenwood (1967), as reviewed recently by Kerrick (1974).

Calculations may be hampered by inadequate thermochemical data. Ramberg initiated a program of calorimetric measurement of the thermochemical properties of silicate minerals, and results were published by his students at the University of Chicago during the 1950s. The combination of calorimetry and phase equilibrium studies is developing into a powerful approach through the recent development of new methods for high temperature solution calorimetry applicable to the refractory rock-forming minerals (Kleppa and Newton, 1975). An example is the paper by Newton et al. (1974), in which equilibrium relations involving cordierite and other phases in the system $MgO–Al_2O_3–SiO_2$ were defined with a clarity impossible from the single approaches of either experiment or thermochemistry. Data for silicate liquids are less reliable, but a paper by Marsh and Carmichael (1974) is an example of how a difficult problem, the origin and magmatic history of andesites in subduction zones, can be examined by consideration of experimental phase equilibrium data, thermodynamic calculations, and geophysical fluid dynamics. Ramberg's (1967) experiments on centrifuged models are similarly supported by theoretical dynamic analysis.

The application of high pressure phase equilibrium studies to petrogenetic problems had contributed a great deal to our understanding of rocks, despite the preceding remarks about uncertainties. For elucidation of processes of magma generation of great depths it is still the most productive approach. Let us therefore take the measure of the earth's nether regions by examining the products from crucibles after they have been inserted into a simulated subduction zone.

As we noted early in this chapter, the subsidence of the granite controversy began with publication of Tuttle and Bowen's (1958) experimental work on the melting relationships in feldspar–quartz–water systems. Their results extended to 4 kbar, pressures corresponding to almost halfway through the average continental crust. We can begin our entry into a subduction zone by reviewing the phase relationships in the granite system, $NaAlSi_3O_8–KAlSi_3O_8–SiO_2–H_2O$, which were recently extended to pressures greater than 30 kbar, corresponding to a depth of more than 100 km in the earth.

Figure 1 shows the effect of pressure on the solvus crest for the alkali feldspars, and a series of univariant curves that fall into three groups. The fusion curves for dry minerals and mineral assemblages have steep positive slopes. In the presence of excess water, the temperatures of the fusion curves are strongly depressed by application of a few kilobars pressure, through solution of H_2O in the silicate liquids, and less so with increasing pressure. The slopes of these hydrothermal fusion curves change from negative to positive where the low pressure feldspar structures undergo phase transitions into the more dense minerals jadeite and

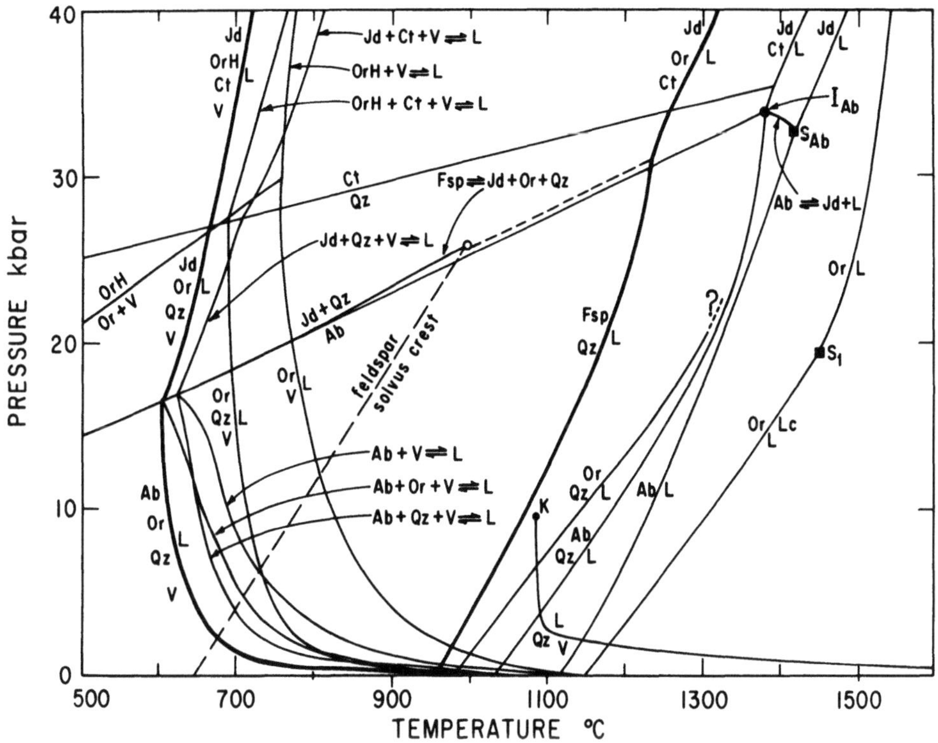

Fig. 1. Univariant melting curves for most mineral combinations in the systems Ab–Or–Qz and Ab–Or–Qz–H_2O, based on experimental results and published data cited in text. *Two heavy lines:* the minimum solidus temperatures in the dry system, and in the system with excess H_2O (Huang and Wyllie, 1975). Ab: albite; Or: sanidine; Qz: quartz; Ct: coesite; Jd: jadeite; OrH: $KAlSi_3O_8.H_2O$, sanidine hydrate; Fsp: feldspar solid solution; *L:* liquid; *V:* vapor; *I, S:* invariant and singular points

sanidine hydrate. These phase transitions, along with quartz–coesite, comprise the third group of reactions.

Figure 2A shows the effect of pressure on the composition of the H_2O-saturated liquids coexisting with three minerals (only two at low pressures above the feldspar solvus crest) at the minimum temperature on the liquidus. This is the solidus reaction for the system, shown by the heavy line at lowest temperatures in Figure 1A. The eutectic liquid composition moves away from SiO_2, as shown to 4 kbar by Tuttle and Bowen (1958), and extended to 10 kbar by Luth et al. (1964). Note the changes in primary minerals around the eutectic at 10, 20, and 30 kbar. The water contents of these saturated liquids are approximately 15, 20, and 25 wt%, respectively.

Figure 2B compares the field boundaries with excess water at 5 and 30 kbar from Figure 2A with the estimated positions of the dry field

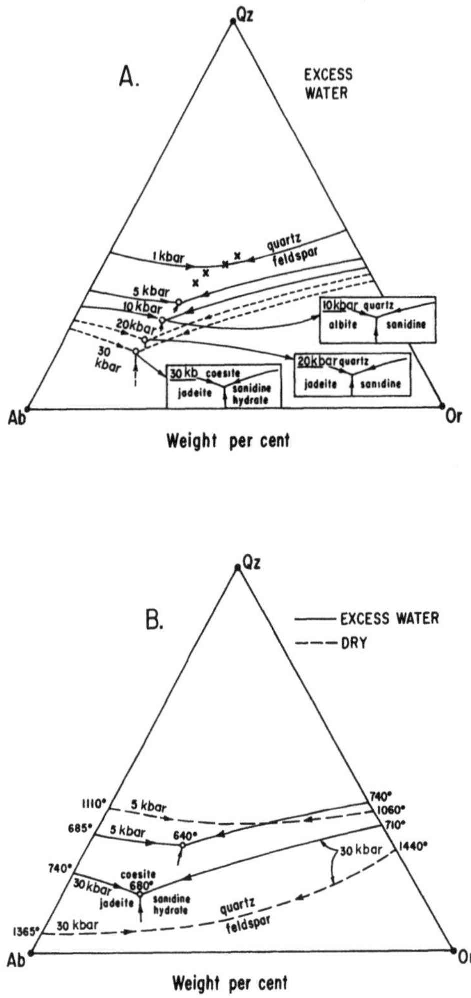

Fig. 2A, B. (A) Effect of pressure on H_2O-saturated liquidus field boundaries in the system Ab–Or–Qz–H_2O. Cf. Figure 1. Data to 10 kbar from Luth et al. (1964), and boundaries for 20 and 30 kbar estimated as described by Huang and Wyllie (1975). (B) Comparison of field boundaries in systems Ab–Or–Qz–H_2O and Ab–Or–Qz at two pressures, 5 and 30 kbar. Data and estimates as described by Huang and Wyllie (1975)

boundaries between quartz and feldspar solid solution at these pressures. Neither the shapes of these boundaries nor the compositions of their temperature minima are known precisely. However, Steiner et al. (1975) have measured a similar pair of boundaries for 4 kbar, with disposition similar to those depicted in Figure 2B for 5 kbar.

In the complete tetrahedron, with H_2O added as an apex, each pair of isobaric field boundaries is connected by a surface giving the compositions of water-undersaturated liquids coexisting with quartz (or coesite) and feldspars (or jadeite, sanidine hydrate). The compositions of liquids on these surfaces, expressed in terms of anhydrous components, change less at 5 kbar than at 30 kbar.

The isobaric surfaces for coexisting feldspars and quartz (or high pressure equivalent minerals) show the range of liquids produced during fusion in the presence of a trace of water. At 30 kbar, the first liquid produced at 680°C would be water-saturated. With progressive fusion and solution of all of the water, the undersaturated liquid migrates across the isobaric surface towards the anhydrous base with little change in composition (anhydrous components), until one of the crystalline phases is exhausted. Steiner et al. (1975) have worked out paths of equilibrium and fractional crystallization on the surface for 4 kbar.

Figure 2 shows that the composition of the liquid coexisting with alkali feldspar and quartz (or high pressure equivalent minerals), either dry or in the presence of water, moves progressively away from SiO_2 with increasing pressure. At 30 kbar, corresponding to 100 km depth, the liquid generated from partial fusion of a subducted feldspathic sandstone would not correspond to that of a granite.

4 SUBDUCTION

In a paper published before plate tectonics was formalized, Hamilton and Myers (1967) concluded that batholiths had their source at mantle depths, and that deep crustal material was involved to various extents by melting or assimilation. Differentiation, assimilation, and zone melting could be so extensive that very little of the final high level magma need represent material present in the initial mantle-derived melt. The concept of subducted oceanic lithosphere added a new dimension to the problem of the origin of batholiths by providing a relatively high-silica source material for magma generation at depts of 100 km or more.

When Hamilton (1969) discussed the origin of batholiths and volcanic island arcs from magmas generated in subducted zones, he incorporated his previous interpretations, and concluded that these magmas "underwent profound modification by differentiation and by reactions with the mantle and crust through which they rose." In their first experimentally based model for the origin of the calc-alkaline magmas from subducted ocean crust Green and Ringwood (1968) concluded, in contrast, that primary andesite magmas generated by partial melting of quartz eclogite rose through the mantle without substantial interaction. Marsh and Carmichael (1974) similarly considered the andesitic magma as primary material from subducted oceanic crust, metamorphosed to quartz eclogite.

Armstrong (1971) and Noble et al. (1975) both published papers about "isotopic and chemical constraints" on magma genesis in subduction zones. They concluded that andesites were unlikely to represent liquids derived directly from subducted ocean-floor basalt (see also Stern and Wyllie, 1973; Stern, 1974). For similar reaons, Ringwood (1975) abandoned the simple one-stage model and formulated a complex scheme involving generation of more silicic liquids from hydrous quartz eclogite, their reaction with overlying mantle peridotite to produce pyroxenites, diapiric uprise of the pyroxenites, followed by partial melting to produce hydrous magmas with compositions ranging from basalt to andesite. He found support for the model in experimental data of Green (1972) and Nicholls and Ringwood (1973).

Batholiths may include material derived from magmas whose genesis was initiated in three distinct environments: subducted oceanic lithosphere, including metamorphosed sea-floor basalt and sediments; mantle peridotite overlying the subducted oceanic slab; and the base of the continental crust or volcanic island arc.

In order to trace the magmatic processes associated with subduction zones, we must know the structural position and rate of movement of the subducted lithosphere slab, the composition of the material subducted, its phase relationships, and the temperature distribution through the slab. In addition, we need information about the physics of migration and concentration of volatiles and magmas at high pressures.

The broad process of progressive melting of the subducted ocean crust as it moves to deeper, hotter levels, with inward melting into the lithosphere slab, is understood in general terms (Oxburgh and Turcotte, 1970; Fitton, 1971). We now have much detailed information about the phase relationships of the materials. However, there are many factors influencing the thermal structure of a subducted slab, and a variety of solutions has been published. Without a specific temperature distribution, particularly within the upper 5 km of the slab covering the former oceanic crust, it is impossible to use the detailed phase relationships to trace a quantitative sequence of magma generation at various depths. There is much to be said for Ringwood's approach (1975, p. 195), which is to assume a level of heating "that leads to a satisfactory petrological model." If only everyone could agree as to what is "satisfactory!"

4.1 Temperatures

It is generally agreed that shear strain heating occurs along narrow zones of relative movement between the downgoing slab and surrounding mantle, but the magnitude of this effect is debated. Melting begins along the upper friction surface of the subducted slab, and progresses into the cooler interior as the material moves to deeper levels.

Figure 3 shows three estimates of the temperature along the upper friction boundary. Note that a depth of 100 km, there is a difference of 500°C between the maximum and minimum estimates.

At any particular depth, the temperature gradient into the interior of the lithosphere slab is not well established. Figure 5 compares a temperature curve for the friction boundary with a calculated curve for the layer 1 km below the surface of the slab. The difference is 100 to 200°C. The temperature at the base of the oceanic crust, 5 km thick, could, therefore, be several hundred degrees lower than that of the original surface of the oceanic crust, as illustrated by Fitton (1971), using geotherms calculated by Oxburgh and Turcotte (1970).

4.2 Water

Water is generally considered to play a significant role in the genesis of calc-alkaline magmas. Certainly, water is carried down with the subducted lithosphere slab. It occurs in the pore fluids of sediments and basaltic crust, and structurally bound in hydrous minerals. Pore fluids may be partially squeezed out as the rocks are carried deeper, or they may be trapped and carried down to levels where they initiate partial melting. The hydrous minerals eventually reach temperatures where they dissociate, releasing water, or where they participate directly in melting reactions. The depths at which reactions occur may be estimated by comparing the experimentally measured thermal stabilities of the minerals with temperature distributions in subducted slabs.

Figure 3A shows the maximum stability ranges of most of the hydrous minerals that could be present in the lithosphere before subduction begins. Serpentine and talc occur in the mantle peridotite–H_2O system, and bodies of serpentinite occur in the oceanic crust. The curve for

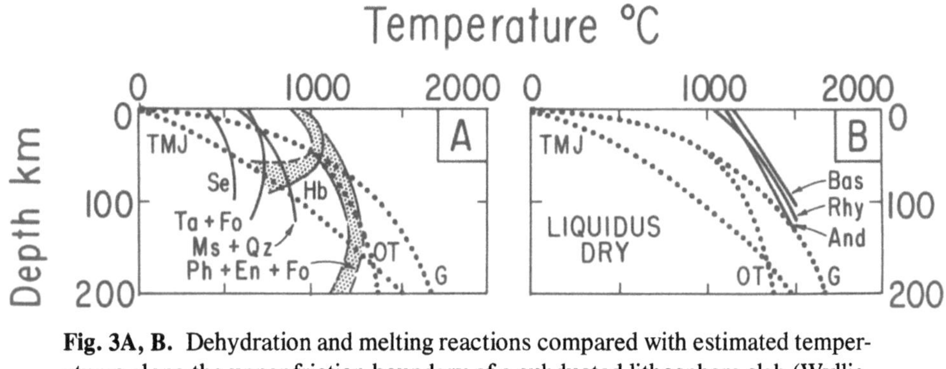

Fig. 3A, B. Dehydration and melting reactions compared with estimated temperatures along the upper friction boundary of a subducted lithosphere slab (Wyllie, 1973). Se: serpentine; Ta: talc; Fo: forsterite; Ms: muscovite; Qz: quartz; Ph: phlogopite; En: enstatite; Hb: amphibole; Bas: basalt; And: andesite; Rhy: rhyolite. TMJ: Toksöz et al. (1971). QT: Oxburg and Turcotte (1970)

muscovite + quartz can represent the micas produced by the metamorphism of subducted sediments. Amphibole is believed to be a prominent mineral in metasomatized ocean crust. The most refractory mineral shown is phlogopite, which is stable to somewhat higher temperatures than any biotite that might be produced in basic rocks of the oceanic crust.

The dotted lines for estimated temperature distributions show that along the upper friction boundary of a subducted slab, only phlogopite could be carried deeper than 100 km. Temperatures are lower within the slab, and the minerals could survive to greater depths (see the pair of curves for surface and 1-km depth in Fig. 5). The maximum depths to which hydrous minerals could be carried within subducted slabs are indicated in Figure 4.

Figure 4 is based on the mineral stabilities in Figure 3A and two calculated thermal structures reviewed elsewhere (Wyllie, 1973). Given a cross section with isotherms, each point in the section is defined by a depth (pressure) and temperature. For each point, the minerals that are stable can be read directly from Figure 3A.

If we assume that the initial lithosphere peridotite was hydrated to the maximum extent, the distribution of serpentine, talc and phlogopite is as shown in Figure 4. The maximum depths to which these minerals could be carried are given by the boundaries to their areas of stability. Amphibole in the peridotite could be carried no deeper than 80–100 km whatever the temperature distribution, because it breaks down through a pressure-sensitive reaction (Fig. 3A).

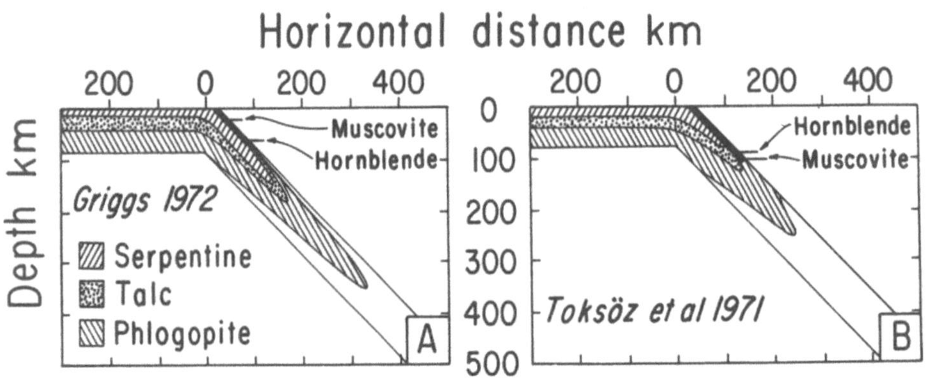

Fig. 4A, B. Schematic cross-sections through subducted slab, showing oceanic crust and maximum depths to which hydrous minerals can be carried, if the host-rock composition is appropriate, according to two temperature distributions (Wyllie, 1973)

The 5 km-thick subducted ocean crust is represented in Figure 4 by the heavy line at the upper edge of the sinking slab. Neither amphibole nor muscovite are carried deeper than 100 km. There is no experimental evidence on epidote to indicate that this would be carried any deeper. Phlogopite, if present, would not reach 100 km depth according to Figure 4A, but could reach about 175 km according to Figure 4B. Biotite in the oceanic crust would break down at shallower depths, probably less than 150 km.

Figure 4 shows the maximum depths to which hydrous minerals could survive during subduction, if they were present initially in the starting material. The mantle peridotite of the lithosphere slab, depleted in basaltic and volatile components during chemical fractionation processes beneath an ocean ridge, is not the kind of peridotite likely to carry phlogopite. Neither is it likely to contain much serpentine. Anhydrous residual peridotite formed beneath an ocean ridge would be at too high a temperature to become serpentinized. When it has migrated away from the active ridge region, the ocean crust probably protects it from serpentinization by sea water, and serpentinization associated with rising magmas of oceanic volcanoes would be localized and probably trivial.

Let us assume no phlogopite in the depleted mantle peridotite, but allow some serpentine. Then, according to Figure 4B, all hydrous minerals in the subducted slab react by a depth of 100 km. It is argued that a faster-moving slab remains cooler in the center to greater depths, and this therefore would permit water to be carried deeper. The result in Figure 4 is based on one of the coolest-interior models yet published. Serpentine, if present in the depleted lithosphere peridotite 10 to 20 km below the ocean crust, reaches only 150 km. Amphibole, serpentine and muscovite of the subducted ocean crust break down by 100 km depth; biotite if present may reach about 150 km.

We can conclude, therefore, on the basis of Figures 3 and 4, that hydrous minerals carried down in subducted lithosphere slabs can reach depths of 100 km, with serpentine in mantle peridotite or biotite in oceanic crust reaching 150 km under favorable conditions. According to this conclusion, magma generation at depths greater than 100–150 km does not involve water. However, if a satisfactory petrological model demands that water be released from the oceanic crust down to depths of 300 km (Ringwood, 1975, p. 295), we have to find a mechanism to retain water in the subducted slabs.

There is a possibility that serpentine may dissociate into dense hydrated magnesium silicates (DHMS) and other minerals at high pressures, and the DHMS could carry water down to greater depths. This has been reviewed by Ringwood (1975, p. 294–7), who concluded that discrete bodies of original serpentinite in the oceanic crust, converted to DHMS, could continue to release water to depths of 300 km. Confirmation of this proposal requires more experimental data and close examination. Serpentinite bodies at the base of the oceanic crust would have to stay below 600°C until they reached a depth of about 150 km. On the basis of presently available evidence, it appears that the slab would already be dehydrated before the pressure became high enough to stabilize DHMS minerals. Dense high pressure hydrates can serve as storage sites for water only if water is present where they become stable.

4.3 Sediments

There are divergent opinions about the extent to which subducted oceanic sediments participate in magma generation, including the view that they are not subducted at all, but are scraped off and accreted laterally to continental margins or island arcs.

Moore (1975) pointed out that the biogenic oozes that initially cover young oceanic crust, calcareous ooze followed by siliceous ooze, become lithified during transportation to subduction zones. These tough rocks, with densities similar to the underlying basalt, would tend to be subducted whereas the overlying pelagic-terrigenous muds and ocean trench clastic units would tend to become deformed and accreted. Mass balance calculations for trace elements and lead isotope compositions for andesites and oceanic sediments led Armstrong (1971) to conclude that, on the average, sediment contamination is required to explain the chemistry of magmas generated in subduction zones (but not necessarily in every subduction zone).

Results from deep drilling on Leg 37 of the Deep Sea Drilling Project provide strong evidence that some sediments are subducted along with gabbroic crust. Abundant unconsolidated and lithified foram-bearing nannofossil ooze occurs within the basalt sequence of ocean crust layer 2 (Geotimes, 1974). Some recrystallized ooze persists as veins and irregular zones in basalt to at least 544-m depth.

Figure 5 illustrates what could happen to pelagic or ocean trench sediments if these should be trapped and subducted with the ocean crust. Many of these sediments when metamorphosed could produce a rock composed dominantly of mica, quartz, and two feldspars. The known phase relationships for a muscovite-granite can therefore be used as a guide for the melting behavior of such sediments, and to illustrate the processes during subduction (Huang and Wyllie, 1973).

The line a–b is the solidus curve for muscovite-granite in the presence of excess water. It corresponds very closely to the minimum melting curve in the synthetic granite-water system (Fig. 1). The muscovite-granite contains 0.6% H_2O structurally held in muscovite. The line a'–b' is the solidus for the muscovite-granite with no additional water; it gives the conditions where the muscovite reacts with the other minerals to produce a small amount of water-undersaturated liquid. More liquid would be generated in a similar rock with more muscovite. The rhyolite liquidus shows the temperature required to produce a rhyolite liquid if only 0.6% water is available.

The dotted line *a–a′* shows the temperature/pressure path for the subducted lithosphere surface calculated by Toksöz et al. [1971 (see Fig. 3a)]. Consider the subduction of sediment along this path. If interstitial pore fluid is trapped within the sediment, melting begins with the generation of a water-saturated liquid at *a*, about 60 km deep. The pore fluid dissolves within a very narrow temperature interval, and with progres-

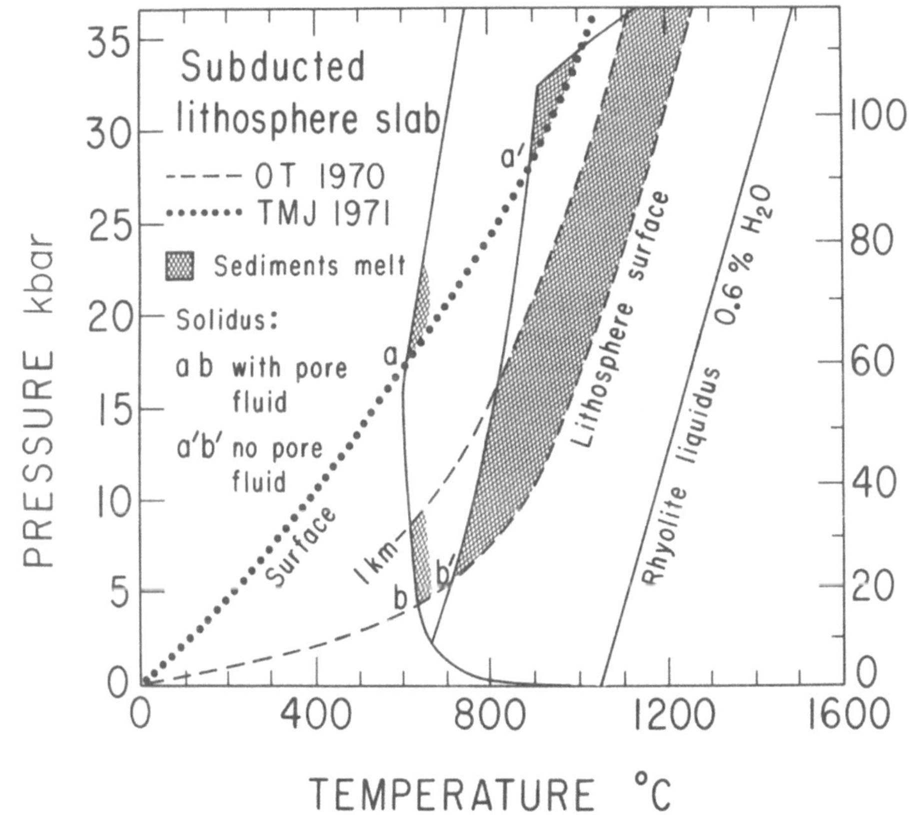

Fig. 5. A model for anatexis of a subducted oceanic sediments (micaceous) in pressure (depth)—temperature projection showing two episodes of melting. The curves b–a and b′–a′ are the solidus curves. *Dashed curves:* published estimates of temperature distributions along the surface, and 1 km below the surface of a subducted lithosphere slab, according to Oxburgh and Turcotte (1970). *Dotted curve:* another estimate along the surface of a subducted lithosphere slab published by Toksöz et al. (1971). See text for details (Huang and Wyllie, 1973)

sive fusion the liquid becomes water undersaturated. The magma would rise from the slab into the overlying mantle, carrying the dissolved pore fluid with it. The remaining dry sediment would then move down with no further melting until the muscovite reacted to produce water-undersaturated liquid at a', near 90 km.

According to Oxburgh and Turcotte (1970), the surface of the subducted lithosphere slab reaches higher temperatures (Fig. 3A), and these two episodes of melting begin at much shallower depths b and b', respectively, where the slab reaches depths of about 17 and 21 km. The shaded bands illustrate the way the fusion zones would migrate from the lithosphere surface into its interior, if there were a layer of sediments 1 km thick—a purely hypothetical example. The first episode of melting would begin at b, 17 km, and it would have reached the base of a 1 km-thick layer at a depth of 32 km.

If the magma generated in the shaded zone extending from b' should escape into the overlying mantle, the remaining dehydrated sediments would yield no more liquid until temperatures reached the solidus for the dry rock; this would probably exceed 1200°C at 100-km depth (Fig. 1). Continued melting would therefore depend upon the addition of water escaping from deeper within the subducted slab.

Figure 5 illustrates some of the complexities of magma generation in a subducted slab. The shaded zones relate to hypothetical (and unrealistic) sediment layers 1 km thick, but the same principles apply to the fusion of a layer of amphibole-quartz-eclogite. Consider the 1-km layer at the top of the slab, at a depth of 34 km. The base of the layer at this depth is just beginning to produce water-saturated melt at a temperature of 630–650°C. The top of the layer is about 2 km directly above the base, and at 33–34 km depth water-undersaturated magma is being generated at temperatures rising from 750°C in the center of the layer to 850°C near its surface. The magma from the lower part of the layer will be cooler than the partially melted rock into which it rises.

The composition of the melt generated from this kind of sediment is similar to that in the synthetic granite system, as shown in Figure 2. Winkler et al. (1975) have emphasized that with addition of calcium, or normative anorthite, initial liquid compositions are significantly enriched in potassium compared with the eutectic liquids in Figure 2A. Figure 2 shows that with increasing pressure, the composition of the liquid changes progressively away from that of granite. At depths corresponding to 100 km, the composition is nearer to that of a syenite. For a more calcic sediment, it appears that the liquids might approximate diorites more closely than granites, although this requires experimental confirmation.

Subsolidus and melting relationships in the system CaO–SiO_2–Co_2 have been measured to 30 kbar by Huang, Wyllie, and Nehru (manuscript in preparation), and Figure 6 shows some of the reactions that could affect subducted calcareous sediments. Two of the estimated temperature paths for the surface of a subducted slab have been transferred from Figure 3A, and these are compared with the solidus temperatures of dry peridotites.

The fusion curve for calcite (Irving and Wyllie, 1973, 1975) remains well above the temperatures in the subducted slab, suggesting that pure limestones could be carried to considerable depths without change. For siliceous limestones, the two heavy curves meeting at invariant point I_2 are relevant. If temperatures were to exceed these curves, calcite and quartz would react to produce wollastonite at depths down to 63 km, and at greater depths they would melt to produce a liquid, releasing CO_2 simultaneously. According to the estimated temperatures for the friction surfaces of subducted slabs, it appears that siliceous limestones also could be carried down to considerable depths in the mantle without reaction. This conclusion needs modification if additional components are introduced into the limestones either by magmatic activity, or by migration of water from deeper within the slab.

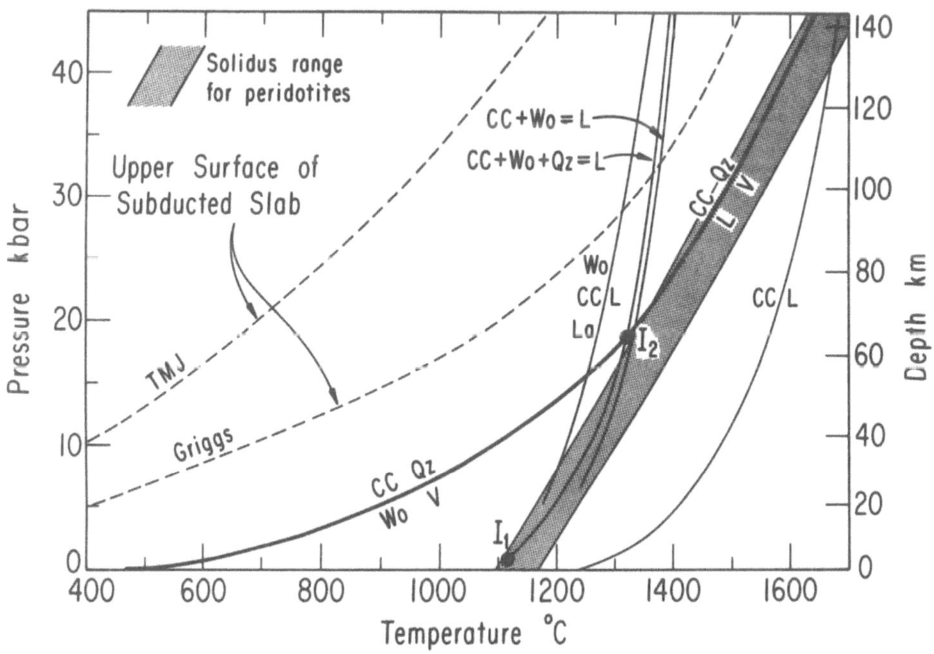

Fig. 6. Decarbonation and melting reactions in the system CaO–SiO_2–CO_2 (Huang, Wyllie and Nehru, in prep.) compared with temperatures for beginning of melting of peridotite and subduction temperatures (see Fig. 3). CC: calcite; Qz: quartz; Wo: wollastonite; La: larnite; *L:* liquid; *V*–CO_2 vapor

Figure 7 shows the equilibrium liquidus diagram for the system at 27 kbar. If the temperature should exceed the fusion curve in Figure 6 at 27 kbar (about 90 km), the calcite and quartz would react to produce liquid *k* in Figure 7, plus carbon dioxide. This silicate-carbonate liquid contains about 45% carbonate.

The influence of water is indicated in Figure 8, which reproduces the calcite-quartz reactions from Figure 6, the granite-water solidus curve from Figure 5, and portions of the muscovite and amphibole curves along with the slab surface temperatures from Figure 3A. The muscovite curve shows the depths where water would be released from any subducted clay minerals, and the amphibole curve shows where most of the water would be released from the main oceanic crust. If any water diffused into subducted limestones, then dissociation would begin at much lower temperatures on the divariant surface shown by the contours for constant P_{CO_2}. This surface is intersected by the pressure/temperature paths for the lithosphere surface.

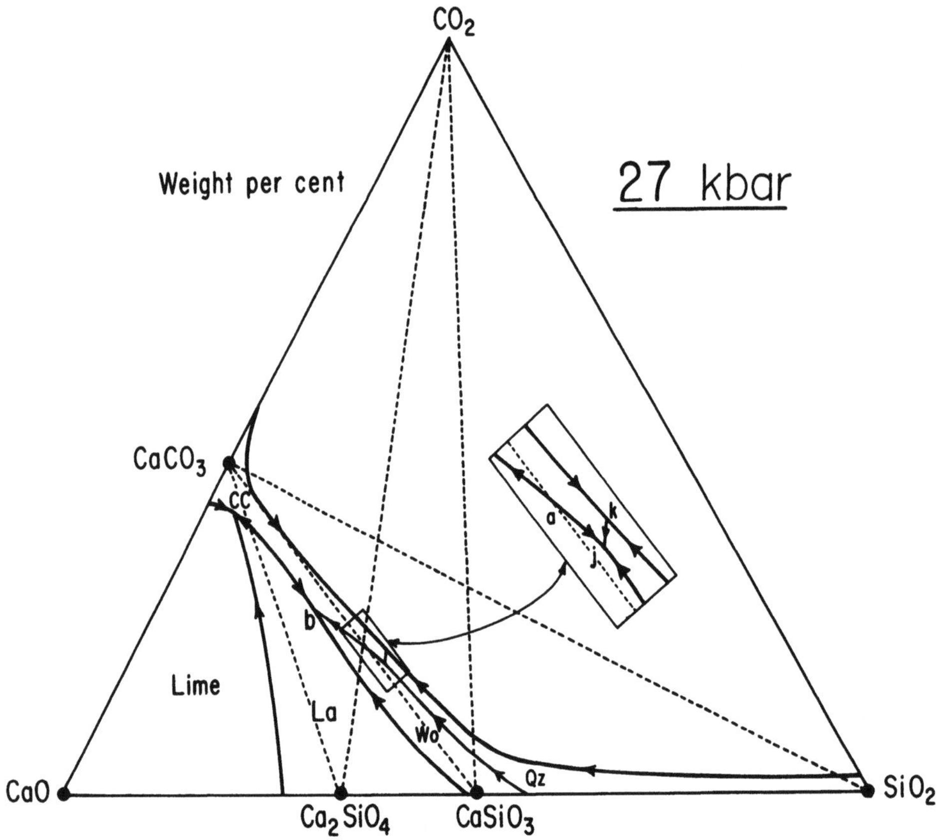

Fig. 7. Liquidus for the system $CaO–SiO_2–CO_2$ at 27 kbar (Huang, Wyllie and Nehru, in prep). For abbreviations see Figure 6

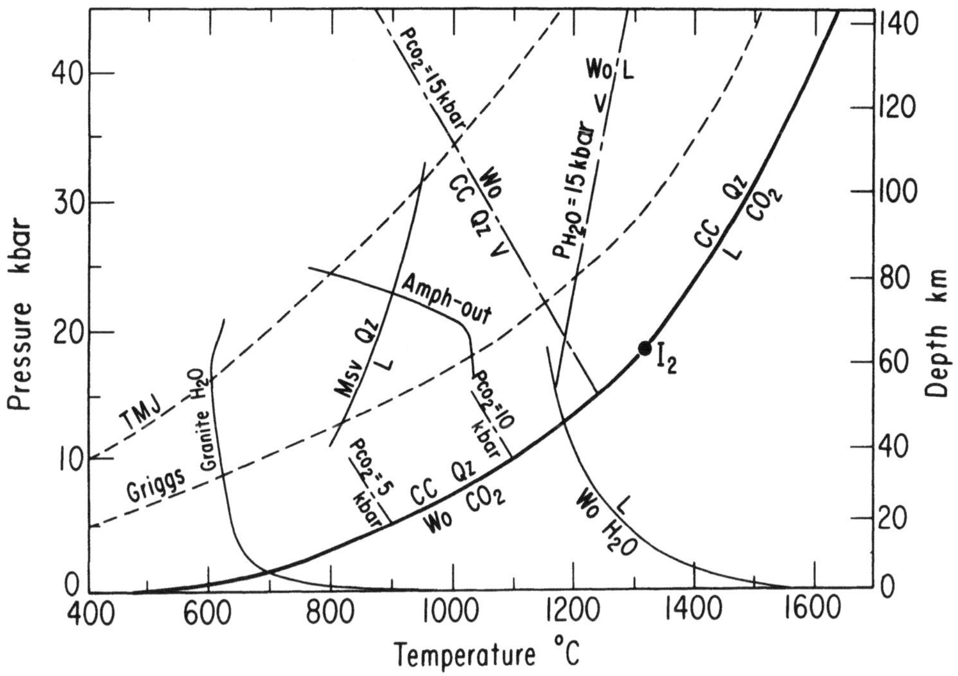

Fig. 8. Reaction curves for calcite-quartz assemblages transferred from Figure 6, with contours to illustrate the shape of the divariant reactions in the presence of CO_2–H_2O mixtures. Solidus from Figure 5, and dehydration reactions from Figure 3a

Dissociation of siliceous limestone through addition of water would leave an assemblage of calcite + wollastonite, or wollastonite + quartz. The fusion curve for calcite + wollastonite is shown in Figure 6; this reaction would occur somewhere between 100 and perhaps 200 km, depending upon the temperature, with liquid composition corresponding to *a* in Figure 7, very similar to liquid *k* that would be produced from calcite + quartz.

If water is added to the assemblages undergoing fusion, the reactions are represented by divariant surfaces with shapes different from the dissociation reaction, as indicated by the single contour for the wollastonite-vapor melting reaction in Figure 8. At the depths where partial fusion of siliceous limestones could begin, there is little water left in the subducted slab to lower the melting temperatures. There is a very good prospect that the reacting limestones would be able to force the diffusing fluid compositions toward CO_2 enrichment, thus permitting some of the carbonate to escape complete dissociation and melting, and to be carried down to considerable depths within the mantle, possibly for long-term storage of CO_2.

4.4 Ocean Crust and Magma Generation

Although some sediments may escape lateral accretion and be captured in subduction, their contribution to magma generation must be minor compared with melting of the basaltic ocean crust. Fundamental for our understanding of this magmatic system is a thorough knowledge of the phase relationships for the rock series basalt–andesite–rhyolite–H_2O, through the pressure interval from the depths where the metamorphosed basaltic crust yields its melts, to the near-surface levels where plutonic quartz diorites and granites are emplaced and chemically comparable andesites and rhyolites are erupted. This problem has been reviewed in considerable detail elsewhere, so the present discussion can be limited to selected items. In particular, Ringwood (1975) recently reviewed the experimental studies from Canberra (Green, Ringwood, Nicholls). Many of my former students and associates have attempted to build up a comprehensive experimental framework for the complete system basalt–andesite–rhyolite–H_2O from 0.5 kbar to 35 kbar (Piwinskii, 1968, 1973a,b, 1975; Piwinskii and Wyllie, 1968, 1970; Boettcher and Wyllie, 1968; Lambert and Wyllie, 1968, 1972, 1974; Gibbon and Wyllie, 1969; Robertson and Wyllie, 1971a,b; Stern and Wyllie, 1973; Huang and Wyllie, 1973; Stern, 1974; Stern et al., 1975; Maaløe and Wyllie, 1975).

For representation of the system we need four variables, pressure, temperature, the water content X_{H_2O} and a variable for rock composition such as SiO_2 content X_{SiO_2}. With one variable held constant, we can examine the effects of the other three in a diagram with three axes. These three-dimensional models are often difficult to visualize, and sections or projections are usually more convenient.

Robertson and Wyllie (1971b) outlined the phase relationships in a P–T–X_{H_2O} model for a single rock, that is, with X_{SiO_2} held constant. Figure 9 shows some of the relationships within P–T–X_{H_2O} models for each of three compositions, with X_{SiO_2} corresponding to rhyolite, andesite, and basalt. The dotted lines show estimated temperatures along the upper friction boundary of subducted slabs transferred from Figure 3. The shaded bands in the left-hand diagrams are the crystallization intervals with excess water. These diagrams also show the dry liquidus boundaries (0% H_2O) and the liquidus curves for each rock with 2% H_2O. The dry liquidus boundaries are compared in Figure 3B. With increasing pressure, the liquidus curves for andesite and rhyolite cross so that the liquidus for rhyolite becomes higher than that for andesite, as illustrated also in Figure 11. This feature was discovered and discussed by Green and Ringwood (1968).

The right-hand diagrams show the melting intervals for each rock with X_{H_2O} constant, and low enough that there is no free vapor within the stability fields of the minerals muscovite and amphibole. All of the

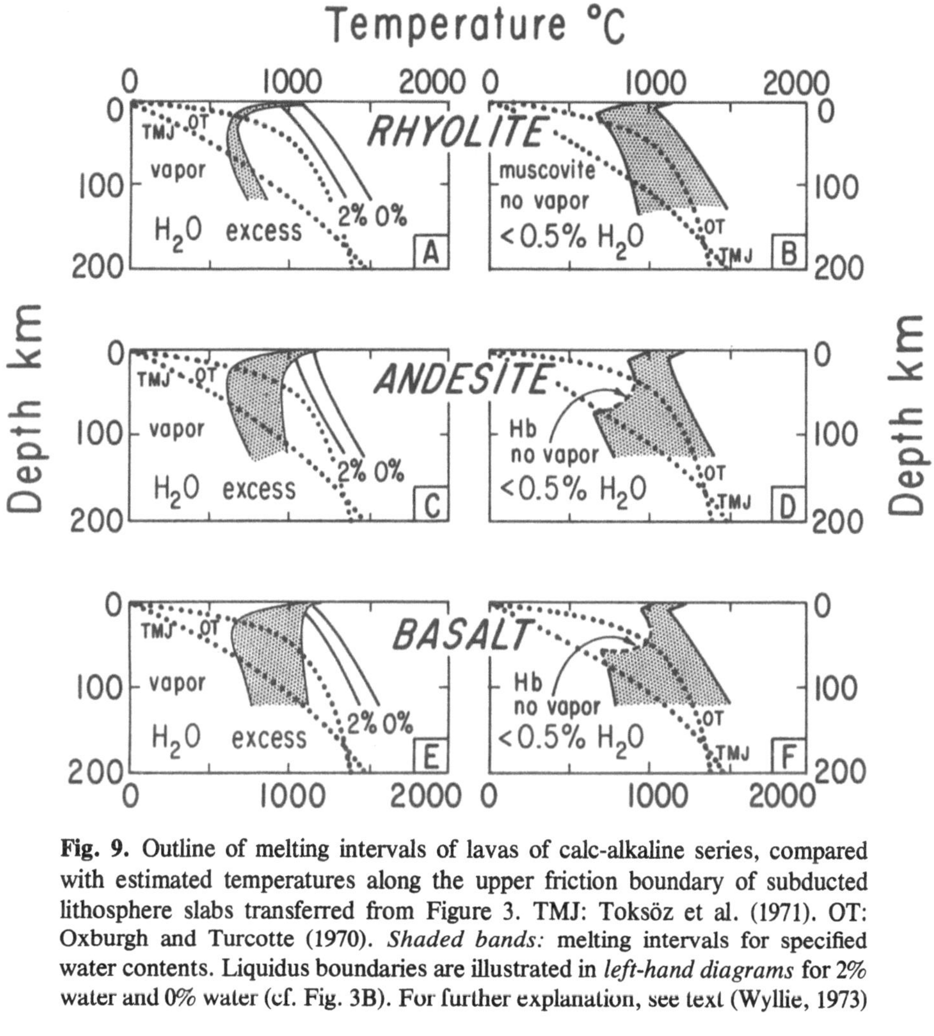

Fig. 9. Outline of melting intervals of lavas of calc-alkaline series, compared with estimated temperatures along the upper friction boundary of subducted lithosphere slabs transferred from Figure 3. TMJ: Toksöz et al. (1971). OT: Oxburgh and Turcotte (1970). *Shaded bands:* melting intervals for specified water contents. Liquidus boundaries are illustrated in *left-hand diagrams* for 2% water and 0% water (cf. Fig. 3B). For further explanation, see text (Wyllie, 1973)

water is then bound within the hydrous minerals, and melting begins only where a hydrous mineral reacts, releasing its water to water-undersaturated liquids. Note the high liquidus temperatures and the correspondingly wider melting intervals for these low water contents.

The left-hand diagrams show that if free water is present in subducted ocean sediments and oceanic crust, melting would begin at shallow depths, ranging from 20 to 80 km depending on the temperature at the surface of the slab (compare Fig. 5). Erupted lavas give no indication of the high water contents required to saturate the magmas (see Figs 12, 13 and 14; Stern and Wyllie, 1973); it is likely that water-saturated liquids generated at relatively shallow depths would rise into the overlying mantle and crystallize there, releasing water vapor.

The application of Figure 9B to partial melting of subducted sediments was reviewed with Figure 5. Figure 9F outlines the conditions for fusion of subducted oceanic basalt, metasomatized and metamorphosed into amphibole eclogite. The two estimates of friction boundary temperatures illustrate the two alternatives for magma generation. Melting may begin directly where the amphibole breaks down or, if temperatures are lower, the amphibole breaks down under subsolidus conditions, and the water vapor released rises into the overlying mantle causing diapiric uprise and partial melting (Wyllie, 1971a), with the separation of hydrous tholeiitic magmas. Ringwood (1975) now prefers the latter alternative, with additional water being supplied by bodies of serpentinite or DHMS for partial fusion of dehydrated quartz eclogite at depths greater than 100 km. Allen et al. (1975) determined the stability of amphiboles in andesite and basalts in the presence of excess water with variable, controlled oxygen fugacities, and they concluded that magmas generated in the subducted slab were derived from amphibole-bearing crust, and that some andesitic magmas are derived from basaltic parents through amphibole-liquid equilibria.

The petrology of erupted andesites is generally considered to indicate the involvement of water, and the estimated temperatures in Figure 9C require the solution of at least 2% water for andesitic liquids to be generated in the depth interval 75–100 km. This is consistent with Eggler's (1972) experimentally based estimate of water content for an erupted andesite. Marsh and Carmichael (1974) calculated a value of 0.75% water dissolved in andesite generated at these depths. Figure 9C shows that according to the TMJ temperature estimate, andesitic liquid generated at 100 km would have to be saturated with water; Figure 13 shows that this would require about 25% dissolved water. This very high value suggests that the TMJ temperature curve between 50 and 150 km is too low. Petrology and experimental petrology together provide constraints for geophysical calculations of temperature distribution.

Another convenient way to examine the phase relationships in this system is using isobaric models with axes T–X_{H_2O}–X_{SiO_2}. Figure 10 is a section through the model for 30 kbar at a constant X_{H_2O} of 5%. The lowest point on the liquidus corresponds to a rhyodacite with about 70% SiO_2. At this point, the rock-series composition-join through the multicomponent natural system is pierced by a liquidus field boundary for the coprecipitation of garnet and quartz. There is another point on the liquidus at a basaltic andesite composition with about 52% SiO_2 where the clinopyroxene-out curve meets the garnet-out curve. This point probably corresponds to a piercing point for a garnet-clinopyroxene field boundary, although the primary field for clinopyroxene has not yet been distinguished from the coprecipitation of clinopyroxene and garnet on or near the liquidus in experiments completed so far for basaltic compositions.

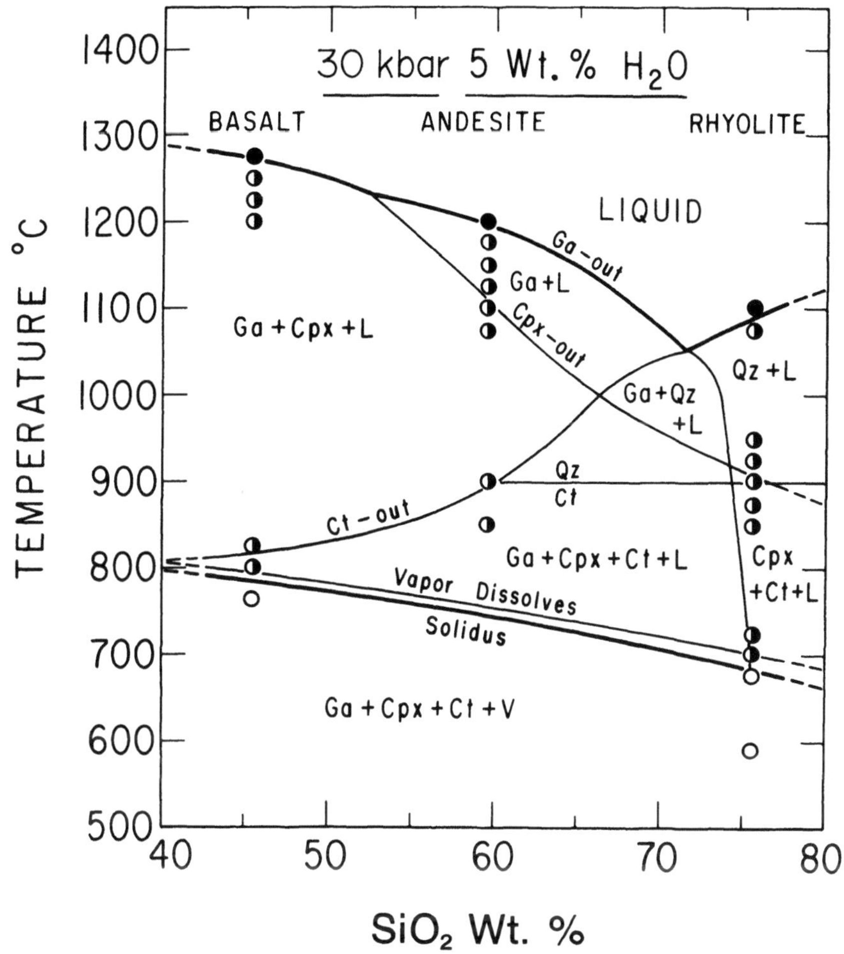

Fig. 10. Isobaric diagram of the melting relations of calc-alkaline rocks as a function of silica content at 30 kbar and with 5 wt% H_2O. Ga: garnet; Cpx: clinopyroxene; Ct: coesite; Qz: quartz; *V:* vapor, *L:* liquid

From experimental studies of the rocks with varying water contents, other diagrams similar to Figure 10 but with different water contents can be constructed. The liquidus relationships for a series of these are superimposed in Figure 11, which shows a projection of the water-undersaturated liquidus surface from within the 30 kbar T–X_{H_2O}–X_{SiO_2} model. The dry liquidus is constructed from the data of Green and Ringwood (1968) and Green (1972). The other lines are contours on the liquidus surface for specific water contents. The liquidus temperature is drastically lowered by addition of water for all compositions, but more so for acid than basic compositions. The liquidus field for garnet, narrow near the dry andesite composition, increases considerably at the expense of both quartz and clinopyroxene. The composition of the low-melting garnet–quartz piercing point

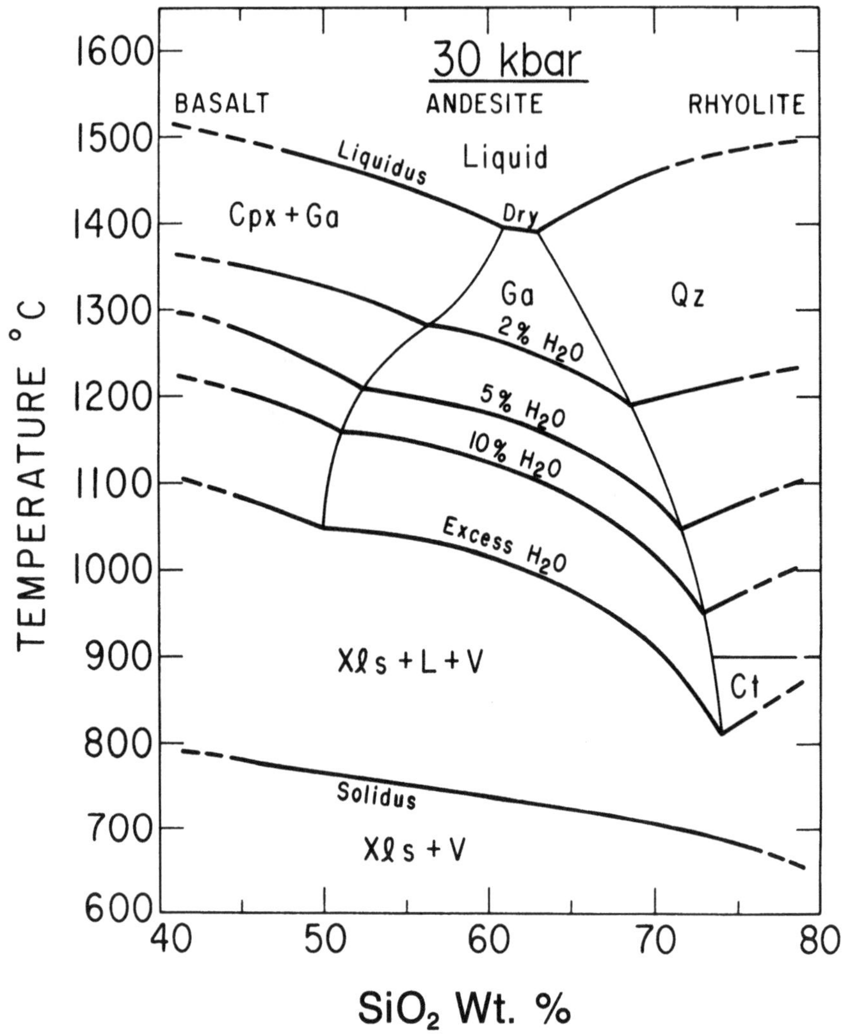

Fig. 11. Isobaric diagram of the liquidus temperatures and primary phases of calc-alkaline rocks with varying amounts of water as a function of silica content. Excess H_2O varies from 20 to 25 wt% with increasing SiO_2 content. Solidus is with excess H_2O. Abbreviations: see Figure 10

becomes enriched in SiO_2 with increasing water content and decreasing temperature. The solidus line remains unchanged regardless of water content, and the field for crystals + liquid + vapor extends from this line to the liquidus for excess water.

In order to illustrate the effect of pressure on the water-undersaturated liquidus surface displayed in Figure 11 we must return to P–T–X_{H_2O} models for specific values of X_{SiO_2}. Figures 12, 13, and 14 show projections of the surfaces from within models for basalt, andesite, and rhyolite. Each surface extends from the excess-water liquidus curve to the dry liquidus for 0% water; these curves are shown in the left-hand diagrams of Figure 9. Water solubilities along the excess water liquidus curves increase from 0% at 1 bar to about 27% at 30 kbar (Stern and Wyllie, 1973b; Green, 1973). Each surface is contoured by lines of constant water content (compare the 2% contours with the 2% lines in Fig. 9). The maps of the surfaces also show the fields for liquidus and near-liquidus minerals. The phases on the surfaces at 30 kbar in Figures 12, 13, and 14, as well as the water contours, can be compared directly with the liquidus surface in Figure 11 for the corresponding SiO_2 values.

It is important to appreciate the fact that the phase diagrams illustrated in Figures 9 through 14 do not provide a complete representation of the phase relationships. They illustrate the sizes and distribution of multicomponent phase fields intersected by the single composition join basalt–andesite–rhyolite in the presence of various amounts of H_2O. There are only two compositional variables represented in this multicomponent

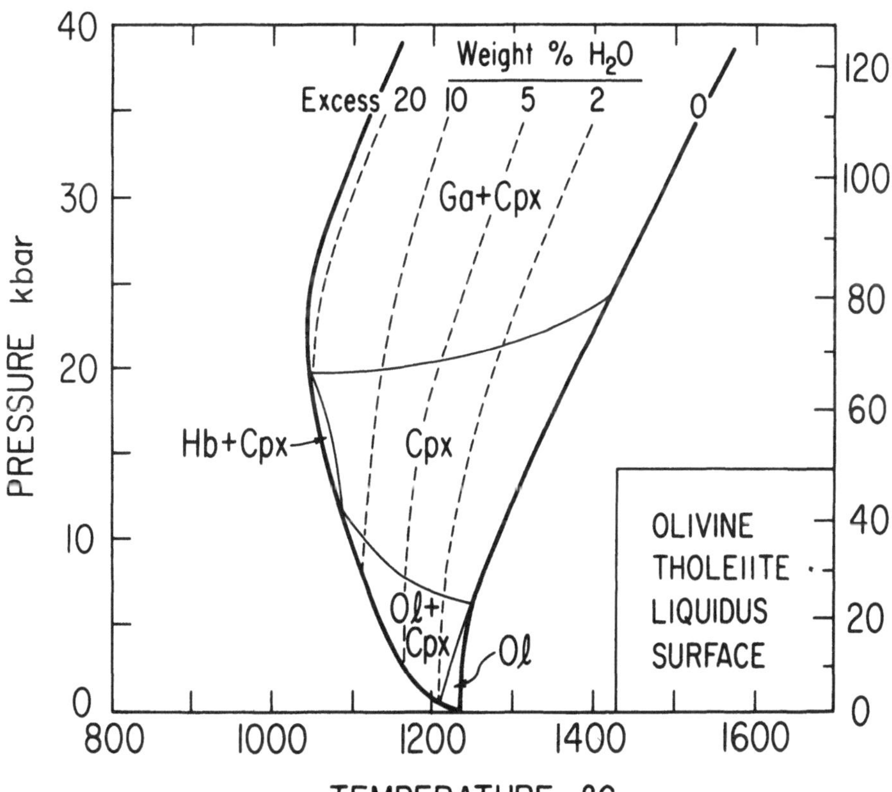

Fig. 12. Contoured map of water-undersaturated liquidus surface for tholeiite showing liquidus and near-liquidus minerals. Pl: plagioclase; Qz: quartz; Ct: coesite; Ol: olivine; Px: pyroxene; Cpx: clinopyroxene; Ga: garnet; Hb: amphibole; (Stern et al., 1975)

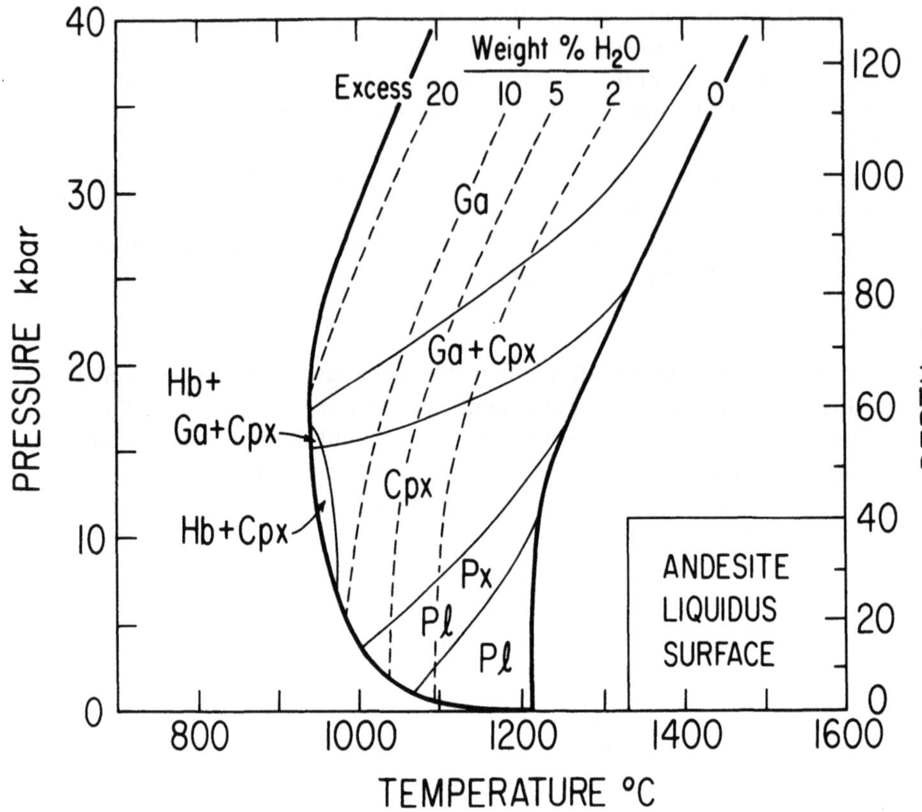

Fig. 13. Contoured map of water-undersatured liquidus surface for andesite, showing liquidus and near-liquidus minerals. For abbreviations see Figure 12 (Stern et al., 1975)

system X_{SiO_2} and X_{H_2O}. Neither the compositions of minerals nor the compositions of liquids coexisting with them are represented on any of the diagrams. Only the liquid compositions in one-phase fields are given by the composition axes. Consequently, as soon as a mineral begins to crystallize from one of these liquids, the liquid composition departs from the diagram. A critical question for petrogenesis is how far do the liquids depart.

Green and Ringwood (1968) concluded that the thermal valley in the anhydrous liquidus profile for basalt–andesite–rhyolite at high pressure (Fig. 11) prevented the fractionation of a dry basalt to liquids more SiO_2-rich than andesite (with water present, the thermal minimum occurs nearer rhyodacite compositions). Stern and Wyllie (1973) argued that this conclusion appears to assume that the thermal minimum on the basalt–andesite–rhyolite join is a terminus for crystallizing liquids, as in a binary system. In fact, Figure 11 shows that there are two piercing points near the thermal minimum.

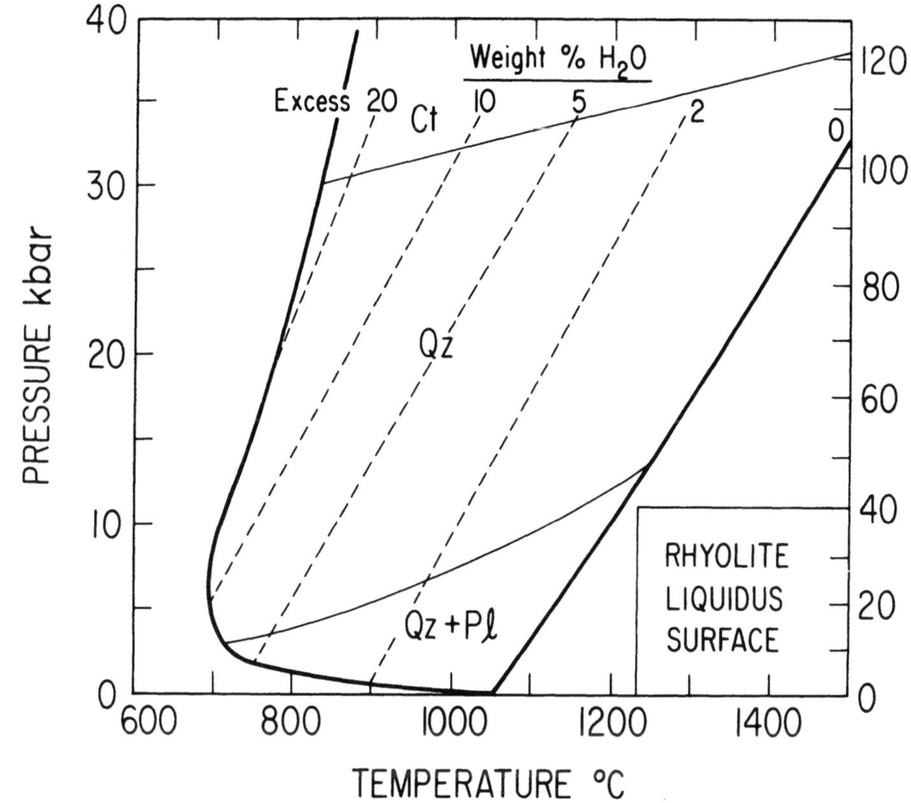

Fig. 14. Contoured map of water-undersaturated liquidus surface for rhyolite, showing liquidus and near-liquidus minerals. For abbreviations see Figure 12 (Stern et al., 1975)

Similarly, Green and Ringwood (1968) stated that with a small degree of partial melting of quartz eclogite, andesitic magma would be formed because of this thermal minimum. They concluded that the magma would not move from the andesite minimum until all the quartz had been incorporated into the liquid (Ringwood, 1975, p. 269) generating perhaps 50% andesite liquid.

These applications of binary phase equilibria to a multicomponent join are not valid. Even if the coexisting minerals happen to constrain the liquid composition close to the join, as appears likely at 30 kbar during coprecipitation of clinopyroxene and garnet at temperatures above the thermal minimum, there is not necessarily anything special about the liquid composition at the minimum. The available data allow the possibility of continued SiO_2-enrichment of crystallizing dry liquids if the clinopyroxene–garnet field boundary moves obliquely away from the composition join, as suggested by Stern (1974). Similarly, the first liquid developed by partial fusion of quartz eclogite would be at a near-invariant point for coexisting quartz–clinopyroxene–garnet that could occur at considerably lower temperatures and with liquids richer in SiO_2 than the thermal minimum on the composition join. The extent of SiO_2-enrichment still needs experimental determination; it could be little, or it could be significant.

Consideration of near-liquidus phases in primary magmas and residual minerals in source rocks suggests that andesite of Figure 13 is a possible derivative of quartz eclogite, with lower water contents at depths greater than 80 km, and water-rich between 50 km and 80 km, and with amphibole in the residue near 50–60 km. However, Stern and Wyllie (1973) and Stern (1974) concluded that although partial melting of subducted oceanic crust at 100 km depth produces broadly acid and intermediate liquids, these liquids depart from the calc-alkaline basalt–andesite–rhyolite join. At shallower depths, where amphibole is involved, as shown in Figures 12 and 13, Allen et al. (1975) reviewed the experimental evidence supporting the generation of andesites from amphibole–quartz–eclogite and by the crystallization of amphibole-bearing basalt at mantle depths.

The high-pressure liquidus mineralogy of rhyolite in Figure 14 is dominated by quartz and coesite, and quartz (or coesite) crystallizes alone for a wide temperature interval below the liquidus (Stern et al., 1975). Rhyolite, therefore, can not be a primary magma from subducted quartz eclogite (Fig. 10). Nor could it be a primary magma from subducted sediments, for similar reasons discussed above (Fig. 2).

4.5 Mantle Peridotite

One consequence of subduction is dehydration of the subducted lithosphere slab (Figs 3A and 4). The amount of carbon dioxide driven from the plate would be negligible compared with water. The water may contribute to magma generation within the subducted oceanic crust, as discussed above, or it may migrate into the overlying wedge of mantle peridotite and lower its melting temperature.

Determination of the compositions of liquids generated from mantle peridotite under various conditions has been subject to much experimental attention (for reviews see Wyllie, 1971a,b, Chapter 8; Ringwood, 1975, Chapts. 4 and 7). Basalts are generated by partial fusion of mantle peridotite, and basaltic magma and its plutonic derivatives are certainly associated with some batholiths (e.g., Gastil, 1975), but this is not an adequate explanation for the origin of the batholiths.

There is evidence that peridotite in the presence of excess water can produce quartz-normative magmas, and it has been argued that andesitic parent magmas can be derived in this way. This question is still controversial, and the literature is extensive, and repetitive.

In outline, Kushiro (1969) concluded that his experimental data indicated that water-saturated andesitic liquids could be generated from peridotite at mantle pressures. Green (1969) countered that such liquids are nephelinitic, arguing that andesitic liquids could not coexist with olivine at water pressures greater than 10 kbars. In a symposium volume, Kushiro (1973) summarized additional experimental data for the generation of water-saturated andesitic or dacitic magmas from peridotite at pressures to 25 kbar; Green (1973) reviewed additional data demonstrating that andesitic or dacitic liquids were not equilibrium partial melting products of peridotite at pressures above 10 kbar; and Boettcher (1973) reviewed experimental results suggesting that water-saturated andesitic liquids could be generated from mantle peridotite at pressures up to 20 kbar.

The cases for and against primary andesites from mantle peridotite were argued in discussions by Mysen et al. (1974). Nicholls (1974) presented a detailed experimental study confirming the Canberra conclusion that andesitic liquids could be generated only at water pressures less than 10 kbar, and that these liquids would be chemically distinct from typical natural andesites. Mysen and Boettcher (1975) presented a detailed experimental study confirming the Geophysical Laboratory-Pennsylvania State University conclusion that andesitic liquids are generated from hydrous mantle peridotite to pressures of at least 25 kbar. Nehru and Wyllie (1975) studied the compositions of hydrous glasses from partially melted peridotite, emphasized the problem of iron loss from sample to capsule wall, and the related problems of attaining and demonstrating equilibrium in the experiments. We concluded that we were "not altogether surprised to note that there is no consensus on the compositions of liquids produced under various conditions — even admitting that other investigators may have been more successful in solving or compensating for the problems."

As far as the origin of batholiths is concerned, I consider the composition of water-saturated liquids generated from peridotites to be of little more than academic interest. Figure 13 shows that in the pressure range 10–20 kbar (40–70 km depth), water-saturated andesite contains 15–20% dissolved water. If magmas of this type were generated in mantle peridotite, they would begin to crystallize and evolve vapor as soon as they began to rise upwards (Fig. 13), and it seems unlikely that they could penetrate far into the crust without undergoing considerable fractionation. I am aware of no evidence for the existence of magmas with such high water contents in batholiths, or as eruptive andesites. Therefore, I conclude that material contributed to batholiths from mantle peridotite consists of basaltic magma rather than andesitic magma.

5 BATHOLITHS

Batholiths are emplaced in the crust in complicated tectonic environments with relatively high geothermal gradients. Whatever the material contribution may be from mantle peridotite or subducted ocean crust, there can be no doubt that anatexis of crustal rocks is involved. Some of the experimental studies leading to this conclusion were outlined earlier.

The dominant rocks of the deep crust consist of the calc-alkaline plutonic rock series gabbro–diorite–granite and their metamorphic equivalents, together with metamorphosed sedimentary rocks whose mineralogy overlaps with that of the plutonic rocks. For most of these rocks in the presence of aqueous pore fluid, melting begins at a moderate temperature. Whatever alkali feldspar is present reacts with the quartz and plagioclase to generate a liquid of granite composition. Progressive melting along grain boundaries has been studied in a series of experiments by Mehnert et al. (1973) and Büsch et al. (1974).

Figure 15 shows the phase relationships in the presence of excess water for the plutonic rock series of the Wallowa Batholith (Piwinskii and Wyllie, 1970). The composition variable in this diagram is the differentiation index instead of X_{SiO_2}, but the diagram is directly comparable with the excess water part of the isobaric T–X_{SiO_2}–X_{H_2O} model for 30 kbar shown in Figure 11. Figure 15 shows that for rocks with compositions ranging from diorite to granite in the presence of excess water, the granitic components are melted within a few tens of degrees of the solidus. The granite liquid then coexists with residual biotite, amphibole, and more calcic plagioclase through a wide temperature interval with little change in composition. Piwinskii (1975) presented similar diagrams for pressures of 1 and 10 kbar.

The small amount of water available in pore fluids is insufficient to saturate the liquids produced by fusion except very close to the solidus. In order to examine processes in systems with low water contents, we need P–T–X_{H_2O} models for rocks of fixed composition (X_{SiO_2} or differentiation index).

Figure 16 shows an isobaric section through such a model for a biotite-granite (Maaløe and Wyllie, 1975). This diagram shows that a rock at fixed pressure may have three different solidus temperatures, depending on its water content (compare Fig. 9). We have discussed the solidus in the presence of a vapor phase, at constant temperature regardless of how little vapor may be present. Ramberg (1952) noted that most of the water in high grade metamorphic rocks would be bound in hydrous minerals. If there is only enough water present to constitute minerals such as epidote, muscovite, biotite and amphibole, then there will be no

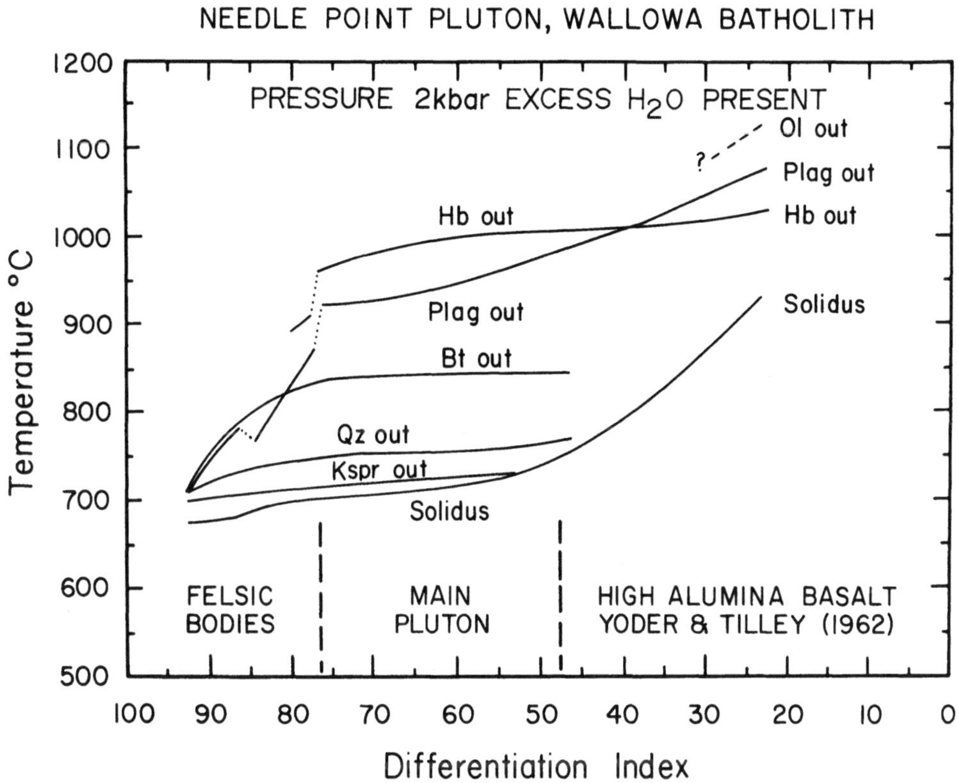

Fig. 15. Isobaric *T*-Differentiation Index diagram for the rock series of the Needle Point pluton with excess water vapor (Piwinskii and Wyllie, 1970). Compare Figures 10 and 11

free vapor phase (in the absence of other volatiles such as CO_2). Melting will begin at somewhat higher temperature, where the hydrous mineral breaks down in a melting reaction. If the rock is completely anhydrous, then the solidus is at much higher temperatures. The applications of the second type of melting for anatexis have been discussed by Brown and Fyfe (1970), Robertson and Wyllie (1971a,b), Fyfe (1973), and Maaløe and Wyllie (1975).

Figure 16 shows that for a metamorphic rock with a reasonable pore fluid content, less than 1%, the water-saturated liquid produced by initial anatexis becomes water-undersaturated within a few degrees of the solidus. This was illustrated by Whitney (1975a) for a series of rocks at several pressures. With progressive fusion, the liquid coexists with alkali feldspar, plagioclase and quartz through a very wide temperature interval. This interval corresponds to the movement of liquid compositions across the water-undersaturated surface illustrated in Figure 2B for the synthetic granite system. During this melting interval, the liquid composition does not change very much in terms of anhydrous components.

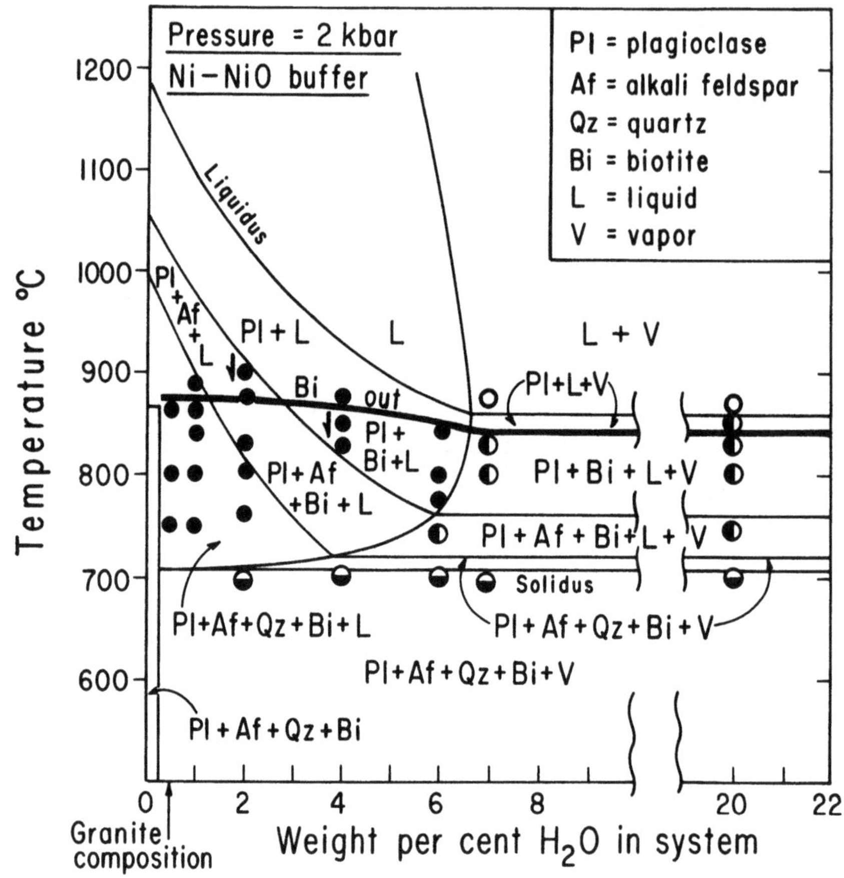

Fig. 16. Crystallization interval of a biotite–granite as a function of H_2O content at 2 kbar pressure (Maaløe and Wyllie, 1975). Note the boundary between the phase fields with vapor *(run circles open or partly open)* and those without vapor *(run circles closed)*. *Heavy line:* upper stability limit of biotite; this intersects the phase boundaries for quartz and the feldspars in the vapor-absent region

Fyfe (1973) reviewed the melting products of biotite–amphibole-bearing granitic and dioritic rocks under vapor-absent conditions, and presented a diagram showing liquid compositions labeled tonalite and diorite at about 900°C in the pressure interval 8–11 kbar. According to Figure 15, liquids of this composition even under conditions of water-saturation require temperatures of 1000°C at 2 kbar, and Figure 13 shows that this falls to about 950°C at 10 kbar. But the amount of water required to saturate the liquid at this pressure is about 15% (Fig. 13), and partial fusion of a vapor-absent biotite–amphibole–diorite can produce liquids with only small water contents.

I cannot accept Fyfe's (1973; Fig. 3) diagram showing the generation of tonalite and diorite liquids from crustal rocks at 900°C. According to Figure 13, a tonalite with 2% water would be completely molten only at

1100°C at pressures to 10 kbar. Figure 14 shows that for complete melting, granite with 2% water requires temperatures of 950–1000° C in the pressure range 5–10 kbar (compare Fig. 16). I conclude that the partial fusion of crustal rocks produces water-undersaturated liquids of granite composition, becoming granodioritic at highest temperatures, but that tonalite and diorite liquids cannot be generated by anatexis of the crust under conditions of normal regional metamorphism.

Figure 16 shows that in the vapor-absent region of a rock-water system, the steeply sloping phase boundaries for anhydrous silicates intersect the nearly isothermal phase boundary for biotite; boundaries for muscovite and amphibole have similar attitudes (Huang and Wyllie, 1973). At each point of intersection, the sequence of crystallization changes. Completion of phase diagrams like Figure 16 for a series of rock types at various pressures will provide grids potentially useful in defining limits for the water content of magmas.

For the biotite-granite of Figure 16, Maaløe and Wyllie (1975) concluded from petrographic study that the sequence of crystallization was plagioclase, microcline, quartz, and finally biotite. This sequence was reproduced in the experiments at 2 kbar only with water content less than 1.2%. These results, together with a review of other data, led Maaløe and Wyllie (1975) to conclude that the water content of large bodies of batholithic magma is unlikely to exceed about 1.5%.

For a large batholithic magma body, consisting of a mixture of liquid and crystals, the water content of the liquid may reach much higher levels than that of the whole magma. Uprise and crystallization of the magma produces water-saturated liquids in the upper regions and margins of the magma chamber, as illustrated by Whitney (1975b).

Figures 9A and 15 show that with excess water at crustal pressures, a granite liquid crystallizes within a narrow temperature interval. Figure 16 shows that for more realistic compositions, a granite with 1–2% water, the crystallization interval is several hundred degrees. This is true even for the ideal "eutectic" granite, as shown in Figures 1 and 2B.

6 CONCLUSIONS

Some batholiths are produced from crustal rocks as a normal consequence of regional metamorphism, with the formation of water-undersaturated granite liquid and mobilized migmatites (Hutchison, 1970). Chappell and White (1974) have distinguished two contrasting types of batholiths that they interpret as being derived by partial melting of either igneous (I-type) or sedimentary (S-type) source material. The results in Figures 12 to 16 relate directly to the I-type. Experiments with pelitic sediments by Green (1976) and with muscovite–granite by Huang and Wyllie (1973) relate to the S-type.

Batholiths may also contain contributions of material and heat from less siliceous magmas generated in subducted ocean crust and mantle peridotite. Magmas with intermediate-SiO_2 contents rising from subducted oceanic crust could differentiate into tonalite and granite by the time they enter the crust, or within the crust. Basaltic magma from the mantle could also differentiate to produce tonalite liquids, or it could hybridize with anatectic granite liquids to produce magmas of intermediate composition.

Is it possible that the emplacement of basaltic magma could raise deep crustal temperatures high enough to generate anatectic liquids of tonalite composition? These conditions appear to be required by the models of Presnall and Bateman (1973) for the Sierra Nevada plutons, and the reconstruction of Gastil (1975) for the Peninsular Ranges batholith. Gastil's interpretation states that liquids from mantle and subducted oceanic crust were intruded into preexisting crust where they "gradually generated huge welts of tonalitic magma" from which plutons rose as diapirs. Andesitic volcanoes had their sources in the tonalite magma, so largely liquid tonalites are implied.

In many batholiths, true granites are rare and alkali feldspar is subordinate to plagioclase and quartz. This is particularly well documented for the Coast Plutonic Complex of British Columbia, which is composed largely of diorite, quartz diorite, tonalite, and granodiorite rather low in alkali feldspar (Roddick and Hutchison, 1974). Similarly, diorites and tonalites predominate over granites in the Peninsular Ranges of southern California and northern Baja California (Gastil, 1975) and in the Coastal Batholith of Peru (Pitcher, 1974).

According to the anatectic model, a crystal mush of water-undersaturated granite liquid and residual crystals is mobilized and intruded upwards in the early stages of the long history of a batholith. If temperatures become high enough the liquid composition may reach that of a granodiorite, but not that of tonalites. The abundance of quartz and plagioclase and paucity of alkali feldspar in many batholiths presents difficulties for this model.

The evidence presented by Hutchison (1970) and Roddick and Hutchison (1974) for the Coast Plutonic Complex of British Columbia appears to be consistent with the mobilization and intrusion of metamorphic rocks with variable proportions of interstitial water.undersaturated liquid. To a first approximation, one can use the alkali feldspar content as a guide to the maximum amount of liquid present in a mushy magma; this would be three times the percentage of alkali feldspar. This suggests that there was little granite liquid within the diorite and tonalite plutons of British Columbia, and an abundance of quartz in the source rocks. This is consistent with the conclusion of L. S. Hollister that the bulk composition of the rocks in the deep metamorphic zones of this region appears to approximate that of tonalite, or andesite (pers. comm., 1975).

I suggest that the exposed plutons rooted in the migmatite complex have lost much of the interstitial liquid that initially caused their mobilization. Figure 16 shows that water-undersaturated granite liquid can exist through a very wide temperature interval. As migmatite domes were slowly intruded upwards, the interstitial granite liquid could be expected to percolate upwards at somewhat higher rate, leaving residual masses of recrystallized plagioclase, quartz and mafic minerals, and producing smaller plutons enriched in alkali feldspar at higher levels, now largely removed by erosion.

Fyfe (1973) reviewed the time scales for the accumulation of crustal melts, their separation from residual crystals, and the dimensions of rising magma "bubbles." There is abundant field evidence for plutonic intrusion of migmatite domes from which the interstitial liquid presumably did not escape (e.g. Haller, 1971). Yet, Ramberg's (1967, Chap. 11) centrifuge experiments with scale models indicate that despite the high viscosity of granite liquids, the liquids would rise through the crust rapidly compared with the uprise of low-density crystalline domes. I conclude that rising migmatite domes can leave behind residual metamorphic rocks of the granulite facies. At some later stage the rising crystal mush can lose its more mobile component, the liquid, leaving a pluton of recrystallized tonalite, with variable proportions of trapped liquid. The distribution of liquid is indicated by the distribution of alkali feldspar, which may be quite irregular in plutonic tonalites and granodiorites.

In many batholiths, there is evidence that magmas derived from mantle or subduction zone are intruded into and through the region of deep crustal anatexis. The associaton of mantle-derived gabbro with batholiths is clearly displayed, for example, in southern California (Gastil, 1975) and in Peru (Pitcher, 1974). This produces a situation with several different types of energetic bodies rising toward the surface at different rates.

Crystal-mush domes and magmatic "bubbles," derived from different sources at different temperatures, transport material and energy to progressively higher levels. Magma bodies become trapped temporarily in chambers, depositing crystals as they differentiate. Liquid fractions may escape from these chambers, or from crystal-mush domes, to rise as independent "bubbles." There are opportunities for hybridism at all stages of these complex processes. At low pressures near the surface, the small initial water contents have become concentrated to the stage of saturation, and the separation of a vapor phase is accompanied by crystallization which temporarily halts the uprise of the magma (see Figs 13 and 14). The granitic liquids are erupted, often explosively, from these magma chambers.

The Coastal Batholith of Peru is an example of the upper levels of a batholith (Pitcher, 1974), and the subvolcanic environment is illustrated by a series of centred complexes within the batholith. Magma was continuously emplaced through a period of 80 million years. During this time there was a gross order of intrusion from gabbro through intermediate to granitic rocks. Superimposed on this main cycle are minor cycles, where the most basic component of an intrusive event, diorite, or tonalite, or granodiorite, is associated with a series of rocks varying towards granite. Stoping and gas coring is an important mechanism of emplaement at these high crustal levels.

The process of subduction carries water deep into the mantle, and concentrates thermal energy in great linear belts. This generates magma in three different environoments, the subducted oceanic crust, the overlying mantle peridotite, and the silicic crust. Gravity forces the magmas to rise, transferring energy into and through the crust in a series of plutons, until they "spread out laterally at shallow depth . . . many of them . . . crystallize beneath a cover of their own volcanic ejecta" (Hamilton and Myers, 1967). Ramberg's (1967) scaled centrifuge models show that the effect of gravity can produce many of the features of batholiths described by Hamilton and Myers.

ACKNOWLEDGMENTS

The experimental data reviewed here was gathered by many research students and associates cited in the text. This research was supported by the Earth Sciences Section, National Science Foundation NSF Grant DS 73-00191 AO1. We have also benefitted from support of the Materials Research Laboratory by the National Science Foundation.

REFERENCES

Allen, J. C., Boettcher, A. L., Marland, G.: Amphiboles in andesite and basalt: I. Stability as a function of P-T-f_{O_2}. Am. Mineral. *60,* 1069–1085 (1975)

Armstrong, R. L.: A model for the evolution of strontium and lead isotopes in a dynamic earth. Rev. Geophys. *6,* 175–199 (1968)

Armstrong, R. L.: Isotopic and chemical constraints on models of magma genesis in volcanic arcs. Earth Planet. Sci. Lett. *12,* 137–142 (1971)

Biggar, G. M., O'Hara, M. J.: Reversibility and equilibrium in atmospheric pressure experiments. In: Progress in Experimental Petrology. Nat. Envir. Res. Counc. Publ. Ser. D, No. 2, London, 1972

Boettcher, A. L.: Volcanism and orogenic belts—the origin of andesites. Tectonophysics *17*, 223–240 (1973)

Boettcher, A. L.: Experimental igneous petrology. Rev. Geophys. Space Phys. *13*, 75–79 (1975)

Boettcher, A. L., Wyllie, P. J.: Melting of granite with excess water to 30 kilobars pressure. J. Geol. *76*, 135–244 (1968)

Bowen, N. L., Tuttle, O. F.: The system $NaAlSi_3O_8$-$KAlSi_3O_8$-H_2O. J. Geol. *58*, 489–511 (1950)

Brown, G. C., Fyfe, W. S.: The production of granitic melts during ultrametamorphism. Contr. Mineral. Petrol. *28*, 310–318 (1970)

Büsch, W., Schneider, G., Mehnert, K. R.: Initial melting at grain boundaries, Part II: Melting in rocks of granodioritic, quartzdioritic and tonalitic composition. N. Jb. Miner. Mh. *8H*, 345–370 (1974)

Carmichael, I. S. E., Turner, F. J., Verhoogen, J: Igneous Petrology. New York: McGraw-Hill, 1974, 739 p

Chappell, B. W., White, A. J. R.: Two contrasting granite types. Pacific Geol. *8*, 173–174 (1974)

Eggler, D. H.: Water-saturated and undersaturated melting relations in a Paricutin andesite and an estimate of water content in the natural magma. Contr. Mineral. Petrol. *34*, 261–271 (1972)

Faure, G., Powell, J. L., Strontium Isotope Geology. New York: Springer, 1972

Fitton, J. G.: The generation of magmas in island arcs. Earth Planet. Sci. Lett. *11*, 63–67 (1971)

Fyfe, W. S.: Hydrothermal synthesis and determination of equilibrium between minerals in the subliquidus region. Geol. *68*, 553–566 (1960)

Fyfe, W. S.: The generation of batholiths. Tectonophysics *17*, 273–283 (1973)

Gastil, R. G.: Plutonic zones in the Peninsular Ranges of southern California and northern Baja California. Geology *3*, 361–363 (1975)

Geotimes: Deep sea drilling project, 16–18, December 1974

Gibbon, D. L., Wyllie, P. J.: Experimental studies of igneous rock series: The Farrington Complex, North Carolina and the Star Mountain Rhyolite, Texas. J. Geol. *77*, 221–239 (1969)

Green, D. H.: The origin of basaltic and nephelinitic magmas in the earth's mantle. Tectonophysics *7*, 409–422 (1969)

Green, D. H.: Contrasted melting relations in a pyrolite upper mantle under mid-oceanic ridge, stable crust and island arc environments. Tectonophysics *17*, 285–297 (1973)

Green, T. H,: Crystallization of calc-alkaline andesite under controlled high-pressure conditions. Contr. Mineral. Petrol. *34*, 150–166 (1972)

Green, T. H.: Experimental generation of cordierite- or garnet-bearing granitic liquids from a pelitic composition. Geology *4*, 85–88 (1976)

Green, T. H., Ringwood, A. E.: Genesis of the calc-alkaline igneous rock suite. Contr. Mineral. Petrol. *18*, 105–162 (1968)

Greenwood, H. J.: Mineral equilibria in the system MgO-SiO_2-H_2O-CO_2. In: Researches in Geochemistry. Abelson, P. H. (ed.). New York: Wiley, 1967, Vol. II, pp 542–547

Haller, J.: Der "Zentrale Metamorphe Komplex" von Nordöstgronland. Teil I. Die geologische Karte von Suess Land, Gletscherland und Goodenoughs Land. Medd. Grønland *73(I)*, 1–174 (1955)

Haller, J.: Geology of the East Greenland Caledonides. New York: Interscience, 1971

Hamilton, W.: Mesozoic California and the underflow of Pacific mantle. Geol. Soc. Am. Bull. *80,* 2409–2430 (1969)

Hamilton, W., Myers, W. B.: The Nature of Batholiths. U.S. Geol. Survey Prof. Paper *554-C,* 1967, 30 p.

Huang, W. L., Wyllie, P. J.: Melting relations of muscovite-granite to 35 kbar as a model for fusion of metamorphosed subducted oceanic sediments. Contr. Mineral. Petrol. *42,* 1–14 (1973)

Hutchison, W. W.: Metamorphic framework and plutonic styles in the Prince Rupert region of the Central Coast Mountains, British Columbia. Can. J. Earth Sci. *7,* 376–405 (1970)

Irving, A. J., Wyllie, P. J.: Melting relationships in CaO–CO_2 and MgO–CO_2 to 36 kilobars, with comments on CO_2 in the mantle. Earth Planet. Sci. Lett. *20,* 220–225 (1973)

Irving, A. J., Wyllie, P. J.: Subsolidus and melting relationships for calcite, magnesite, and the join $CaCO_3$-$MgCO_3$ to 36 kilobars. Geochim. Cosmochim. Acta *39,* 35–53 (1975)

Kerrick, D. M.: Review of metamorphic mixed volatile (H_2O–CO_2) equilibria. Am. Mineral. *59,* 729–762 (1974)

Kleppa, O. J., Newton, R. C.: The role of solution calorimetry in the study of mineral equilibria. Fortschr. Mineral. *52,* 3–20 (1975)

Kushiro, I.: Discussion of the paper "The origin of basaltic and nephelinitic magmas in the earth's mantle" by D. H. Green. Tectonophysics *7,* 427–436 (1969)

Kushiro, I.: Origin of some magmas in oceanic and circum-oceanic regions. Tectonophysics *17,* 211–222 (1973)

Lambert, I. B., Wyllie, P. J.: Stability of hornblende and a model for the low velocity zone. Nature (London) *219,* 1240–1241 (1968)

Lambert, I. B., Wyllie, P. J.: Melting of gabbro (quartz eclogite) with excess water to 35 kilobars, with geological applications. J. Geol. *80,* 693–708 (1972)

Lambert, I. B., Wyllie, P. J.: Melting of tonalite and crystallization of andesite liquid with excess water to 30 kilobars. J. Geol. *82,* 88–97 (1974)

Luth, W. C., Jahns, R. H., Tuttle, O. F.: The granite system at pressures of 4 to 10 kilobars. J. Geophys. Res. *69*, 759–773 (1964)

Maaløe, S., Wyllie, P. J.: Water content of a granite magma deduced from the sequence of crystallization determined experimentally with water-undersaturated conditions. Contr. Mineral. Petrol. *52,* 175–191 (1975)

Marsh, B. D., Carmichael, I. S. E.: Benioff zone magmatism. J. Geophys. Res. *79,* 1196–1206 (1974)

Mehnert, K. R., Büsch, W., Schneider, G.: Initial melting at grain boundaries of quartz and feldspar in gneisses and granulites. N. Jb. Mineral. Mh. *4H,* 165–183 (1973)

Moore, J. C.: Selective subduction. Geology *3,* 530–532 (1975)

Mysen, B. O., Boettcher, A. L.: Melting of a hydrous mantle: II. Geochemistry of crystals and liquids formed by anatexis of mantle peridotite at high

pressures and temperatures as a function of controlled activities of water, hydrogen, and carbon dioxide. J. Petrol. *16*, 549–593 (1975)
Mysen, B. O., Kushiro, I., Nicholls, I. A., Ringwood, A. E.: A possible mantle origin for andesitic magmas. Discussion of a paper by Nicholls and Ringwood. 1. Opening discussion, 2. Reply to opening discussion, 3. Comments on the reply of Nicholls and Ringwood, and 4. Final reply. Earth Planet. Sci. Lett. *21*, 221–229 (1974)
Nehru, C. E., Wyllie, P. J.: Compositions of glasses from St. Paul's peridotite partially melted at 20 kilobars. J. Geol. *83*, 455–471 (1975)
Newton, R. C., Charlu, T. V., Kleppa, O. J.: A calorimetric investigation of the stability of anhydrous magnesium cordierite with application to granulite facies metamorphism. Contr. Mineral. Petrol. *44*, 295–311 (1974)
Nicholls, I. A.: Liquids in equilibrium with peridotitic mineral assemblages at high water pressures. Contr. Mineral. Petrol. *45*, 289–316 (1974)
Nicholls, I. A., Ringwood, A. E.: Effect of water on olivine stability in tholeiites and the production of silica-saturated magmas in the island-arc environment. J. Geol. *81*, 285–300 (1973)
Noble, D. C., Bowman, H. R., Hebert, A. J., Silberman, M. L., Heropoulos, C. E., Fabbi, B. P., Hedge, C. E.: Chemical and isotopic constraints on the origin of low-silica latite and andesite from the Andes of central Peru. Geology *3*, 501–504 (1975)
Orville, P. M.: Alkali exchange betwen vapor and feldspar phases. Am. J. Sci. *261*, 201–237 (1963)
Oxburgh, E. R., Turcotte, D. L.: Thermal structure of island arcs. Geol. Soc. Am. Bull. *81*, 1665–1688 (1970)
Pitcher, W. S.: The Mesozoic and Cenozoic batholiths of Peru. Pacific Geol. *8*, 51–62 (1974)
Piwinskii, A. J.: Experimental studies of igneous rock series: Central Sierra Nevada batholith, California. J. Geol. *76*, 548–570 (1968)
Piwinskii, A. J.: Experimental studies of igneous rock series, central Sierra Nevada batholith, California: Part II. Neues Jahrb. Mineral. Monatsh. 6 193–215 (1973a)
Piwinskii, A. J.: Experimental studies of granitoids from the Central and Southern Coast Ranges, California. Tschermaks Mineral. Petrogr. Mitt. *20*, 107–130 (1973b)
Piwinskii, A. J.: Experimental studies of granitoid rocks near the San Andreas fault zone in the Coast and Transverse ranges and Mojave Desert, California. Tectonophysics *25*, 217–231 (1975)
Piwinskii, A. J., Wyllie, P. J.: Experimental studies of igneous rock series: a zoned pluton in the Wallowa batholith, Oregon. J. Geol. *76*, 205–234 (1968)
Piwinskii, A. J., Wyllie, P. J.: Experimental studies of igneous rock series: "Felsic Body Suite" from the Needle Point pluton, Wallowa batholith, Oregon. J. Geol. *78*, 52–76 (1970)
Presnall, D. C., Bateman, P. C.: Fusion relations in the system $NaAlSi_3O_8$–$CaAl_2Si_2O_8$–$KAlSi_3O_8$–SiO_2–H_2O and generation of granite magmas in the Sierra Nevada batholith. Geol. Soc. Am. Bull. *84*, 3181–3202 (1973)
Ramberg, H.: The Origin of Metamorphic and Metasomatic Rocks. Chicago: Univ. Chicago, 1952
Ramberg, H.: Gravity, Deformation and the Earth's Crust. London: Academic Press, 1967

Ramberg, H., DeVore, D. G. W.: The distribution of $Fe^{\cdot\cdot}$ and $Mg^{\cdot\cdot}$ in coexisting olivines and pyroxenes. J. Geol. *59,* 193–210 (1951)
Ringwood, A. E.: Composition and Petrology of the Earth's Mantle. New York: McGraw-Hill, 1975
Robertson, J. K., Wyllie, P. J.: Experimental studies on rocks from the Deboullie stock, northern Maine, including melting relations in the water-deficient environment. J. Geol. *79,* 549–571 (1971a)
Robertson, J. K., Wyllie, P. J.: Rock-water systems, with special reference to the water-deficient region. A. J. Sci. *271,* 252–277 (1971b)
Roddick, J. A., Hutchison, W. W.: Setting of the Coast Plutonic Complex, British Columbia. Pacific Geol. *8,* 91–108 (1974)
Saxena, S. K.: Thermodynamics of Rock-Forming Crystalline Solutions. New York: Springer, 1973
Steiner, J. C., Jahns, R. H., Luth, W. C.: Crystallization of alkali feldspar and quartz in the haplogranite system $NaAlSi_3O_8$–$KAlSi_3O_8$–SiO_2–H_2O at 4 kb. Geol. Soc. Am. Bull. *86,* 83–98 (1975)
Stern, C. R.: Melting products of olivine tholeiite basalt in subduction zones. Geology *2,* 227–230 (1974)
Stern, C. R., Huang, W. L., Wyllie, P. J.: Basalt–andesite–rhyolite–H_2O: crystallization intervals with excess H_2O and H_2O-undersaturated liquidus surfaces to 35 kilobars, with implications for magma genesis. Earth Planet. Sci. Lett. *28,* 189–196 (1975)
Stern, C. R., Wyllie, P. J.: Melting relations of basalt–andesite–rhyolite–H_2O and a pelagic red clay at 30 kilobars. Contr. Mineral. Petrol. *42,* 313–323 (1973b)
Stern, C. R., Wyllie, P. J.: Effect of iron absorption by noble-metal capsules on phase boundaries in rock-melting experiments at 30 kilobars. Am. Mineral. *60,* 681–689 (1975)
Toksöz, M. N., Minear, J. W., Julian, B. R.: Temperature field and geophysical effects of a downgoing slab. J. Geophys. Res. *76,* 1113–1138 (1971)
Tuttle, O. F., Bowen, N. L.: Origin of granite in the light of experimental studies in the system $NaAlSi_3O_8$–$KAlSi_3O_8$–SiO_2–H_2O. Geol. Soc. Am. Memoir *74,* 153 p. (1958)
Walker, D., Kirkpatrick, R. J., Longhi, J., Hays, J. F.: Crystallization history of lunar picritic basalt sample 12002: phase equilibria and cooling rate studies. Geol. Soc. Bull. *87,* 646–656 (1976)
Wegmann, C. E.: Tectonic patterns at different levels. Geol. Soc. South Africa, annexure to *66,* 1–78 (1963)
Whitney, J. A.: The effects of pressure, temperature and X_{H_2O} on phase assemblages in four synthetic rock compositions. J. Geol. *83,* 1–31 (1975a)
Whitney, J. A.: Vapor generation in a quartz monzonite magma: a synthetic model with application to porphyry copper deposits. Econ. Geol. *70,* 346–358 (1975b)
Winkler, H. G. F.: Petrogenesis of Metamorphic Rocks, 2nd ed. New York: Springer, 1967
Winkler, H. G. F., Boese, M., Marcopoulos, T.: Low temperature granitic melts. N. Jb. Miner. Mh. *6H,* 245–268 (1975)
Wyllie, P. J.: The role of water in magma generation and initiation of diapiric uprise in the mantle. J. Geophys. Res. *76,* 1328–1338 (1971a)
Wyllie, P. J.: The Dynamic Earth. New York: Wiley, 1971b

Wyllie, P. J.: Experimental petrology and global tectonics: a preview. In: Experimental Petrology and Global Tectonics. Tectonophysics *17,* 189–209 (1973)
Wyllie, P. J., Tuttle, O. F.: Hydrothermal melting of shales. Geol. Mag. *98,* 56–66 (1961)
Yoder, H. S., Tilley, C. E.: Origin of basalt magmas: an experimental study of natural and synthetic rock systems. J. Petrol. *3,* 342–532 (1962)
Zen, E-an: The stability relations of the polymorphs of aluminum silicate: a survey and some comments. Am. J. Sci. *267,* 297–309 (1969)

17

E. J. OLSEN

Abstract Thermodynamic calculations on meteorite mineral assemblages have been applied in two ways: (1) subsolidus equilibria that pertain to conditions within or on parent objects, and (2) condensation of solids from a cooling gaseous nebula of solar composition. The subsolidus calculations fall into two categories: (1) oxidation-reduction equilibria in ordinary chondrites, iron meteorites, pallasites, enstatite chondrites, and carbonaceous chondrites; (2) cation distributions between coexisting phases, olivine-orthopyroxene, orthopyroxene-clinopyroxene, olivine-chromite, and troilite-sphalerite. The subsolidus calculations show that ordinary chondrites and iron meteorites equilibrated under approximately the same oxidation conditions, while enstatite chondrites are five to six orders of magnitude more reduced. The chondrites equilibrated at temperatures in the lower igneous range, 900–985°C. The iron meteorites equilibrated at relatively low pressures, 200–3100 bars, commensurate with parent planets $\leq$ 200 km in radius. The condensation calculations have been successful in predicting mineral assemblages that have been observed in the primitive, unaltered carbonaceous chondrites. Further refinement of condensation calculations, along with additional thermochemical data, promise to explain the earliest processes in planetary formation.

EQUILIBRIUM THERMODYNAMIC CALCULATIONS APPLIED TO METEORITE MINERAL ASSEMBLAGES

1 INTRODUCTION

Hans Ramberg utilized thermodynamic calculations to make inferences and obtain insights into the processes that existed during the formation of metamorphic rock facies. It is the purpose of this short chapter to illustrate how the same approach has been applied to understanding the conditions that existed in the formation of several major groups of meteorites.

The direct application of equilibrium thermodynamic calculations to pertinent reactions among mineral phases in meteorites has not been practiced on as large a scale as one might expect. One of the reasons for this is that some of the significant groups of meteorites are clearly agglomerates and/or disequilibrated assemblages. There are, on the other hand, some well-equilibrated meteorites, principally the majority of the ordinary chondrites. Because most of the ordinary chondrites can be considered to be reduced soda-rich peridotites (with metallic Fe–Ni alloy and albite-rich plagioclase, as opposed to magnetite and calcic-plagioclase in terrestrial peridotites) many of the reactions and calculations applied to peridotites that appear in the terrestrial petrological literature will also apply to many of the equilibrated stone meteorites. In this chapter, however, this latter approach will not be discussed.

Perhaps the first time the approach of Ramberg was applied to a meteorite mineral assemblage was the statement made by Prior (1916), what has come to be called "Prior's Rule." This rule states that the smaller the quantity of Fe–Ni alloy in chondritic stone meteorites the richer the alloy is in Ni content and the richer in FeO contents are the coexisting ferromagnesian silicates. Prior never attempted to quantitatively treat this rule nor did he explicitly write out the pertinent reactions. Nevertheless, his observation was stated in terms that suggested a quantitative oxidation-reduction relationship. Mueller (1963) showed the pertinent reactions suggested by the rule are:

$$\tfrac{1}{2}\, Mg_2SiO_4 + FeSiO_3 \rightleftharpoons \tfrac{1}{2}\, Fe_2SiO_4 + MgSiO_3 \qquad (a)$$
$$2\, MgSiO_3 + 2\, Fe + O_2 \rightleftharpoons Mg_2SiO_4 + Fe_2SiO_4 \qquad (b)$$
$$FeSiO_3 + Fe + \tfrac{1}{2}\, O_2 \rightleftharpoons Fe_2SiO_4\,. \qquad (c)$$

By making certain assumptions on the nature of the three solid solutions represented in these equations, olivine, pyroxene, and metal, he was able to utilize observed chemical analyses to calculate the theoretical distribution relationship between the mole fraction, Fe/(Fe + Ni) in metal and the mole fraction, FeO/(FeO + MgO) in the coexisting silicates. The agreement between the theoretically calculated curve and a curve previously fitted to a group of empirical data points by Ringwood (1961) is remarkably close.

Later, Mueller, and Olsen (1967) pointed out a necessary corollary to Prior's Rule is indicated in the equations above: as the oxidation state of the meteorite assemblage increases, the quantity of olivine must also increase relative to pyroxene and metallic iron. In this paper they derived a rigorous parametric equation to express the extent of the oxidation–reduction reactions in terms of the FeO content of the ferromagnesian silicates and the quantity of metallic iron. Relatively satisfactory agreement is attained between the theoretically calculated relationship and a plot of empirical points. The paper made it clear that while Prior's Rule does have validity it does not necessarily imply a genetic oxidation–reduction relationship between the several types of equilibrated chondritic meteorites; that is, the more oxidized chondrites are not formed by oxidation of the more reduced types of chondrites, or visa versa.

Another early attempt to apply calculated equilibria to meteoritic problems was Urey (1952). In this book (especially pp. 114–157) Urey calculates a group of about fifty homogeneous and heterogeneous equilibria involving oxides, silicates, hydroxides, sulfides, noble gases, methane, ammonia, and water in an attempt to delineate the kinds of phases that would condense under the conditions of a cooling solar nebular gas at various points in that nebula. Many of the equations are highly interdependent and although tentative conclusions could be drawn an accurate calculated equilibrium condensation sequence was not to come until better thermodynamic data and high speed computers were available two decades later (discussed later in this chapter).

During the 1960s and early 1970s a moderate number of papers appeared that approached meteorite phase equilibria through thermodynamic calculations. These fall into two main categories: (1) subsolidus equilibria that apply to conditions that obtained within or on solid parent bodies, and (2) condensation calculations that apply to the conditions of formation of the parent bodies or planetesimals. The first category falls into several subclasses and these will be treated separately.

2 SUBSOLIDUS EQUILIBRIA

2.1 Oxidation–Reduction Equilibria

2.1.1 Ordinary Chondrites. The important oxidation–reduction relationship embodied in Prior's Rule has been discussed already. In addition, there are a number of other papers that have examined subsolidus oxidative equilibria among coexisting meteorite phases.

In two papers Mueller (1964, 1965) examined several important oxidation-reduction equilibria between silicate–oxide–metal phases in equilibrated ordinary chondrites. The boundary curve between the fields of ferrite (Mg-bearing magnetite) and metal marks a change in the behavior of ferro-magnesian solid solution compositions, as indicated by the reactions:

$$2\,MgSiO_3 + 2\,Fe + O_2 \rightleftharpoons Mg_2SiO_4 + Fe_2SiO_4 \quad \text{(d)}$$
$$2\,MgSiO_3 + \tfrac{2}{3}\,Fe_3O_4 \rightleftharpoons Mg_2SiO_4 + Fe_2SiO_4 + \tfrac{1}{3}\,O_2. \quad \text{(e)}$$

In the limiting case, when X^{ol}_{Mg}, X^{px}_{Mg}, and $X^{ferrite}_{Fe}$ each goes to unity, the equilibrium constants reduce to

$$K_d = (1 - X^{ol}_{Mg})^2/p(O_2), \text{ and } K_e = (1 - X^{ol}_{Mg})^2\, p(O_2)^{1/3}/(a^{ferrite}_{Fe})^{2/3}. \quad (1)$$

On either side of the ferrite-Fe boundary the Mg/(Mg + Fe^{2+}) ratio of the ferromagnesian silicate increases as $p(O_2)$ increases (in the ferrite field), or decreases (in the metallic iron field), respectively, at constant temperature. The compositions of the silicates show a different dependence on the partial pressure of oxygen in the two cases, such that the compositional sensitivity to $p(O_2)$ is less when the silicate is in equilibrium with ferrite than when it is in equilibrium with metal.

Mueller (1964) examined in detail the quantitative equations relating the fugacity of oxygen to the observed FeO/(FeO + MgO) mole ratio in equilibrated chondrites. The equations incorporate the observed distribution relation of FeO between the coexisting olivine and orthopyroxene. The final relationship is given by:

$$\log_{10} f(O_2) = 2\log_{10}\frac{(X^{ol}_{Fe})(X^{ol}_{Mg})}{(X^{px}_{Mg})(a^{m}_{Fe})} - \log_{10} K_d \quad (2)$$

where ol = olivine, px = pyroxene, m = metal, K_d is the equilibrium constant for reaction (d), written out earlier in this chapter. Olivine and pyroxene are considered ideal solutions over the compositional range of interest.

For the case where the metal is pure Fe, Table 1 gives some representative values of $f(O_2)$, for two fayalite contents of the olivine. Fa 22 is an

Table 1 Table illustrates the dependence of $\log_{10} f(O_2)$ [in atmospheres] on the composition of olivine in a chondritic meteorite and on temperature (from Mueller, 1964)

T°(K)	$\log f(O_2)$ for Fa 22	$\log f(O_2)$ for Fa 1
900	−26.0	−28.7
1100	−20.2	−22.9
1300	−16.1	−18.8

approximate average composition of olivines in most ordinary chondrites.

It is clear that these numbers approach, in magnitude, the reciprocal of Avogadro's Number. For mechanistic reasons then the oxidation state of the internal atmosphere in these meteorites must be controlled by gaseous species other than only oxygen. Because the metal will contain carbon, oxidative equilibria will involve the reactions:

$$C + CO_2 \rightleftharpoons 2\,CO \tag{f}$$
$$CO + \tfrac{1}{2}\,O_2 \rightleftharpoons CO_2. \tag{g}$$

Using calculated $f(O_2)$ values from Table 1 it is possible to calculate the corresponding values of $f(CO)$ and $f(CO_2)$. Similarly, from the reaction it was possible for Mueller to calculate the values of $\log_{10}[f(H_2O)/f(H_2)]$ for the various temperatures and olivine compositions given in Table 1. A portion of these calculated data are given in Table 2.

These data demonstrate that the dominant gaseous species present during the equilibration of the silicates in equilibrated ordinary chondrites are CO and H_2. At the time these conclusions were at variance with a commonly held notion that the dominant gases were CO_2 and H_2O, by analogy with terrestrial peridotites. Subsequent years of research on both meteorite mineral assemblages and returned lunar rocks have demonstrated the validity of Mueller's conclusions.

Williams (1972) approached the question of oxidation state in chondrites from a different point of view. By fitting empirical activity-composition data for the olivine solid solution from a group of sources in the experimental literature he obtained an expression for the activity coefficient of the fayalite component as a function of T, P, and X_{Fa}^{ol} . His results are in close agreement with those calculated by Mueller. Williams' value of $\log_{10}f(O_2)$ at 1200°K for the ordinary chondrites is given in Table 3.

Table 2 Table illustrates the dependence of $\log_{10}f(CO)$ [in atmospheres], $\log_{10}f(CO_2)$ [in atmospheres], and $\log_{10}[f(H_2O)/f(H_2)]$ on the composition of olivine in a chondritic meteorite and on temperature (from Mueller, 1964)

	Fa 22			Fa 1		
T°(K)	log f(CO)	log f(CO_2)	log[f(H_2O)/f(H_2)]	log f(CO)	log f(CO_2)	log[f(H_2O)/f(H_2)]
900	−1.9	−3.0	−1.5	−3.3	−5.7	−2.9
1100	−0.2	−1.4	−1.2	−1.6	−4.1	−2.6
1300	1.0	−0.2	−1.0	−0.3	−2.9	−2.3

Table 3 Calculated values of $\log_{10}f(O_2)$ [in atm] for several meteorite types at 1200°K (temperature chosen arbitrarily for the purpose of comparison]

Meteorite type	$\log f(O_2)$
Enstatite chondrites (Olsen and Fuchs, 1967)	−24.2
Enstatite chondrites (Larimer, 1968a)	−25.5
Ordinary chondrites (for Fa 22, Mueller, 1964)	−18.0
Ordinary chondrites (for Fa 19, Williams, 1972)	−17.9
Pallasites (with phosphate)	−17.3
Iron meteorites (with ferrous phosphates)	−16.6
Iron meteorites (with chlorapatite and silicates)	−17.6

2.1.2 Iron Meteorites and Pallasites. The state of oxidation of chondritic meteorites can be calculated because of a sufficient number of phases that participate in oxidation-reduction equilibria. For the major phases in iron meteorites and pallasites a sufficient number of phases relative to the number of components are generally not present and calculations are not possible. Some irons, however, contain trace phases that are indicators of oxidation state. Olsen and Fredriksson (1966) and Olsen and Fuchs (1967) calculated oxidation conditions as a function of T for coexisting phosphate-phosphide phases which are found as trace inclusions within some irons and pallasites. For irons which contain Ca-free sarcopside or graftonite (both $Fe_3(PO_4)_2$), or pallasites that contain farringtonite, $Mg_3(PO_4)_2$, and schreibersite with phosphorus dissolved in metal, the reactions below apply:

$$3\,Fe + 2\,[P]^{Fe} + 4\,O_2 \rightleftharpoons Fe_3(PO_4)_2 \qquad \text{(h)}$$
$$3\,Mg_2SiO_4 + 4\,[P]^{Fe} + 8\,O_2 + 6\,Fe \rightleftharpoons 2\,Mg_3(PO_4)_2 + 3\,Fe_2SiO_4 \qquad \text{(i)}$$

where $[P]^{Fe}$ indicates phosphorus dissolved in metal.

In addition, some irons contain traces of chlorapatite and the silicates olivine and orthopyroxene, along with schreibersite and P dissolved in metal. In all cases sufficient phases are present to calculate the states of oxidation of such meteorites for comparison with the chondritic meteorites discussed earlier. To make the calculations an insufficiency of thermochemical data and knowledge of some activity coefficients for the several solid solutions involved made it necessary to make simplifying assumptions. Because oxidation-reduction equilibria are so energetic the effects of the assumptions on the final calculations are demonstrably small. Representative numbers are given for comparative purposes in Table 3.

Buseck and Goldstein (1969) made calculations from the equation

$$Fe + \tfrac{1}{2}\,[Ni_2SiO_4]^{ol} \rightleftharpoons Ni + \tfrac{1}{2}\,[Fe_2SiO_4]^{ol}. \qquad \text{(j)}$$

By taking the solid solutions, metal and olivine, to be ideal in the temperature range of interest they were able to calculate the equilibrium distribution of Ni and Fe between coexisting olivine and metal as a function of T. On the basis of the observed compositions of coexisting metal and olivine in the pallasites they were able to conclude that equilibrium did obtain for most of the pallasites with equilibration having taken place between 1200°C and 800°C, and perhaps lower for some individual meteorites.

2.1.3 Enstatite Chondrites. From the point of view of oxidation-reduction equilibria the enstatite chondrites are interesting subjects. They are the most reduced meteorites known and contain such exotic reduced phases as CaS (oldhamite), TiN (osbornite), Si_2ON_2 (sinoite), 1.1 to 1.6% Si dissolved in kamacite metal, etc. Calculation of the oxidation state is possible by several means. Olsen and Fuchs (1967) used a reaction involving known coexisting phases in some enstatite chondrites,

$$2\,CaMg(SiO_3)_2 + 2\,FeS \rightleftharpoons 2\,CaS + 2\,MgSiO_3 + 2\,SiO_2 + 2\,Fe + O_2 \qquad \text{(k)}$$

Larimer (1968a) used both experimental runs and thermodynamic calculations to determine the equilibrium,

$$\tfrac{1}{2}\,CaS + O_2 \rightleftharpoons \tfrac{1}{2}\,CaSO_4 \qquad \text{(l)}$$

Anhydrite does not occur in any enstatite chondrite. Therefore, this reaction marks an upper limit of $f(O_2)$ for enstatite chondrites. Other equilibria are required to delimit the oxidation state before the appearance of anhydrite. Larimer uses several equilibrium calculations:

$$Si + O_2 \rightleftharpoons SiO_2 \qquad \text{(m)}$$
$$2\,CaSiO_3 + S_2 \rightleftharpoons 2\,CaS + 2\,SiO_2 + O_2 \qquad \text{(n)}$$
$$2\,Fe + S_2 \rightleftharpoons 2\,FeS \qquad \text{(o)}$$

On the basis of his work the oxidation state of enstatite chondrites can be calculated. As indicated in Table 3 his results and those obtained by Olsen and Fuchs from Equation (k) are in good agreement.

2.1.4 Carbonaceous Chondrites. The Type I carbonaceous chondrites (= C1) present the opposite end of the oxidation–reduction spectrum from the enstatite chondrites. C1 meteorites contain such phases as goethite (FeOOH), several hydrated Mg-sulfates (principally epsomite), sulfur, breunnerite, and gypsum along with magnetite and some troilite. C1 meteorites are the most oxidized group known. The matrix phases of these meteorites consist of at least two polymorphs of a septechlorite (probably ferric chamosite), which has low thermal stability (Fuchs, et al., 1973). Boström and Fredriksson (1966) examined a number of equilibria for $T \leqslant 40°C$ in the presence of liquid water. For example,

$$24\ FeOOH + FeS \rightleftharpoons 8\ Fe_3O_4 + Fe^{++} + 12\ H_2O + SO_4^{=}.$$

Utilizing activity data for the hydrous solute ions at several pH values in equilibrium with breunnerite, epsomite, and gypsum they calculated Eh-pH diagrams. Based on their calculations the oxidation state was too high to be accounted for in a closed system. They concluded that open access to free oxygen under low temperature hydrothermal conditions was necessary to account for the coexisting oxidized phases observed. They were not, unfortunately, able to make quantitative calculations of $f(O_2)$.

2.1.5 Conclusions on Oxidation States of Meteorites. Based on thermodynamic calculations it is possible to compare the relative oxidation states of several groups of meteorites. Table 3 gives values at 1200 K only, however, because $\log f(O_2) - 1/T$ curves are parallel to each other and nonintersecting, the relative oxidation states are preserved at any chosen temperature. From the calculations it is apparent that iron meteorites and pallasites are somewhat more oxidized than the ordinary chondrites, although both are within calculation error of each other. In terms of parent bodies it can be concluded that irons and ordinary chondrites occupied the same oxidation–reduction regime. This could be interpreted in two ways: (1) Irons occurred as masses distributed throughout a stony body, or (2) Irons occurred as cores in parent bodies that were small enough that the oxidation state was uniform throughout.

Enstatite chondrites, on the other hand, are five or six orders of magnitude more reduced. Application of these data to the interpretation of genetic relationships among these types of meteorites is still forthcoming with no clear consensus at the present time.

2.2 Subsolidus Cation Distributions between Coexisting Phases

2.2.1 Olivine–Orthopyroxene. Ramberg and Devore (1951) examined analytically the distribution of Mg^{2+} and Fe^{2+} between coexisting olivine and orthopyroxene as a function of temperature. In principle it is possible to establish a thermometer that would register the last temperature at which cation distribution equilibration took place. Calibration of the thermometer remained a problem. Experimental studies (Larimer, 1968b; Medaris, 1969) indicated the temperature dependence is too small to provide a useful thermometer. A number of attempts to calibrate the thermometer from the compositions of naturally coexisting pairs in both meteorites and terrestrial rocks proved fruitless. The issue was further complicated when experimental Mössbaur spectroscopy (for example, Virgo and Hafner, 1969) demonstrated a degree of sublattice site occupancy nonideality to the orthopyroxene solid solution. Analytical treatment of the cation distribution between two phases, one of which has multiple, nonequivalent sublattice cation sites, was provided by Grover and Orville (1969) and Mueller (1962) for the case where the sublattice partitioning is ideal.

In terms of meteorites, in which olivine and orthopyroxene commonly coexist, the apparent lack of temperature sensitivity of this thermometer is unfortunate. Olsen and Bunch (1970), on the other hand, utilized the tightly clustered, observed compositions of Mg-rich olivine–orthopyroxene pairs as they occur as trace phases in iron meteorites to estimate the degree of nonideality of the olivine solid solution, which had been observed experimentally by Nafziger and Muan (1967). Assuming orthopyroxene nonideality on the basis of measured site-occupancy data of Virgo and Hafner (1969), it was possible to calculate the activity coefficients of Mg^{2+} and Fe^{2+} in the olivine solid solution, using the Gibbs-Duhem equation, for $1.00 \leqslant X_{Mg}^{ol} \leqslant 0.91$. The results gave activity coefficients from unity at $X_{Mg}^{ol} = 1$, rising to 1.02 at $X_{Mg}^{ol} = 0.91$. Most interesting is the value of the activity coefficient of Fe^{2+} which extrapolates to 2.00 at infinite dilution (i.e., as $X_{Fe}^{ol} \rightarrow 0$). This is in good agreement with the calculation of Williams (1972) in which he obtains a value of 1.96. Both calculations are for $T = 1100°C$.

At the present time it appears that the use of an olivine-orthopyroxene thermometer for meteorites is a dead issue.

2.2.2 Orthopyroxene–Clinopyroxene. Kretz (1963) established the theoretical basis for a mineral thermometer based on the distribution of Fe^{2+} and Mg^{2+} between coexisting Ca-poor orthopyroxene and Ca-rich clinopyroxene. He calibrated the thermometer empirically on measured compositions in equilibrated terrestrial rocks. The temperature dependence is distinct. He also concluded that the thermometer would be essentially independent of pressure. For the reaction:

$$MgSiO_3 + CaFeSi_2O_6 \leftrightharpoons FeSiO_3 + CaMgSi_2O_6 \tag{q}$$

the distribution constant K_D is

$$\log_{10}K_D = -363.6/T + 0.1178. \tag{3}$$

Later, McCallum (1968) provided a modified calibration. For the purposes of this discussion the Kretz calibration will be used.

Using Kretz's model, Van Schmus and Koffman (1967) calculated that all the ordinary chondrites equilibrated at approximately 820°C, irrespective of the chondrite group, H, L, or LL. Later, Blander (1972) questioned Kretz's model and established equations for a model in which the two nonequivalent cation sites M1 and M2 in both pairs show unequal distribution of Mg^{2+} and Fe^{2+}. In addition, because the Ca^{2+} ion in clinopyroxene goes preferentially into the M2 site a significant percentage of those sites are blocked for occupancy by Fe^{2+} or Mg^{2+}. In his model the distribution of Fe^{2+} and Mg^{2+} between coexisting orthopyroxene–clinopyroxene pairs is a function not only of T, but also of mole fraction of $CaSiO_3$ in the clinopyroxene. Blander developed a group of equations to calculate these dependences and concluded that the Mg–Fe distribution was an extremely sensitive function of the Ca content (i.e., Wo content) of the clinopyroxene, such that accurate calibration of the thermometer would be impossible.

In 1974 Bunch and Olsen published superior analyses of coexisting orthopyroxene-clinopyroxene pairs in well equilibrated H, L, and LL group chondrites—mainly the same chondrites Van Schmus and Koffman (1967) had used in their work. They found, contrary to the results of Van Schmus and Koffman, that three different equilibration temperatures were obtained when the data were calculated in Equation 3, above: H group = 900°C, L group = 940°C, LL group = 984°C. If, however, the Wo content of the clinopyroxenes were taken into account, within the framework of Blander's group of equations, only a single equilibration temperature is implied; there is no way, however, to know what this temperature is. Later, Olsen and Bunch (1975) attempted to calibrate the thermometer within the framework of Blander's equations and found that it was possible only to delineate temperature ranges on a fairly crude scale: low temperatures of 700–750°C, higher temperatures of approximately 1000°C, and highest temperatures of approximately 1200°C. The three groups of chondrites, on this scale, equilibrated at around 1000°C. In terms of the thermal histories of meteorite parent bodies it is desirable to know if silicate cation equilibration took place in the igneous or low temperature metamorphic temperature regimes. At this time it appears that they equilibrated in the lower igneous range. This is very reasonable in that cation exchange would have to have taken place in a completely water-free environment, as opposed to terrestrial crustal rocks. It is unlikely that diffusional equilibrium could have been achieved at temperatures down in the 700 to 800°C range under absolutely dry conditions.

At the moment the Blander model appears uncertain. Kretz's model gives very reasonable temperatures when applied to a variety of equilibrated terrestrial rocks and this cannot be ignored. Blander's model assumes that *all* the Ca^{2+} is in the M2 site, however, structurally it only prefers that site. It is entirely likely that Ca^{2+} ions blocking some of the M1 sites balance the effect of Ca^{2+} ions blocking the M2 sites such that the thermometer works along the lines established by Kretz after all. If so, then the three groups of ordinary chondrites did equilibrate at three different temperatures, as discussed earlier, and these were in the lower range of igneous temperatures, 900 to 984°C. Saxena (1973, p. 146), however, questions the simplicity of the Kretz model and believes that variation of Ca content in both orthopyroxene and clinopyroxene with temperature may have a pronounced effect on the Fe–Mg distribution dependence on temperature. He concludes that more information is required on the ternary solid solution, En, Fs, Wo before the validity of the thermometer can be determined.

2.2.3 Other Thermodynamic Treatments of Meteorite Mineral Assemblages. Another thermometer based on cation exchange between olivine and chromite, which commonly coexist in many meteorites, was developed theoretically by Jackson (1969). Applications to meteorites, however, have neither been widespread nor successful. The complexity of the chromite solid solution render the model problematical.

Barton and Toulmin (1966) developed the theoretical basis for a mineral barometer based on the FeS content of sphalerite coexisting with troilite. Schwarcz et al. (1975) developed the barometer to apply to those iron meteorites that have traces of sphalerite in troilite. From the initial condition,

$$\frac{dX_{\mathrm{FeS}}^{\mathrm{sphal}}}{dP} = \frac{[\bar{V}_{\mathrm{FeS}}^{\mathrm{sphal}} - V_{\mathrm{FeS}}^{0}]\, X_{\mathrm{FeS}}^{\mathrm{sphal}}}{RT}, \tag{4}$$

where $\bar{V}_{\mathrm{FeS}}^{\mathrm{sphal}}$ is the partial molal volume of FeS dissolved in sphalerite and V_{FeS}^{0} is the molal volume of pure troilite, they derived

$$\ln X_{\mathrm{FeS}}^{\mathrm{sphal}} = \frac{\ln [1.5T/10^4 + 0.394] - P[5.803 - 4.806T/10^4] - 5.22P^2/10^3}{8.206T/10^2}. \tag{5}$$

Using measured FeS contents of sphalerite inclusions in troilite they were able to calculate that the iron meteorites, for which they had data, formed under pressure regimes of 200 to 3100 bars. This indicates an origin for irons within stony planetary objects of radius $\leq$ 200 km. This is in accord with the conclusions on the oxidation states of irons and stones discussed earlier in this chapter.

Ringwood and Kaufman (1962) utilized thermodynamic data on the Fe–Ni system, along with observed data on molar volumes of Fe–Ni alloys as a function of temperature and composition, to calculate phase diagrams for the Fe–Ni subsolidus system at high pressures. They conclude that iron meteorites formed under pressures of 30 000 to 60 000 bars. The results are in marked conflict with the calculations of Schwarcz, et al., cited just above. Other lines of evidence, beyond the scope of this chapter, strongly indicate that the low pressures calculated by Schwarcz et al. (1975) are correct.

3 THERMODYNAMIC CONDENSATION CALCULATIONS

Using thermodynamic calculations to illuminate the processes by which molecular species and solid matter in the solar system initially formed has been a long sought goal by cosmochemists. In recent years this approach has reached a new level of elegance and predictions on the behavior of what is a very complex system have been satisfyingly accurate. Before describing recent advances in this field a brief historical review is appropriate.

As noted in the introduction to this chapter, Urey (1952) made calculations based on a large group of reactions and attempted to delineate a set of conditions that would apply to the condensation of matter in the primitive solar system. At the time, however, he was not able to handle the complex interdependence of the reactions and the large number of components necessary to obtain directly applicable results. That required high speed computers which were not available at the time. Thus, his conclusions were necessarily tentative and limited. Later, Wood (1963) made a simplified calculation based on the single reaction

$$\text{Fe(gas)} + H_2O\text{ (gas)} \leftrightarrows \text{FeO (liquid in silicate solution)} + H_2\text{ (gas)}. \qquad \text{(r)}$$

By a series of simplifying assumptions he was able to conclude that condensation of gaseous Fe into molten silicate droplets at temperatures around 2000 K would require a total gas pressure (dominantly H_2 gas pressure) of 100 to 10 000 atm. On the basis of astrophysical lines of reasoning total pressures of this magnitude are out of the question (e.g., see Hoyle, 1960). The most reasonable H_2 pressure regime is in the range of 10^{-5} to 10^{-2} atm.

In that pressure range condensation of solid phases takes place directly from the gaseous state. The problem has been treated by a host of workers including Lord (1965), Eck et al. (1966), Larimer (1967, 1973), Blander and Katz (1967), Anders (1968, 1971, 1972), Griffiths et al. (1972), and Lewis (1972), just to cite a representative few. Most workers have treated the problem by considering condensation of solids from a neutral solar nebular gas. Arrhenius and Alfven (1971), however, have examined the case of condensation of solid particles from an ionized gas that is not in thermal equilibrium with the solids.

For the purpose of illustrating the approach of applying thermodynamic calculations to the condensation of matter in a neutral solar nebular gas the more recent and largely successful calculations of Grossman (1972) will be used. For a very thorough review of the subject, with an exhaustive bibliography, Grossman and Larimer (1974) is recommended.

The physical constraints on the system are set by various theoreticians, most notably Cameron and Pine (1973). Condensation takes place mainly in a solar nebular gas disk in which temperature-pressure gradients range from 2000 K and 10^{-1} atm, at the center, to 100 K and 10^{-5} atm at 10 AU (astronomical units, each approximately 1.5×10^8 km).

The calculation is based on the stability, with falling temperature, of each element relative to its several gaseous molecules, and the stabilities of both elements and molecules relative to condensed solids. The list of possible condensed solids must be chosen aforehand. These include phases found to be abundant in primitive chondrites as well as phases not necessarily known in meteorites but ones which are simply reasonable, such as any known highly refractory phase. What phases appear to condense out in the calculation at what temperatures depends on the bulk composition of the gas and the choice of condensed phases whose thermodynamic data are entered into the computational program. As phases condense the composition of the gas changes. Certain phases, which condense early may, at lower temperatures, become unstable with respect to the changed gas composition and be resorbed into the gas again, thereby affecting its composition and, in turn, affecting the stability of other phases in the process of reaching saturation. Questions of nucleation and degree of subcooling (supersaturation) are not considered in the equilibrium case, although it is recognized that these factors will affect the ultimately observed mineralogy and petrography of the resulting accreted bodies (i.e., meteorites). The starting bulk composition of the gas is largely hydrogen and helium. The hydrogen mainly sets the total pressure regime. The remainder of the gas consists of neutral atoms of the major elements in their relative abundances as known from astrophysical compilations (Cameron, 1973).

The condensed solids are conceived of as dust-size grains of the kinds known to occur in interstellar nebular clouds. Grain growth, overgrowth of later condensates nucleated on earlier condensates, and reaction between later condensates and earlier condensates finally produce particles large enough to accrete into planetesimals and ultimately parent bodies of minor to major planetary dimensions. With the condition that $p(\text{total}) \approx p(H_2)$ and $p(H_2) \leqslant 10^{-2}$, the ideal gas law obtains and

$$p(H_2) = n_{H_2}RT, \text{ in which } n_{H_2} = \text{moles of } H_2 \text{ per liter} \quad (6)$$

$$2n_{H_2} \approx n_H, \text{ in which } n_H = \text{total moles of H per liter} \quad (7)$$

where n_H is a function of P and T. Thus,

$$n_H = 2p(H_2)/RT. \tag{8}$$

For a gaseous element E, its abundance relative to H is

$$\frac{n_E}{n_H} = \frac{E}{H}. \tag{9}$$

For each element E, a linear mass balance equation can be written based on all the gaseous species in which E can appear. Oxygen, for example, would appear thus,

$$n_O(\text{total}) = n_O + n_{H_2O} + n_{CO} + 2n_{CO_2} + n_{MgO} + n_{CaO} + \ldots, \text{etc.} \tag{10}$$

Equilibrium between a condensed phase and vapor can be illustrated for the case of corundum.

$$Al_2O_3(\text{solid}) \rightleftarrows 2\ Al\ (\text{gas}) + 3\ O\ (\text{gas}) \tag{11}$$

which has an equilibrium constant $K = (p_{Al})^2\,(p_O)^3/a_{Al_2O_3}$.

For a pure solid the activity of corundum, $a_{Al_2O_3}$ is unity. It is assumed that in a tenuous gas the fugacities of the gaseous elements are equal to their partial pressures and that these are, in turn, directly proportional to their abundances.

As the gas saturates with those elements relative to a solid phase that is composed of them the solid condenses. The gas is thus sharply diminished in the components of that solid phase at the equilibrium condensation temperature. Since the condensed solid is considered to keep in chemical equilibrium with the gas, the linear mass balance equation for each element involved in that particular solid must have an additional term added for the number of moles in that solid per liter of space. For a meaningful calculation 20 elements, 134 gaseous species, and 75 solids had to be considered simultaneously. For each total gas pressure considered the process had to be repeated.

The results of the calculations are gratifying for temperatures from 1700 K down to 1000 K. Early, high temperature phases, such as hibonite, spinel, melilite, Cr-bearing iron-nickel alloys, olivine, Ca-poor pyroxene, diopside, and calcic feldspars, as observed in certain primitive stone meteorites, conform mineralogically to the predicted sequence. Certain phases, however, show petrographic relationships that do not conform to the predicted sequence. By repeating calculations under different sets of boundary conditions it is hoped these problems can be resolved. It may also be that kinetics of nucleation, and large degrees of subcooling, as suggested by Blander and Abdel-Gawad (1969) can affect the equilibria in ways that account for the observed petrographic relationships (Blander and Fuchs, 1975). In any event, accurate thermodynamic calculations of equilibrium condensation provides the best starting point from which to explain the chemical-mineralogical features observed in primitive meteorites.

4 CONCLUDING COMMENTS

In the decades of the 1940's and 1950's Hans Ramberg pioneered an approach to calculating chemical equilibria in complex metamorphic petrological systems with a view to gaining insights into the physical–chemical conditions that existed at the time metamorphism took place. The approach became universal to the extent that it has now become an integral part of the education of graduate petrology students. Twenty-five to thirty years ago thermochemical data for rock-forming minerals and solid solutions were sparse. Because of the impetus Ramberg provided, many more data are now available from calorimetric laboratories, such as the U.S. Geological Survey laboratory of Richard Robie, one of Ramberg's former students (see Robie and Waldbaum, 1968). Today it is possible to make many calculations based on good thermochemical data, although in certain cases only approximate calculations can be made where data are lacking, especially activity–composition data. One must, nevertheless, agree with the statement of Mueller (1964), "Numerical approximations are, in any event, always preferable to qualitative speculations."

In meteoritics, as well as in all branches of mineralogy and petrology, this statement embodies the spirit of Hans Ramberg's teachings.

APOLOGIA

In the literature of meteoritics many able research workers have made significant contributions over the past seven decades. It was not possible to cite the works of most of these individuals in this review. The content of this chapter is restricted to only those papers that approached meteorite mineral equilibria via thermodynamic calculations—the approach Ramberg applied to terrestrial metamorphic rocks. Thus, many major contributions to meteoritics by experimentation and analysis are not included, except where they were needed to support a particular calculation discussed in the text. To any individual who feels his work was slighted the author offers his most sincere apology.

REFERENCES

Anders, E.: Chemical processes in the early solar system as inferred from meteorites. Acc. Chem. Res. *1*, 289–298 (1968)

Anders, E.: Meteorites and the early solar system. Ann. Rev. Astron. Astrophys. *9*, 1–34 (1971)

Anders, E.: Physico-chemical processes in the solar nebula as inferred from meteorites. In: Symposium on the Origin of the Solar System. Reeves, H. (ed.), Paris; Centre Nat. de la Rech. Sci., 1972, pp. 179–195

Arrhenius, G., Alfven, H.: Fractionation and condensation in space. Earth Planet. Sci. Lett. *10*, 253–267 (1971)

Barton, P. B., Jr., Toulmin, P. III: Phase relations involving sphalerite in the Fe-Zn-S system. Econ. Geol. *61*, 815–849 (1966)

Blander, M.: Thermodynamic properties of orthopyroxenes and clinopyroxenes based on the ideal two-site model. Geochim. Cosmochim. Acta *36*, 787–799 (1972)

Blander, M., Abdel-Gawad, M.: The origin of meteorites and the constrained equilibrium condensation theory. Geochim. Cosmochim. Acta *33*, 701–716 (1969)

Blander, M., Fuchs, L. H.: Calcium–aluminum-rich inclusions in the Allende meteorite: evidence for a liquid origin. Geochim. Cosmochim. Acta *39*, 1605–1619 (1975)

Blander, M., Katz, J. L.: Condensation of primordial dust. Geochim. Cosmochim. Acta *31*, 1025–1034 (1967)

Boström, K., Fredriksson, K.: Surface conditions of the Orgueil meteorite parent body as indicated by mineral associations. Smithsonian Misc. Coll. *151*, 1–39 (1966)

Bunch, T. E., Olsen, E.: Restudy of pyroxene-pyroxene equilibration temperatures for ordinary chondrite meteorites. Contr. Mineral. Petrol. *43*, 83–90 (1974)

Buseck, P., Goldstein, J.: Olivine compositions and cooling rates of pallasitic meteorites. Bull. Geol. Soc. Am. *80*, 2141–2158 (1969)

Cameron, A. G. W.: Abundances of the elements in the solar system. Space Sci. Rev. *15*, 121–146 (1973)

Cameron, A. G. W., Pine, M. R.: Numerical models of the primitive solar nebula. Icarus *18*, 377–406 (1973)

Eck, R. V., Lippincott, E. R., Dayhoff, M. O., Pratt, Y. T.: Thermodynamic equilibrium and the inorganic origin of organic compounds. Science *153*, 628–633 (1966)

Fuchs, L. H., Olsen, E., Jensen, K. J.: Mineralogy, mineral-chemistry, and composition of the Murchison (C2) meteorite. Smithsonian Contrib. to the Earth Sci. *10*, 1–39 (1973)

Griffiths, P. R., Brown, C. W., Lippincott, E. R., Dayhoff, M. O.: Thermodynamic models in cosmochemical systems. Geochim. Cosmochim. Acta *36*, 109–128 (1972)

Grossman, L.: Condensation in the primitive solar nebula. Geochim. Cosmochim. Acta *36*, 597–619 (1972)

Grossman, L., Larimer, J. W.: Early chemical history of the solar system. Rev. Geophys. Space Phys. *12*, 71–101 (1974)

Grover, J., Orville, P.: The partitioning of cations between coexisting single- and multi-site phases with application to the assemblages: orthopyroxene–clinpyroxene and orthopyroxene–olivine. Geochim. Cosmochim. Acta *33,* 205–226 (1969)

Hoyle, F.: On the origin of the solar nebula. Quart. J. Roy. Astron. Soc. *1,* 28–55 (1960)

Jackson, E. D.: Chemical variation in coexisting chromite and olivine in chromite zones of the Stillwater Complex. Econ. Geol., Mono. *4,* 41–71 (1969)

Kretz, R.: Distribution of magnesium and iron between orthopyroxene and calcic pyroxene in natural mineral assemblages. J. Geol. *71,* 773–785 (1963)

Larimer, J. W.: Chemical fractionations in meteorites, 1: Condensation of the elements. Geochim. Cosmochim. Acta *31,* 1215–1238 (1967)

Larimer, J. W.: An experimental investigation of oldhamite, CaS, and the petrologic significance of oldhamite in meteorites. Geochim. Cosmochim. Acta *32,* 965–982 (1968a)

Larimer, J. W.: Experimental studies on the system Fe–MgO–SiO_2–O_2 and their bearing on the petrology of chondritic meteorites. Geochim. Cosmochim. Acta *32,* 1187–1207 (1968b)

Larimer, J. W.: Chemical fractionations in meteorites, 7: Cosmothermometry and cosmobarometry. Geochim. Cosmochim. Acta *37,* 1603–1623 (1973)

Lewis, J. S.: Low temperature condensation from the solar nebula. Icarus *16,* 241–252 (1972)

Lord, H. C., III: Molecular equilibria and condensation in a solar nebula and cool stellar atmospheres. Icarus *4,* 279–288 (1965)

McCallum, I. S.: Equilibrium relationship among the coexisting minerals in the Stillwater Complex, Montana. Ph.D. dissertation, University of Chicago, Chicago, IL. (1968)

Medaris, L. G., Jr.: Partitioning of Fe^{++} and Mg^{++} between coexisting synthetic olivine and orthopyroxene. Am. J. Sci. *267,* 945–968 (1969)

Mueller, R. F.: Energetics of certain silicate solid solutions. Geochim. Cosmochim. Acta *26,* 581–598 (1962)

Mueller, R. F.: A comparison of oxidative equilibria in meteorites and terrestrial rocks. Geochim. Cosmochim. Acta *27,* 273–278 (1963)

Mueller, R. F.: Phase equilibria and the crystallization of chondritic meteorites. Geochim. Cosmochim. Acta *28,* 189–207 (1964)

Mueller, R. F.: The system Fe–MgO–SiO_2–O_2 with applications to terrestrial rocks and meteorites. Geochim. Cosmochim. Acta *29,* 967–976 (1965)

Mueller, R. F., Olsen, E.: The olivine, pyroxene, and metal content of chondritic meteorites as a consequence of Prior's rule. Mineral. Mag. *36,* 311–318 (1967)

Nafziger, R., Muan, A.: Equilibrium phase compositions and thermodynamic properties of olivines and pyroxenes in the system MgO-"FeO"-SiO_2. Am. Mineral. *52,* 1364–1385 (1967)

Olsen, E., Bunch, T.: Empirical derivation of activity coefficients for the magnesium-rich portion of the olivine solid solution. Am. Mineral. *55,* 1829–1842 (1970)

Olsen, E., Bunch, T.: An attempt to recalibrate the pyroxene–pyroxene thermometer. Meteoritics *10,* 468 (1975)

Olsen, E., Fredriksson, K.: Phosphates in iron and pallasite meteorites. Geochim. Cosmochim. Acta *30,* 459–470 (1966)

Olsen, E., Fuchs, L. H.: The state of oxidation of some iron meteorites. Icarus *6,* 242–253 (1967)

Prior, G. T.: On the genetic relationship and classification of meteorites. Mineral. Mag. *18*, 26–44 (1916)

Ramberg, H., Devore, G.: The distribution of Fe^{++} and Mg^{++} in coexisting olivines and pyroxenes. J. Geol. *59*, 193–210 (1951)

Ringwood, A. E.: Chemical and genetic relationships among meteorites. Geochim. Cosmochim. Acta *24*, 159–197 (1961)

Ringwood, A. E., Kaufman, L.: The influence of high pressure on transformation equilibria in iron meteorites. Geochim. Cosmochim. Acta *26*, 999–1009 (1962)

Robie, R. A., Waldbaum, D. R.: Thermodynamic properties of minerals and related substances at 298.15 K (25.0°C) and one atmosphere (1.013 bars) pressure and at higher temperatures. U.S. Geol. Surv. Bull. *1259*, 1–256 (1968)

Saxena, S. K.: Thermodynamics of Rock-Forming Crystalline Solutions. New York-Heidelberg-Berlin: Springer, 1973, pp. 1–188

Van Schmus, W. R., Koffman, D. M.: Equilibration temperatures of iron and magnesium in chondritic meteorites. Science *155*, 1007–1011 (1967)

Schwarcz, H. P., Scott, S. D., Kissin, S. A.: Pressures of formation of iron meteorites from sphalerite compositions. Geochim. Cosmochim. Acta *39*, 1457–1466 (1975)

Urey, H. C.: The Planets. New Haven, Conn.: Yale Univ. (1952)

Virgo, D., Hafner, S.: Fe^{2+}, Mg order-disorder in heated orthopyroxenes. Mineral. Soc. Am. Spec. Paper *2*, 67–81 (1969)

Williams, R. J.: Activity-composition relations in the fayalite–forsterite solid solution between 900° and 1300°C at low pressures. Earth Planet. Sci. Lett. *15*, 296–300 (1972)

Wood, J. A.: On the origin of chondrules and chondrites. Icarus *2*, 152–180 (1963)

18

S. K. SAXENA

Abstract Using the data available on the orthosilicates of divalent cations, an equation has been obtained which relates ionic characters such as sublimation energy, ionization potential and oxide thermochemical radius to the enthalpy. From the thermochemical and estimated values of enthalpies, the electronegativity χ of divalent and tetravalent cations has been determined from Pauling's equation:

$$-\Delta H_f^0 = 23\ \Sigma(\chi_{SiO_4} - \chi_M)^2$$

where χ_{SiO_4} is 3.885 determined arbitrarily from enthalpy data on oxide and orthosilicate of beryllium. χ_M for trivalent cations are calculated from enthalpies of oxides and have been added to the values determined from orthosilicates to complete the list.

Thermodynamic considerations of the relationship between partition coefficients involving ideal or nearly ideal solutions and χ_M lead to the following scheme of element fractionation:

$$K_{(A)}^{\alpha-\beta} > K_{(B)}^{\alpha-\beta} \ldots > K_{(N)}^{\alpha-\beta}$$

where Ks are partition coefficients, A to N are cations increasing in electronegativity, and α and β represent more and less polymerized silicate structures, respectively. Element fractionations in Skaergaard minerals and between Cl chondrites and lunar and terrestrial rocks have been discussed using the new electronegativity scale.

A NEW ELECTRONEGATIVITY SCALE FOR GEOCHEMISTS

1 INTRODUCTION

Goldschmidt's (1937) rules regarding the distribution of trace elements have been critically discussed by Ahrens (1952), Shaw (1953), Ringwood (1955) and Burns and Fyfe (1966). As discussed by McIntyre (1963), Burns and Fyfe (1966), and Whittaker (1968), the distribution of trace elements in a system must be considered on the basis of the thermochemical properties of the coexisting phases. Since the enthalpies and entropies of a large number of silicates are not available and cannot be easily determined, geochemists have attempted to use atomic properties such as ionic potential, ionization potential and electronegativity to assess indirectly the effect of variation in the thermochemical properties. It is obviously unrealistic to consider that any one of these atomic characters can be directly correlated to the variation in the free energy of the silicates. Theoretical estimation of the entropy of many silicates is possible by considering molar volumes of silicates (Saxena, 1976) and by other methods (Latimer, 1951). Estimates of enthalpy, however, are more difficult, and an attempt will be made in this paper to calculate enthalpies from available thermochemical data. In many qualitative discussions, enthalpy alone may be useful in determining the sequence of trace element distributions. Therefore, we need a suitable atomic property which can be correlated with the enthalpies of a large number of silicates. For halides, this property may be recognized as Pauling's (1939) electronegativity. No attempt has been made so far to determine an electronegativity scale which is related to the thermochemical properties of oxides and silicates in the same way as Pauling's scale is related to the enthalpies of halides. The purpose of this paper is to present a new electronegativity scale for use in geochemistry. Since our purpose is to obtain electronegativity values preferably from enthalpies of silicates and since only a few of these are available, the first part of this paper deals with a method to estimate enthalpies of orthosilicates.

2 ENTHALPY OF FORMATION OF ORTHOSILICATES

The enthalpy of formation of a crystal MX (M = metal cation, X = anion group) can be calculated from the energy terms in a hypothetical reaction path described by the Born–Haber cycle. It involves first subliming the metal to give gaseous metal atoms and dissociating the anion to give gaseous atoms to interact as indicated in the diagram below

$$\begin{array}{ccccc}
\mathrm{M(cryst)} + \tfrac{1}{2}\mathrm{X_2}(g) & \xrightarrow{S + \frac{1}{2}D} & \mathrm{M}(g) & + & \mathrm{X}(g) \\
\Delta H_f^0 \downarrow & & \downarrow + I_{\mathrm{M}} & & \downarrow -E_{\mathrm{X}} \\
\mathrm{MX(cryst)} & \xleftarrow{-U_L} & \mathrm{M}^+(g) & + & \mathrm{X}^-(g)
\end{array}$$

where S is the sublimation energy of the metal, D is the dissociation energy of X_2, I_M is the ionization potential and E_x is the electron affinity. + means heat absorbed, − means heat evolved, and the terms nRT have been neglected. From Hess' law we have

$$\Delta H_f^0 = S + \tfrac{1}{2} D + I_M - E_x - U_L \tag{1}$$

In orthosilicate, we may consider the interaction of M^{2+} cation and anion SiO_4, the latter being regarded as a single unit. For an orthosilicate M_2SiO_4, we have

$$\Delta H_f^0 = 2S + D + 2\sum_{1\rightarrow 2} I_M - 4\,E_x - U_L \tag{2}$$

The thermochemical data on sublimation energy S and ionization potential I_M are easily available (Samsonov, 1968). The data on lattice energy U_L for orthosilicates is generally unavailable. The direct application of Equation (2) is, therefore, not possible. We may, however, use this equation empirically as discussed in the following section using enthalpy data on silicates that are available.

Equation (2) may be rewritten as

$$H_f^0 - 2S - 2\sum_{1\rightarrow 2} I_M = a - U_L \tag{3}$$

The quantity a represents the energy of formation of SiO_4 anion from elements and is unknown.
The lattice energy is given by

$$U_L = \frac{N_0\, A\, z_1 z_2 e^2}{r_0}\left(1 - \frac{\rho}{r_0}\right) \tag{4}$$

where N_0 is Avogadro's number, A is Madelungs constant, z_1 and z_2 are charges, r_0 is the equilibrium internuclear distance and ρ the Born repulsion exponent. Application of Equation (4) is difficult because of the lack of data. A semiempirical approach was used by Kapustinskii (in Greenwood, 1970) to calculate lattice energy for a number of compounds. His equation is

$$U_L = 287.2\,\frac{n z_1 z_2}{r_c + r_a}\left(1 - \frac{0.345}{r_c + r_a}\right) \tag{5}$$

where n is the number of ions in the formula and r_c and r_a are the formal 6-coordinated cation radii and anion radii in ångstrom units. Since our present interest is to obtain a suitable expression with two or more constants, a compromise between Equation (5) and an extended form of Equation (4) due to Ladd and Lee (in Greenwood, 1970) was used. The equation is

$$U_L = \frac{b}{r_c + r_a}\left(1 - \frac{.345}{r_c + r_a}\right) + \frac{c}{(r_c + r_a)^6}\left(1 - \frac{6 \times .345}{r_c + r_a}\right) + \frac{d}{(r_c + r_a)^8}\left(1 - \frac{8 \times .345}{r_c + r_a}\right) \tag{6}$$

The first term is similar to the term in Equation (5). The second and third terms are designed to account for London attractive energy and quadrupole interactions respectively. ρ has been accepted as 0.345 and r_0 as $(r_c + r_a)$. It is obvious that the constants b, c, and d to be determined statistically may not have a direct physical significance. Substituting Equation (6) for U_L in Equation (3) we have

$$\Delta H_f^0 - 2S - 2\sum_{1\to 2} I_M = a - \frac{b}{r_c + r_a}\left(1 - \frac{0.345}{r_c + r_a}\right) - \text{etc. as in Equation (6)} \tag{7}$$

Available data on enthalpy of orthosilicates (Kubaschewski et al., 1967; Robie and Waldbaum, 1968) with single divalent metal cation are listed in Table 1. Kubaschewski et al. (1967) have also listed data on Be_2SiO_4 and Pb_2SiO_4 which they consider as of doubtful accuracy. Data on ionization potential and sublimation energy of cations are from Samsonov (1968). Cation radii are available in Muller and Roy (1974). Greenwood (1970) has listed r_{SiO_4} as 2.4 Å. These data may be used for a simultaneous solution of seven equations (7) to determine the constants a, b, c, and d. Calculations involving the Shannon and Prewitt radii (in Muller and Roy, 1974) lead to rather high errors in calculated enthalpy.

Table 1. Thermochemical data*

	$-\Delta H_f^0$	I_M	S	†Calculated $-\Delta H_f^0$	Error
$-Ca_2SiO_4$	553.973	414.60	42.090	554.248	−0.275
Fe_2SiO_4	353.544	555.30	99.780	355.140	−1.596
Mg_2SiO_4	520.370	522.60	35.190	518.059	2.311
Mn_2SiO_4	413.520	532.00	70.086	415.631	−2.111
Zn_2SiO_4	391.140	630.80	31.253	391.544	−0.404
Ba_2SiO_4	560.270	350.70	41.690	561.447	−1.177
Sr_2SiO_4	556.050	385.60	39.092	554.955	1.095
Be_2SiO_4	—	634.80	77.680	487.219	—
Cd_2SiO_4	—	597.00	26.790	355.112	—
Pb_2SiO_4	—	517.65	46.891	354.408	—
Co_2SiO_4	—	574.50	101.580	341.532	—
Ni_2SiO_4	—	594.50	101.281	339.090	—
Cu_2SiO_4	—	646.00	80.884	297.779	—
Hg_2SiO_4	—	672.90	14.657	272.448	—

*All data in kcal/mol ΔH_f^0, enthalpy of formation. I_M, ionization potential summation. S, sublimation energy. For source of data see text.
†Enthalpy calculated using Equation (9) in which the constants a, b, c, and d have been obtained from the enthalpy data on the seven orthosilicates for which thermochemical enthalpies are available.

For example the errors in calculated H_f^0 for Mg_2SiO_4, Mn_2SiO_4, and Fe_2SiO_4 are 38.09, −12.29, and −27.26 kcal/mol, respectively. Similar results are obtained if any other set of ionic radii are used.

Two of the parameters that may be improved in Equation (7) are $(r_c + r_a)$ and ρ (= .345). Without the necessary physical data on all orthosilicates, ρ's cannot be determined. However we may consider improving $(r_c + r_a)$ by substituting 'thermochemical radius' of Kapustinskii for r_c calculated from the following equation

$$U_L\,(\text{oxide}) = 287.2\,\frac{nz_1z_2}{r'_c + 1.40}\left(1 - \frac{0.345}{r'_c + 1.40}\right) \tag{8}$$

where r'_c is the thermochemical radius. This procedure may be justified on the basis of Ramberg's (1954) arguments on the relative stabilities of oxides and silicates as discussed in the next section. The lattice energies for most oxides may be easily calculated from thermochemical data on elements (Samsonov, 1968). Table 1 also shows the result of using the oxide thermochemical radii. Considering the uncertainty in the thermochemical values of enthalpies, the estimated enthalpies are satisfactory. The following equation has been used:

$$\Delta H_f^0 = 2S + 2\sum_{1\to 2} I_{\mathrm{M}} - 525 - \frac{2630}{r'_c + 2.4}\left(1 - \frac{0.345}{r'_c + 2.4}\right) - \frac{878}{(r'_c + 2.4)^6}\left(1 - \frac{6 \times 0.345}{r'_c + 2.4}\right) + \frac{725}{(r'_c + 2.4)^8}\left(1 - \frac{8 \times .345}{r'_c + 2.4}\right) \tag{9}$$

where r'_c is the thermochemical radius determined from Equation (8). Using Equation (9) it is possible to estimate ΔH_f^0 for the orthosilicates of divalent cations assuming that this equation also applies to other elements of atomic numbers higher than that of Fe. For this work, this equation need only yield enthalpy data which changes systematically even if not correct in the absolute sense. Table 1 shows the calculated ΔH_f^0 for many divalent orthosilicates for which the thermochemical data on enthalpy do not exist.

3 ELECTRONEGATIVITY SCALE

Electronegativity values that could be useful in oxide and silicate geochemistry must show a simple correlation between enthalpy and the numbers assigned as electronegativity to the atom forming the oxide or silicate. As noted in Figure 1, Pauling's scale does not qualify in this respect. This is true also of the other electronegativity scales that are currently in use.

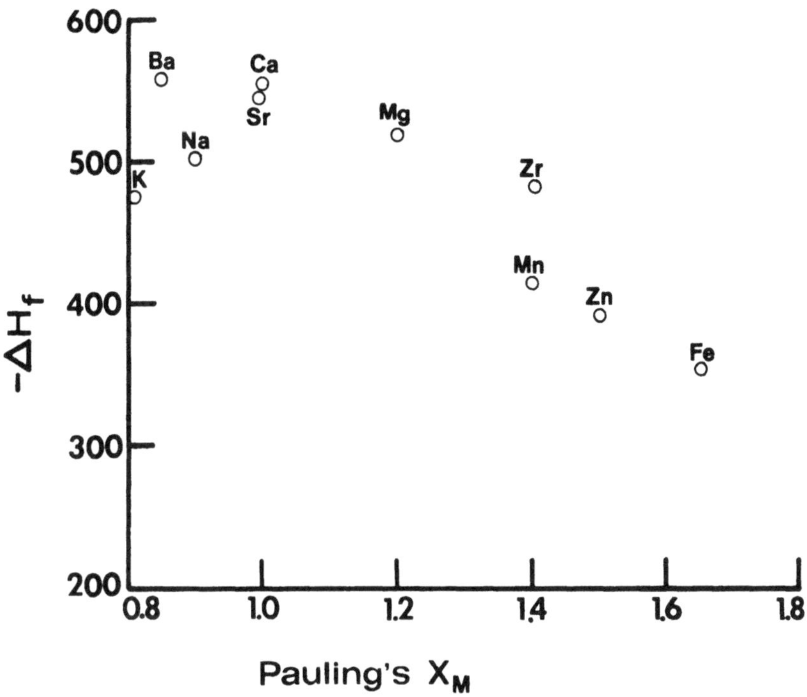

Fig. 1. Pauling's electronegativity plotted against the enthalpy of orthosilicate. Scatter shows that this scale is not satisfactory for use in geochemistry

3.1 Electronegativity from Enthalpies of Oxides

Since oxygen is shared between strongly contrapolarizing Si ions in quartz, Ramberg (1954) argued that the Si–O bond is usually weaker in quartz than the Si–O bond in orthosilicates where none of the oxygens is shared: "The M–O link is weaker in a silicate than in the metal oxide. The strengthening of the Si–O bond represents a decrease in energy thus tending to stabilize the silicate whereas the weakening of the M–O bond represents an increase in energy thus working against the formation of the silicate. The heat of formation of the silicate from the constituent oxides is then the difference between these two bond-energy changes." If therefore the strengthening of the M–O bond is represented by decreasing oxide enthalpy, and considering that the formation of a silicate takes place when an equilibrium between the strengths of the M–O bond is reached, there should be a significant correlation between the enthalpies of oxides and silicates. This argument may be verified by plotting the enthalpies of oxides against enthalpies of silicates as done in Figure 2.

Using Pauling's postulate (Pauling, 1939; Gutman and Mayer, 1972), namely that the 'enthalpy of formation ΔH_f^0 of a compound in standard state from the elements in the standard state is equal to the heat of

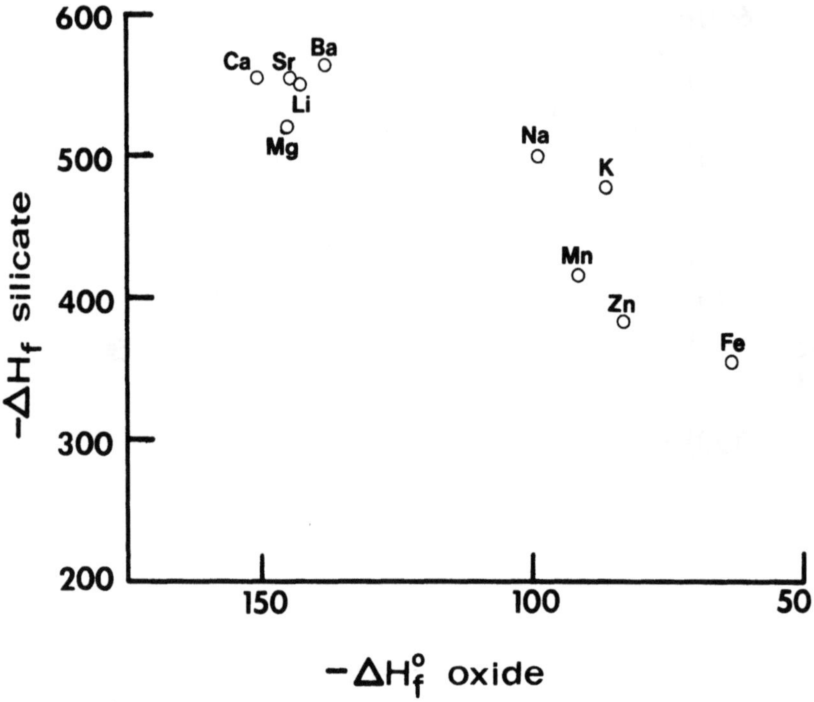

Fig. 2. Enthalpies of oxides plotted against enthalpies of orthosilicates of univalent and divalent cations.

formation of the gaseous compound from the gaseous element molecules,' we have

$$-\Delta H_f^0 = \Sigma\Delta \tag{10}$$

where Δ is given by Pauling's equation

$$\Delta = 23\ \Sigma(\chi_A - \chi_B)^2 \tag{11}$$

where χ_A and χ_B are the electronegativities of atoms A and B, respectively. For compounds involving oxygen, Equation (11) becomes

$$-\Delta H_f^0 = 23\Sigma(\chi_O - \chi_M)^2 - 26n_o \tag{12}$$

where O and M represent oxygen and metal cation, respectively, and n is the number of oxygens. Using Equation (12), it is possible to calculate χ_M from $-\Delta H_f^0$ of oxides. Figure 2 shows that while the correlation between enthalpy of silicates and χ_M is good for most divalent and univalent cations, it is not good for group IIA cations.

3.2 Electronegativity of Divalent Cations

Estimated enthalpies of orthosilicates as determined from Equation (9) may be used to obtain the electronegativity of divalent cations from the relation

$$-\Delta H_f^0 = 23\Sigma(\chi_{SiO_4} - \chi_M)^2 \tag{13}$$

where χ_{SiO_4} and χ_M represent the electronegativity of the anion SiO_4 and the cation M, respectively. χ_{SiO_4} is determined as follows. χ_{Be} may be assumed to be the same when determined from the enthalpies of BeO and Be_2SiO_4. The choice of Be is arbitrary. From Equation (12) χ_{Be} is determined as 1.583. Substituting this value in Equation (13) and the estimated enthalpy of Be_2SiO_4 ($H_f = -487.22$ kcal/mol), χ_{SiO_4} is determined to be 3.885.

3.3 Electronegativity of Tetravalent Cations

While an empirical estimation of enthalpy of orthosilicates of divalent cations could be possible, a similar method cannot be developed for orthosilicates of tetravalent cations because only one enthalpy value is available ($ZrSiO_4$ −483.7 kcal/mol, JANAF tables). The use of an equation similar to Equation (6) i.e.,

$$\Delta H_f^0 - S - \Sigma I_M = a - U_L$$

requires the knowledge of the constant a and the lattice energy U_L. For the present purpose let us assume that the Kapustinskii equation in the modified form

$$U_L = \frac{287.2\, n z_1 z_2}{r'_c + 2.4}\left(1 - \frac{0.345}{r'_c + 2.4}\right) \tag{14}$$

where $n = 2$, $z_1 = z_2 = 4$, and r'_c represents the thermochemical radius, would yield systematically changing values of 'lattice energies'. For $ZrSiO_4$ we have

$$\underset{\Delta H_f^0}{-483.7} - \underset{S}{145.772} - \underset{\underset{1\to 4}{\Sigma I_M}}{1787.9}$$

$$= a - \underset{U_L}{2575.02}\ (r'_c = 0.782 \text{ from } \Delta H_f^0 \text{ of } ZrO_2) \tag{15}$$

which yields $a = 157.65$ kcal. Using Equation (15) with $a = 157.65$ and U_L determined from Equation (14), it is possible to obtain 'enthalpies' of orthosilicates of tetravalent cations. These 'enthalpies' obviously cannot be useful in an absolute sense but being derived from the atomic properties of sublimation and ionization potential and from oxide enthalpy (yielding r'_c), we can be certain that their variation would be systematic and of sufficient accuracy for the present purpose. From such orthosilicate enthalpies using Equation (13), χ_M for tetravalent cations can be determined.

3.4 A Combined Scale of Electronegativity

As discussed before, a scale of electronegativity is most suitable for geochemical work if it is based on enthalpies of silicates. Not all elements, however, form silicates. While most divalent and tetravalent cations could be considered to combine with SiO_4 anion and form silicates, trivalent cations, unless in combination with other divalent cations, do not form silicates. One exception to this rule is Al_2SiO_5. A combination of all three different methods as discussed in the preceding sections has to be used in the final formulation of the electronegativity scale. The success of this scale will depend on how systematically the electronegativities vary with the thermochemical data on enthalpies of silicates of divalent and tetravalent cations and how well it represents the enthalpies of silicates with cations of different valencies. The com-

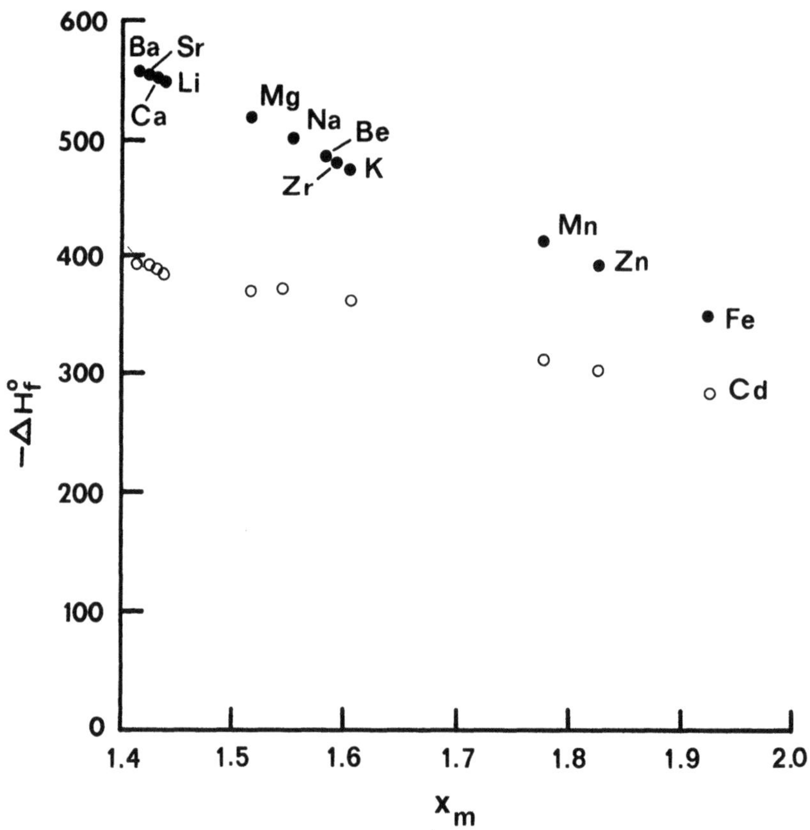

Fig. 3. Enthalpies of orthosilicates and metasilicates plotted against the new electronegativity (χ_M). Filled circles: orthosilicates, open circles: metasilicates.

Table 2 Electronegativity scale for use in geochemistry*

Univalent	Divalent	Trivalent	Tetravalent
Li 1.433	Bc (1.583)	Al 1.638	Ge 2.315
Na 1.553	Mg 1.507 (1.512)	Fe 2.087	Sn 2.015
K 1.603	Sr 1.427 (1.429)	Mn 2.009	Te 2.272
	Ca 1.431 (1.431)	Sc 1.549	Ti 1.878
	Ba 1.417 (1.415)	B 1.837	Zr 1.592
	Mn 1.765 (1.760)	As 2.195	Hf 1.509
	Zn 1.823 (1.823)	Bi 1.941	Th 1.376
	Fe 1.925 (1.920)	Cr 1.906	
	Cd (1.920)	Sb 2.161	
	Pb (1.922)	V 1.864	
	Co (1.958)		
	Ni (1.965)		
	Cu (2.086)		
	Hg (2.164)		

*Electronegativity for univalent and divalent cations is calculated from the enthalpy of orthosilicates. When estimated enthalpy [Eq. (9)] is used the values are placed in parentheses. Electronegativity for trivalent ions are calculated from the oxide enthalpies. For tetravalent ions refer to text.

bined scale is presented in Table.2. Figure 3 shows a plot of χ_M against ΔH_f^0. Note that a linear relationship exists only between $(\chi_{SiO_4} - \chi_M)^2$ and ΔH_f^0. The plot shows rather linear trend because χ_{SiO_4} is assumed as constant resulting in only small deviation from a linear relationship between χ_M and ΔH_f^0.

4 APPLICATIONS

4.1 Trace Element Distributions; Thermodynamic and Crystal–Chemical Considerations:

Distribution of cations $A, B, \ldots, N$ between coexisting phases such as an orthosilicate $(A, B, \ldots, N)\ Si_{.5}O_2$ and a metasilicate $(A, B, \ldots, N)\ SiO_3$ at a certain P and T may be represented by a set of equations on one cation basis as follows

$A Si_{1/2}O_2$	$+ \frac{1}{2}SiO_2$	$\rightleftharpoons$	$ASiO_3$
Orthosilicate Solid Solution			Metasilicate Solid Solution
. . .	. . .		. . .
. . .	. . .		. . .
$\frac{1}{2}N Si_{1/2}O_2$	$+ \frac{1}{2}SiO_2$	$\rightleftharpoons$	$NSiO_3$

Equilibrium constants K's for these reactions are

$$K_a \exp\left(\frac{-P\Delta V_a}{RT}\right) = \frac{a_{ASiO_3}}{a_{ASi_{1/2}O_2}\, a_{SiO_2}^{1/2}} = \frac{X_{ASiO_3}}{X_{ASi_{1/2}O_2}} \text{(ideal solution)} = K_{(A)}$$

and similarly others, where a represents activity and may be replaced by X the mole fraction for ideal or closely ideal solutions, R is the gas constant, P is pressure, ΔV represents change in molar volume in the reaction identified by the subscript, T is temperature and $K_{(i)}$ represents the partition coefficient for the ion i. Note that a clear distinction should be made between the partition coefficient $K_{(i)}$ as defined above and the distribution coefficient with the symbol K_D referring to an ion-exchange equilibrium between coexisting phases (see also Perchuk 1977). K_D for the ion-exchange equilibrium for i and j is given by the ratio of the partition coefficients $K_{(i)}/K_{(j)}$. Since K_D is a ratio, in general it will not be a function of the composition variation in other coexisting phases. a_{SiO_2} is unity when quartz is present. In absence of quartz, it is unknown. For equilibrium constant K, we have

$$RT \ln K = -\Delta G^0 + P\Delta V = -\Delta H^0 + T\Delta S + P\Delta V. \tag{16}$$

The equilibrium constant K is a function of P, T, and composition of the phases. The latter effect is determined by the solution properties of the phases. The sign of ΔG^0 determines whether a cation A is enriched in the orthosilicate or metasilicate. At room temperature, $T\Delta S$ and $P\Delta V$ may be neglected and the sign of ΔH^0 determines the partitioning of cations.

From Figure 3 we note that the slope of the line relating enthalpies of orthosilicate with electronegativity is greater than the slope of a similar line for metasilicate. In other words the difference between enthalpy of ortho- and meta-silicates of the divalent cations decreases as the electronegativity of the cations increases. In Figure 4, ΔH^0 for the reaction

$$\tfrac{1}{2}\, M_2SiO_4 + \tfrac{1}{2}\, SiO_2 \rightleftharpoons MSiO_3$$

for silicates of various divalent cations have been plotted to illustrate the effect of increasing electronegativity. The ΔH_f^0 of orthosilicates are as determined by Equation (9) while the ΔH_f^0 of metasilicates are from the thermochemical tables. The scatter is attributed to the irregular variation in the ΔH_f^0 of some metasilicates (for example $\Delta H^0_{SrSiO_3}$) with electronegativity. In view of the otherwise good correlation between χ and ΔH_f^0 a 'smoothening' of the enthalpy data on metasilicates should be regarded as realistic.

From Figures 3 and 4, the conclusion is that the partition coefficients for cations $A, B, \ldots, N$ which are increasing in χ_M from A to N should be in the following order

$$K_{(A)}^{\alpha-\beta} > K_{(B)}^{\alpha-\beta} > \ldots > K_{(N)}^{\alpha-\beta}$$

where α represents a metasilicate and β an orthosilicate and both β and α are ideal or nearly ideal solutions. Note that the above scheme applies to those cases also where quartz (SiO_2) is not present along with the two silicates. If SiO_2 is insufficient for the reactions (a) to (n) at a certain

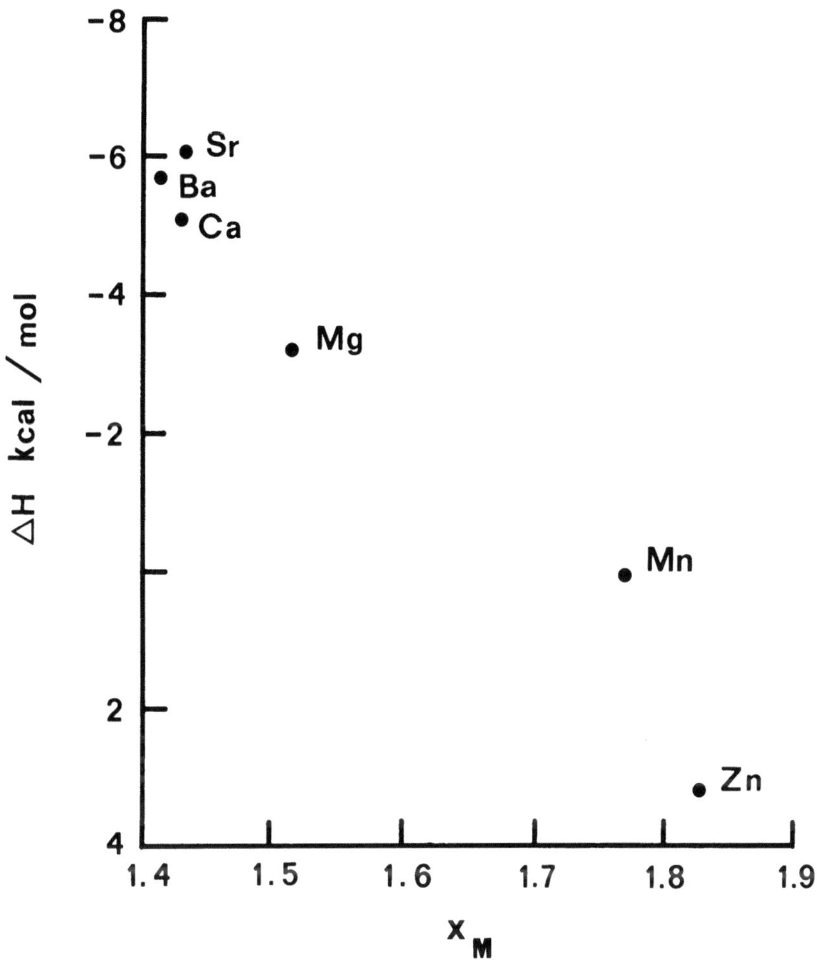

Fig. 4. ΔH^0 for reaction $MSi_{1/2}O_2 + \frac{1}{2}\ SiO_2 = MSiO_3$ plotted against χ_M to show the effect of increasing χ_M on ΔH_f^0

temperature and pressure, only part of the reaction will take place. In such a case the partition coefficients will be different but their scheme of variation as presented above will not change. Note that while the presence or absence of quartz affects the absolute values of the partition coefficients $K_{(A)}^{\alpha-\beta}$, etc., it does not affect the distribution coefficients which are ratios of the partition coefficients.

The enthalpies of crystalline silicates have been estimated here but the enthalpy of the silicate melt is unknown. In this regard, we may use the following rule: the enthalpy of silicates when normalized on one Si basis (see below) increases with increased sharing of the oxygen anion. The origin of this rule may be found in Ramberg's (1952, 1954) discussion of the distribution of Fe and Mg between coexisting silicates and may be illustrated with the following data on K-silicates (Kubaschewski et al., 1967) (all kcal/mol):

K_4SiO_4 −478.75, K_2SiO_3 −368.95,
½$K_2Si_2O_5$ −298.30, ¼ $K_2Si_4O_9$ −259.6.

Note how the enthalpy considered on one Si basis increases from the least polymerized orthosilicate to the most polymerized framework silicate. We expect that the isolated Si–O tetrahedral units and K ions in the melt would represent the continuation of the decreasing enthalpy effect and therefore the enthalpy of a melt component would be lower than that of the corresponding orthosilicate. Simply stated, the normalized enthalpy of formation of silicates of a cation varies according to the scheme

melt < SiO_4 < SiO_3 < $SiO_{3.5}$, etc.

In view of the discussion above, we may generalize our cation distribution scheme as follows

$$K_{(A)}^{\alpha-\beta} > K_{(B)}^{\alpha-\beta} > \ldots > K_{(N)}^{\alpha-\beta}$$

where α represents a more polymerized silicate structure than β. Thus α could be an orthosilicate while β a melt. This scheme of distribution will hold as long as the phases are ideal solutions or else the components except one are in trace amounts (generally parts per million) so that Nernst's law is applicable. Note that in some cases even in the range of extreme dilution, the cation may not be suitable for a particular site in a crystal. This should be regarded as a nonideality effect. The nonideality effects cannot be predicted without additional information on solution behavior. Although the scheme of change of $K_{(i)}$ as a function of electronegativity is based on enthalpies only, the term $T\Delta S$ may not significantly affect this scheme. ΔS_{298} for a reaction of the type (a) for Ba to Mn shows a minimum to maximum variation of −0.1 to 0.26 gibbs/mol. Therefore, at a fixed temperature the neglect of the term $T\Delta S$ may affect the scheme only when the two cations are closely similar in electronegativity such as Ca and Sr. The effect of the term $P\Delta V$ may be evaluated approximately by considering the ΔV at room temperature. For silicates of Mg, Fe, Ca, and Mn in reaction of the type (a), the ΔV (orth- and metasilicates) are −0.043, −0.038, −0.023 and −0.012 cal/mol, respectively. At 10 kbar, neglecting the compressibility coefficients, the difference between $-RT \ln K_{(\mathrm{Mg})}^{Cpx-Opx}$ and $-RT \ln K_{(\mathrm{Mn})}^{Cpx-Opx}$ may be of the order of 310 cal, which is not significant.

4.1 Trace Element Distributions in Skaergaard Minerals

Although the importance of distribution data has been realized for many decades, reliable data on partition coefficients in natural assemblages are scarce. Wager and Brown (1967) have presented the trace element data on various minerals in zones *A* to *F* as vertical bars in their Figure 122 (p. 180–182). The data for mineral separates of olivine and pyroxenes are examined here. The pyroxene data probably represent the data on both ortho- and clino-pyroxenes. The partition coefficients vary according to the following scheme:

A $K^{Px-01}_{(Mn)} > K^{Px-01}_{(Co)} > K^{Px-01}_{(Ni)} > K^{Px-01}_{(Cu)}$

B $K^{Px-01}_{(Mn)} > K^{Px-01}_{(Cu)} > K^{Px-01}_{(Ni)} > K^{Px-01}_{(Co)}$

D $K^{Px-01}_{(Cu)} > K^{Px-01}_{(Mn)} > K^{Px-01}_{(Co)} > K^{Px-01}_{(Ni)}$

E,F $K^{Px-01}_{(Cu)} > K^{Px-01}_{(Co)} > K^{Px-01}_{(Mn)} > K^{Px-01}_{(Ni)}$ (?,Ni < detectable limit)

Note that in zone *A*, which represents rocks of the marginal border group (MBG), olivine and pyroxenes crystallized in equilibrium from the melt as shown by the systematic variation of *K*'s according to the electronegativity of the cations. In *B* and *D* rocks, $K_{(Cu)}$ is greater than $K_{(Ni)}$ or $K_{(Co)}$ while in the late stage zones *E* and *F* the scheme is reversed. This reversal may be recognized as due to the occurrence of cumulus olivine and pyroxene with increasing amounts of intercumulus or late stage pyroxene.

4.2 Distribution of Elements in Planetary Rocks

From the theory discussed in the previous sections, it is possible to consider that the regularity of the scheme of cation partition which exists between less polymerized and more polymerized silicates and between a silicate crystal and a silicate melt continues in the distribution of cations between a gaseous solution and a melt or directly between a gaseous solution and a crystal. Therefore it is expected that the equilibrium distribution of cations between a cooling gas phase and a condensing phase will vary systematically according to electronegativity. Thus the partition coefficients for all cations incorporated from the solar gas into a high temperature condensate such as a melilite will decrease with increasing electronegativity provided both melilite and the gas phase do not deviate highly from ideal solution for those components whose concentration is not stoichiometric.

Unfortunately compositional data on minerals from meteorites is scarce. Grossman (1973) and Grossman and Ganpathy (1976) have determined compositions of the Ca–Al rich inclusions in Allende meteorite. These inclusions are regarded as the high temperature condensates from the Solar gas but these are also not monomineralic. It is, however, possible to consider the distribution of cations between a melt solution representing a terrestrial magma and Solar gas. Irrespective of what models of origin we may consider, the distribution of elements can be studied to understand chemical equilibrium and petrogenetic relations between two planetary bodies. The choice of elements to be studied must depend on the nature of solution and stoichiometry of the phases. While gas and melt solutions can be assumed to be ideal and variable in most components, the crystalline phases are stoichiometric for some components and could be significantly nonideal in the variation of other components. For a comparison of two planetary bodies we may define partition coefficients as

$$K = \frac{\text{ppm of an element in a planetary body } A}{\text{ppm of the element in a planetary body } B}.$$

If natural logarithms of K's are plotted against χ_M and the slope is found to be significantly negative, we may expect that the condition for the formation of these two bodies in chemical equilibrium is approximately satisfied and the planetary body B may be regarded as more 'primitive' than body A. The situation is akin to the crystallization of olivine (body A) from a basaltic magma (body B) when those components which mix ideally are considered.

Figure 5 shows plots of ln K given by

$$\ln K = \ln \frac{\text{ppm of element in a planetary rock}}{\text{ppm of element in C 1 chondrites}}$$

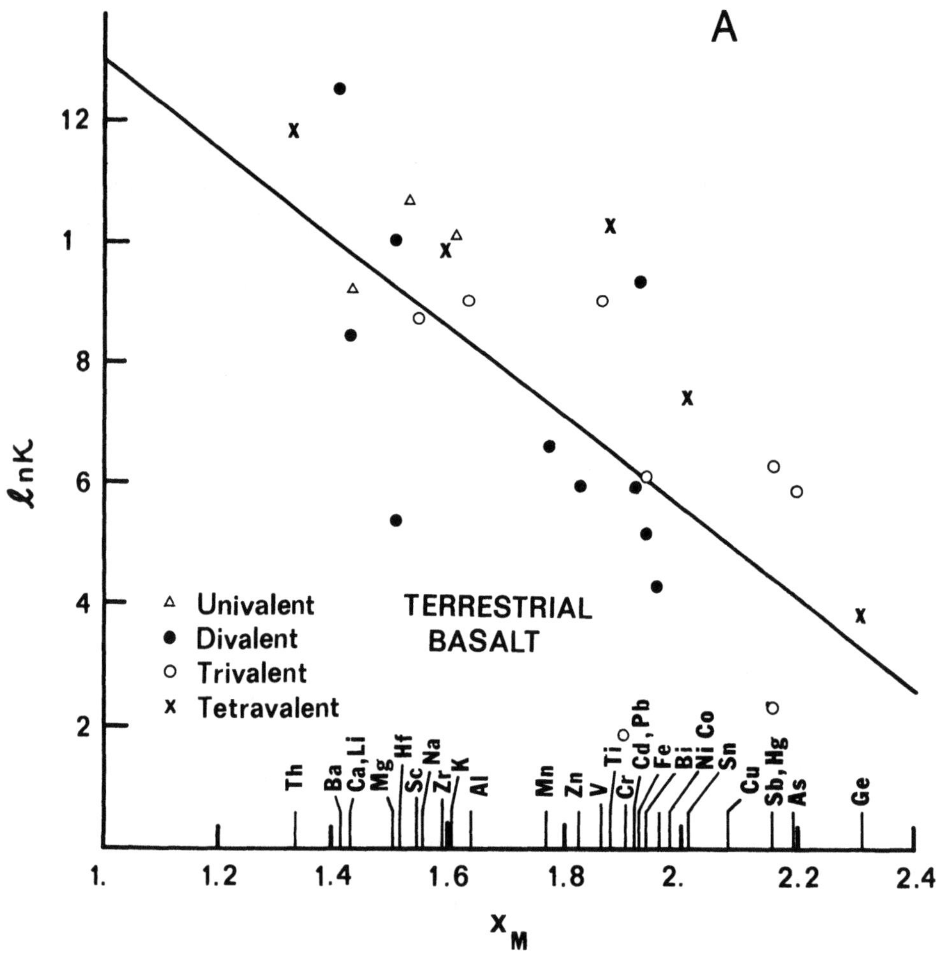

Fig. 5A–C. ln K for partioning of cations between C 1 chondrites and various other compositions plotted against χ_M. (A) Basalt (USGS-BCR-1) (B) Apollo 15 green glass (C) Model pyrolite

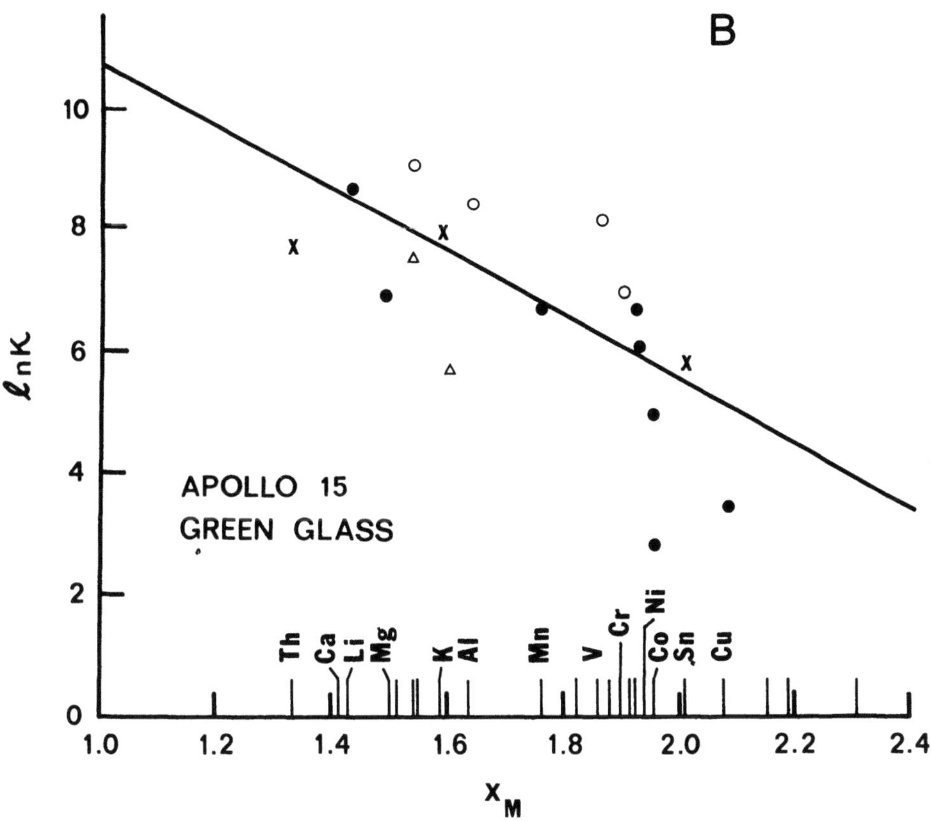
B
APOLLO 15
GREEN GLASS
ℓnK
X_M
Th Ca Li Mg K Al Mn V Cr Ni Co Sn Cu

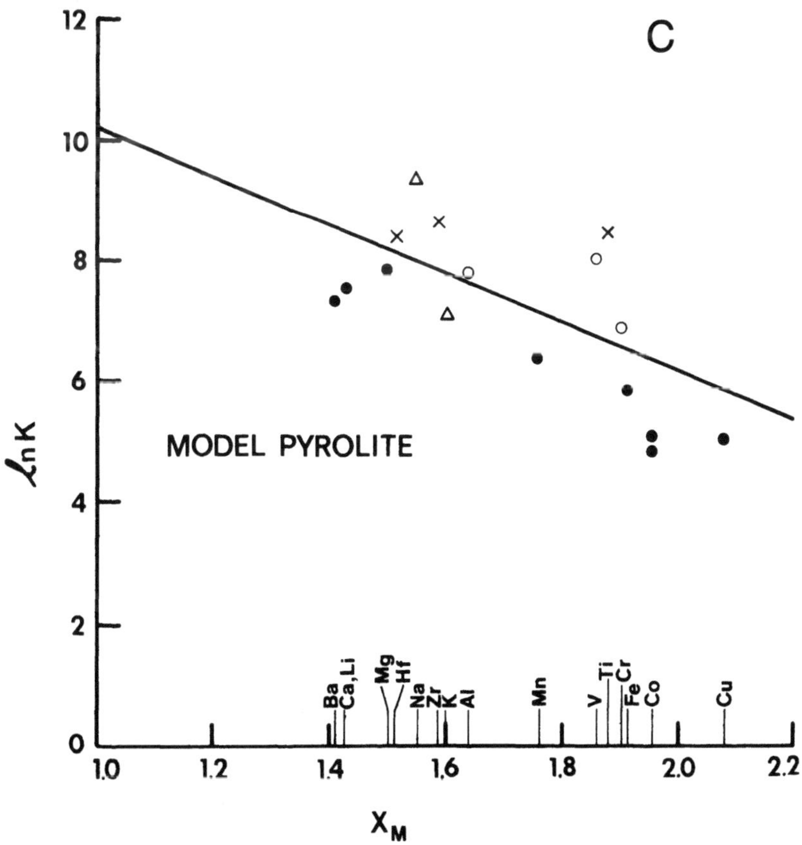
C
MODEL PYROLITE
ℓnK
X_M
Ba Ca,Li Mg Hf Na Zr K Al Mn V Ti Cr Fe Co Cu

against χ_M. The data on mean concentrations in C 1 chondrites are taken from Baedecker and Wasson (1975, Table 6) and Wasson (1975). The planetary rocks under consideration are average basalt (USGS-BCR-1, Flanagan, 1972), lunar glass (green glass, Apollo 15, Taylor, 1974) and pyrolite (Brett, 1971). The results of Figure 5 may be discussed statistically by the following least squares trends:

Basalt: $20.126 - 7.215\chi_M = \ln K$, $(\text{error})^2 = 5.461$, F ratio $= 11.92$
Apollo 15: $15.642 - 5.078\chi_M = \ln K$, $(\text{error})^2 = 1.887$ F ratio $= 7.02$
Pyrolite: $13.828 - 3.827\chi_M = \ln K$, $(\text{error})^2 = 1.019$ F ratio $= 6.65$.

Note that the slope and the scatter continuously change from basalt to pyrolite. The slope for bulk chondrite analysis will be close to zero. Unfortunately the statistics are not strictly comparable because some data for the Apollo 15 glass and for pyrolite are missing. However a continuous change in slope from the less evolved pyrolite to the more evolved basalt with Apollo 15 glass in the middle is quite apparent.

ACKNOWLEDGMENTS

Thanks are due to J. Ganguly, M. J. Drake, N. Pingitore, S. J. Louisnathan, and R. Pizer for critical review. However the author alone bears the responsibility of any errors in calculations.

REFERENCES

Ahrens, L. H.: The use of ionization potentials. Geochim. Cosmochim. Acta *2,* 155–156 (1952)

Baedecker, P. A., Wasson, J. T.: Elemental fractionations among enstatite chondrites. Geochim. Cosmochim. Acta *39,* 735–766 (1975)

Brett, R.: The earth's core: Speculations on its chemical equilibrium with the mantle. Geochim. Cosmochim. Acta *35,* 203–221 (1971)

Burns, R. G., Fyfe, W. S.: Distribution of elements in geological processes. Chem. Geol. *1,* 49–56 (1966)

Flanagan, F. J.: 1972 values for international geochemical reference samples. Geochim. Cosmochim. Acta *37,* 1189–1200 (1973)

Goldschimidt, V. M.: The principles of distribution of chemical elements in minerals and rocks. J. Chem. Soc. 655–672 (1937)

Greenwood, N. N.: Ionic Crystals, Lattice Defects and Nonstoichiometry. New York: Chemical, 1970

Grossman, L.: Refractory trace elements in Ca-Al-rich inclusions in the Allende meteorite. Geochim. Cosmochim. Acta *37,* 1119–1140 (1973)

Grossman, L., Ganpathy, R.: Trace elements in the Allende meteorite-I, Coarse-grained, Ca-rich inclusions. Geochim. Cosmochim. Acta *40,* 331–344 (1976)

Gutman, V., Mayer, U.: Thermochemistry of the chemical bond. In: Structure and Bonding, eds. P. Hemmerich, C. K. Jørgensen, J. B. Neilands, R. S. Nyholm, D. Reinen, R. J. P. Williams. Berlin-Heidelberg-New York: Springer, 1972, Vol. X, pp. 127–151

Janaf Thermochemical Tables. Suppl. 41. Compiled by the Thermal Res. Lab., Dow Chem. Co., Midland, Michigan (1974)

Kubaschewski, O., Evans, E. Li., Alcock, C. B.: Metallurgical Thermochemistry. London: Pergamon Press, 1967

Ladd, M. F. C., Lee, W. H.: Lattice energy and related topics. In: Progress in Solid State Chemistry. Reiss, H. (ed.). New York: MacMillan, 1964, Vol. 1

Latimer, W. M.: Methods of estimating the entropy of solid compounds. Am. Chem. Soc. J. *73,* 1480–1482 (1951)

McIntyre, W. L.: Trace element partion coefficients. Geochim. Cosmochim. Acta *27,* 1209–1264 (1963)

Muller, O., Roy, R.: The Major Ternary Structural Families. Crystal Chemistry of Non-metallic Materials. Heidelberg: Springer, 1974, Vol. IV

Pauling, L.: The Nature of Chemical Bond. Cornell Univ., Ithaca, New York 1939

Perchuk, L. L.: Thermodynamic control of metamorphic processes. In *Energetics of Geological Processes,* eds. S. K. Saxena and S. Bhattacharji. Springer-Verlag, New York, p. 285–352 (1977)

Ramberg, H.: The Origin of Metamorphic and Metasomatic Rocks. Chicago: Univ. Chicago, 1952

Ramberg, H.: Relative stabilities of some silicates as related to the polarization of oxygen ions. Am. Mineral. *39,* 256–271 (1954)

Ringwood, A. E.: The principles governing trace element distribution during magmatic crystallization. Part 1. The influence of electronegativity. Geochim. Cosmochim. Acta *7,* 189–202 (1955)

Robie, R. A., Waldbaum, D. R.: Thermodynamic properties of minerals and related substances at 298°K (25.0°C) and one atmosphere (1.013 Bars) pressure and at higher temperatures. U.S. Geol. Surv. Bull. *1259,* p. 1–256 (1968)

Samsonov, G. V.: Handbook of the Physicochemical Properties of the Elements. New York: Plenum, 1968

Saxena, S. K.: Entropy estimates for some silicates at 298°K from molar volumes. Science, *193,* 1241–1242 (1976)

Shaw, D. M.: The camouflage principle of trace element distribution in magmatic minerals. J. Geol. *61,* 142–151(1953)

Taylor, S. R.: Lunar Science: A Post-Apollo View. London: Pergamon Press, 1974

Wager, L. R., Brown, G. M.: Layered Igneous Rocks. San Fransisco: Freeman, 1967

Wasson, J. T.: Meteorites. In: Minerals and Rocks. Wyllie, P. J. (ed.) Berlin-Heidelberg-New York: Springer, 1974, Vol. X

Whittaker, E. J. W.: Factors affecting element ratios in the crystallization of minerals. Geochim. Cosmochim. Acta *31,* 2275–2288 (1967)

INDEX

MIX
Papier aus verantwortungsvollen Quellen
Paper from responsible sources
FSC® C105338

If you have any concerns about our products,
you can contact us on
ProductSafety@springernature.com

In case Publisher is established outside the EU,
the EU authorized representative is:
Springer Nature Customer Service Center GmbH
Europaplatz 3, 69115 Heidelberg, Germany

Printed by Libri Plureos GmbH
in Hamburg, Germany